CODEX778

EXPERIMENTAL ECOLOGY

University Library
Stamp

Date Due

EXPERIMENTAL ECOLOGY

Issues and Perspectives

Edited by

William J. Resetarits Jr.

Joseph Bernardo

WITHDRAWN
FAIRFIELD UNIVERSITY
LIBRARY

MAR 0 1 2000

New York   Oxford

Oxford University Press

1998

Oxford University Press

Oxford   New York
Athens   Auckland   Bangkok   Bogota   Bombay
Buenos Aires   Calcutta   Cape Town   Dar es Salaam
Delhi   Florence   Hong Kong   Istanbul   Karachi
Kuala Lumpur   Madras   Madrid   Melbourne
Mexico City   Nairobi   Paris   Singapore
Taipei   Tokyo   Toronto   Warsaw

and associated companies in
Berlin   Ibadan

Copyright © 1998 by Oxford University Press, Inc.

Published by Oxford University Press, Inc.
198 Madison Avenue, New York, New York 10016

Oxford is a registered trademark of Oxford University Press

All rights reserved. No part of this publication may be reproduced,
stored in a retrieval system, or transmitted, in any form or by any means,
electronic, mechanical, photocopying, recording, or otherwise,
without the prior permission of Oxford University Press.

Library of Congress Cataloging-in-Publication Data
Experimental ecology : issues and perspectives /
edited by William J. Resetarits Jr. and Joseph Bernardo.
p.   cm.
Includes bibliographical references and index.
ISBN 0-19-510241-X
1. Experimental ecology.   I. Resetarits, William J.
II. Bernardo, Joseph, 1963–   .
QH541.24.E955   1998
577'.07'24—dc21   97-16559

9 8 7 6 5 4 3 2 1

Printed in the United States of America
on acid-free paper

This book is dedicated to the pioneers and evangelists of experimental ecology
—WJR and JB

and to Josh, Emlyn, and Cheryl, all of whom have shown incredible patience with me in their own distinct ways

—WJR

and to Pamela, who has given my life deep meaning
—JB

Preface

Experimentation played a pivotal role in the early development of ecology as a science. However, despite the seminal contributions of Tansley, Turesson, Gause, Clausen, Keck, and Heisey, Park, Huffaker, Connell, Paine, and others, experimentation has been rather slow to emerge as a primary approach to ecological investigation. Although many critical breakthroughs in evolutionary ecology have resulted from application of experimental approaches to either testing ecological theory or to answering specific questions derived from field patterns, it is only in the last two decades that experiments have come to occupy a principal role in ecological research.

In spite of this recent emergence, the experimental paradigm is now a dominant force in forging ecology into a more rigorous discipline, and it continues to expand rapidly, with both the number and quality of experiments increasing. Perhaps as a result of this rapid ascension, the application of experiments to ecological questions has not developed in an integrated fashion, nor has the role of experiments in ecology been the subject of much self-reflective discussion (at least in the literature). Until recently, experiments have had either proponents or detractors, with proponents strongly advocating the use of experiments and detractors seeking to stem their rising tide. As experiments have become more pervasive and more fully and irrevocably integrated into the fabric of ecology, proponents–practitioners of experimental ecology now enjoy the luxury of perhaps being less evangelical and more reflective on their craft. Ecology has arrived at a point where it is both possible and prudent to examine critically the role of experiments and the experimental paradigm within the larger arena of ecological investigation and to lay a foundation for the next phase in the evolution of experimental ecology.

As with most emerging paradigms, the development of experimental ecology has proceeded primarily through the efforts of individual investigators and laboratories, using diverse systems and approaches. Because of the lack of formal opportunities for conscious reflection, practitioners of experimental ecology have had relatively few fo-rums in which to discuss more philosophical ideas and exchange viewpoints. One goal of this project was to provide a framework to bridge that diversity of systems and approaches in order to synthesize a cogent statement about the role of experimentation as an analytical paradigm for ecology and to foster its conscious and thoughtful incor-poration into the fabric of ecological research.

Such an appraisal seems timely for ecology. With the continuing growth of exper-imentation and the increasing focus on experiments in general and experimental design in particular, it seems at least as important to discuss the more philosophical issues relating to the role of experiments in ecology. Our hope is that this volume will provide, for the first time, an effective conceptual and philosophical overview of experimentation as an ecological tool and will more fully establish its relationship to the other principal paradigms in ecology: theory and descriptive empiricism.

We were fortunate to assemble a distinguished group of ecologists whose research spans diverse axes (e.g., ecological/evolutionary questions, experimental approaches, taxa, and habitats). They were assembled with no agenda other than a shared belief that experiments play an important role in ecological research and that the role of experi-ments needs to be more critically debated and clearly defined. Our authors, in addition to being strong advocates of experimental approaches, are also extremely thoughtful and introspective individuals. We could not have imagined how well they would em-brace our vision for this project, sharing both their expertise and their individual view-points and philosophies on the role of experiments in ecology. Hence, we believe this book reflects the current state of the art and sets the tone for future directions for experimentation in ecology. We hope this synthesis will better integrate the diversity of opinion in the field and serve as a focal point for the continuing exchange of ideas.

The diverse group of practicing experimental ecologists represented here was largely drawn from the ''animal'' side of ecology. There are certainly as many from the ''plant'' side who could have made significant contributions to this volume, as well as other ''animal'' people we might have included. Clearly, many others have already contributed to the dialogue on experimental ecology. The ''animal'' emphasis was due partly to the appearance of the original symposium at the American Society of Zool-ogists meetings, but more so it resulted from our feeling that the practice of experi-mental ecology has always presented somewhat different problems to those ecologists who attempt to manipulate and assay the responses of mobile organisms. Consequently, plant ecology is in many ways far ahead of us in the practice of experimental ecology and evolutionary biology, and we individually, and animal ecology collectively, have certainly ''cut our teeth'' on lessons of experimental plant ecology. Given that frame-work, we tried to cover as wide a range of issues and perspectives as possible, in the belief that the insights of ''animal'' ecologists should be valuable to all ecologists. We sincerely hope that the emphasis on animal systems is not seen as reinforcing the separation of plant and animal ecology; rather, we believe this book reflects common issues in the practice of experimentation in ecology.

We brought to this venture our enthusiasm for the subject and our vision that this sort of synthesis was needed at this point in the evolution of experimental ecology. Given the contributors to this volume and the quality of the papers they have produced, it is easy for us now to stand back and say to the readers ''have at it.'' However, it is our most sincere hope that readers will not simply select the specific level of organization, ecological system, or taxa that appeals directly to them; the goal of bringing together a diverse group was not to provide ''something for everybody.'' Although each contribution certainly has important lessons to impart individually, we believe that the true value of this volume will lie in the emergent insights that arise from consideration of the entire work. Every student of ecology, whether a practitioner of experiments or critic of experimental approaches, will profit from careful consideration of the issues raised and the perspectives presented in this volume. Experiments are here to stay, and it is critical for all of us, proponents and opponents alike, to understand their potential contributions, as well as their limitations, within the practice of ecology.

Champaign, Illinois   W. J. R.
Philadelphia, Pennsylvania   J. B.
September 1996

Acknowledgments

In a project such as this there are always lots of people to thank—and lots of people who contributed who never get thanked. We will try to minimize the latter. First and foremost, we thank all the participants in the symposium and all of our contributors to the book for helping us make this project a reality. Although organizing a symposium and editing a book are undeniably a pain, that pain paled in comparison with the pleasure of interacting with this fantastic group of people. From the initial invitations to the final manuscripts, we could not have asked for a more enthusiastic and cooperative group. Their willingness to entertain and embrace our vision for this project is largely responsible for its success.

We thank ASZ Ecology Program Chair Art Dunham, who encouraged us to use the ASZ meetings as the forum for the symposium, and the American Society of Zoologists, for providing a fine venue and waiver of registration fees for the symposium participants. We also thank those who attended the symposium sessions and participated in the many stimulating discussions. We want to express our sincerest gratitude to our infinitely patient acquiring editor at Oxford, Kirk Jensen, for his unwavering support and encouragement throughout. Cheryl Rogers provided helpful editorial advice and prepared the index. Deb Corti took copious notes for us on every symposium presentation over two full days.

Perhaps the most critical component of any editorial operation is the review process. The quality of a publication is dependent in large measure on the quality of the reviewers, and this one is no exception. Over 60 individuals helped us immeasurably by reviewing one or more manuscripts; however, since anonymity would be the first (but perhaps not the last!) casualty if we listed their names, we will simply offer them all our heartfelt thanks.

Finally, we thank those who got us here in the first place—those individuals who are responsible for our passion for experimental ecology. We are extremely fortunate in having had the opportunity to interact with gifted and stimulating ecologists with diverse perspectives. We especially thank Janis Antonovics, Henry Wilbur, Mark Rausher, and Nelson Hairston, Sr., who formed the core of both our doctoral committees, for sharing their diverse perspectives on the nature, role, and goals of experiments in ecology. We can't imagine better inspiration for budding experimental ecologists; their guidance and example are the primary reason we could even conceive of this project. Similarly, what we learned from our "extended committee," the POPGEN group at Duke: Janis, Henry, Mark, Rob Brandon, Marcy Uyenoyama, and their students and postdocs, continues to astound us, and we perpetually benefit from their impact on our thinking.

Since I (WJR) first entertained the possibility of being an experimental ecologist, two people have served as my inspiration; I have other idols (several of whom are in this book), but two have always had special significance for me; Nelson Hairston and Bob Paine. They, too, are a large part of why this book came about, and I thank them for their inspiration.

Art Dunham opened a door for me (JB) onto ecology as a career and onto field-oriented studies of natural populations as an approach. I also thank Art, along with Bill Resetarits, for having always challenged me to think analytically and to argue from a logical perspective.

Contents

Contributors

JOSEPH D. BARNES
Department of Biology, Vanderbilt University, Nashville, Tennessee

STEVEN J. BEAUPRE
Department of Biological Sciences, University of Arkansas, Fayetteville, Arkansas

JOSEPH BERNARDO
School of Environmental Science, Engineering and Policy, Center for Biodiversity and Conservation, Drexel University, Philadelphia, Pennsylvania

STEVEN C. BLUMENSHINE
Department of Biological Sciences, University of Notre Dame, Indiana

JAMES H. BROWN
Department of Biology, University of New Mexico, Albuquerque, New Mexico

WILLIAM E. DIETRICH
Department of Geology and Geophysics, University of California, Berkeley, California

ARTHUR E. DUNHAM
Department of Biology, University of Pennsylvania, Philadelphia, Pennsylvania

JOHN E. FAUTH
Department of Biology, College of Charleston, Charleston, South Carolina

SALLY J. HOLBROOK
Department of Ecology, Evolution and Marine Biology and Coastal Research Center, Marine Science Institute, University of California, Santa Barbara, California

STEPHEN D. HURD
Department of Biology, Vanderbilt University, Nashville, Tennessee

CHRISTOPHER TODD JACKSON
Department of Biology, Vanderbilt University, Nashville, Tennessee

SHARON P. LAWLER
Department of Entomology, University of California, Davis, California

JOHN H. LAWTON
NERC Centre for Population Biology, Imperial College, Silwood Park, Ascot, UK

MATHEW A. LEIBOLD
Department of Ecology and Evolution, University of Chicago, Chicago, Illinois

DAVID M. LODGE
Department of Biological Sciences, University of Notre Dame, Notre Dame, Indiana

ROBERT J. MARQUIS
Department of Biology, University of Missouri–St. Louis, St. Louis, Missouri

ELIZABETH A. MARSCHALL
Aquatic Ecology Laboratory, Department of Zoology, The Ohio State University, Columbus, Ohio

J. A. MONGOLD
Center for Microbial Ecology, Michigan State University, East Lansing, Michigan

PETER J. MORIN
Department of Ecology, Evolution and Natural Resources, Rutgers University, New Brunswick, New Jersey

BARBARA L. PECKARSKY
Department of Entomology, Cornell University, Ithaca, New York, and Rocky Mountain Biological Laboratory, Crested Butte, Colorado

PETER S. PETRAITIS
Department of Biology, University of Pennsylvania, Philadelphia, Pennsylvania

CATHERINE A. PFISTER
Department of Ecology and Evolution, University of Chicago, Chicago, Illinois

GARY A. POLIS
Department of Biology, Vanderbilt University, Nashville, Tennessee

MARY E. POWER
Department of Integrative Biology, University of California, Berkeley, California

WILLIAM J. RESETARITS JR.
Center for Aquatic Ecology, Illinois Natural History Survey, Champaign, Illinois

DAVID N. REZNICK
Department of Biology, University of California, Riverside, California

BERNADETTE M. ROCHE
Department of Biology, Loyola College in Maryland, Baltimore, Maryland

FRANCISCO SANCHEZ-PIÑERO
Department of Biology, Vanderbilt University, Nashville, Tennessee

RUSSELL J. SCHMITT
Department of Ecology, Evolution and Marine Biology and Coastal Research Center, Marine Science Institute, University of California, Santa Barbara, California

KATHLEEN O. SULLIVAN
Weyerhaeuser Company, Tacoma, Washington

ALAN J. TESSIER
W. K. Kellogg Biological Station and Department of Zoology, Michigan State University, Hickory Corners, Michigan

JOSEPH TRAVIS
Department of Biological Science, Florida State University, Tallahassee, Florida

A. J. UNDERWOOD
Institute of Marine Ecology, Marine Ecology Laboratories, University of Sydney, NSW, Australia

YVONNE VADEBONCOEUR
Department of Biological Sciences, University of Notre Dame, Notre Dame, Indiana

JAMES D. WAGNER
Department of Entomology, University of Kentucky, Lexington, Kentucky

EARL E. WERNER
Department of Biology, University of Michigan, Ann Arbor, Michigan

CHRISTOPHER J. WHELAN
Midewin National Tallgrass Prairie, Illinois Natural History Survey, Wilmington, Illinois

DAVID H. WISE
Department of Entomology, University of Kentucky, Lexington, Kentucky

J. TIMOTHY WOOTON
Department of Ecology and Evolution, University of Chicago, Chicago, Illinois

EXPERIMENTAL ECOLOGY

Ecological Experiments and a Research Program in Community Ecology

EARL E. WERNER

It is not surprising that in a relatively young science such as ecology there will be a backdrop of discussion concerning appropriate methodologies and approaches. This self-consciousness is reflected in a long series of introspective articles over the last several decades (e.g., Weins 1977; Simberloff 1981, 1983; Lehman 1986; Peters 1991; Keddy 1992; Pickett et al. 1994; Weiner 1995). This discourse has brought with it much acrimony as individuals or groups have taken extreme stances on the merits or shortcomings of different approaches (see Strong et al. 1984). It is often clearer in hindsight that the excesses of various approaches need to be sanctioned, not the approaches themselves.

The questions before us are not whether we should do experiments in community ecology or to what extent. We should, and in abundance. Nor is the question whether experiments are the only way to contribute toward a predictive ecology. They are not. Questions concerning the proper protocol for conducting and analyzing experiments certainly are important, but these questions too often have been asked out of context of a larger research program in community ecology. The relevant concerns, it seems to me, are: How do experiments fit into a research program? What role should they play, and how does experimental work relate to descriptive and theoretical studies? That is, how does a research program in community ecology shape the use and design of the experimental protocol that we employ? The strategy, philosophy, and logistics of the research program are topics that have been relatively little emphasized in the discussions of methodology in ecology.

In this essay, I detail a particular research program directed at the study of ecological communities. It is a personal testimony to an approach that I have found useful in satisfying my own curiosities about the nature of ecological systems. I provide illustrations from my own and colleagues' work with fish and amphibian assemblages that have been instructive to me in developing this perspective. I am not attempting to outline what I think is the only way to do our science. There are many useful approaches to extract scientific understanding of the systems that we choose to study (for an extreme statement see Feyerabend 1978), and this diversity is needed if a science is to progress. The research program that I outline in the following discussion is a productive way to approach certain questions in community ecology, but certainly for many other questions it will not be feasible or productive. The art form here is to maintain perspective on the limitations of different approaches and the inferences that can be drawn from them.

Background

In recent years, two broad approaches have been important in the pursuit of understanding ecological communities. The approach that came to be closely associated with Robert MacArthur in the 1960s and 1970s consisted of comparing patterns in the natural world to those predicted by specified processes. Essentially this method took a hypothetico-deductive approach to developing mathematical theory and then attempted to use the theory to predict or account for patterns in the natural world. Many processes can contribute to community development, and the rationale was that such pattern analysis should be able to discriminate between these processes, since each should correspond to a different predicted pattern (Inchausti 1994).

The pitfalls and misuses of this approach have been the whipping boys of many ecologists over the last several decades (e.g., Connell 1980, Simberloff 1983, Strong et al. 1984). First, many dynamics can lead to similar patterns, so correspondence between pattern and theory does not necessarily identify the correct causal process. Second, quite vague concordance of predictions and pattern was often taken uncritically as support for the theory and the underlying hypothesis of the process (Simberloff 1983). The latter was a large source of the "excesses of the orthodoxy" (Colwell 1984) of the 1960s and 1970s that led to a considerable backlash to the approach in the 1980s. This backlash certainly helped foster the increasingly experimental nature of community ecology over the last 15–20 years. The problem, however, was not inherent in the method itself so much as in the liberties taken with the method. There is much to be gained from attempts to compare predicted patterns from theory with those of the natural world, but there are no shortcuts to understanding complex ecological systems (e.g., the promotion of measures such as overlap in resource use for competitive effects or simple measures of who eats whom to understand the dynamics of food webs). Reevaluation of the manner in which community ecology was conducted in the 1960s and 1970s certainly had a salutary effect on the rigor with which most ecological studies were conducted. Experimental studies became de rigueur, and there was much discussion of appropriate methodology (e.g., Hurlbert 1984, Scheiner and Gurevitch 1993, Underwood 1995).

The other major approach to questions concerning the processes responsible for community structure employed perturbation experiments. Though there is a long history of experimental work in ecology (Jackson 1981, Paine 1994), many feel that the modern era of experimental community ecology begins with the work of Connell (1961) and Paine (1966). In this case, some aspect of community composition, the environment of a community, or a chosen segment of a community is manipulated. A significant response to this manipulation is interpreted as an indication that this factor is important in determining the structure of the community. This approach certainly is more powerful in determining causality than pattern analysis, and this power has led to its dominant status in current community studies. In fact, strong adherents to the approach feel that there is no other avenue to unraveling the factors responsible for the structure of ecological communities (Paine 1994).

The limitations of this approach are less publicized, though certainly recognized by many ecologists. The results of perturbation experiments in complex systems are neither so clear, precise, nor easily interpretable as generally suggested (Bender et al. 1984, Diamond 1986, Yodzis 1988, Inchausti 1994). For example, the indirect effects inherent in complex systems make it difficult to pinpoint causal pathways, time scales of direct and indirect effects can differ greatly, and there are correlated effects of manipulations that introduce artifacts into the results. Moreover, in community ecology such experiments fall far short of providing a predictive basis for ecological systems because of the lack of integration with a strong conceptual or theoretical foundation. These experiments also are rarely directed at uncovering mechanisms. In other words, this approach has largely focused on a search for factors, not the construction of a coherent conceptual framework for predicting the consequences of interactions of these factors.

In the remainder of this essay, I will discuss a research program that integrates aspects of both the experimental approach described previously and pattern analysis. I hope to convince the reader of the power of focusing on mechanism and integrating aspects of the preceding two approaches. Viewed in this context, the chauvinisms of the practitioners of different approaches become moot. Each approach has its place, its advantages, and its limitations. The issue is how to most effectively use the strengths of each approach so that they enhance each other and thereby increase our overall effectiveness at generating understanding in community ecology.

Elements of the Program

Community ecologists are interested in understanding features of natural assemblages such as species composition, relative abundances, diversity, and food web structure as they are influenced by species interactions. The research program that I discuss here emphasizes the importance of integrating descriptive/comparative, experimental, and theoretical work to approach these questions. It is the iteration among theory, experiments, and a specific field pattern that is so valuable. Much has been written on the practice and merits of each of these elements of the research program. What is lacking in most discussions is a clear view of how to integrate different aspects of the scientific process into a successful research program.

Biological communities may be described as hierarchical systems, and my approach is explicitly mechanistic in that it seeks explanatory factors of a specific phenomenon in entities and processes that belong to a lower level (Schoener 1986a, Inchausti 1994, Pickett et al. 1994). Mechanistic approaches to community ecology often employ individual-level ecological concepts—for example, those of behavioral ecology, physiological ecology, and ecomorphology—as a basis for understanding community patterns. This mechanistic approach embodies the attempt to define and interrelate theory at each level in the hierarchy: individual, population, and community. (What is viewed as mechanistic, of course, depends on where one stands in the hierarchy; all models of phenomena are at once mechanistic and phenomenological depending on one's perspective.) Though typically I have sought causality in lower level mechanism, causality is bidirectional in these systems and I briefly discuss "downward causality" later.

The Descriptive/Comparative Component

There are many advantages if a research program is focused on a clear and dramatic pattern in the field, the description of which is part of the program. First, the focus on broad, repeatable patterns ensures that we can infer something general about ecological systems from our studies (Tilman 1989). All studies perforce will be narrow and specific to the system that we choose. The broader and more repeatable the pattern that the phenomena in our system represent, the more likely that we can emerge from the uniqueness of our study system with abstractions that are general in their implications. Second, description/quantification of the field patterns (and the comparative analyses this avails), provides the inferences that guide the development of theory and directs experimental work. Third, and very important, making explicit predictions from the experimental and theoretical work to a particular pattern in the study system provides the self-correcting element of the program. This exercise constantly realigns the questions asked and the experiments and theory considered. Unfortunately, much experimental ecology is conducted without this sort of context, and this can result in experimental programs directed at phenomena suggested by experimental artifacts. In the absence of the context provided by the field pattern and the self-correcting steps it avails, it is easy for a program to be deflected into unproductive avenues that can consume a research career. Last, but not least, the presence of a clear and striking pattern generally indicates the presence of strong causative factors. This knowledge helps to maintain our motivation when we encounter dead ends in theoretical and experimental work directed at understanding the pattern. There is at least the sense that factors can be eliminated and we can go back to the drawing board somewhat wiser.

Several striking patterns have been the backdrop for our studies. In our work on fish assemblages, the obvious patterns in habitat segregation of species and differences in species composition or relative abundances across lakes served as this backdrop. Our work on amphibian assemblages similarly has focused on the strong patterns of species' replacement along the environmental gradient of permanent to ephemeral ponds and certain patterns of change in species composition over larger expanses of space and time. In both cases, the patterns are clear, repeatable in most taxa within these groups, and representative of broad patterns across a wide range of taxa. All our studies are examined eventually in the context of what each can tell us about the particular patterns in the field systems that we study. In the following discussion, I illustrate some of these points.

The patterns of habitat partitioning in the fish assemblages naturally suggested that competitive interactions might be important. This prompted a series of small pond experiments that elucidated the basic mechanisms responsible for habitat partitioning among these fish (Werner and Hall 1976, 1977, 1979). We documented strong niche shifts under competition that conformed to the field pattern and showed that these shifts were an example of the compression hypothesis of MacArthur and Pianka (1966). This work documented that species, in the presence of competitors, often retreated to non-preferred habitats where they realized a relative foraging advantage. Thus, we were able to elucidate a general mechanism proposed to account for habitat partitioning in animals (Schoener 1986a,b).

Attempts to document the habitat partitioning in natural lakes and to take predictions from these experimental studies back to the lakes, however, alerted us to patterns that demanded a much deeper understanding. For example, while conducting descriptive studies to document patterns in habitat use suggested by the pond experiments we found that different size classes of the same species were found in different habitats (Hall and Werner 1977, Werner et al. 1977). Either resource partitioning was on a much finer scale than expected or other factors were unaccounted for. It was not apparent from the mechanisms documented in the experimental ponds (with similar-sized individuals) why size-specific partitioning of habitat would occur.

This puzzle motivated in large part the work that we conducted on interactions in size-structured populations (Werner and Gilliam 1984, Werner 1986) and caused us to examine the classical ideas on resource partitioning in a different light. Laboratory work on the scaling of foraging relations with body size (Werner 1977, Mittelbach 1981) enabled us to predict some of these habitat use patterns. Further, this prompted comparative work across lakes on the consequences of size-specific predation risk to species habitat use (Werner and Hall 1988), and colleagues have examined the ramifications for species interactions in this system (Mittelbach and Chesson 1987, Osenberg et al. 1994). Ultimately, these studies led to the identification of consequences of nonlethal effects of the presence of predators on habitat use of, and interactions among, species (Mittelbach 1981, Werner et al. 1983b, Werner and Hall 1988). Study of the behavioral responses to predators has subsequently become a very active area of ecological research (Lima and Dill 1990). Models of optimal habitat shifts as a function of size, developed in this context (Gilliam 1982), additionally led to speculation on the evolution of complex life cycles (Werner 1986, 1988).

What was clearly so valuable in this endeavor was the iteration among theory, experiment, and the specific field patterns. Experimental advances and theoretical developments were intimately tied to iterations with the descriptive phase. The patterns directed the thrust of the research, and the research also caused us to look more closely at the patterns. When either an experiment or theoretical work is only vaguely related to an explicit pattern (e.g., descriptions of such patterns in the literature) the intimate iteration with the descriptive work that guides the research is largely lost. For example, much of experimental community ecology conducted in laboratory containers, cattle tanks, and gardens focuses on a process. These studies are usually motivated by general patterns in the literature or the known importance of the particular processes. These are extremely valuable, but the limitations of inferences from the results of such experiments are not as readily apparent as when predictions can be taken directly back to a specific field pattern. In the long run, this iteration greatly enriches a research program. The more explicitly that predictions can be taken back to the pattern, the more specific the directions to future work will be.

Another example of the insight gained from very primitive sorts of descriptive studies comes from our amphibian work, in this case at very different spatial and temporal scales. Simple presence or absence data on 14 species of amphibians across 32 ponds on the University of Michigan's E. S. George Reserve collected in 1967–1972 by Collins and Wilbur (1979) and by us from 1988 to 1992 (Skelly et al. forthcoming) indicated that the average species experienced roughly 50% turnover of breeding ponds between surveys. These dramatic changes, however, led to little net change in the amphibian species composition on a regional scale: local (pond) extinctions (37) were balanced by colonizations (40). These data and data retrieved from aerial photographs suggested (1) that the changes in the amphibian assemblage in part were directional due to forest succession around the ponds and (2) that some of the changes were due to metapopulation structure of the species' populations. These very minimal descriptive studies on an expanded spatial and temporal scale have generated a host of hypotheses about the dynamics of amphibian populations on the reserve and the mechanisms of changes in community structure due to succession and metapopulation dynamics. Some of these inferences have been supported by recent experimental tests (Werner and Glennemeier forthcoming), which in turn suggest more refined and directed studies of the distributional properties of these species on the reserve. It would be very difficult to convince most ecologists to fund this type of survey work today, and yet in the context of a specific research program it becomes a bonanza of ideas directing research.

The preceding discussion illustrates how descriptive/comparative work can be an essential element of a research program and not a separate approach to the study of ecological communities practiced in isolation. Criticisms of the way descriptive work was employed in the 1960s and 1970s have caused prominent ecologists to bare their souls in the confessional of book introductions to justify practicing it (e.g., Brown 1995). Yet little can be more exciting or important than the first comprehension of a pattern in the field and the glimmerings of its implications. And little can be more captivating than the opportunity that such a pattern avails to grasp at generality, to attempt to generate an explanation or possible alternative explanations responsible for the pattern. The excesses of the 1960s and 1970s in using descriptive approaches to test ecological theory are subject to criticism, but we have to guard against setting aside one of our most critical tools for the wrong reasons.

The Experimental Component

A research program clearly should have a strong experimental component. There is no need to justify the value of experiments for understanding patterns in community structure. Experimentation is our most powerful tool for uncovering causality, and experimental manipulations at the community level have been the centerpiece of my own research program. Next I focus on how we can effectively integrate experiments into a larger research program. But first I will note some of the shortcomings that I perceive concerning the current practice of experimental community ecology.

Modern experimental community ecology is dominated by perturbation experiments. These studies are often conducted in the field or in more controlled surrogates for the field (e.g., gardens, cages, small ponds, and cattle tanks). These are the sorts of experiments that, for example, form the basis for recent reviews of the effects of competitors or predators on communities (Connell 1983, Schoener 1983, Sih et al. 1985, Hairston 1989, Goldberg and Barton 1992).

The power and importance of such experiments are unquestioned. What is generally overlooked are their limitations and the fact that other types of experiments are often suggested, or analyses subtly changed, when experiments are viewed as integral parts of a larger research program. For example, greater attention may be given to experiments that are used to parameterize theory or to test lower level theory that provides the mechanistic basis for our inferences. Further, greater attention may be given to developing an explicit model that will underpin an experimental design, as opposed to simply accepting the traditional analysis of variance (ANOVA) model. Examined in the context of a larger research program, experimental community ecology takes on a different flavor than is represented, for example, in the reviews mentioned previously.

What are some of the limitations of the general perturbation experiment approach? As noted earlier, the results of such experiments in complex systems are neither so clear, precise, nor easily interpretable as they are generally given credit for being. As Bender et al. (1984), Yodzis (1988), and others have noted, the role of indirect effects in the larger community and the differing time scales of direct and indirect effects complicate the interpretation of field perturbation experiments. Further, there always are correlated effects of manipulations that can cause strong artifacts in ecological experiments. There also are logistical limits on the number of factors that can be manipulated in a given experiment, and there are critical ecological questions that reside at scales at which it is impossible for us to conceive of conducting experiments (e.g., Brown 1995).

In addition, and perhaps more important, the practice of field perturbation experiments isolated from descriptive and theoretical work reduces their power. For example, suppose a perturbation experiment is conducted manipulating the presence of a predator to demonstrate that the predator and, by inference, predators in general have an effect on the community of interest. If the predator does have an effect, what is the next step? We may have a qualitative understanding of the impact of predators on community properties, but too often the research stalls at this point because directions to further work are missing. This sort of experimentation leads to catalogs of examples (such as abundantly represented in the reviews noted previously), with little conceptual development of the field. But when conducted in the context of a natural pattern and theoretical development, such experiments prompt conceptual developments, tests of mechanisms, and the experimental derivation of parameter estimates that permits more quantitative tests of the theory.

To take a specific example, we have documented very strong indirect effects of a fish predator on the distribution of bullfrogs (*Rana catesbeiana*) and green frogs (*R. clamitans*) along an environmental gradient (Werner and McPeek 1994). Manipulating the presence of fish indicated that they facilitate the presence of the bullfrog by eliminating invertebrate predators. Thus, we have a qualitative understanding of how direct and indirect effects cause differences in community composition along the environmental gradient. However, we are in no better position than before to predict how average relative abundances of these species change along the gradient or to interpret the extensive variation in relative abundances. Nor are we in a good position to go to a new system and predict whether such effects will be important there. The same topology of food web will give very different results as strengths of interaction coefficients change (Werner and McPeek 1994). Yet this is where most experiments in community ecology are left. In other words, we do not understand the mechanisms responsible for the strength of the interaction coefficients and what generalizable species attributes are associated with these. It is the latter that provide much of the direction for further research.

We explored the mechanisms responsible for these interactions in a series of laboratory experiments. These studies indicated that general activity level and body size affected competitive ability, predation risk, and the trade-offs involved (Skelly and Werner 1990, Werner 1991, Skelly 1992, Werner and McPeek 1994, Anholt and Werner 1995). We then developed models of the adaptive responses of anurans under this trade-off (Werner and Anholt 1993) which we are now testing. This theory should allow us to quantitatively predict the changes in strength of interactions in different environments along the gradient. Focusing on the mechanisms and theory development identified individual traits that are general and that determine the strength of interactions of many taxa with their competitors and predators (Werner and Anholt 1993, Wellborn et al. 1996). This approach further prompted new research directions as it became apparent that a consequence of adaptive responses would be strong trait-mediated indirect effects (a higher order interaction; Abrams 1995, Werner and Anholt 1996, Peacor and Werner 1997). If the strengths of interactions can be predicted using this approach, this will lead us back to the field for further refinement of the natural pattern.

It is the separation of theoretical and experimental traditions in community ecology that is part of the problem with the standard practice of perturbation experiments. This separation has several important consequences. First, in the absence of explicit theoretical predictions to guide experimentation, there is a preoccupation with simply demonstrating that a factor is "important." Thus, the experimental program is not effectively harnessed to adjudicate theory or major conceptual frameworks or to measure quantities that can be employed with the theory to make more specific predictions for further tests.

Second, theory is not used as an effective guide to experimental studies. Theory frames very explicit hypotheses concerning the nature of the interactions occurring, and we can then tailor the experiment or analysis to examine these hypotheses. That is, along with the preoccupation of simply demonstrating that a factor is important comes a kind of unhealthy reliance on the standard designs of the ANOVA. Although the ANOVA is a powerful method for analyzing the effects of different factors and providing a basis for designing competent experiments, it can be a kind of mental straitjacket in the sense that standard designs easily become a surrogate for thinking critically about how we conceptualize a process or interaction—that is, developing a model of the system that we envision. Attributes of communities result from the joint operation of a number of causative chains that interact, and what we are fundamentally aiming for is a general statement of functional relationships in such systems. The ANOVA is not a good tool for elucidating these functional relationships unless our systems are perfectly additive (i.e., causes independent; Levins and Lewontin 1985; Chapter 4). We often expect the interactive effects to dominate in our systems, and yet these are the effects poorly handled by the ANOVA (see Levins and Lewontin 1985 and Wade 1992 for analyses of similar considerations in genetics). The confusion that can arise when there is no explicit model guiding the experimentation is evident in the debates in the behavioral ecology literature as to whether an interaction term in an ANOVA really indicates that animals are balancing resource gain and predation risk (e.g., Cirri and Fraser 1983, Horat and Semlitsch 1994) and the debates as to whether an interaction term represents a higher order interaction when a third (or more) species is added to a community (e.g., Worthen and Moore 1991, Billick and Case 1994, Wootton 1994). In both cases, it is clear that a significant interaction term may or may not indicate these phenomena—often the standard ANOVA simply is not the appropriate model or the appropriate test (Billick and Case 1994, Wootton 1994). Each ANOVA represents a particular model that is being tested—something usually ignored, as is the fact that the model changes (e.g., from additive to multiplicative) with transformations of the data (Billick and Case 1994, Wootton 1994). If a theoretical model of the system is guiding the experimentation, experiments often will be conceived or analyzed differently and a number of these problems circumvented.

Further, preoccupation with only demonstrating that a factor is important results in catalogs of examples that in turn foster analyses of, for example, whether a phenomenon is caused by this or that factor or the relative importance of factors (is it competition or predation?, top-down or bottom-up?). In interacting systems, this is a fruitless enterprise as it is conceptually impossible to assign quantitative values to specific causal factors or separate them in this way (Levins and Lewontin 1985). The emphasis should instead be on how such factors interact and the nature of the interconnectedness.

Third, perturbation experiments in community ecology are too rarely designed to build on what we know and extend it or even to measure those parameters of theory that supposedly represent what we think we know. There is an extensive tradition of theory in ecology concerning the effects of competition between species. Despite the shortfalls of this theory, especially the Lotka-Volterra system, it identifies several issues that are critical to the evaluation of competitive interactions. For example, the ratio of intra- to interspecific effects is critical, and the theory requires some estimate of per capita effects. Yet experiments continue to be conducted that simply look for an effect of competition and rarely quantify the above parameters (see Connell 1983, Schoener 1983, Goldberg and Barton 1992). Experiments should be designed to provide measures of the parameters of the theory and to determine functional relations inherent in the interactions. Consequently, experiments or observations often should be nested within larger perturbation experiments to document mechanisms and to quantify important functional relations (see also Leibold and Tessier, this volume). A closer integration of theory and experiments would lead to better experiments and better theory.

Integrating experiments into a larger research program changes the character of the experiments we conduct in other ways. For example, because explanation is often found by going to lower levels of organization to uncover mechanisms, experiments perforce are conducted over a range of different levels and scales. This integration across levels offers two advantages. First, new phenomena emerge at different levels or scales. Second, the strengths of studies at one level can be used to increase the effectiveness of studies at other levels.

Conducting experiments at different levels and scales confronts the issue of trade-offs between control and realism. This point has been made a number of times in relation to conducting experiments in the field versus the laboratory (e.g., Diamond 1986, Hairston 1989, Morin, this volume). In most such discussions, the case is made that experiments at different scales are intrinsically better at answering different questions. I take a slightly different emphasis. If this inevitable trade-off is acknowledged, how should experiments of different types or scales be integrated into a research program in community ecology to maximize progress? Here the emphasis is on how to use the power of experiments at different scales to attack common, not different, questions and how this integrated approach changes the emphasis of experiments conducted at different levels of realism and precision compared to their conduct in isolation.

The experiments that we have conducted on fish and amphibian assemblages lie on certain countergradients of control and realism (essentially Morin's precision and realism gradient). These gradients are represented by studies conducted at scales ranging from laboratory containers through pens and cattle tanks to small ponds and finally the field. Experiments conducted at these various scales also traverse levels of organization from the individual to the community. The art form is to use these countergradients in such a way that the drawbacks of going one direction on the gradient can be countered as much as possible by the advantages of going the other direction. Specifically, studies must be integrated in such a way that the artifacts introduced by experimental control are eventually canceled out by the guiding hand of reality, and the intractableness of reality must be whittled away by the surgical knife of control.

At the one extreme, we have conducted highly controlled studies in laboratory aquaria or small containers. These studies have the potential to suffer from many artifacts but enable us to examine the specific mechanisms related to interactions between species or to parameterize the models that are used to make predictions concerning interactions in the field. The results of these experiments have to be extrapolated to the natural community to reveal both their predictive power and their artifacts. Such experiments have been very useful for us. For example, Mittelbach (1981) conducted laboratory experiments to parameterize foraging models that enabled him to predict optimal diet and habitat use of bluegill sunfish (*Lepomis macrochirus*) in the field (see also Werner et al. 1983a, Osenberg and Mittelbach 1989). Similarly, we have conducted laboratory studies to determine the activity responses of amphibians that in conjunction with theory have enabled us to interpret the outcome of competitive and predator–prey interactions and determine the presence of trait-mediated indirect effects in these communities (Werner and McPeek 1994, Anholt and Werner 1995, Werner and Anholt 1996, Peacor and Werner 1997).

An example of a case where artifacts evidently rendered laboratory studies inaccurate is work that I have conducted on competitive mechanisms in wood and leopard frog larvae (Werner 1992a). These experiments were conducted under highly controlled conditions with measured resource levels to elucidate the role of size and activity level on the outcome of interactions. The laboratory results were clear and appeared to fit the patterns of the influence of activity level on competitive interactions—that is, leopard frogs were more active and competitively dominant. Contrarily, every experiment that we or others have conducted in more natural settings has yielded the opposite conclusion: that wood frogs are the clear and asymmetrically dominant competitors (Smith-Gill and Gill 1978, Werner and Glennemeier forthcoming). There is no a priori way to anticipate the adequacy of these laboratory experiments to reflect natural phenomena. Therefore, the attempt to take predictions from the controlled studies back to more natural conditions is critical to prevent deflections of a program in unproductive directions.

In order to take the results of highly controlled laboratory work to a more natural context, we have made extensive use of small cage or container experiments (i.e., moving up the gradient of increased reality, Morin's hybrid experiments). This level of control is widely practiced in experimental community ecology. In our case, these experiments are conducted in either cattle watering tanks or cages on the scale of 1 to 3 m² constructed of screen and set in ponds and lakes. Thus, resources and spatial relations are more realistic. These experiments have been very useful in testing the predicted interactions proposed from individual behavior (e.g., Mittelbach 1988, Skelly 1992, Werner 1994, Werner and Anholt 1996). Interactions can be examined at this scale or level of control that are impossible to study at other scales or higher levels, but these experiments are not without their associated artifacts. Tanks, cages, limno-corrals, and so on can introduce unnatural environments or favor certain species. Also, such experiments obviously do not capture the appropriate spatial scale for interactions in many natural communities.

The next scale of experiments along the gradient of increasing reality that we have employed is that of reasonably large experimental ponds (15 to 30 m in diameter and 1 to 2 m deep) with natural substrates. These ponds have an established natural fauna and flora and further increase the spatial reality of the interactions in question while still allowing us to retain considerable control over species composition of the communities. This has been an extraordinarily productive scale on which to explore questions of the consequences of species interactions, and most of these results appear to translate to interactions in natural communities. As noted earlier, for fish assemblages we used this scale to elucidate the basic mechanisms of resource partitioning (Werner and Hall 1976, 1977, 1979) and the role of predation risk in affecting the size-specific habitat use of fish (Werner et al. 1983b). Both phenomena are important in determining the interactions and distributions of species in natural lakes (Mittelbach 1981, Werner and Hall 1988) and have been found to be important in many other systems (Power et al. 1985, Schoener 1986b, Brown 1988, Kohler and McPeek 1989, Kotler et al. 1991, Persson 1993). We also have used the ponds to illustrate how indirect effects are responsible for differences in the distribution of several ranid frogs along an environmental gradient (Werner and McPeek 1994).

The final scale of study on the gradient toward greater reality is the actual field situation (lakes in the case of fish and natural ponds in the case of amphibians). This level represents the ultimate reality check on understanding the phenomena generated in the more controlled arenas. Experiments are certainly feasible at this level (e.g., Carpenter et al. 1987, Carpenter and Kitchell 1993), although we have primarily conducted descriptive/comparative studies at this scale. First, with both the fish and amphibian assemblages, comparative studies over both space and time set the critical questions and drove the more controlled experimental work in profitable directions. For example, comparative studies of the fish assemblages of a series of lakes guided our studies, and subsequently those of colleagues, on the competitive interactions between species (Hall and Werner 1977, Mittelbach 1988, Osenberg et al. 1988), the role of habitat structure and resource abundances in affecting relative species abundances (Osenberg et al. 1994; see also Leibold and Tessier 1991), and the role of predation risk in size class interactions in species (Werner and Hall 1988). Second, as previously discussed, these descriptive studies provide the patterns that we compare with predictions generated from the experimental and theoretical work. These comparisons have been critical in directing research by indicating factors left out of the explanatory framework developed within the more controlled arenas. It is impossible to overestimate the importance of this guidance for the experimental program.

By moving from the field through ponds and cages to the lab we were able to decompose complex systems into manageable subcommunities, populations, or individuals to conduct controlled studies of mechanisms. The field studies provided the ideas and guiding patterns that, in turn, provided the context for these experimental studies. Going the other direction, from highly controlled laboratory studies to studies subject to the uncontrolled variation in the natural environments, provided the correcting force on the program of experimental work, weeding out the artifactual and irrelevant results and ideas and guiding subsequent experimentation. In the process of attempting to make explicit predictions from each level to the adjacent, the coherence of the predictive framework is constantly checked and refined. It is not possible to go far afield if the results at one level of control are constantly judged on the basis of whether they can predict results at the adjacent levels or at least are consistent with results at adjacent levels. Again, the focus on mechanism and relating models of phenomena up and down levels of organization or scales casts experimental studies in a slightly different light, and this is one of the important roles of theory. The theoretical component is the glue that holds these studies at various levels together. Theory should integrate results from, for example, the individual through the population to the community.

The Theoretical Component

As an essential element in a research program, theory is used to embody the perceived generalities of a field in a coherent statement and can therefore be our aid to extrapolation, generalization, and understanding (e.g., Pickett et al. 1994). Theory forces us to make assumptions clear and unambiguous. It gives coherence to facts and observations. Theory helps us see through the complexities of our systems to the essential elements. Thus, theory can be used to clarify hypotheses, identify key elements in a system, suggest the critical experiments, and provide a forum for presenting and examining the logical consequences of new ideas.

Theory therefore has great advantages in clarifying our thinking about a phenomenon. It is puzzling that often those who vehemently complain about the simplifications of theory are quite willing to take vague verbal models as their guide. Verbal models rarely allow very specific predictions and generally are as simplistic as most mathematical theory. Theory is also very useful in providing a guide for data standardization that then enables the comparison of effects across different systems and generalization of these effects (Kareiva 1989).

A close dialogue between theoretical and empirical components of a research program strengthens both enterprises. When theory is developed in the context of an experimental system, it tends to be framed in operational terms and is directed to questions that are feasible to address experimentally. Further, experiments can then be used to parameterize the theory so that explicit quantitative tests can be performed. Kareiva (1989) makes the plea that more effort be devoted to devising models that include concrete, directly measurable parameters, noting that this often leads to less general models, but models that will receive much greater attention by biologists. (Of course, some theory is not intended to be operational and directly testable, but to develop ideas or explore limiting cases and the implications of ideas.) A dialogue with theory in a research program perforce increases the generality of the program. The current structure of ecological work is not optimal for a profitable interaction between theory and experiment. Experimental ecologists often harbor a distrust of theory and to a large extent practice their craft in the absence of a direct dialogue with it. Similarly, theorists often are not attentive to empirical traditions and do not formulate their theories in operational terms. The lack of a constructive dialogue limits both theoretical and empirical advances in ecology (Kareiva 1989, Pickett et al. 1994).

Theory is an essential tool in prompting new research directions. First, theory can motivate the collection of data or conduct of experiments that would not otherwise occur to the investigator. An example from my own experience concerns the role of higher order interactions or, more specifically, trait-mediated indirect effects (Abrams in press). Theory that we developed to predict the adaptive activity responses of individual organisms to a growth/risk trade-off (Werner and Anholt 1993) suggested that these responses should result in higher order effects in communities. That is, changes in activity alter competitive ability and the ability to avoid predators, and therefore the presence of predators should modify competitive interactions among prey and presence of competitors should modify predator–victim interactions. In these cases, addition of a third species to a community alters the per capita effects of two other species on each other. Experiments motivated by this theory document that both of these effects are quite strong in an amphibian system (Werner 1991, Werner and Anholt 1996, Peacor and Werner 1997).

Second, using theory to take predictions from lower level mechanisms back to higher levels or field patterns also is an important means of discovery. The traditional modus operandi in testing theory is to work at just one level (or to drop a level to see if mechanisms are consistent). For example, the individual-level theory that we have employed obviously needs to be tested at the level of the individual, and this is the agenda of behavioral ecology. A research program is greatly enhanced, however, if predictions of the implications of individual behavior are applied to higher levels and more complex situations. As noted earlier, attempts to take predictions of foraging models that were reasonably successful both in the laboratory and in simple pond communities (Werner and Hall 1974, Werner et al. 1983a) to a natural lake implicated the role of predation risk in foraging behavior (Mittelbach 1981). The larger individuals of a bluegill population closely approximated the predictions of habitat use that maximized foraging rates, but the small individuals violated the predictions and remained in cover. This study and those of Sih (1982) clearly identified the important role of predators in altering behavior of prey in an ecological context (Lima and Dill 1990). These results have influenced conceptual advances both at the level of the theory of individual behavior (Gilliam 1982, Houston et al. 1993) and at the level of the community (reviewed in Werner 1992b). Constant reference to the larger system also keeps a research program focused and in context. This assures that foolish questions are not posed at the lower level. Thus, deviations from theoretical predictions are a powerful way of indicating important factors that we have not incorporated in our explanatory structure and therefore indicating new directions for theoretical and empirical work. Kareiva (1989) makes a similar point about the role of "bottle experiments" in population biology illustrating the need to incorporate spatial processes.

The close iteration of theory, experiment, and descriptive fieldwork has facilitated advances on all fronts in our work. For example, in an attempt to explain a pattern of size selection of prey in bluegill sunfish (Hall et al., 1970), we developed a model of optimal foraging behavior (Werner and Hall 1974). We tested this model in the laboratory (Werner and Hall 1974), in the experimental ponds (Werner et al. 1983a), and in the field (Mittelbach 1981). This iteration of theory, experiment, and field test led us in several important directions. Deviations from predicted patterns led to studies of the role of learning and sampling in foraging behavior (Ehlinger 1990) and ultimately to evidence for individual differences related to intraspecific morphological differentiation that have implications for speciation theory (Ehlinger and Wilson 1988). Tests of the theory also necessitated that parameters be measured as a function of predator size, and this opened questions concerning ontogeny and size-structured interactions (Werner and Gilliam 1984). Tests in the field identified the role of predation risk in foraging behavior culminating in the development of the μ/g rule predicting size-specific adaptive behavior under the trade-off between foraging rates and predation risk (Gilliam 1982, Werner and Gilliam 1984). The tests in more controlled environments (laboratory and ponds) gave us confidence that the theory was predictive at some level, and the field tests showed where it was incomplete. The theory was further used to develop a general model of adaptive activity responses in mobile animals (Werner and Anholt 1993). Thus, the iteration of theory, experiment, and field predictions was very important to both advances in the theory and new directions for our research program.

One might argue that such a close interaction between theory and experiments might result in very system-specific theory. We have not found this to be the case. For example, the foraging theory developed in the context of understanding size selection of prey by the bluegill was quite general. Attempting to understand why species of sunfish undergo niche shifts at certain sizes during their ontogeny prompted Gilliam's (1982) μ/g model. Conceptualized in this manner, the theory also turns out to be quite general (e.g., see Houston et al. 1993, McNamara and Houston 1994). The μ/g theory also has provided a mechanistic tool for understanding when ontogenetic niche shifts occur in many taxa and what the implications are to species interactions (Werner and Gilliam 1984). Even more broadly, the μ/g results suggested a way that we could conceptualize metamorphosis in animals (Werner 1986, Ludwig and Rowe 1990, Rowe and Ludwig 1991) and then more generally a hypothesis about the evolution of complex life cycles (Werner 1988, Moran 1994). Thus, the attempt to explicitly relate the individual foraging behavior of bluegill to ontogenetic niche shifts so that we could predict species interactions across ontogeny in a particular fish assemblage spawned a theory that was very general in applicability. This illustrates that conceptual relations that are developed in the context of making predictions for a system do not have to be specific to that system. It is almost always possible to identify the biological features key to the phenomena under investigation that then can be abstracted in general statements.

In this context, it is useful to consider why have we focused on the individual and the consumer/resource link in the development of a mechanistic approach to communities. The consumer/resource link is a fundamental dynamic link between species in ecological systems, forming the basis for analysis of competitive, predator-prey, host-parasite, and some mutualistic interactions. We reasoned that if we could predict resource use of species then we could predict the manner in which these interactions changed as a function of resources and predators (as well as the effect of predation on resource species). In order for this approach to be useful, such models must incorporate species characteristics that are responsible for differences in foraging effectiveness or predator avoidance (e.g., Werner 1977, Mittelbach 1981, Osenberg and Mittelbach 1989, Werner 1994). These relations to species characteristics are the abstractions that then should allow us to generalize to other species.

The consumer–resource link makes a clear connection between food web, community, and ecosystem properties in one direction and the evolution of morphology, behavior, physiology, and life histories in the other. As a consequence, it is a natural conduit for integrating across levels of organization in ecology. Further, the connection to evolutionary theory is important. For example, we have employed adaptive individual-level models to predict behavior, life history attributes, and activity levels, all of which have direct import to the interactions among species in communities. The major advantage of working at this level is the ability to formulate a priori predictions to initiate the iteration of theoretical and empirical work. In part, the generality inherent in these exercises arises from the identification of fundamental trade-offs. Organisms typically face trade-offs in dealing with environmental constraints (factors that limit reproduction or increase mortality), and this is a fundamental force behind the generation of biotic diversity (e.g., Tilman 1989).

Thus, the role of theory in a research program is multifaceted. Theory has a critical role in directing empirical research, generalizing that research, and enabling predictions elaborated from one level of research to be linked to other levels. It is a preeminent tool in discovery both through plausibility arguments and by identifying systematic deviations from predictions. It is also clear that theory development and judgments on the validity of theoretical constructs are greatly enhanced when done in the context of the empirical component of a research program.

Mechanistic Studies in Community Ecology

In this final section, I attempt to put the research program that I have discussed into a larger perspective. Much of what I have advocated can be viewed as a reductionistic approach to community ecology, and I now address some of the advantages and disadvantages of this approach (see also Schoener 1986a, Persson and Diehl 1990, Inchausti 1994). It is always critical to keep in clear view the limitations of any approach. Community properties arise due to the reciprocal interaction of the community components (in our case, typically species) and the greater whole of which species are a part (e.g., Levins and Lewontin 1985). The naive reductionist, for example, who treats species as the components of the community would give the whole no more properties than the properties of these species and their interactions. That is, the description of the community is without causal reality (Levins and Lewontin 1985). We will not arrive at an adequate description of the phenomena at the community level without considering the reciprocal interactions of systems at different levels. It is clear that causality operates bidirectionally in the biological hierarchy, and next I discuss how some of our results illustrate the reciprocal interaction of component species and the community context. Knowledge of the mechanisms of interactions arrived at by a reductionistic approach nonetheless allows considerable understanding of community patterns and, in some cases, a means of discovery of the reciprocal effects.

What are some of the general advantages of the mechanistic approach that we have taken? First, it allows one to establish from individual ecology quantitative predictions about community structure that are constructed independent of the community-level phenomena (Schoener 1986a). In this way, we are able to map the phenomenological consequences of processes (e.g., competition) to attributes of the organism (e.g., body size, mouth or bill size, and behavior), and vice versa. Our work at the individual level that focuses on some of these attributes greatly aided our understanding of the role of species interactions in creating community patterns (Werner 1977, Mittelbach 1981, Werner and Gilliam 1984, Werner 1994, Werner and McPeek 1994, Werner and Anholt 1996).

Second, by identifying mechanisms we vastly increase our power to resolve and discriminate processes responsible for community patterns over the use of phenomenological models. Since many phenomena in ecology have multiple causes, we are in a much better position to discriminate these if we have explicit predictions from a mechanism. For example, predators as well as competitors can be responsible for patterns in habitat partitioning (Holt 1977, 1984, Jefferies and Lawton 1984; Holt and Lawton 1994). In the sunfish system, our studies of the mechanisms of responses to resources and predators enabled us to clearly document that patterns of size-specific habitat use within species often were due to predators (Mittelbach 1981, Werner et al. 1983b, Werner and Hall 1988). Similarly, in the amphibian system knowledge of the mechanism of the responses to resources and predators enabled us to propose that density-dependent predation often can arise from the responses of the prey rather than those of the predator (Anholt and Werner 1995).

Third, we can begin to rebuild the community from these various theories at the individual level. I think that we can go a considerable way in understanding community patterns by employing this approach. There are limitations, however. The naive reductionist would not consider some of the reciprocal effects that come back the other direction from the higher levels. As I discussed previously, systematic deviations from the predictions of lower level theory taken back to the field can alert the investigator to certain phenomena that are unaccounted for in the current rendition of the explanatory framework. Thus, this iteration can provide a tool for discovering such properties if one is alert to the importance of the larger context in influencing interactions of the component parts. In fact, in some cases individual-level theory has allowed us to anticipate such properties (see discussion of higher order effects following).

Fourth, understanding mechanisms enables generalization of results. The difficulty with phenomenological approaches is that the analysis is local and cannot be easily transported to the next system. The estimation of parameters is often very system-specific (e.g., estimates of alpha for competitive effects in the Lotka-Volterra system;

Tilman 1987). Phenomenological models assume that conditions do not change and that the phenomena that go into the model sample the causal pattern of interest (Koehl 1989). Mechanistic models, on the other hand, may allow predictions of much greater scope. For example, the inclusion of a few simple mechanisms of competitive interactions for nutrients that seemingly would be less general than Lotka-Volterra type systems enabled Tilman (1982) to make predictions of phytoplankton assemblages in a wide range of lakes. Mechanistic forest simulators have been developed for communities with over a hundred tree species (Pacala 1989). (Again, as noted earlier, the mechanistic/phenomenological distinction depends on where one stands in the hierarchy of ecological systems; the optimal point to work along this gradient must be determined empirically [Tilman 1989].)

What are some of the difficulties and drawbacks of the mechanistic approach? First, attempts at explanation of community-level phenomena can be extremely complex. If we consider the information necessary to parameterize individual-level models and to build a community with a number of species, the process begins to appear overwhelming. It is not clear, however, that this is any more demanding than the alternative of predicting system behavior with phenomenological models that require estimates of highly system-specific parameters. It is also possible that developing simple rules for higher level theory based on results of mechanistic studies will help reduce dimensionality. Perhaps the reductionist program will aid in the process by providing direction on how representative macromodels at the population and community levels are when used across different environments (Gross 1989). In other words, individual-level models may not only form the basis for more empirical macrodescriptors of system behavior at larger scales but also provide a means to test the robustness of these descriptors across different environments.

Second, the reductionist approach encourages the tendency to ignore the influence of unmeasured factors. In order to operationalize the approach we generally isolate components of the community and concentrate on the phenomena of interest. This to some varying extent removes the community from its larger environmental and historical context. That is, we have to simplify the community to a considerable extent in order to provide the appropriate detail on mechanisms. As discussed earlier in the context of conducting experiments at different scales or levels, the investigator simply has to be cognizant of these limitations and constantly iterate predictions from lower level studies back to the natural system to expose these limitations.

Thus far, I have focused on the power of examining lower levels of organization to understand component interactions and therefore whole-system properties. That is, we essentially have assumed that causality operates in one direction. The components that we have abstracted to constitute our system at the lower levels are presumed to be sufficiently robust and immutable that concepts or models organized around them predict the higher level phenomena under different conditions. But the larger system context can influence the nature of our components in a way that alters the qualitative or quantitative nature of their interactions—that is, there are reciprocal effects or ''downward causality'' (Campbell 1974, Wimsatt 1976, Mayr 1982). It is well recognized that genes or species may have different properties on different backgrounds (e.g., Mayr 1982). In the case of community ecology, the specific issue is whether species A always has the same per capita effect on species B regardless of the background of other species in the system. Do the traits that determine the nature of interactions between species remain the same when species composition changes? This has been a central issue of the long-standing concern with higher order interactions in community ecology (Vandermeer 1969, Neill 1974, Abrams 1983, Wilbur and Fauth 1990).

If species are adaptively responding to their environment, then their capabilities as competitors or predators inevitably change with the composition of the community. In our case, this recognition was generated by work on individual behavior and prompted experiments that manipulated nonlethal presence of a predator to assess the effect on competitive abilities of prey independent of the numerical impact of the predator. This framework also suggested that the presence of a competitor by reducing resources could increase the activity rate of other species and thereby render them more vulnerable to predators. We have tested both propositions with the amphibian system and found that these higher order (or in this case trait-mediated indirect effects; Abrams 1995) can have quantitatively important influences on per capita interaction strengths (Werner 1991, Anholt and Werner 1995, Werner and Anholt 1996, Peacor and Werner 1997). Such conclusions could be reached with more phenomenological experiments that manipulate combinations of different species to test for additive effects (e.g., Wilbur and Fauth 1990), but focusing on the mechanism is a much more powerful way to proceed. By focusing on the mechanism we can generate an a priori expectation that suggests explicit experimental designs or analyses and it is clear what is causing the deviations from additivity. This example illustrates how the conduct of a mechanistic research program does not have to ignore the causal effects of the larger system on component parts.

In conclusion, a research program that integrates comparative/descriptive, experimental, and theoretical work directed at understanding mechanism can be a very powerful approach to the analysis of community patterns. In addition, the interplay of studies at very different levels of organization and at different scales or levels of control enriches this enterprise. The examples used in this essay I think point to the manner in which employing a number of the different tools at our disposal in the context of a research program greatly enhances the power of each. This is not the only, or sometimes the best, approach to develop understanding of community phenomena, nor will it be sufficient in and of itself. We clearly need a diversity of approaches to the problems in community ecology, and we simply must be aware of the strengths of the different approaches and vigilant of their limitations. A more balanced perspective on these issues would have circumvented much of the acrimony in the debates on approaches to community ecology over the last several decades and moved us forward on the work at hand.

ACKNOWLEDGMENTS   It will be obvious to the reader the debt that I owe to the succession of students who have been in my laboratory and became wonderful colleagues. They have clarified my thinking, made fundamental contributions to both ideas and method, and contributed elements to the program that I did not have the talents to accomplish. In particular, I would like to acknowledge with pleasure the contributions of (in chronological order) Gary Mittelbach, Jim Gilliam, Tim Ehlinger, Craig Osenberg, Matthew Leibold, Mark McPeek, David Skelly, Gary Wellborn, Rick Relyea, Scott Peacor, and Kerry Yurewicz.

Postdoctoral students also contributed to my thinking in important ways, especially David Hart, Andy Sih, Carol Folt, Steve Kohler, Shahid Naeem, Brad Anholt, Josh VanBuskirk, Andy McCollum, and Peter Eklov. The manuscript was greatly improved by the comments of Peter Eklov, Deborah Goldberg, Mathew Leibold, Andy McCollum, Gary Mittelbach, Scott Peacor, Les Real, Rick Relyea, Bill Resetarits, Josh VanBuskirk, John Vandermeer, and Kerry Yurewicz. The work represented here has been generously supported by the National Science Foundation since 1972.

Literature Cited

Abrams, P. A. 1983. Arguments in favor of higher order interactions. American Naturalist 121:887–891.

———. 1995. Implications of dynamically variable traits for identifying, classifying and measuring direct and indirect effects in ecological communities. American Naturalist 146:112–134.

Anholt, B. R., and E. E. Werner. 1995. Interaction between food availability and predation mortality mediated by adaptive behavior. Ecology 76:2230–2234.

Bender, E. A., T. J. Case, and M. E. Gilpin. 1984. Perturbation experiments in community ecology: theory and practice. Ecology 65:1–13.

Billick, I., and T. J. Case. 1994. Higher order interactions in ecological communities: what are they and how can they be detected? Ecology 75:1529–1543.

Brown, J. H. 1995. Macroecology. University of Chicago Press, Chicago, Illinois.

Brown, J. S. 1988. Patch use as an indicator of habitat preference, predation risk, and competition. Behavioral Ecology and Sociobiology 22:37–47.

Campbell, D. 1974. Downward causation in hierarchically organized systems. Pages 179–186 in F. Ayala and T. Dobzhansky (eds.), Studies in the Philosophy of Biology: Reduction and Related Problems. University of California Press, Berkeley.

Carpenter, S. R., and J. F. Kitchell (eds.). 1993. The Trophic Cascade in Lakes. Cambridge University Press, Cambridge.

Carpenter, S. R., J. F. Kitchell, J. R. Hodgson, P. A. Cochran, J. J. Elser, M. M. Elser, D. M. Lodge, D. Kretchmer, and X He. 1987. Regulation of lake primary productivity by food web structure. Ecology 68:1863–1876.

Cirri, R. D., and D. F. Fraser. 1983. Predation and risk in foraging minnows: balancing conflicting demands. American Naturalist 121:552–561.

Collins, J. P., and H. M. Wilbur. 1979. Breeding habits and habitats of the amphibians of the Edwin S. George Reserve, Michigan, with notes on the local distribution of fishes. Occasional Papers Museum of Zoology, University of Michigan, no. 686:1–34.

Colwell, R. K. 1984. What's now? Community ecology discovers biology. Pages 387–396 in P. W. Price, C. N. Slobodchikoff, and W. S. Gaud (eds.), A New Ecology: Novel Approaches to Interactive Systems. Wiley, New York.

Connell, J. H. 1961. The influence of interspecific competition and other factors on the distribution of the barnacle *Chthamalus stellatus*. Ecology 42:710–723.

———. 1980. Diversity and coevolution of competitors, or the ghost of competition past. Oikos 35:131–138.

———. 1983. On the prevalence and relative importance of interspecific competition: evidence from field experiments. American Naturalist 122:661–696.

Diamond, J. M. 1986. Overview: laboratory experiments, field experiments, and natural experiments. Pages 3–22 in J. Diamond and T. J. Case (eds.), Community Ecology. Harper and Row, New York.

Ehlinger, T. J. 1990. Habitat choice and phenotype-limited feeding efficiency in bluegill: individual differences and trophic polymorphism. Ecology 71:886–896.

Ehlinger, T. J., and D. S. Wilson. 1988. Complex foraging polymorphism in bluegill sunfish. Proceedings of the National Academy of Science of the USA 85:1878–1882.

Feyerabend, P. 1978. Against Method. Verso, London.

Gilliam, J. F. 1982. Habitat use and competitive bottlenecks in size-structured fish populations. Ph.D. dissertation, Michigan State University, Lansing.

Goldberg, D. E., and A. M. Barton. 1992. Patterns and consequences of interspecific competition in natural communities: a review of field experiments with plants. American Naturalist 139:771–801.

Gross, L. J. 1989. Plant physiological ecology: a theoretician's perspective. Pages 11–24 in J. Roughgarden, R. M. May, and S. A. Levin (eds.), Perspectives in Ecological Theory. Princeton University Press, Princeton, New Jersey.

Hairston, N. G., Sr. 1989. Ecological Experiments: Purpose, Design, and Execution. Cambridge University Press, Cambridge.

Hall, D. J., and E. E. Werner. 1977. Seasonal distribution and abundance of fishes in the littoral zone of a Michigan lake. Transactions of the American Fisheries Society 206: 545–555.

Hall, D. J., W. E. Cooper, and E. E. Werner. 1970. An experimental approach to the production dynamics and structure of freshwater animal communities. Limnology and Oceanography 15:839–928.

Holt, R. D. 1977. Predation, apparent competition, and the structure of prey communities. Theoretical Population Biology 12: 197–229.

———. 1984. Spatial heterogeneity, indirect interactions, and the coexistence of prey species. American Naturalist 124: 377–406.

Holt, R. D., and J. H. Lawton. 1994. The ecological consequences of shared natural enemies. Annual Review of Ecology and Systematics 25:495–520.

Horat, P., and R. D. Semlitsch. 1994. Effects of predation risk and hunger on the behavior of two species of tadpoles. Behavioral Ecology and Sociobiology 34:393–401.

Houston, A. I., J. M. McNamara, and J. M. C. Hutchinson. 1993. General results concerning the trade-off between gaining energy and avoiding predators. Philosophical Transactions of the Royal Society of London B 341:375–397.

Hurlbert, S. H. 1984. Pseudoreplication and the design of ecological field experiments. Ecological Monographs 54:187–211.

Inchausti, P. 1994. Reductionist approaches in community ecology. American Naturalist 143: 201–221.

Jackson, J. B. C. 1981. Interspecific competition and species' distributions: the ghost of theories and data past. American Zoologist 21:889–901.

Jeffries, M. J., and J. H. Lawton. 1984. Enemy-free space and the structure of ecological communities. Biological Journal of the Linnean Society 23:269–286.

Kareiva, P. 1989. Renewing the dialog between theory and experiments in population ecology. Pages 68–88 in J. Roughgarden, R. M. May, and S. A. Levin (eds.), Perspectives in Ecological Theory. Princeton University Press, Princeton, New Jersey.

Keddy, P. A. 1992. A pragmatic approach to functional ecology. Functional Ecology 6:621–626.

Koehl, M. A. R. 1989. Discussion: From individuals to populations. Pages 39–53 in J. Roughgarden, R. M. May, and S. A. Levin (eds.), Perspectives in Ecological Theory. Princeton University Press, Princeton, New Jersey.

Kohler, S. L., and M. A. McPeek. 1989. Predation risk and the foraging behavior of competing stream insects. Ecology 70:1811–1825.

Kotler, B. P., J. S. Brown, and O. Hasson. 1991. Factors affecting gerbil foraging behavior and rates of owl predation. Ecology 72:2249–2260.

Lehman, J. A. 1986. The goal of understanding in limnology. Limnology and Oceanography 31:1160–1166.

Leibold, M. A., and A. J. Tessier. 1991. Contrasting patterns of body size for *Daphnia* species that segregate by habitat. Oecologia 86:342–348.

Levins, R., and R. Lewontin. 1985. The Dialectical Biologist. Harvard University Press, Cambridge, Massachusetts.

Lima, S. L., and L. M. Dill. 1990. Behavioral decisions made under the risk of predation: a review and prospectus. Canadian Journal of Zoology 68:619–640.

Ludwig, D., and L. Rowe. 1990. Life history strategies for energy gain and predator avoidance under time constraints. American Naturalist 135:686–707.

MacArthur, R. H., and E. R. Pianka. 1966. On optimal use of a patchy environment. American Naturalist 100:603–609.

Mayr, E. 1982. The Growth of Biological Thought: Diversity, Evolution, and Inheritance. Harvard University Press, Cambridge, Massachusetts.

McNamara, J. M., and A. I. Houston. 1994. The effect of a change in foraging options on intake rate and predation rate. American Naturalist 144:978–1000.

Mittelbach, G. G. 1981. Foraging efficiency and body size: a study of optimal diet and habitat use by bluegills. Ecology 62:1370–1386.

———. 1988. Competition among refuging sunfishes and effects of fish density on littoral zone invertebrates. Ecology 69:614–623.

Mittelbach, G. G., and P. L. Chesson. 1987. Indirect effects on fish populations. Pages 315–332 in W. C. Kerfoot and A. Sih (eds.), Predation: Direct and Indirect Impacts on Aquatic Communities. University Press of New England, Hanover, New Hampshire.

Moran, N. A. 1994. Adaptation and constraint in the complex life cycles of animals. Annual Review of Ecology and Systematics 1994: 573–600.

Neill, W. E. 1974. The community matrix and interdependence of the competition coefficients. American Naturalist 108:399–408.

Osenberg, C. W., and G. G. Mittelbach. 1989. The effects of body size on predator-prey interaction between pumpkinseed sunfish and gastropods. Ecological Monographs 59: 405–432.

Osenberg, C. W., E. E. Werner, G. G. Mittelbach, and D. J. Hall. 1988. Growth patterns in bluegill (*Lepomis macrochirus*) and pumpkinseed (*L. gibbosus*) sunfish: environmental variation and the importance of ontogenetic niche shifts. Canadian Journal of Fisheries and Aquatic Sciences 45:17–26.

Osenberg, C. W., M. A. Olsen, and G. G. Mittelbach. 1994. Stage-structure in fishes: resource productivity and competition gradients. Pages 151–170 in D. J. Stouder, K. L. Fresh, and R. J. Feller (eds.), Theory and Application of Fish Feeding Ecology. University of South Carolina Press, Columbia.

Pacala, S. W. 1989. Plant population dynamic theory. Pages 54–67 in J. Roughgarden, R. M. May, and S. A. Levin (eds.), Perspectives in Ecological Theory. Princeton University Press, Princeton, New Jersey.

Paine, R. T. 1966. Food web complexity and species diversity. American Naturalist 100:65–75.

———. 1994. Marine Rocky Shores and Community Ecology: An Experimentalist Perspective. Ecology Institute, Oldendorf/Luhe, Germany.

Peacor, S. D., and E. E. Werner. 1997. Trait-mediated indirect interactions in a simple aquatic food web. Ecology 78:1146–1156.

Persson, L. 1993. Predator-mediated competition in prey refuges: the importance of habitat-dependent prey resources. Oikos 68:12–22.

Persson, L., and S. Diehl. 1990. Mechanistic, individual-based approaches in the population/community ecology of fish. Annales Zoologici Fennici 27:165–182.

Peters, R. H. 1991. A Critique for Ecology. Cambridge University Press, Cambridge.

Pickett, S. T. A., J. Kolasa, and C. G. Jones. 1994. Ecological Understanding: The Nature of Theory and the Theory of Nature. Academic Press, San Diego.

Power, M. E., W. J. Mathews, and A. J. Stewart. 1985. Grazing minnows, piscivorous bass, and stream algae: dynamics of a strong interaction. Ecology 66:1448–1456.

Rowe, L., and D. Ludwig. 1991. Size and timing of metamorphosis in complex life cycles: time constraints and variation. Ecology 72:413–427.

Scheiner, S. M., and J. Gurevitch (eds.). 1993. Design and Analysis of Ecological Experiments. Chapman and Hall, New York.

Schoener, T. W. 1983. Field experiments on interspecific competition. American Naturalist 122:240–285.

———. 1986a. Mechanistic approaches to community ecology: a new reductionism? American Zoologist 26:81–106.

———. 1986b. Resource partitioning. Pages 91–126 in J. Kikkawa and D. Anderson (eds.), Community Ecology: Pattern and Process. Blackwell Scientific, Oxford.

Sih, A. 1982. Foraging strategies and the avoidance of predation by and aquatic insect, *Notonecta hoffmanni*. Ecology 63:786–796.

Sih, A., P. Crowley, M. McPeek, J. Petranka, and K. Strohmeier. 1985. Predation, competition, and prey communities: a review of field experiments. Annual Review of Ecology and Systematics 16:269–312.

Simberloff, D. 1981. The sick science of ecology: symptoms, diagnosis, and prescription. Eidema 1:49–54.

———. 1983. Competition theory, hypothesis-testing, and other community ecological buzz-words. American Naturalist 122:626–635.

Skelly, D. K. 1992. Larval distributions of spring peepers and chorus frogs: regulating factors and the role of larval behavior. Ph.D. dissertation, University of Michigan, Ann Arbor.

Skelly, D. K., and E. E. Werner. 1990. Behavioral and life historical responses of larval American toads to an odonate predator. Ecology 71:2313–2322.

Skelly, D. K., E. E. Werner, and S. A. Cortwright. Forthcoming. Forest Succession and the distributional dynamics of an amphibian assemblage. Ecology.

Smith-Gill, S. J., and D. E. Gill. 1978. Curvilinearities in the competition equations: an experiment with ranid tadpoles. American Naturalist 112:557–570.

Strong, D. R., J. D. Simberloff, L. G. Abele, and A. B. Thistle (eds.). 1984. Ecological Communities: Conceptual Issues and the Evidence. Princeton University Press, Princeton, New Jersey.

Tilman, D. 1982. Resource Competition and Community Structure. Princeton University Press, Princeton, New Jersey.

———. 1987. The importance of the mechanisms of interspecific competition. American Naturalist 129:769–774.

———. 1989. Discussion: population dynamics and species interactions. Pages 89–101 in J. Roughgarden, R. M. May, and S. A. Levin (eds.), Perspectives in Ecological Theory. Princeton University Press, Princeton, New Jersey.

Underwood, A. J. 1995. Experimental Design. Blackwell Scientific, Oxford.

Vandermeer, J. H. 1969. The competitive structure of communities: an experimental approach with protozoa. Ecology 50:362–371.

Wade, M. J. 1992. Sewall Wright: gene interaction and the shifting balance theory. Pages 35–62 in J. Antonovics and D. Futuyma (eds.), Oxford Surveys of Evolutionary Biology. Vol. 6. Oxford University Press, Oxford.

Weiner, J. 1995. On the practice of ecology. Journal of Ecology 83:153–158.

Weins, J. A. 1977. On competition and variable environments. American Scientist 65:590–597.

Wellborn, G. A., D. K. Skelly, and E. E. Werner. 1996. Mechanisms creating community structure across a freshwater habitat gradient. Annual Review of Ecology and Systematics 27:337–363.

Werner, E. E. 1977. Species packing and niche complementarity in three sunfishes. American Naturalist 111:553–578.

———. 1986. Amphibian metamorphosis: growth rate, predation risk, and the optimal size at transformation. American Naturalist 128:319–341.

———. 1988. Size, scaling, and the evolution of complex life cycles. Pages 60–81 in B. Ebenman and L. Persson (eds.), Size-Structured Populations: Ecology and Evolution. Springer-Verlag, Berlin.

———. 1991. Nonlethal effects of a predator on competitive interactions between two anuran larvae. Ecology 72:1709–1720.

———. 1992a. Competitive interactions between wood frog and northern leopard frog larvae: the influence of size and activity. Copeia 1992:26–35.

———. 1992b. Individual behavior and higher-order species interactions. American Naturalist 140:S5–S32.

———. 1994. Ontogenetic scaling of competitive relations: size-dependent effects and responses in two anuran larvae. Ecology 75:197–213.

Werner, E. E., and B. R. Anholt. 1993. Ecological consequences of the tradeoff between growth and mortality rates mediated by foraging activity. American Naturalist 142:242–272.

———. 1996. Predator-induced behavioral indirect effects: consequences to competitive interactions in anuran larvae. Ecology 77:157–169.

Werner, E. E., and J. F. Gilliam. 1984. The ontogenetic niche and species interactions in size-structured populations. Annual Review of Ecology and Systematics 15:393–425.

Werner, E. E., and D. J. Hall. 1974. Optimal foraging and the size-selection of prey by the bluegill sunfish (*Lepomis macrochirus*). Ecology 55:1042–1052.

———. 1976. Niche shifts in sunfishes: experimental evidence and significance. Science 191:404–406.

———. 1977. Competition and habitat shift in two sunfishes (Centrarchidae). Ecology 58:869–876.

———. 1979. Foraging efficiency and habitat switching in competing sunfishes. Ecology 60:256–264.

———. 1988. Ontogenetic habitat shifts in the bluegill sunfish (*Lepomis macrochirus*): the foraging rate-predation risk tradeoff. Ecology 69:1352–1366.

Werner, E. E., and M. A. McPeek. 1994. Direct and indirect effects of predators on two anuran species along an environmental gradient. Ecology 75:1368–1382.

Werner, E. E., and K. S. Glennemeier. Forthcoming. The influence of forest canopy cover and succession on the breeding pond distributions of several amphibian species. Copeia.

Werner E. E., D. J. Hall, D. R. Laughlin, D. J. Wagner, L. A. Wilsmann, and F. C. Funk. 1977. Habitat partitioning in a freshwater fish community. Journal of the Fisheries Research Board of Canada 34:360–370.

Werner, E. E., G. G. Mittelbach, D. J. Hall, and J. F. Gilliam. 1983a. Experimental tests of optimal habitat use in fish: the role of relative habitat profitability. Ecology 64:1525–1539.

Werner, E. E., J. F. Gilliam, D. J. Hall, and G. G. Mittelbach. 1983b. An experimental test of the effects of predation risk on habitat use in fish. Ecology 64: 1540–1548.

Wilbur, H. M., and J. E. Fauth. 1990. Experimental aquatic food webs: interactions between two predators and two prey. American Naturalist 135:176–204.

Wimsatt, W. 1976. Reductive explanation: a functional account. Pages 671–710 in R. Cohen, C. Hooker, A. Michalos, and J. van Evra (eds.), PSA 1974. Reidl, Dordrecht.

Wootton, J. T. 1994. Putting the pieces together: testing the independence of interactions among organisms. Ecology 75:1544–1551.

Worthen, W. B., and J. L. Moore. 1991. Higher-order interactions and indirect effects: a resolution using laboratory *Drosophila* communities. American Naturalist 138:1092–1104.

Yodzis, P. 1988. The indeterminacy of ecological interactions as perceived through perturbation experiments. Ecology 69:508–515.

2

Ecological Experiments

Scale, Phenomenology, Mechanism, and the Illusion of Generality

ARTHUR E. DUNHAM & STEVEN J. BEAUPRE

In attempting to understand the distribution and abundance of organisms, ecologists often seek general organizing principles to aid in understanding and predicting the responses of populations to environmental variation. The principle goal of this essay is to clarify the concept of generality in ecology. We are motivated by the perception that many ecologists attempt to generalize both theory and experimental results beyond specific experimental systems. In some cases, this generalization may be justified; in others, it is not. We believe that ecology as a science will benefit from an open discussion regarding the nature of and limits to generality. To this end, we (1) discuss the nature of ecological processes and their implications for epistemology and generality; (2) discuss concepts of generality and provide a mechanism-based definition; (3) offer a definition and several examples of mechanism, a concept that is central to generality; and, finally, (4) illustrate our position with a discussion of the role of mechanisms in ecological inquiry and with several examples. We emphasize at the outset that our discussions of specific research programs are meant solely to foster constructive discussion of issues related to generality in ecology.

The Nature of Ecological Processes

In the process of studies directed at providing an objective basis for understanding patterns in the distribution and abundance of organisms, ecologists have established three principles. First, the fundamental unit of ecology is the individual organism. Second, most ecological patterns result from multiple simultaneously acting processes (i.e., multiple causality). Third, very few generalizations apply to all ecological systems and remain valid regardless of spatial, temporal, and organismal scales. In the following, we discuss each of these principles and their implications for the process of ecological inquiry.

All ecological processes are ultimately transduced through individual organisms. Regardless of the scale that ecologists use to study ecological processes, these processes result from energy and mass exchanges among individual organisms or between individuals and the physical environment. Therefore, the fundamental unit of ecology is the individual organism and every individual is potentially unique. The uniqueness of individuals results from the nature of inheritance and from the fact that each individual has a unique history of interaction with its environment (including other organisms) throughout its life. The uniqueness of individuals means that individual organisms cannot necessarily be aggregated in ecological theory by imposition of the law of large numbers or by the assertion that individuals can be treated as if the differences among them are inconsequential. Nonetheless, the vast majority of models in population biology are based on this assertion. We submit that this is a dangerous assertion because the fundamental units of ecology do not comprise homogenous sets. It may be the case that for certain questions, particularly those that involve spatial or temporal patterns on large scales, this assumption may be safely made. However, the validity of imposing such an assumption should be tested before the assumption is made.

Most, if not all, of the patterns that ecologists seek to explain result from several, if not many, simultaneously acting and potentially interacting processes (Quinn and Dunham 1983). Thus, ecological patterns are complex, and that complexity should be reflected in both theory and empiricism. Because multiple interacting causal mechanisms may often produce the patterns that ecologists seek to understand, the potential for multiple causal mechanisms must be incorporated into the construction of ecological theory and into the design of ecological experiments.

Ecologists have also established that very few general principles apply to all ecological systems and remain valid irrespective of spatial, temporal, or organismal scales. Examples of such general principles are the first and second laws of thermodynamics (the law of conservation of energy and the law that conversion efficiency among energy forms must be less than 100%, respectively). Energy and mass balance must hold in all ecological systems regardless of the system boundary. However, most processes or principles that ecologists use to understand the patterns they study are not general because they are valid only over a restricted range of spatial, temporal, or organismal scales (= the domain of generality of a given process or principle). The domain of generality of a given process is process-specific.

The preceding considerations have at least two important implications for ecology: in the design of ecological experiments and in the generalization of experimental results. The first implication is that ecological theory cannot generally be done in a manner that ignores the differences among individuals. The second implication is that because multiple interacting causal mechanisms often produce the patterns that ecologists seek to understand, the potential for multiple causal mechanisms should be incorporated into the construction of ecological theory and into the design of ecological experiments.

The Concept of Generality and the Importance of Mechanism

When ecological theory is constructed such that differences among individuals are ignored or assumed negligible and/or such that a single mechanism is assumed to produce a particular pattern, a coarse and usually unrealistic body of theory (e.g., the large literature on Lotka-Volterra dynamics in community ecology) is the result. Many ecologists refer to such simple, unstructured theory as "general," but in reality these theoretical constructs often lack logical generality and are best described as extremely simple and crude. A general theory is one that has many special cases and holds true for all of them, not one that fails to hold for any special case even though it may capture a few gross features. The recent increase in the use of physiologically structured, individual-based models (e.g., Adams and DeAngelis 1987, DeAngelis and Gross 1992, DeAngelis et al. 1993, Dunham 1993, Dunham and Overall 1994) in population biology probably reflects the realization that the preceding implications are important.

Several ecologists (e.g., Foster 1990, Hurlburt 1984, Underwood 1986, Underwood and Fairweather 1986) have suggested that part of the difficulty in understanding patterns in ecology arises because much of the evidence that supports putative causal mechanisms is poor. In part, this is because single-factor causality is often assumed and methods and experimental designs used in many studies do not adequately explore alternative hypotheses that involve the potential effects of multiple interacting mechanisms. In addition, the patterns themselves are often not well documented. For example, Foster (1990) examined patterns of zonation in intertidal macroalgal assemblages and noted that most research done on these assemblages had been conducted at a few protected sites where patterns of zonation are distinct. Foster surveyed a variety of algal assemblages at wave-exposed sites and demonstrated considerable variation in assemblage structure. Foster argues convincingly that explanations (e.g., mussel–algal interactions) for zonation in macroalgal assemblages developed at only a few sites may be inadequate to explain patterns in macroalgal assemblages at all or even most sites. Simply put, the commonly accepted causes of algal zonation may not apply generally due to environmental heterogeneity. We suggest that the problems described by Foster (1990) are more pervasive in ecology than is commonly appreciated. Ecology needs a set of objective criteria for judging the domain of generality of theory and of the results of experiments designed to test theory.

Generality—A Definition

Foster provides one definition of generality applicable to the macroalgal assemblages he studied: "By generality, I mean over what proportion of the coast in some defined geographic region does a particular organization apply" (1990: 22). This definition is in the right spirit, but we suggest the following alternative definition: *By generality we mean the range of spatial, temporal, and organismal scales over which a particular mechanism or set of mechanisms applies.* This definition differs from Foster's in two important respects. First though different ecological mechanisms and processes may operate over different temporal, spatial, and organismal scales, Foster's definition explicitly incorporates only the spatial scale, whereas our definition incorporates all relevant scales. Second, Foster's definition attempts to define generality in terms of ecological organization rather than the processes or mechanisms that produce patterns in ecological organization, whereas our definition, since it is clear that, similar patterns of community organizations may be produced by different mechanisms, is in terms of the set of temporal, spatial, and organismal scales over which the mechanisms that produce a particular pattern in nature are valid.

Mechanism—A Definition

In order to infer the domain of generality of a particular causal explanation for an ecological pattern, we must be explicit about what we mean by *mechanism*. There is some disagreement among ecologists with regard to the meaning of mechanism (e.g., Peters 1991, Schoener 1986). *By* mechanism *we mean an appropriate level of reductionism that provides a causal explanation of the functional relationship among a set of variables.* In discussions of mechanism with our colleagues we have heard the argument that ''one person's mechanism is another's phenomenological description'' in many forms. Note that by our definition, the relationship between pattern and mechanism is scale-dependent and hierarchical in nature. The key in this distinction is the determination of what the ''appropriate level of reductionism'' is in any attempt to understand a particular ecological pattern. For a particular set of functional relationships to qualify as a mechanism to explain a particular pattern, the functional relationships must be quantified independently of the pattern of interest and at a lower level of hierarchy. We clarify the hierarchical nature of the relationship between pattern and mechanism by example in the next section. Mechanisms are generally described by a set of parameters that allow prediction of the functional relationships among the variables of interest. This set of parameters is subject to the following constraints: (1) they cannot be derived from the variables under consideration and (2) they must be objective and measurable. The first constraint precludes regression models and other such models based solely on curve-fitting procedures (for example) from being mechanisms and stands in contrast to ''instrumentalist'' approaches to ecology (e.g., Peters 1991). The second constraint ensures that hypotheses that involve ecological mechanisms are empirically testable. The domain of generality of a particular mechanism or set of mechanisms is simply the range of temporal, spatial, and organismal scales over which the set of variables and the associated parameter set remain invariant.

Mechanism—An Example

There is often disagreement among ecologists about the importance of mechanism, as well as about what constitutes a mechanism. To illustrate our definition of mechanism as a level of reductionism subject to the above constraints we provide the following example from physiological ecology. The set of complex mechanistic interactions that describe the mass and thermal energy exchanges between an individual organism and its environment are well-known and have been described in detail elsewhere (e.g., Dunham et al. 1989, Gates 1980, Porter and Gates 1969, Porter and Tracy 1983). This set of interactions can be summarized by a set of coupled thermal energy–mass balance equations (Porter and Tracy 1983). A simplified version of these equations is shown in Figure 2-1. The set of interactions which result in the rates of change of temperature and mass of an individual organism at any particular time and which are described by the set of equations depicted in Figure 2-1 constitutes an ecological mechanism. This mechanism is general in that it applies on any spatial and temporal scales relevant to an individual organism and can be applied to any individual organism irrespective of the type of organism or environment. In order to apply this mechanism to a particular organism, a number of physical characteristics and functional relationships characteristic

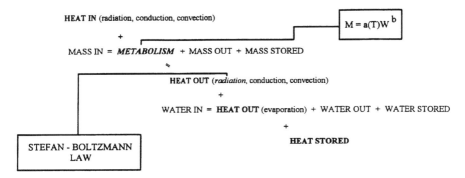

Figure 2-1. Simplified version of the coupled heat and mass balance equations relating variation in operating environments to body temperature and net allocatable resources. Modified from Dunham (1993), Dunham et al. (1989), and Porter and Tracy (1983).

of the organism must be supplied. For illustrative purposes, we examine two terms from this complex set of equations in detail (Fig. 2-1). The first term specifies the rate (Q_{IR}) of heat loss from the surface of the organism due to infrared radiation. The physical relationship that describes the rate of heat loss due to emission of infrared radiation is known as the Stefan-Boltzmann law and may be written $Q_{IR} = \sigma \varepsilon K^4$, where Q_{IR} is the rate of heat loss due to infrared radiation (Watts), σ is the Stefan-Boltzmann constant (5.67×10^{-8} Watts·m^{-2}·K^{-4}), ε is the emissivity of the organism's surface in the infrared wavelengths (dimensionless), and K is the surface temperature of the organism (° K) (equation 1).

The second term describes the dependence of heat produced by metabolism on body temperature of the organism and on the organism's mass. For a small ectothermic vertebrate this relationship may be written $M = a(T_b)W^b$, where M is the rate of heat production due to metabolism (Watts), T_b is body temperature (°C), a is a function of T_b, and b is a fitted constant (equation 2). Note that the relationship described by equation 2 involves a fitted function (a which depends on T_b and a fitted constant (b) and that the relationship described by equation 1 involves a measured quantity (ε). The fitted function, fitted constant, and measured quantity render the mechanism embodied by the set of thermal–mass balance equations (Fig. 2-1) specific to the organism under consideration. However, this does not affect the generality of the thermal–mass balance equations as a mechanism responsible for the rates of change of temperature and mass of any particular individual in any environment.

Physics and physiology dictate the terms and form of the thermal–mass balance mechanism, and those are general irrespective of the type of organism or environment. In contrast, consider the term describing the dependence of the heat produced by metabolism on the body temperature of the organism and the organism's mass. Equation 2 describes the form of this relationship for a small ectothermic vertebrate, but the fitted function, $a(T_b)$, and constant, b, represent empirical fits to laboratory data. The empirically estimated function, $a(T_b)$, and constant, b, and the estimate of metabolic heat production provided by equation 2 constitute an acceptable estimate of the appropriate term in the set of thermal–mass balance equations. However, equation 2 provides no information about the actual physiological mechanism that relates body temperature and mass to metabolic heat production because the parameter estimates are derived empirically by a curve-fitting procedure from data on the response variable M taken at different levels of the experimental variables T_b and W. Note that the empirically derived relationship between M and the fitted function $[a(T_b)]$ and estimated parameter (b) can serve as a component of the mechanism described by the coupled mass–energy balance equations. However, the same relationship by our definition cannot be a mechanism of metabolic heat production because, at this level of hierarchy, the estimated parameters are derived from the variables of interest. As a consequence of the specificity of the relationship between M, $a(T_b)$, and b, the domain of generality associated with equation 2 is precisely the organism and ranges of the values of the experimental variables used in the experiment to relate measurements of M to experimentally controlled values of T_b and W.

Mechanism—A Practical Example

We provide an explicit example that illustrates the importance of mechanism in developing hypotheses about the causes of an ecological pattern in a system potentially involving multiple interacting causes. There is an interesting and perplexing pattern of geographic variation in several life history characters among three populations of the canyon lizard (*Sceloporus merriami*) that occur along a steep elevational gradient in Big Bend National Park, Texas (Dunham et al. 1989; Grant and Dunham 1988, 1990). These populations are Maple Canyon (MC, 1609-m elevation), Grapevine Hills (GV, 1036 m), and Boquillas Canyon (BQ, 560 m). The pattern of variation in life history characteristics and the environmental differences among these sites are discussed in detail by Dunham et al. (1989) and Grant and Dunham (1990). For the purposes of this example, we consider only the differences among these populations with regard to individual growth rates of yearling lizards (Fig. 2-2a). The pattern of geographic variation in individual growth rates for other age classes is similar and is discussed in detail elsewhere (Dunham et al. 1989, Grant and Dunham 1990). The perplexing aspect of the pattern of among-population variation in individual growth rates is that individuals from the population at the intermediate elevation (GV) have higher average growth rates than do individuals from either the low-(BQ) or high-(MC) elevation population. Within the GV population, individual growth rates depend on food resource availability such that individual growth rates are higher during periods of high resource abundance (Dunham 1978). Several environmental gradients that operate in this system potentially influence individual growth rates. In this system, both primary productivity and associated prey availability depend on rainfall and, because precipitation increases with elevation, prey availability also increases with elevation (Fig. 2-2b) (Dunham 1993, Dunham and Overall 1994, Dunham et al. 1989, Grant and Dunham 1990). Biophysically imposed thermal constraints limit the amount of time an individual can be out of refugia (crevices, etc.), and foraging on each day throughout the active season and the fraction of the day during which an individual lizard can forage increases with elevation (Fig. 2-2b) (Grant and Dunham 1990). These considerations lead to the prediction that there should be a pattern in which individual growth rates increase monotonically from the low-elevation population (BQ) to the high-elevation population (MC) because food

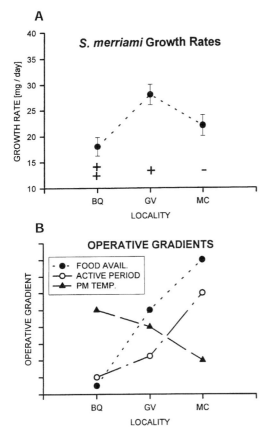

Figure 2-2. (**A**) Pattern of geographic variation in individual growth rates of yearlings among three populations of the lizard *Sceloporus merriami*. (**B**) Pattern of variation in three environmental gradients (food availability, length of daily activity period, and average operative temperature, (T_e) during scotophase among three populations of the lizard *S. merriami* from Big Bend National Park, Texas. BQ is the Boquillas Canyon population (elevation 560 m); GV is the Grapevine Hills population (elevation 1036 m); and MC is the Maple Canyon population (elevation 1609 m).

availability and time available for foraging are lowest at low elevations and increase monotonically with elevation (Fig. 2-2b). Clearly, this prediction is not met and some other set of factors must be involved.

Growth rate can be limited by any mechanism whereby the amount of assimilated resources available for allocation to growth is limited. In the present system, such limitation can occur in two different ways: (1) the rate of prey ingestion is limited by availability of prey and/or time during which foraging can occur, and (2) the rate at which ingested prey can be digested and the resulting nutrient resources assimilated is limited by the nature of digestive physiology (Beaupre and Dunham 1995, Beaupre et al. 1993, Dunham et al. 1989). The second mechanism is termed *process limitation* (Dunham et al. 1989). In *S. merriami*, as in other ectotherms, digestive physiology is strongly temperature-dependent (Beaupre and Dunham 1995, Beaupre et al. 1993). At naturally occurring temperatures in these populations, passage rates are such that much of the digestion of prey items ingested during the day takes place at night, when these lizards are normally in refugia (Beaupre and Dunham 1995, Beaupre et al. 1993). There is significant variation among these populations in the average operative environmental temperatures available to these lizards such that at any particular time of day the average temperature declines with increasing elevation (Fig. 2-2b) (Dunham et al. 1989, Grant and Dunham 1990). Metabolizable energy (ME) and passage time are two temperature-dependent variables that influence the rate at which ingested food is digested and as-similated (Beaupre and Dunham 1995, Beaupre et al. 1993). Metabolizable energy is a measure of the amount of the energy ingested that is assimilated and available for allocation to processes such as growth. Passage time is a measure of the amount of time it takes the digestive tract to completely process ingested food. In *S. merriami*, for a given meal size, ME decreases significantly with increasing temperature (largely due to increasing uric acid production) over the range of temperatures normally exhib-ited by field-active lizards (Beaupre and Dunham 1995, Beaupre et al. 1993). In *S. merriami*, passage time decreases significantly with decreasing temperature over the range of temperatures normally exhibited by field-active lizards (Beaupre et al. 1993). In addition, rate of food consumption by *S. merriami* decreases significantly with de-creasing temperature over the range of temperatures normally exhibited by field-active lizards (Beaupre et al. 1993).

The interaction among the thermal dependence of ME, passage time, and consump-tion rate, and the environmentally imposed temperature gradient (Fig. 2-2b) suggests a mechanistic hypothesis for the lower growth rate of individuals in the high-elevation (MC) population. That is, lower individual growth rates in the MC population may result from process limitation in which the resources assimilated per unit time that could be allocated to growth are reduced relative to the GV population because of decreased passage time and consumption rate due to the lower temperatures at which digestion must take place in the MC population. The lower individual growth rates seen in the BQ population relative to the GV population are hypothesized to be due to lower prey availability due to lower rainfall and resulting primary productivity interacting with a biophysically imposed constraint that greatly reduces the time available for foraging in the BQ population. Thus, the overall hypothesis that explains the pattern of geographic variation in individual growth rates in this system involves multiple interacting mech-anisms (process limitation due to complex nonlinear temperature dependencies of di-gestive performance and resource limitation due to lower prey availability and biophysically imposed thermal constraints on time available for foraging) and interac-tions among several environmental gradients (temperature, precipitation, productivity, and food availability). This complex causal scenario can be tested experimentally in several ways. The most obvious approach involves a resource supplementation exper-iment in which prey availability is artificially increased in all populations. Under the current causal hypothesis, a strong positive growth rate response is predicted in the BQ population, a smaller response in the GV population, and no increase in growth rate in response to increased prey availability in the MC population.

Mechanism in Ecological Experiments

Currently most ecological experiments are not formulated and carried out with ex-plicit mechanisms as the alternative hypotheses being tested. That is, most ecological experiments are phenomenological or mechanism-vague (or mechanism-free). For ex-ample, most density manipulation experiments which test for a significant effect of the density of one species on some response variable (e.g., growth rate or density) of another species, when considered alone, provide no means for inferring the causal mechanism whereby a response to density manipulation is produced. Phenomenological experiments may be used to suggest hypotheses of causality but are rarely adequately designed to test potentially important alternative causal mechanisms or even the "mechanisms" that the experimenter claims to be testing. Typically, such mechanism-vague experiments: (a) involve mechanisms (in the sense defined previously) that are unspecified or unknown, (b) involve manipulated variables that bear no unambiguous relationship to a mechanism or set of mechanisms, and (c) produce outcomes that are not generalizable to other systems or even to the experimental system in which they are carried out when the range of experimental variables falls outside the range of manipulations already accomplished or when initial conditions vary. In such experiments, manipulated and response variables generally have no clear mechanistic connection and multiple interacting causes may be responsible for the observed response to experimental manipulation.

Statistical Considerations

The nature of experimental design and statistical analysis may pose problems for generalizing experimental results to natural systems. The purpose of doing an experiment is to test a statistical hypothesis. Basic principles of experimental design involve establishment of orthogonal contrasts among experimental variables. Establishment of orthogonal contrasts allows one to test hypotheses of main effects in factorial designs independently but may not reflect naturally occurring covariation among the experimental (= predictor) variables. For example, consider a two-factor analysis of variance with three levels of each factor (Fig. 2-3). Natural covariation between these two factors may be reflected in only a limited subset of treatment combinations. Suppose that, in nature, there is negative covariation between the magnitude of factor 1 and the magnitude of factor 2. In the example diagrammed in Figure 2-3, treatments where both factors are high or where both factors are low do not occur naturally. Most experimental designs currently employed by ecologists yield yes-no answers (there is or there is not a statistically significant effect of experimental variable x on response variable y). In a case such as that presented in Figure 2-3, significant treatment effects may be due to combinations of treatment levels that do not occur naturally. In addition, the results and interpretations of analysis of variance or similar linear model analyses are dependent on which predictive factors (treatments) are included in the experiment and, therefore, in the analysis. In such an analysis, the magnitude of the main effect of a

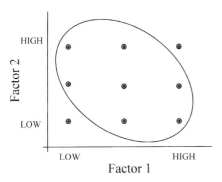

Figure 2-3. Diagram of a 3 × 3 factorial design ecological experiment. Assume that the ellipse encloses the treatment combinations that occur in nature. To the extent that significant treatment main effects or interaction effects are due to treatment combinations that never occur in nature (e.g., low-low or high-high combinations), extrapolation of the results of such experiments to natural systems is problematic.

factor (e.g., density of another species) may be large in a controlled experiment, yet if the effect is reanalyzed in the presence of other factors (e.g., variation in the biophysical environment) statistical significance may be lost or the magnitude of the effect may change dramatically. As a result of these two issues, it may often be impossible to understand the dynamic behavior of natural systems based on experimental results (for an example see Petraitis, this volume).

On the Relationships among Mechanism, Scale, and Generality

All studies in ecology begin with a specific phenomenon or set of phenomena that requires explanation (Fig. 2-4). It is usually the case that many potential mechanisms may be responsible for a given pattern. In order to understand a particular ecological pattern, one must identify the subset of potential mechanisms that is actually responsible for the pattern of interest. It is also the case that each potential mechanism may act over a unique set of scales. For example, population density may be limited by physiological constraints on reproduction (e.g., developmental rates, pelvic girdle size, etc.) which may be common to all members of a species. Alternatively, population densities may be limited by food availability or predation, both of which may vary among populations. Each potential mechanism, therefore, is valid over a set of spatial, temporal, and organismal scales. The set of scales over which a given mechanism (or set of mechanisms) is valid directly determines the appropriate spatial, temporal, and organismal scales for critical experimental tests of the mechanisms in question. Thus, as depicted in Figure 2-4, for any epistemological sufficient phenomena, potential mechanisms may be listed ($M_1, \ldots, M_n$ and critical experiments may be designed (EXP_1, $\ldots$, EXP_n) each with its own spatial, temporal, and organismal scale ($SCALE_1, \ldots,$ $SCALE_n$). Execution of the designed experiments should lead, ultimately, to a subset of mechanisms (M^*) to which the original phenomenon is attributed. The set of mechanisms included in M^* collectively exhibits a unique set of spatial, temporal, and organismal scales. It is the relevant scales of M^* that determine the domain of generality over which the mechanisms in M^* have explanatory and predictive power. Thus, if M^* is known, an unambiguous statement regarding the generality of experimental results may be made.

Observation and Mechanism in Ecology, Illustration by Example

The fundamental question that concerns us is: How do we recognize the generality of a particular principle derived from observations taken in the context of ecological study? Using examples from the present volume and the primary literature, we illustrate in the following discussion how the concepts of mechanism and scale determine directly the domain of generality over which a given experimentally derived principle applies. Our choice of particular studies as examples is not meant as criticism; rather, we chose these examples because of their utility for illustrating epistemological limitations shared to a greater or lesser degree by all ecological studies. Prior to discussion of specific

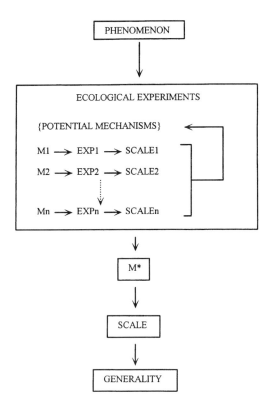

Figure 2-4. Relationships among mechanisms (alternative hypotheses of causality), mechanism-explicit ecological experiments, the spatial, temporal, and organismal scales of ecological experiments, and the domain of generality over which the results of experimental experiments apply.

examples, it is useful to consider the variety of contexts under which observations are made in ecological studies.

Observation is the fundamental tool of all scientists, and studies differ only in the degree to which the observer exerts control over the circumstances under which observations are made. In ecology, we define three broad classes of observation: (1) natural history; (2) mechanism-free (phenomenological) experiments, and (3) mechanism-explicit experiments. We note that these categories are not necessarily mutually exclusive—that is, it is possible to test critically some mechanisms with natural history (i.e., uncontrolled) observations, and the distinction between mechanism-free and mechanism-explicit experiments will depend to some extent on the level of reductionism implied by each hypothesis and associated experiment. Nevertheless, these distinctions are useful for illustrating the relationships among mechanism, scale, and generality.

The primary purpose of natural history is to observe and document patterns of variation in natural phenomena. The documentation of pattern in nature produces ecological questions at the most fundamental level, and it is toward answering these questions that the activities of ecologists are generally directed. At their core, natural history observations may establish pattern and often suggest a potentially broad range of mechanisms that may be responsible for an observed pattern. In some cases, competing or alternative mechanistic hypotheses may be tested through further uncontrolled observation of natural history. Natural history observations can falsify some mechanisms, but they generally cannot unambiguously support any particular mechanism as responsible for pattern in nature. Typically therefore, identification of relevant mechanisms cannot be accomplished without imposing greater control on conditions of observation. Thus, because exact mechanisms cannot be clearly established, the scale over which a particular pattern obtains and also its domain of generality remain ambiguous if only natural history observations are utilized.

In mechanism-free or mechanism-vague experiments, a manipulation is performed with the aim of testing a specific prediction. However, either a specific mechanism is not stated or more than one mechanism (usually unspecified) may produce the predicted result. Investigators focus on particular factors without a clear statement of the hypothesized relationship(s) between manipulated independent and observed dependent measures. For example, many studies have manipulated densities of potential competitors to study "competition" or predation. Competition, as used in this sense, is actually a family of mechanisms, including potentially a variety of inter-and intraspecific interactions (e.g., competition for food, interference, and a host of corollary effects attendant to changes in species density). Careful attention to experimental detail is required to differentiate among the variety of mechanisms embodied in "competition." Without knowledge of the exact mechanisms that produce an experimental result, it is difficult to determine the scale over which the result may obtain in nature, and thus it is impossible to specify the domain of generality of the principle being tested. That mechanism-free experiments are impossible to generalize does not necessarily diminish their utility. Mechanism-free experiments play a critical role in refining hypotheses and in suggesting potential mechanisms for further, more controlled, studies.

Mechanism-explicit experiments are among those that are most tightly controlled. The experimenter has identified a single or very few mechanisms and has carefully designed the experiment to test critically the mechanisms of interest as hypotheses to account for some observed behavior. Mechanism-explicit experiments have the greatest potential for unambiguously identifying mechanisms responsible for a particular pattern. Tightly controlled experiments designed to test particular mechanisms give the experimenter confidence that the tested mechanism or set of interacting mechanisms (if not falsified) is responsible for pattern in nature. Once appropriate mechanisms are identified, the scale over which each mechanism operates can be inferred, and an explicit statement concerning the domain of generality of the principle being tested may be made. For example, suppose an experimenter determines that two species compete with each other through direct interference of one of the species with the other. Interference competition operates on the spatial scale determined by the amount of space required for one individual of each species to interact and on the temporal scale which determines how often individuals of these two species will occur in syntopy. The domain of generality over which interference competition is likely to influence population dynamics of the species in question is, therefore, all populations of these species that occur in syntopy and that exhibit interference behavior.

We emphasize here that it is not our intention to assign greater or lesser value to any of the three observational contexts we have identified. Clearly, the advancement of ecology as a science requires all three kinds of observation. We maintain only that a concise statement of the generality of a particular experimental result hinges on a clear understanding of the mechanism(s) responsible for that result. In our opinion, only mechanism-explicit experiments can provide the required level of understanding for unambiguous generalization. Our arguments have implications for the interpretation of experimental results and may aid in determining the degree to which particular experimental results are generalizable. In the remainder of this essay, we illustrate the relationships among mechanism, scale, and generality with specific examples.

Studies of Ectotherms in Big Bend

The mechanistic explanation of geographic variation in life history has been the focus of our research on ectothermic vertebrates in the Big Bend region of Texas. Both *S. merriami* (as just described) and the mottled rock rattlesnake (*Crotalus lepidus*) exhibit similar variation in growth and size along an elevational gradient (Beaupre 1995a, Grant and Dunham 1990). Our approach has been to quantify the relevant fluxes of mass and energy through individuals by quantifying interactions between the properties of organisms (physiological and behavioral) and their environment (thermal distributions, seasonality, and productivity). Our goal in each case is to narrow the set of potential mechanisms $(M_1, \ldots, M_n)$ by (1) making critical observations in nature that can falsify some mechanisms and (2) designing critical experiments to test those mechanisms that cannot be falsified by observation. As such, the process is iterative, and it is our hope to eventually understand the natural complexity of these systems at a mechanistic level.

We are the first to acknowledge that we are far from a complete mechanistic understanding of environmental effects on life history. However, our studies have documented complex trade-offs among a number of mechanisms that affect the patterns of interest. For example, variation in metabolism, growth rate, and adult body size of *C. lepidus* on an elevational gradient may be the complex result of simultaneous variation in environmental thermal distributions (that affect body temperature); prey capture success, which varies with productivity; and time available for foraging (Beaupre 1993, 1995a, b, 1996). We have already described such complex and interacting effects on growth and size of *S. merriami* (see above). These processes qualify as mechanisms by our definition, because they operate at a lower level of hierarchy (individuals) than the patterns we wish to explain (populations). As the set of mechanisms and their likely interaction is narrowed, specific experimental tests of mechanistic hypotheses can be designed. An example of such a test is the supplemental feeding experiment proposed here for *S. merriami*. Through the iterative process of observation and experimentation, each employed as appropriate, we hope to attribute much of the variation in this system to sets of interacting mechanisms (M^*). Understanding these mechanisms will allow an unambiguous statement regarding the domain of generality (on organismal, temporal, and spatial scales) of the processes that we study. We are not to the point of making such an unambiguous statement, and we are aware of the implications for generalization posed by the complexity of our systems.

Mechanism, Higher Order Effects, and Higher Order Interactions   Recently there has been a great deal of concern over the existence and interpretation of higher order effects and higher order interactions in ecological systems and associated experiments (e.g., Abrams 1983; Adler and Morris 1994; Billick and Case 1994; Fairweather 1990;

Wootton 1993, 1994). This concern is motivated by the following question: To what extent can the complex dynamics of communities and ecosystems be predicted with knowledge of pairwise species interactions? Clearly, this question cannot be addressed with only experimental data on pairwise interactions because the experiments and data are at the wrong scale to address the fundamental issue. For illustrative purposes we consider the typical approach to answering this question. Typically, investigators conduct a set of pairwise and generally mechanism-free experiments and then model the resulting pairwise interactions using simple mathematical constructs like ordinary differential equations or difference equations. The next step is constructing a larger model that combines the previously derived pairwise models in additive fashion and makes some predictions about the dynamics of the complete system based on simple pairwise interactions. This set of predictions is then compared to data from an experiment that involves all relevant species. The comparison of model-based prediction with experimental results can only have two outcomes: either there is agreement between the predictions of the model and the results of the experiment or there is no agreement. A lack of agreement may occur for many reasons, including: (a) error in the data, (b) incorrect pairwise or additive models, (c) indirect effects that must be modeled at a finer scale, and (d) nonlinear "higher order" effects that cannot be predicted by simple linear combinations of pairwise interactions. The indirect effects mentioned in (c) are effects of one species on another mediated through a third species. This kind of interaction is easily incorporated into simple models because it occurs by the only "mechanism" normally incorporated into such models (numbers of organisms of each type, experimentally estimated rates of increase, and interaction coefficients). As an example of the "higher order" effects described in (d), consider a situation where, say, species A influences the number of refugia available to species B by eating macrophytes and species B has fewer places to hide from species C, which consequently eats disproportionately more of species B than it would if species A were not present. Wootton (1993, 1994) and others would argue that this kind of effect could not be predicted based on pairwise comparison and that experiments that involved all three species would be required to elucidate the "higher order" term that describes this interaction. Although we agree for the comparison involving these relatively simple experimental systems and models, a critical issue is being overlooked that is fundamental to issues of scale, mechanism, and generality in ecology. The distinction that Wootton and others raise between indirect effects and higher order effects is artificial because it arises solely due to the inadequacy of the original model with respect to mechanism. These models are generally implemented using empirically derived relationships among numbers of organisms of each type, experimentally estimated rates of increase, and experimentally estimated interaction coefficients which attempt to incorporate density dependence. Note that these empirically derived relationships are not mechanisms under our definition because they are derived by a curve-fitting procedure from variables at the same level of organization as the pattern they are attempting to explain. It is simple to incorporate interactions that act directly on numbers of organisms into these models because the indirect effect can be expressed directly in terms of the currency of the model numbers of each species. In the preceding example of a "higher order" effect, an investigator that considers only change in numbers of each species will not be able to explain the disproportionate decrease in species B. However, it seems reasonable that careful attention to mechanism (as we define it) in this system might lead to the realization that the supposedly "nonlinear" effect was actually the complex result of several interacting linear effects. For example, species A causes a linear decrease in macrophyte density with the result that refugia eventually become limiting, and there is a linear increase in the number of species B exposed to predation with a concomitant linear increase in the number of species B captured by species C. This chain of causality could occur with no change in the density of species A. Such a simple causal scenario could explain what would look like a complex nonlinear interaction, as the result of a series of interacting linear functions. In this case, as with most of the debate over higher order effects and higher order interactions, the lack of mechanism results in an inability to generalize to more realistic cases. Thus, we believe that much of the argument over higher order interactions derives from (1) failure to explicitly incorporate mechanism into theory and (2) attempts to generalize the results of mechanism-free experiments.

Experimental Exclosures   Competition among guilds of seed-eating rodents in the Chihuahuan desert has been the subject of long-term studies involving experimental exclosures (Brown, this volume, Brown and Munger 1985, Brown et al. 1986, Heske et al. 1994). Brown and coworkers have fostered a paradigm in ecological field studies. Their experiments have documented increases in population density of small granivorous rodents in response to removal and exclusion of large granivorous rodents of the genus *Dipodomys*. These density increases of small granivores have been attributed to competitive release. The pattern of increase in granivores is consistent with the hypothesis of some generalized competitive release. However, the exact mechanisms (interference, exploitation, etc.) that govern the response remain unknown. It is also the case that the experimental manipulations may not rule out some alternative explanations. For example, *Dipodomys* ssp. were excluded from experimental plots by gates which are too small to allow free movement of kangaroo rats. Such gates are also likely to restrict free movement of large viperid snakes that may consume many rodents on an annual basis. Heske et al. (1994) attribute immediate increases in granivore density on *Dipodomys* removal plots to migration of granivores from surrounding areas and active selection of *Dipodomys*-free microhabitats. Are migrating granivorous rodents responding to decreased density of *Dipodomys* or to decreased density of snake predators? Heske et al. (1994) note that insectivorous rodents (genus *Onychomys*) show no response to *Dipodomys* removal and argue that this observation supports the competitive release hypothesis rather than a decrease in predators. However, alternative explanations may exist for the failure of *Onychomys* to respond (e.g., insectivorous rodents may differ in their behavior and in their propensity to move among sites).

These considerations suggest that the *Dipodomys* exclusion experiments are mechanism-free or mechanism-ambiguous experiments. We note that ambiguity of mechanism is largely due to the laudable attempt to produce realism in these experiments. Nevertheless, ambiguous mechanisms lead to an inability to specify the conditions under which a given experimental result will be repeated, and thus it is impossible to specify the domain of generality of the principles of competition tested by these experiments. For example, based on the Chihuahuan desert experiments, what would we predict as the outcome of a similar manipulation in the Sonoran desert? Clearly, any prediction would be based on previously observed pattern, rather than on an explicit mechanistic theory of interaction for the species present in the new system.

Experimental Communities   Lawler and Morin (1993) constructed food chains of protists in microcosms to study how the population dynamics of these protists varied with food chain length and with the presence or absence of omnivorous top predators. Their approach was motivated by the desire to test the predictions of a vast volume of virtually data-free food web theory. Lawler and Morin chose simple and manipulable systems in an effort to match the assumptions of theory to as great a degree as possible.

In their manipulations, food chain elements consisted of an initially similar assemblage of bacteria to serve as prey, two types of bacterivorous ciliates, an omnivore that could persist on either bacteria alone or on bacteria and a bacterivorous ciliate, and a top predator that would eat only ciliates. Their primary manipulations were food chain long (three elements) or food chain short (two elements) and omnivore present or absent. They measured two variables as indicators of system stability: time course of abundance and variance in abundance. Lawler and Morin's results supported the notion that population fluctuations and extinctions will increase with increasing food chain length and that predators feeding on multiple prey species are better buffered from system fluctuations than are specialists. These results were in general agreement with expectations from food web theory.

Lawler and Morin achieved their stated goals and were duly cautious in generalizing their experimental results. They offer the following thoughts in their closing paragraph (Lawler and Morin 1993:682): "Convincing statements about the generality of these patterns will require examination of many more species assembled in various trophic combinations." This quote reflects a pervasive and popular view of generality that is based on the notion that general principles are those that apply to the majority of cases. The show-of-hands concept of generality is also at the core of the meta-analysis approach (Gurevitch et al. 1992). This kind of generality is limited in the sense that (1) it is mechanism-free and determination of "generality" requires endless iteration of experimental permutations and (2) special cases must be explained in post hoc, case-by-case investigation. Furthermore, this view of generality provides only a limited ability to predict the outcome of novel manipulations. We believe these limitations can be avoided by focus on mechanism in experiments and by development of mechanistic theory that can explain a broader range of special cases.

As a demonstration of our position we pose the question, What is the domain of generality of patterns observed in protist food web dynamics? The answer to this question lies in the understanding of mechanisms that govern system stability and the extent to which this understanding is provided by the experimental manipulation. The patterns in time course of abundance and variance in abundance observed by Lawler and Morin were likely governed by many mechanisms; hence their experiments were mechanism-free or mechanism-vague. One uncertainty is whether abundance of food chain elements was governed by predominantly top-down or bottom-up regulation. It is likely that abundances in different treatments were affected by different mechanisms. The stability of food chains should also be affected (at a minimum) by encounter rates, prey capture success rates, and the efficiency and rate of biomass conversion of each participating element. A clear understanding of these mechanisms and their role in producing patterns of abundance in experimental communities would allow a concise statement of the degree of generality of patterns observed in these manipulations (i.e., what properties of natural systems must obtain to exhibit behavior similar to that of experimental communities). Because mechanisms are not precisely known, the domain of generality of these results cannot be exactly specified.

Hormonal Manipulations   Use of hormones to manipulate phenotypes is an experimental technique that is gaining in popularity. Ketterson and Nolan (1992) and Ketterson et al. (1992) outline their rationale and provide an example of hormonal manipulation that has been referred to as "phenotypic engineering." In their own words, "Phenotypic engineering consists of manipulating the phenotype of an organism, quantifying the effects of the manipulation, and relating these effects to performance or fitness. This method permits exploration of the evolutionary significance of phenotypic variation by asking whether a rare or a novel phenotype would increase in frequency, assuming the requisite genetic variation" (Ketterson and Nolan 1992: S41). Furthermore, they suggest that "it is possible to probe the question of why existing phenotypes persist despite the fact that alternative phenotypes are possible" (Ketterson and Nolan 1992:S42).

In the context of these issues of general interest, Ketterson et al. (1992) used testosterone implants to manipulate phenotypes of male dark-eyed juncos during their reproductive season. A sample of male birds were collected while their mates were brooding the first clutch of the season. Half of the male birds received testosterone implants, and the other half received sham implants. Ketterson et al. (1992) made a series of measurements on behavior of both sexes (feeding trips to the nest, time spent brooding by females, and time spent singing by males) and related these behavioral measures to short-term fitness components (eggs laid and hatched in the first and second clutch, number of nestlings at day 10, female mass at day 10, and renesting interval). Testosterone-implanted males exhibited 200% larger home ranges than controls and 300% larger core areas, spent significantly less time at the nest and feeding young than controls, ranged greater maximum distances from the nest than controls, and sang more frequently than controls (Chandler et al. 1994, Ketterson et al. 1992). Clearly, the manipulation produced novel phenotypes whose behavior was far outside the norm for male dark-eyed juncos. Despite this massive behavioral perturbation, effects on short-term fitness components were largely undetectable. Lack of clear fitness effects was likely due to a combination of low statistical power for some measures, compensation by unmanipulated females, and the short duration of the experiments. The authors offer four possible interpretations of the observed results: (1) a wide range of equally fit phenotypes can exist, (2) fitness effects of elevated testosterone were too small to detect, (3) fitness effects were absent in this study but may be detectable at other times or in other environments, and (4) the components of fitness measured were insensitive to the manipulation, but other fitness components may be affected.

In this series of hormone manipulations, the mechanism that governs behavioral changes in male juncos is explicit. There can be little doubt that the increase in testosterone at the particular stage of male life history was directly responsible for shifts in male behavior. For this reason, we consider these manipulations to be mechanism-explicit experiments. The explicitness of mechanism allows a very clear statement of the appropriate scale and domain of generality over which the results of this experiment apply. The appropriate spatial scale for this particular mechanism is on the order of individual males—the mechanism of testosterone increase operates directly on individual patterns of behavior. The appropriate temporal scale for this particular mechanism may be defined most easily by the life history stage during which the manipulation takes place—in adult reproducing males during brooding by females of the first clutch of the year. Therefore, the domain of generality of the results of these experiments (i.e., the set of conditions under which the results of these manipulations have relevance to the stated goal of determining whether a novel phenotype would increase in frequency) includes all possible mutations that produce increases in testosterone similar in magnitude to experimental increases, produce testosterone increases in adult male dark-eyed juncos during brooding of the first clutch of the year by females, and, furthermore, produce no other discernible effects on the phenotype.

Allometric Engineering   In a series of experiments, Sinervo and coworkers have manipulated trade-offs between clutch size and egg size in lizards by applying a variety of techniques, including yolkectomy and hormonal treatments at different stages of the reproductive cycle (Sinervo 1990, 1993; Sinervo and Huey 1990; Sinervo and Licht 1991). Female lizards (*Uta stansburiana*) with large clutches of small eggs were produced by increasing circulating levels of follicle-stimulating hormone (FSH) in vitellogenic females (Sinervo and Licht 1991). Yolkectomy of eggs following oviposition has been used to affect hatchling size (Sinervo 1990, Sinervo and Huey 1990). Radical yolkectomy of oviducal eggs has been used to reduce clutch size in the oviduct and thereby produce smaller clutches of larger eggs (Sinervo and Licht 1991).

The primary purpose of these elegant manipulations has been to investigate the mechanistic basis and fitness consequences of naturally occurring negative covariation between clutch size and egg or hatchling size among populations of *U. stansburiana*. These goals are made clear by the following quotes:

> The experimental manipulations of clutch and egg size address the causal basis of the physiologically based trade-off between clutch size and egg size. (Sinervo 1993: 215)
>
> Our experimental data indicate that these comparative patterns of covariation between clutch size and egg size are governed by the mechanistic bases underlying the regulation of these traits. This experimental confirmation of the comparative patterns has important implications for the adaptive evolution of clutch size and egg size. (Sinervo and Licht 1991: 260)
>
> Given the pervasive nature of the egg size and egg number tradeoff among amniotes . . . it is likely that our results might be generalizable to other groups of vertebrates. (Sinervo and Licht 1991:262)

Such statements may be found throughout the allometric engineering literature, and they imply that Sinervo and coworkers argue that they have directly manipulated the mechanistic basis whereby individual females make clutch size–egg size allocation decisions.

Through allometric engineering, Sinervo and coworkers have succeeded in mimicking the pattern of variation in clutch size–egg size trade-offs observed in among-population comparisons of *U. stansburiana.* Producing variation in the lab that is parallel to variation observed in the field, however, does not imply that the mechanistic basis of variation in the field has been discovered. A cogent statement of the domain of generality of the results of size manipulations requires consideration of the exact mechanisms being manipulated. In fact, each technique employed in this family of experiments arguably represents a different mechanism. Hatchling size manipulation through yolkectomy of recently oviposited eggs is a manipulation that may have little bearing on the actual physiological mechanisms responsible for clutch size–egg size relationships. Fundamental to the interpretation of the results of yolkectomy experiments is the assumption that size-manipulated eggs and hatchlings are equivalent in all meaningful ways with eggs and hatchlings of comparable size produced naturally. This equivalence has yet to be demonstrated (Bernardo 1991). Likewise, reduction of clutch size by radical yolkectomy in the oviduct is a manipulation applied after the female makes the primary clutch size decision. Such a manipulation may only be relevant to the actual mechanisms that establish the clutch size–egg size trade-off if it can be shown that females routinely reduce clutch size through the selective removal of yolk from a subset of available developing follicles. Of all the manipulations employed, increasing circulating FSH alone may be related to the actual mechanisms whereby female lizards make clutch size–egg size allocation decisions. The role of variation in circulating FSH in determining clutch size–egg size relationships in natural populations apparently has yet to be demonstrated and is complicated by the fact that, as yet, no radioimmunoassay for squamate FSH or its analogue is available (Sinervo and Licht 1991). We argue that these procedures do not directly manipulate the actual mechanistic basis of clutch size–egg size trade-offs, but rather, they affect phenotypes after the female has made this critical allocation decision. Even the FSH manipulation is ambiguous with respect to its relationship to actual mechanisms governing clutch size–egg size trade-offs in nature. For these reasons, with respect to the often stated goal of understanding the physiological basis of clutch size–egg size trade-offs, we consider these manipulations to be mechanism-ambiguous.

Despite the apparent mismatch between manipulations employed and the actual mechanisms that govern clutch size–egg size relationships, these experiments produce their results through explicit mechanisms. Thus, we can state explicitly that the domain of generality of the results of these experiments includes all lizard species that affect clutch size through mechanisms that are directly analogous to egg yolkectomy, radical yolkectomy in the oviduct, and variation in levels of circulating FSH.

Summary

For ecology to advance as a science, we must continually evaluate our current understanding of natural processes and the methods by which we arrive at that understanding. The uniqueness of individuals and the potential for multiple causality in patterns of importance to ecologists force a particular structure on ecological theory and experiments. In our opinion, it is the application of simplistic modeling and experimental approaches that has, in part, led to the conception of "general" ecological theory that explains few or no special cases. Ecological theory and experiments are of two basic types: process-explicit (mechanism-based) theory and experiments and phenomenological (mechanism-vague or mechanism-free) theory and experiments. Both types of theory and experiments are useful, but the defensible interpretations that can be drawn from each are fundamentally different. A concept of mechanism is central to the distinction between these two classes of manipulation. We have defined mechanism as a level of reductionism that provides a causal explanation of the functional relationships among a set of variables. Each mechanism operates over a particular set of spatial, temporal, and organismal scales and therefore determines these appropriate scales for experiments designed to test alternative hypotheses of causality in ecological systems. These scales, in turn, determine the domain of generality over which any causal explanation applies or over which the results of any experiment apply. That is, most ecological processes or principles are valid over a restricted range of spatial, temporal, or organismal scales (= the domain of generality of a given process or principle). For ecologists, nature is truly a collection of special cases.

Ecologists frequently overgeneralize the results of observational, theoretical, and experimental studies because of a failure to appreciate the connection between mechanism, scale, and generality. Phenomenological or mechanism-vague experiments are not explicit with regard to mechanism or scale, and therefore the domain of generality associated with these experiments is either ambiguous or zero (results apply only to the experimental system). All ecological experiments are conducted on some set of temporal, spatial, and organismal scales. However, scales of experimentation often do not match the temporal, spatial, and organismal scales over which the processes studied purport to operate. The domain of generality of experimental results cannot be explicitly stated if the experiment was mechanism-free or if the experimental scales and those dictated by the mechanism in question were mismatched. These considerations suggest that discussions of generality should be limited to situations where the set of operating mechanism (M^*) has been unambiguously identified. Clearly, it will take more than experiments alone for ecology to advance as a science. Natural history, phenomenological experiments, and process-explicit experiments all are required to increase our understanding of the natural world. We suggest only that believable statements about generality require knowledge of the relevant mechanisms. A useful practice, and one which we support, might be specifying the domain of generality of theoretical and empirical work in much the same way as we specify the methods we use to acquire data or to construct models as part of the normal reporting process. In addition, application of our definitions may clarify those instances where the domain of generality remains unknown.

Perhaps some of the problems encountered in determining domain of generality in ecology are created by limitations in our current epistemology. Unambiguously identifying the set of mechanisms responsible for a given pattern is problematic. Whereas we have clear criteria for rejecting hypotheses, when do we generally accept that a given mechanism is the correct one? Clearly, among researchers there is wide variance in willingness to attribute ecological patterns to specific mechanisms. This dilemma is at the core of the generality debate. If specification of generality requires knowledge of mechanism and if our criteria for attributing a given pattern in nature to a mechanism or set of mechanisms are to some degree subjective, what then is the prospect for generality in ecology? It is ironic that specifying the generality of a principle may require a degree of specificity that we are epistomologically ill-equipped to obtain.

ACKNOWLEDGMENTS   We are grateful to Peter Petraitis and to the other members of the ecology group at the University of Pennsylvania for fruitful discussions of this topic.

Literature Cited

Abrams, P. A. 1983. Arguments in favor of higher order interactions. American Naturalist 121:887–891.

Adams, S. M., and D. L. DeAngelis. 1987. Indirect effects of early bass-shad interactions on predator population structure and food web dynamics. Pages 102–116 in W. C. Kerfoot and A. Sih (eds.), Predation in Aquatic Ecosystems. University Press of New England, Hanover, New Hampshire.

Adler, F. R., and W. F. Morris. 1994. A general test for interaction modification. Ecology 75:1552–1559.

Beaupre, S. J. 1993. An ecological study of oxygen consumption in the mottled rock rattlesnake, *Crotalus lepidus lepidus*, and the black-tailed rattlesnake, *Crotalus molossus molossus*. Physiological Zoology 66:437–454.

———. 1995a. Comparative ecology of the mottled rock rattlesnake, *Crotalus lepidus*, in Big Bend National Park. Herpetologica 51:45–56.

———. 1995b. Effects of geographically variable thermal environment on bioenergetics of mottled rock rattlesnakes, *Crotalus lepidus*. Ecology 76:1655–1665.

———. 1996. Field metabolic rate, water flux, and energy budgets of mottled rock rattlesnakes, *Crotalus lepidus*, from two populations. Copeia 1996:319–329.

Beaupre, S. J., and A. E. Dunham. 1995. A comparison of ratio-based and covariance analyses of a nutritional data set. Functional Ecology 9:876–880.

Beaupre, S. J., A. E. Dunham, and K. L. Overall. 1993. The effects of consumption rate and temperature on apparent digestibility coefficient, urate production, metabolizable energy coefficient and passage time in canyon lizards (*Sceloporus merriami*) from two populations. Functional Ecology 7:273–280.

Bernardo, J. 1991. Manipulating egg size to study maternal effects on offspring traits. Trends in Ecology and Evolution 6:1–2.

Billick, I., and T. J. Case. 1994. Higher order interactions in ecological communities: what they are and how can they be detected. Ecology 75:1529–1543.

Brown, J. H., and J. C. Munger. 1985. Experimental manipulation of a desert rodent community: food addition and species removal. Ecology 66:1545–1563.

Brown J. H., D. W. Davidson, J. C. Munger, and R. S. Inouye. 1986. Experimental ecology: the desert granivore system. Pages 41–61 in J. Diamond and T. J. Case (eds.), Community Ecology. Harper and Row, New York.

Chandler, C. R., E. D. Ketterson, V. Nolan Jr., and C. Ziegenfus. 1994. Effects of testosterone on spatial activity in free-ranging male dark-eyed Juncos, *Junco hyemalis*. Animal Behaviour 47:1445–1455.

DeAngelis, D. L., and L. J. Gross. 1992. Individual-based models and approaches in ecology. Chapman and Hall, New York.

DeAngelis, D. L., K. A. Rose, L. B. Crowder, E. A. Marschall, and D. Lika. 1993. Fish cohort dynamics: application of complementary modeling approaches. American Naturalist 142: 604–622.

Dunham, A. E. 1978. Food availability as a proximate factor influencing individual growth rates in the iguanid lizard *Sceloporus merriami*. Ecology 59:770–778.

———. 1993. Population responses to global change: physiologically structured models, operative environments, and population dynamics. Pages 95–119 in P. Kareiva, J. Kingsolver, and R. Huey (eds.), Evolutionary, Population, and Community Responses to Global Change. Sinauer, Sunderland, Massachusetts.

Dunham, A. E., and K. L. Overall. 1994. Population responses to environmental change: life history variation, individual based models, and the population dynamics of short-lived organisms. American Zoologist 34:382–396.

Dunham, A. E., B. W. Grant, and K. L. Overall. 1989. The interface between biophysical ecology and the population ecology of terrestrial vertebrate ectotherms. Physiological Zoology 62:335–355.

Fairweather, P. G. 1990. Is predation capable of interaction with other community processes on rocky reefs? Australian Journal of Ecology 15:453–464.

Foster, M. A. 1990. Organization of macroalgal assemblages in the Northeast Pacific: the assumption of homogeneity and the illusion of generality. Hydrobiologica 192:21–33.

Gates, D. M. 1980. Biophysical Ecology. Springer-Verlag, New York.

Grant, B. W., and A. E. Dunham. 1988. Biophysically imposed time constraints on the activity of a desert lizard, *Sceloporus merriami*. Ecology 69:167–176.

———. 1990. Elevational variation in environmental constraints on life histories of the desert lizard, *Sceloporus merriami*. Ecology 71:1765–1776.

Gurevitch, J., L. L. Morrow, A. Wallace, and J. S. Walsh. 1992. A meta-analysis of competition in field experiments. American Naturalist 140:539–572.

Heske, E. J., J. H. Brown, and S. Mistry. 1994. Long-term experimental study of a Chihuahuan desert rodent community: 13 years of competition. Ecology 75:438–445.

Hurlburt, S. H. 1984. Pseudoreplication and the design of ecological field experiments. Ecological Monographs 54:187–211.

Ketterson, E. D., and V. Nolan Jr. 1992. Hormones and life histories: an integrative approach. American Naturalist 140:S33–S62.

Ketterson, E. D., V. Nolan Jr., L. Wolf, and C. Ziegenfus. 1992. Testosterone and avian life histories: effects of experimentally elevated testosterone on behavior and correlates of fitness in the dark-eyed junco (*Junco hyemalis*). American Naturalist 140:980–999.

Lawler, S. P., and P. J. Morin. 1993. Food web architecture and population dynamics in laboratory microcosms of protists. American Naturalist 141:675–686.

Peters, R. H. 1991. A Critique for Ecology. Cambridge University Press, Cambridge.

Porter, W. P., and D. M. Gates. 1969. Thermodynamic equilibria of animals with environment. Ecological Monographs 39:245–270.

Porter, W. P., and C. R. Tracy. 1983. Biophysical analyses of energetics, time-space utilization, and distributional limits. Pages 55–83 in R. B. Huey, E. R. Pianka, and T. W. Schoener (eds.), Lizard Ecology: Studies of a Model Organism. Harvard University Press, Cambridge, Massachusetts.

Quinn, J. F., and A. E. Dunham. 1983. On hypothesis testing in ecology and evolution. American Naturalist 122:602–617.

Schoener, T. W. 1986. Mechanistic approaches to community ecology: a new reductionism? American Zoologist. 26:81–106.

Sinervo, B. 1990. Evolution of maternal investment in lizards: an experimental and comparative analysis of egg size and its effects on offspring performance. Evolution 44: 279–294.

———. 1993. The effect of offspring size on physiology and life history: manipulation of size using allometric engineering. BioScience 43:210–218.

Sinervo, B., and R. B. Huey. 1990. Allometric engineering: an experimental test of the causes of interpopulational differences in performance. Science 248:1106–1109.

Sinervo, B., and P. Licht. 1991. Hormonal and physiological control of clutch size, egg size, and egg shape, in side blotched lizards (*Uta stansburiana*): constraints on the evolution of lizard life histories. Journal of Experimental Zoology 257:252–264.

Underwood, A. J. 1986. The analysis of competition by field experiments. Pages 240–258 in J. Kikkawa and D. J. Anderson (eds.), Community Ecology: Patterns and Process. Blackwell, Melbourne.

Underwood, A. J., and P. G. Fairweather. 1986. Intertidal communities: do they have different ecologies or different ecologists? Proceedings of the Ecological Society of Australia 14:7–16.

Wootton, J. T. 1993. Indirect effects and habitat use in an intertidal community: interaction chains and interaction modifications. American Naturalist 141:71–89.

———. 1994. Putting the pieces together: testing the independence of interactions among organisms. Ecology 75:1544–1551.

3

Realism, Precision, and Generality in Experimental Ecology

PETER J. MORIN

The continuing debate among ecologists over where and how to conduct experiments stems from a failure to appreciate the advantages and disadvantages of different approaches. Much of the current debate revolves not around the value of ecological experiments but around the choice of experimental setting: laboratory, field, or something intermediate. Experiments done in laboratory or field settings differ unavoidably in precision and realism. Hybrid experiments, such as those conducted in various kinds of mesocosms (Morin 1983, Naeem et al. 1994), can be argued to have either the best or worst features of laboratory and field experiments. The distinctions among these experimental settings have been discussed in depth by others (Diamond 1986, Hairston 1989a) and will not be belabored here. Instead, I want to focus on the trade-offs imposed by working with different kinds of organisms and different experimental settings. Richard Levins (1968) made a similar point in a different context, regarding trade-offs encountered in building mathematical models of ecological processes. It is impossible to build a model that is at once realistic, precise, and general. Similar constraints apply to ecological experiments. Nelson Hairston Sr. (1989a), used similar logic in his discussion of experimental design in ecology, emphasizing the inevitable trade-offs between precision and realism, particularly when comparing laboratory and field experiments. Some of these ideas are abstracted in Figure 3-1, which portrays a sort of precision–realism continuum running from laboratory microcosms to field experiments. Laboratory experiments can be much more precise than field experiments because extraneous factors are readily eliminated or controlled and high levels of replication are possible. That precision comes at the expense of ecological realism. Lab experiments typically contain many fewer species than occur in natural communities, and they are often insulated from the disturbances that may drive patterns in nature. Field experiments take place in very realistic settings, but with that realism comes high variation

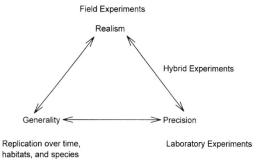

Figure 3-1. Proposed trade-offs among realism, precision, and generality in ecological experiments.

among replicates, together with financial or logistic constraints on the number of experimental units that can be set up and monitored. Precision suffers accordingly.

Unlike the situation described for mathematical models, the problem of generality in ecological experiments arises because no single study, no matter how precise or realistic it is, can claim to yield results that hold for the full diversity of species, communities, and ecosystems. Indeed, some ecologists argue that every situation is so unique that any search for generality is misguided. Others, including myself, are not so pessimistic and feel that general trends can emerge from collections of findings in a variety of systems (e.g., Sih et al. 1985, Gurevitch and Barton 1992, Goldberg et al. 1992). No single study, regardless of its position on the realism-precision continuum, will tell us very much about how another very different system will behave. Instead, we tend to infer generality by comparing the results of many experiments, conducted in a variety of systems, to search for common patterns.

Laboratory and field experiments each provide complementary and essential information about different kinds of organisms and ecological phenomena. Neither experimental setting can or should take primacy over the other, given the unavoidable trade-offs involved. Before arguing that a lab or field experiment is most appropriate and before selecting the system or organisms it is essential to decide what information is needed to answer a particular question. In ecology, the interesting questions are often prompted by theory, although some are suggested by natural history observations or previous experiments. Most ecological theory, with its focus on population dynamics, can be tested best with information obtained from laboratory experiments, since only those studies can provide information about dynamics over sufficiently long time frames to address questions about stability, cycles, chaos, and other dynamic behavior. Field experiments yield information that is either too variable or of insufficient duration relative to the generation times of focal organisms to be of much use in describing population dynamics. Ecologists will have to accept that some systems and settings can yield telling answers to important ecological questions, and others, no matter how intrinsically fascinating or aesthetically appealing, may not. This does not imply that we should not study the latter sort of systems, but we have to accept that there are very real limits to what we can learn from them.

Some of the more ardent proponents of the primary of field experiments feel that if complex natural systems are in any way abstracted or simplified, as invariably happens in laboratory or hybrid settings, then some critically important interactions may be distorted or lost. There is no denying that complex systems can exhibit a fascinating array of dynamics. The problem is that direct study of intact complex systems has done little so far to identify the mechanisms responsible for these dynamics (Lewin 1992).

One advantage of field experiments is their unchallenged realism. Interactions detected in field experiments clearly happen in nature and are not likely to be artifacts as long as the usual controls are used together with sound experimental design and rigorous statistical analysis. Field experiments are often the only feasible approach for studies of large, long-lived, or behaviorally complex organisms that are unlikely to go through their paces in the confines of the laboratory. The chief disadvantage of field experiments is that they usually do not offer the resolution needed to unambiguously test predictions made by theory. Typically, field experiments can tell us whether particular species interact, what the signs of those interactions are, and, in rare cases, whether the strengths of those interactions vary. Because the subjects of most field experiments have relatively long generation times, field experiments can tell us little about population growth rates and long-term dynamics that are central to much ecological theory.

In contrast to field experiments, laboratory experiments that use small organisms with short generation times can often yield sufficiently precise data to directly address ecological theory and are often the only the way that small short-lived organisms can be experimentally manipulated and monitored. The important work of Gause (1934), Park (1962), and Huffaker (1958) simply could not have been done under field conditions. The chief drawback is that laboratory experiments take place in settings that are more highly controlled and greatly simplified than natural communities. This level of control and simplification runs the risk of omitting, either inadvertently or by design, factors of ecological importance in natural systems. Such errors of omission enhance the risk of artifacts in laboratory experiments relative to field experiments. Laboratory systems also typically require intensive and often tedious maintenance. Finally, laboratory experiments are simply inappropriate for organisms of large size that cannot comfortably go about fulfilling their ecological roles within the confines of a laboratory environment.

Hybrid experiments offer some of the advantages and disadvantages of laboratory and field experiments. They are somewhat more realistic variants of laboratory experiments. Like field experiments, hybrid experiments will continue to run with a minimum of heroic intervention, although the need for a high frequency of sampling can add greatly to this. Typically, the organisms in question experience approximately natural regimes of light, temperature, and other physical factors, since the mesocosms are placed in the field. Unlike in natural systems, initial conditions of species composition, density, nutrient levels, and some physical variables can be carefully controlled, often to the great advantage of the experimenter. The chief drawback is that as in laboratory experiments, results may hinge critically on values of initial conditions or features of experimental environments that may not be realistic.

What follows is not intended as a comprehensive review of everything that can be learned from ecological experiments done in different settings. Instead, I use examples of work done by me and my students, mostly in aquatic systems, to illustrate the strengths and weaknesses of different experimental venues. Since this is the work that I know best, it is easy to identify advantages and disadvantages. We have done an assortment of field, hybrid, and laboratory experiments over the years, and along the way we have found that all of these approaches are valuable in asking questions about processes and dynamics that occur at different scales.

Field Experiments

Field experiments tell us what actually happens in nature. However, field experiments typically lack the resolution needed to assay whether natural processes conform to the predictions made by mathematical models. A single field experiment also cannot tell us whether a process operating in one location occurs in different sites or slightly different systems. One example of the limited resolution of field experiments comes from a study of a farm pond in North Carolina (Morin 1984). The experiment tested for the presence of a strong interaction between predatory fish and a group of invertebrate prey, larval dragonflies. A conspicuous natural pattern, rather than a specific mathematical model, motivated the study. There was a striking negative relation between species-specific body size and abundance in an assemblage of larval dragonflies. The pattern was remarkably consistent among years and has been observed in other ponds and lakes (Crowley and Johnson 1992). Similar size-abundance patterns in zooplankton are driven by planktivorous vertebrates that selectively eliminate larger prey (Brooks and Dodson 1965, Dodson 1974). Several fish species, including bluegill sunfish and largemouth bass, were abundant in the farm pond, so size-dependent predation seemed like a reasonable hypothesis for the pattern. A simple field experiment involving predator exclusion tested this idea. Cages of fine mesh screening excluded fish from areas of the littoral zone. Other cages with incomplete walls allowed fish to forage in a comparable experimental setting with an identical sampling area and served as cage controls. Both odonates and fish could move freely between the incomplete cages and the surrounding littoral macrophytes. After a summer of colonization and interaction, all of the pens were removed from the pond, and the odonates within were counted and identified. The results, shown in Figure 3-2, suggest two important points. First, the approach was useful in showing that fish had a strong influence on many, but not all, odonate species. The smallest and largest species were least affected. Second, only rather strong effects are statistically significant. Treatments that differ by about an order of magnitude show up as significant with four replicates per treatment; smaller effects do not. The upshot is that such experiments are useful for showing whether strong interactions occur. Limitation to the detection of effects of large magnitude is a simple consequence of the well-known ways in which the amount of replication and variability among replicates influence the power of a statistical test (Sokal and Rohlf 1981).

Field experiments generally suffer from two limitations: high variability among experimental replicates and low levels of replication. Both reduce the ability of statistical tests to detect modest differences among treatments. A different problem concerns the limited kind of manipulations that are tractable in field settings. Typically, we are limited to comparisons between ambient densities of species and the consequences of reduced densities that result from partial or complete exclusions. Effects of a range of densities, which would be of interest if we wanted to see how some population or community response changed as a continuous function of density, are difficult to test.

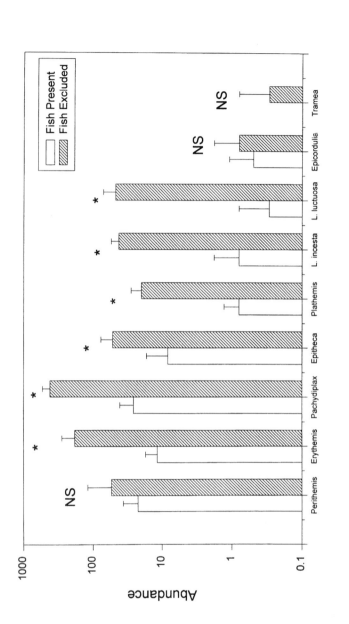
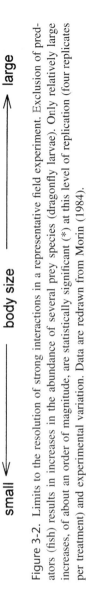

Figure 3-2. Limits to the resolution of strong interactions in a representative field experiment. Exclusion of predators (fish) results in increases in the abundance of several prey species (dragonfly larvae). Only relatively large increases, of about an order of magnitude, are statistically significant (*) at this level of replication (four replicates per treatment) and experimental variation. Data are redrawn from Morin (1984).

For small, sedentary animal species, a range of densities can be established in enclosures. Even then, as the number of density levels increases, the effort and expense of building additional enclosures increase and rapidly become prohibitive.

Had the focus of the study been on the effects of predators on prey population dynamics, a study of much greater (and probably unrealistic) duration would have been required. Depending on the species, the odonates go through only one or two generations per year. This means that a multiyear study of population dynamics would be needed to compare actual dynamics with and without predators.

Field experiments can clearly show when important interactions like competition (e.g., Connell 1961) or predation (Paine 1966) influence communities. Field experiments are, however, very limited in terms of what they can tell us about the precise form of the functional relations between interacting species or about the long-term effects of interactions on population dynamics.

Hybrid Experiments

Hybrid experiments using mesocosms lie somewhere on the continuum between completely natural systems and entirely artificial systems. Hybrid systems share some features of natural systems but remain artificial in the sense that they occur within constructed enclosures and typically consist of a controlled subset of species drawn from a more complex natural community. They may share other features of highly controlled laboratory systems, such as standardized initial values for nutrient levels, spatial heterogeneity, and some physical variables, such as pH.

Although the advantages and disadvantages of mesocosms and hybrid experiments have been much debated (Hairston 1989b, Jaeger and Walls 1989, Morin 1989, Wilbur 1989, Lawton et al. 1993), they can offer special advantages that offset the drawbacks. The mesocosms that I have used are artificial ponds reconstructed in prefabricated containers—in this case, watering tanks designed for livestock (Morin 1981). Many others have used similar systems to successfully explore a diverse array of ecological questions (Hurlbert et al. 1972, Murdoch and Sih 1978, Murdoch and McCauley 1985, Wilbur and Alford 1985, Resetarits and Wilbur 1989, Fauth and Resetarits 1991, Leibold and Wilbur 1992, Harris 1995). The systems offer special advantages for the study of certain groups of aquatic organisms that are difficult to study or manipulate in natural water bodies. For example, often we would like to know how manipulations in the abundance of predators or competitors will influence survival to metamorphosis. Estimation of survival in larval amphibians requires knowledge of the initial number of animals in the population and the final number that survive. Hybrid systems are stocked with an initial number of animals that is known without error. In natural ponds, estimation of the initial size of a cohort may require an exceptional amount of sampling effort that will still yield an error-prone estimate. There are some exceptional situations, like small water-filled rock pools, where direct counts are possible (Smith 1983), but as a rule just estimating the size of a population of animals in a natural setting involves tremendous effort.

The general procedure has been to pick a group of focal organisms, such as larval frogs, and then build many replicate communities of identical species composition. For larval frogs, that is accomplished by collecting and hatching a sufficient number of anuran eggs to stock the requisite number of mesocosms. Depending on the oviposition habits of the anuran species, that is accomplished by either collecting egg masses or collecting amplectant pairs of frogs that are returned to the laboratory where their eggs can be collected. Once the mesocosms have been stocked with a known number of hatching tadpoles, the fate of any cohort can be measured by collecting metamorphosing individuals as they attempt to leave the mesocosm. By counting the survivors and determining the mass and larval period for each it is easy to measure how various experimental manipulations affect species composition, larval performance, and other aspects of the ecology of larvae. The key point is that use of mesocosms provides a known set of initial conditions that permit the subsequent calculation of survival, growth rates, and larval periods.

Some of our first experiments using this system were designed to explore how assemblages of moderate complexity responded to a gradient of predator density and predation intensity. There was considerable interest in how community composition would respond to a gradient in predation intensity. The predator used was the broken-striped subspecies of the common spotted newt, *Notophthalmus viridescens dorsalis*. It was the common polyphagous predator in our natural study ponds. It was relatively simple to establish a gradient of predator density and predation intensity by stocking the artificial ponds with different numbers of newts. Comparable manipulations in natural ponds would have been more difficult and would have required cages stocked with known densities of predators and prey, a situation little different from that used in our mesocosms. Uncaged manipulations of newts at the level of the whole-pond manipulations have other difficulties, because transplanted newts rapidly return to their pond of origin (Gill 1979).

An important result of those experiments is shown in Figure 3-3. Anuran survival, which determines the final composition of prey assemblages, depended critically on predator density. Without *Notophthalmus*, assemblages were dominated by one species, *Scaphiopus holbrooki*, and several species, most noticeably *Pseudacris crucifer*, were underrepresented at metamorphosis. At the other end of the predation gradient, the situation was reversed, with the assemblage being dominated by *P. crucifer*. Using evidence gleaned from observed reductions in growth and survival in the absence of predators, it appeared that some species, like *P. crucifer*, were competitively excluded at low predator densities. These same competitively inferior species were apparently better at persisting with predators. Subsequent work showed that *P. crucifer* is relatively inactive and may be less prone to attack by visually oriented predators (Morin 1986, Lawler 1989).

This result raised other issues. These issues are not specific to mesocosms but apply equally to any single natural site studied in a field experiment. One issue is the extent to which a particular experimental outcome hinges on the initial conditions (say the densities or composition of a set of species) established in mesocosms or found in nature. Another issue is whether the properties of species that yield a particular outcome in one portion of their geographic range are constant across the entire range or whether species evolve to interact in different ways in different locations. This is a very important problem that has been chronically overlooked by experimental ecologists. We often know very little about how well the results obtained in a few well-studied sites or intensively studied systems generalize across locations and taxa.

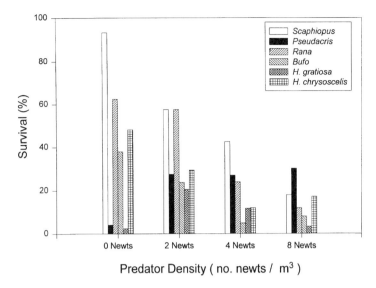

Figure 3-3. Results from a representative hybrid experiment. Variation in predator density creates variation in prey survival, which in turn affects community structure. Establishment of a predation gradient is easy in hybrid experiments based in mesocosms, while manipulations of predators in natural communities are often limited to presence (natural density) versus absence (complete or partial exclusions). Data from Morin (1983).

We have recently addressed this problem (Kurzava 1994, Kurzava and Morin 1994) by using a common environment study conducted in mesocosms to compare the effects of two different subspecies of the newt *Notophthalmus* on similar assemblages of prey. The approach is directly analogous to the common garden experiments that have a long history of use by plant ecologists (Clausen et al. 1948). Mesocosms permit common environment and transplant studies that would be either difficult to monitor or potentially unethical to set up in nature.

The two newt subspecies differ in their geographic distribution, body size (Fig. 3-4), and the array of prey species they are likely to encounter. *N. v. dorsalis* is from our field sites in North Carolina. It is the smaller of the two, its range is restricted to the coastal plain of the Carolinas, and it occurs with a high diversity, approximately 25 species, of anuran prey. *N. v. viridescens* occurs at our study sites in New Jersey. It is roughly twice the mass of *N. v. dorsalis*, it occurs over much of eastern North America, and in New Jersey it encounters perhaps half the number of anuran species seen at our sites in North Carolina. We wondered whether the two subspecies might differ in their impacts on prey, since they differ greatly in body size, gape, and the recent history of evolutionary experience with different prey species.

Our first approach to the problem involved comparing the effects of similar densities of the two different *Notophthalmus* subspecies on a single prey species, the larvae of the toad *Bufo americanus*. The comparison is potentially interesting because *B. americanus* occurs sympatrically with *N. v. viridescens* but does not occur with *N. v. dorsalis* at our study sites in North Carolina. Three other *Bufo* species, including the closely

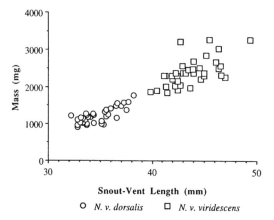

Figure 3-4. An example of geographic variation in the body size of *Notophthalmus*. *N. v. dorsalis* from North Carolina is significantly smaller than *N. v. viridescens* from New Jersey. Differences in size may affect the per capita impact of *Notophthalmus* on prey populations. Redrawn from Kurzava and Morin (1994).

related species *Bufo terrestris*, do occur with *N. v. dorsalis*. We stocked sets of artificial ponds with identical densities of *Bufo* tadpoles (400/m³) and measured their ability to survive predation inflicted by two different densities of the two *Notophthalmus* subspecies. The striking result was that *Notophthalmus* from the population that regularly encountered *B. americanus* inflicted heavy mortality on the tadpoles, while predators from outside the geographic range of *B. americanus* did not (Fig. 3-5). It was as if *N. v. dorsalis* did not recognize *B. americanus* as prey. The result is unlikely to be a simple consequence of differences in body size among the two predator populations, since the small tadpoles of *Bufo* remain vulnerable to predation by *Notophthalmus* right up through metamorphosis. This fascinating result suggests that different populations

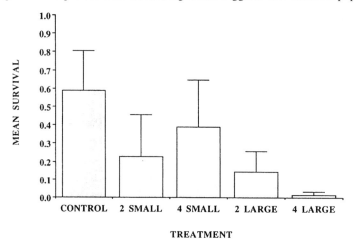

Figure 3-5. Different density-dependent effects of small and large subspecies of *Notophthalmus* on the survival of prey populations of *Bufo americanus* tadpoles. Each *Bufo* population initially contained 400 hatchling tadpoles stocked in a 1 m³ mesocosm with different densities of *Notophthalmus*. Controls contained no *Notophthalmus*. Such common environment studies facilitate investigations of geographic variation in community-level interactions. Redrawn from Kurzava and Morin (1994).

of *Notophthalmus* may have very different effects as predators and different impacts on community organization in different parts of their geographic range.

Kurzava (1994) explored this potential difference further by comparing the impacts of both *Notophthalmus* subspecies on an assemblage of six anuran species. Of the six anuran species used as prey, only *Hyla versicolor* does not occur within the ranges of both predator species, but its morphologically similar congener *H. chrysoscelis* does occur with *N. v. dorsalis*. When a mesocosm experiment similar to the ones described previously was used, it became clear that both newt subspecies have qualitatively similar impacts on community composition, ultimately generating communities dominated by *P. crucifer* (Fig. 3-6). The larger subspecies appears to have somewhat stronger per capita effects, resulting in lower survival of most prey species. The experiment was set up so that two treatments, the ones containing four *N. v. viridescens* and eight *N. v. dorsalis*, contained a similar total biomass of predators. These treatments produced the most similar patterns of prey survival and community composition. This suggests that although some differences in the effects of *Notophthalmus* on novel prey exist, effects on prey that both subspecies experience may scale simply with predator biomass. These studies highlight the power of the mesocosm approach in exploring possible patterns of geographic variation of the functional roles of species in community organization.

Laboratory Experiments

Despite their admitted lack of realism, laboratory studies of organisms with short generation times are the source of most of what we know about the long-term popu-

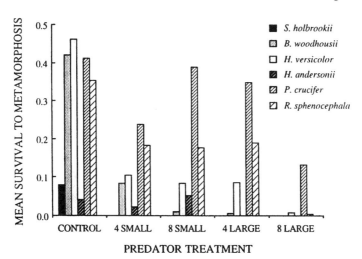

Figure 3-6. Comparisons of the effects of different densities of small and large subspecies of *Notophthalmus* on the survival of six species of larval anurans stocked in experimental mesocosms. The larger subspecies appears to have stronger effects at a given predator density. However, predator populations of similar total biomass (eight small and four large *Notophthalmus*) appear to produce the most similar patterns of prey survival and community composition. Redrawn from Kurzava (1994).

lation dynamics of real organisms in experimental settings (Gause 1934, Huffaker 1958, Park 1962). Most of the large long-lived organisms that field ecologists tend to study yield information about dynamics over periods of tens of years (Elton 1927, Silvertown 1987). However, information about dynamics is absolutely essential for rigorous tests of ecological theory (Harrison 1995). The relative ease with which information about dynamics can be extracted from laboratory systems has prompted theoretical ecologists to call for more and better ''bottle experiments'' (Kareiva 1989). The following three examples illustrate some of the special advantages of experiments conducted with short-lived organisms in laboratory environments.

We have used simple communities assembled from protists and bacteria to attack a range of theoretical issues in population and community ecology. The issues that I will focus on here include experiments designed to test some predictions of three different bodies of theory. The issues concern food web theory, the nonequilibrium maintenance of diversity, and the evolution of sexual reproduction. Protists are particularly useful for experimental tests of such a diverse array of topics because their short generation times facilitate the collection of long-term population dynamic data that are essential for tests of theory. Protists live in the fast lane, with generation times that range from a few hours to 1 or 2 days. Protists are easily cultured in small aquatic microcosms established in glass bottles. In our studies we use microcosms ranging from about 50 to 100 ml in volume. Despite the small volume, such systems can house populations of 10^5 to 10^7 individuals. We have found that such systems are particularly useful for tests of some of the qualitative predictions of food web theory.

Food Web Theory

Simple models show that food chains consisting of a few species arranged in two to four trophic levels will exhibit predictable differences in population dynamics that depend on food chain length (Pimm and Lawton 1977). Long food chains should have longer return times after perturbations than shorter food chains. One consequence of longer return times is greater temporal variation in the abundance of species after a perturbation (see Fig. 3-7). This provides a statistical signature, measurable as a greater standard deviation in the log of abundance over time, that we can use to search for evidence of such dynamics in simple food chains assembled in bottled ecosystems.

Sharon Lawler (Lawler 1993a,b; Lawler and Morin 1993; Morin and Lawler 1995, 1996; Lawler, this volume) came up with the clever idea of assembling simple food chains from bacteria and protists to explore some predictions of food chain theory in bottle ecosystems. Short food chains can be assembled from a nutrient source (a standard growth medium), bacteria, and small ciliates like *Colpidium striatum* or *Tetrahymena thermophila*. The chains can be further lengthened by addition of predators, like *Amoeba proteus* and *Actinosphaerium eichhornii*, that feed on the bacterivorous ciliates. It is then possible to compare the dynamics of the same species in long and short chains to see whether dynamics in longer chains are indeed more variable, as the theory would predict.

Food web theory has been criticized because many of its predictions are based on relatively simple models that are thought to be biologically unrealistic. To our surprise, our findings lend some qualitative support to the predictions of food chain theory,

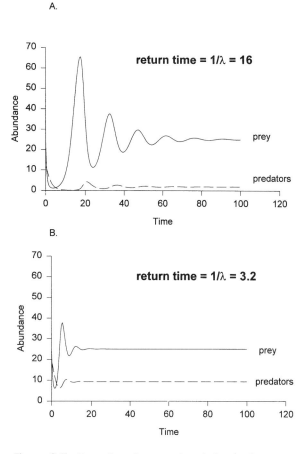

Figure 3-7. Examples of temporal variation in the population abundance in a simple two-species two-level food chain. The chain with the longer return time exhibits damped oscillations of greater amplitude from longer periods of time. The system with the longer return time consequently displays a greater standard deviation in the logarithm of abundance over time.

despite the simplicity of the underlying models. The dynamics seen in replicate microcosms are highly repeatable, an important and comforting attribute of laboratory systems. The dynamics of bacterivorous ciliates in short chains look suspiciously like logistic population growth, with a rapid period of initial growth followed by a prolonged period of relatively constant population size (Fig. 3-8A). This means that the system behaves in a similar fashion to simple models used to model food chains. Comparisons of the dynamics of species in short and longer chains show that dynamics are significantly more variable in longer chains than in shorter ones, for the majority of species combinations examined (Fig. 3-8B–D; Lawler and Morin 1993, Morin and Lawler 1996). This is precisely the statistical signature that we would expect for systems that

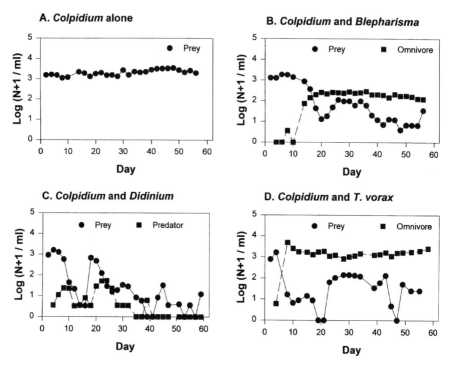

Figure 3-8. Examples of population dynamics in short and long food chains assembled from bacteria and protists in laboratory microcosms (data from Morin and Lawler 1996). Prey dynamics (for *Colpidium*) are much more variable in longer chains than in slightly shorter ones.

differ in return times as the models predict. The pattern holds regardless of whether the bacterivore considered is *Tetrahymena* or *Colpidium*. The dynamics of the predators also appear to be little influenced by the choice of ciliate prey used, which suggests a degree of functional redundancy or similarity in the two bacterivore species used. These findings allow us to return briefly to the issue of generality.

Generality in experiments and theory is attained in different ways. In theory, general models are framed in terms that are not specific to a particular set of species but instead capture the gist of important interactions, such as the positive and negative feedback in predator-prey interactions. In experiments, we strive for generality by determining whether a particular interaction or process occurs repeatedly over a range of species and systems. By comparing the dynamics yielded by simple food chains over a range of different species we can infer that the resulting dynamics are a consequence of food chain configuration rather than the choice of a particular set of species. The same logic should hold true for inferences about the generality of other processes and patterns, whether the results of many studies are digested using a rigorous form of meta-analysis (Gurevitch et al. 1992) or results are simply tallied as the number of experiments that support or refute a particular pattern.

We have not yet taken the next step of estimating the parameters of food chain models to see whether there is a close quantitative agreement between the dynamics-observed in our bottles and the dynamics predicted by mathematical models of food chains. Others have parameterized models of simple predator–prey interactions and obtained good agreement with the patterns seen in laboratory systems (Harrison 1995).

Nonequilibrium Maintenance of Diversity

Laboratory experiments also provide an excellent way to explore the role of distur-bance in molding community composition. The laboratory provides an excellent setting for careful control and manipulation of certain kinds of disturbances. By focusing on species with short generation times, responses to different disturbance regimes can materialize over periods as brief as 1 or 2 months. This contrasts very favorably with the longer time frame required for studies of disturbance and species richness in natural communities where the focus is on relatively long-lived species (Sousa 1979, Pickett and White 1985). Many of our protists live in ephemeral rain pools. Frequent drying and refilling of the pools is a kind of disturbance regime that might promote the co-existence of protist species by interrupting strong interactions before exclusions occur. We know from our field surveys that a single well-studied pool may support up to 28 protist taxa plus a small number of rotifers, tardigrades, and nematodes. It was our impression that when such pools were sampled and the sampled organisms were kept under constant conditions in the laboratory, diversity rapidly declined. This suggested a role for the drying of pools in maintaining the high number of species observed in nature. The underlying mechanism might simply be the intermediate disturbance hy-pothesis (Connell 1978). Another possibility is that a more complex mechanism in-volving the storage effect (Chesson 1986) could promote diversity, since the organisms in ephemeral pools all use some sort of resting stage to endure periods of desiccation.

To learn whether frequent drying promotes the coexistence of species in this system, Jill McGrady-Steed (McGrady-Steed and Morin 1996) assembled microcosms with en-cysted protists from a natural rain pool and then manipulated whether the microcosms dried and refilled or held a constant volume of water without drying. Microcosms that dried were refilled to their initial level and then allowed to dry again. Over the 40-day duration of the study, disturbed microcosms went through an average of five periods of desiccation and refilling. To our surprise, the undisturbed microcosms that retained a constant volume held a slightly greater number of active species over the 40-day duration of the experiment. The number of active species in both disturbed and undis-turbed microcosms tended to decline over time, with peak numbers of species in dis-turbed systems sometimes attaining levels seen in undisturbed systems (Fig. 3-9). This result is a bit misleading, however, since disturbed and undisturbed systems eventually contained rather different sets of species, even though they began with the same species pool. Another way to visualize these patterns is to examine the shapes of the cumulative species richness curves for each microcosm (Fig. 3-10). The curves indicate that num-bers of active species continue to increase for longer periods in undisturbed microcosms, whereas most of the species observed in disturbed microcosms appear during the first few days of community development. The undisturbed microcosms ultimately support a larger number of taxa over a 40-day period because some taxa, the majority of which are predators, only become active well after their populations of prey have become established. These species fail to become active in the repeatedly disturbed microcosms.

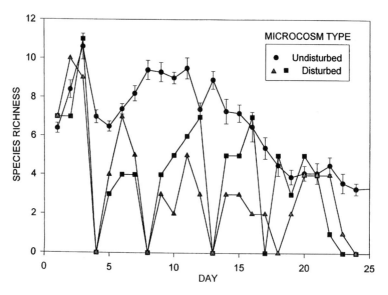

Figure 3-9. Patterns of species richness for protists and small metazoans in disturbed and undisturbed microcosms. Disturbed microcosms are allowed to dry and are then refilled with water to initiate another cycle of community development. Undisturbed microcosms received daily additions of water to maintain a constant volume. The trend for undisturbed systems shows the average number of active species in 10 replicates. Two representative patterns for disturbed microcosms are shown for comparison. The number of active species in disturbed microcosms goes to zero when the microcosms dry and returns to higher levels when the systems are rehydrated and organisms emerge from cysts. Redrawn from McGrady-Steed and Morin (1996).

We are left with the conclusion that the regular disturbance that we imposed does not significantly enhance the number of coexisting protist species in our artificial pools. We cannot rule out the possibility that a more variable hydroperiod regime than the one that we used, with a random alternation of very long and short episodes of community development, might favor greater diversity via the storage effect.

Evolution of Sexual Reproduction

Just as short generation times make it possible to collect multigeneration data on population dynamics, the same properties make viruses, bacteria, and protists well suited for studies of evolutionary ecology. In some protists, the presence or absence of sex can be simply manipulated by varying the number of mating types, the protist equivalent of sexes, present in a microcosm. Asexual reproduction by binary fission occurs in microcosms that contain only one mating type. Conjugation, a form of sexual recombination that is functionally separated from reproduction, occurs in microcosms that contain two or more compatible mating types under appropriate environmental conditions. After conjugation, protists go on to reproduce by fission.

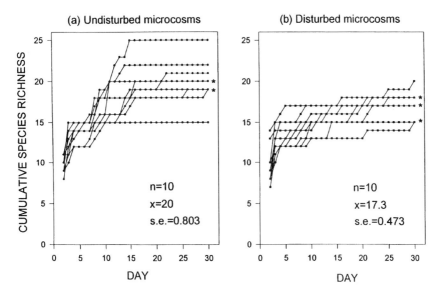

Figure 3-10. Patterns of cumulative species richness, the number of active species observed in a microcosm by a specific day, in undisturbed and disturbed systems. Undisturbed systems support a slightly higher number of active species than do regularly disturbed systems. Redrawn from McGrady-Steed and Morin (1996).

Kondrashov (1994) has pointed out that some 20 different hypotheses have been put forward to explain why sex persists despite its many purported genetic and demographic costs. He also points out the need for clever experiments to test these hypotheses, since models and comparative data alone will not enable the researcher to distinguish among them. While not all of these hypotheses can be easily tested by experiments, some can. Paul McMillan (1995) devised a particularly elegant test of the Tangled Bank Hypothesis (Bell 1982). The Tangled Bank Hypothesis suggests that sex is advantageous when sexually produced progeny disperse into patchy environments where resources vary in quality and quantity among patches. The idea is that genetically variable offspring will use resources in different ways, while genetically uniform offspring will be less plastic. Offspring from a sexual lineage will consequently be able to exploit more kinds of resources and more kinds of habitat patches than offspring from a genetically uniform asexual lineage. When the number of sexually and asexually produced offspring is averaged across an array of different habitat patches, sexuals should be more abundant than asexuals, if sex enhances fitness (the number of offspring produced) in patchy habitats.

McMillan (1995) addressed this question by first isolating two stocks of different mating types of the small hymenostome ciliate *Tetrahymena thermophila*. Either stock grown by itself would reproduce via binary fission without any conjugation (sex) occurring. However, when the two stocks were grown together, conjugation occurred. This simple protocol makes it possible to compare the performance of both the ''parental'' asexual lineages and the sexual ''offspring'' lineage formed from their cross.

Performance (fitness) was measured by the abundance attained by all three lineages in a patchy laboratory environment. Replicated patchy environments were simulated by establishing three different bacterial prey species at two different nutrient concentrations in separate culture dishes, for a total of six different patch types (dishes) per experimental replicate. Each of the three *Tetrahymena* populations, two asexual and one sexual, were introduced into separate replicated sets of the six different patch types. *Tetrahymena* in each dish then grew for 10 days, a minimum of 10 generations. Then densities within each habitat patch type were measured and compared, with the key measure of interest being the average performance of each population, sexual or asexual, across the six patch types. Regardless of whether abundance is measured by geometric or harmonic means of abundance across habitat patches, the sexual lineage significantly outperforms the asexual one (Fig. 3-11). When back transformed to an arithmetic scale, there is a greater than twofold fitness (abundance) advantage for the sexual lineage over its asexual parental lines. This suggests to us that sex may indeed confer a significant demographic advantage in coarse-grained, patchy environments. Certainly, other hypotheses may account for the maintenance of sex in other situations, such as the coevolutionary arms race envisioned in the Red Queen Hypothesis (Lively et al. 1990). Nonetheless, our system has made it possible to detect a very real fitness advantage for sex in patchy environments.

The experiments described here would have been difficult or impossible to conduct with the same organisms in field settings. Protists are well suited for lab experiments, but they literally fall through the cracks in field enclosures. Thus, we are faced with the possibility that we may learn much about protist dynamics in lab settings, without ever being able to confirm those patterns in the field. This may be offset by the knowledge that these or similar systems may be our only hope for obtaining the kinds of population dynamic data needed to test ecological theory. There is also the concern that protists may not be representative of the ecology of "higher" organisms. While this may be so, the converse is likely to be true as well, with field experiments on long-lived "higher" organisms ultimately shedding little light on the ecology of bacteria and protists. Given the prominent role of microbes in nutrient cycling and trophic dynamics, an understanding of their dynamics may be far more critical than an understanding of the ecology of more visually conspicuous but functionally insignificant macroorganisms.

Synthesis

Ecologists often despair over an apparent lack of progress in their field. This malaise is perplexing, because in some very important respects the accomplishments of experimental ecologists far surpass those of other experimental biologists who work on less complex systems found lower in the hierarchy of biological organization. It has been said that the best test of one's understanding of how any system works is whether one can reassemble a functional system from its component parts. Experimental ecologists working with reassembled communities do this as a matter of course! Practitioners of the newly prominent field of restoration ecology also strive to reconstruct functioning ecosystems in the real world. Sometimes the results are a bit like Dr. Frankenstein's monster: functional, but differing from the intended result in unanticipated ways.

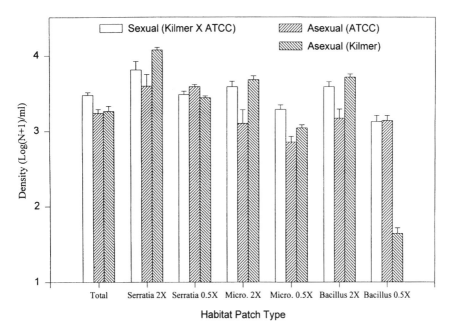

Figure 3-11. Average densities attained by sexual and asexual populations of *Tetrahymena thermophila* in six different experimental habitat patches (e.g., *Serratia* 2× nutrients) and averaged across all six habitat patch types (Total). Although different populations perform better in different settings, sexual populations attain significantly higher densities when averaged over all six habitat types. Redrawn from McMillan (1995).

We continue to need ecological experiments conducted in different systems, with different levels of realism and precision, to enhance our understanding of complex systems. The gap between the amount of ecological theory that has been developed and critical tests of theory is enormous and growing rapidly. If we can test theory in simple laboratory systems, we can at least discard those ideas that fail in simple settings and pursue others that work in the lab and might work in nature. Ultimately, field experiments are essential to show that what can happen in models, bottles, or chemostats can also happen in nature. The point is that the creative use of models together with a variety of experimental approaches will help us to understand far more about the complexities of natural communities than any single approach advocated to the exclusion of others.

ACKNOWLEDGMENTS   Many thanks to Bill Resetarits and Joe Bernardo for inviting me to share my ideas. My research has benefited greatly from the generous support of the National Science Foundation, most recently grants DEB-92-20665 and DEB-94-24494. I benefited from the many thought-provoking discussions stimulated by the 1995 ASZ Symposium on the State of Experimental Ecology. Thanks to Sharon Lawler, Lynn Kurzava, Jill McGrady-Steed, and Paul McMillan for allowing me to include their published and unpublished results here. Comments by Patricia Harris, Mark Laska, Christina Kaunzinger,

Jill McGrady-Steed, Jeremy Fox, Marlene Cole, Jim Baxter, Gabrielle Vivian-Smith, and an anonymous reviewer greatly improved the manuscript.

Literature Cited

Bell, G. 1982. The Masterpiece of Nature. University of California Press, Berkeley.

Brooks, J. L., and S. I. Dodson. 1965. Predation, body size, and composition of plankton. Science 150:28–35.

Byers, R. J., and H. T. Odum. 1993. Ecological Microcosms. Springer-Verlag, New York.

Chesson, P. 1986. Environmental variation and the coexistence of species. Pages 240–256 in J. Diamond and T. J. Case (eds.), Community Ecology. Harper and Row, New York.

Clausen, J., D. D. Keck, and W. M. Heisey. 1948. Experimental studies on the nature of species: III. Environmental responses of climatic races of *Achillea*. Carnegie Institute of Washington Publication 581:1–129.

Connell, J. H. 1961. The influence of interspecific competition and other factors on the distribution of the barnacle *Chthamalus stellatus*. Ecology 42:710–723.

———. 1978. Diversity in tropical rain forests and coral reefs. Science 199:1302–1310.

Crowley, P. H., and D. M. Johnson. 1992. Variability and stability of a dragonfly assemblage. Oecologia 90:260–269.

Diamond, J. M. 1986. Overview: laboratory experiments, field experiments, and natural experiments. Pages 3–22 in J. Diamond and T. J. Case (eds.), Community Ecology. Harper and Row, New York.

Dodson, S. I. 1974. Zooplankton competition and predation: an experimental test of the size-efficiency hypothesis. Ecology 55:605–613.

Elton, C. 1927. Animal Ecology. Methuen, London.

Fauth, J. E., and W. J. Resetarits Jr. 1991. Interactions between the salamander *Siren intermedia* and the keystone predator *Notophthalmus viridescens*. Ecology 72:827–838.

Gause, G. F. 1934. The Struggle for Existence. Williams and Wilkins, Baltimore, Maryland.

Gill, D. E. 1979. Density dependence and homing behavior in the red-spotted newt, *Notophthalmus viridescens* (Rafinesque). Ecology 60:800–813.

Goldberg, D. E., and A. M. Barton. 1992. Patterns and consequences of interspecific competition in natural communities: a review of field experiments with plants. American Naturalist 139:771–801.

Gurevitch, J., L. L. Morrow, A. Wallace, and J. S. Walsh. 1992. A meta-analysis of competition in field experiments. American Naturalist 140:539–572.

Hairston, N. G., Sr. 1989a. Ecological Experiments: Purpose, Designs, and Execution. Cambridge University Press, Cambridge.

———. 1989b. Hard choices in ecological experimentation. Herpetologica 45:119–122.

Harris, P. M. 1995. Are autecologically similar species also functionally similar? A test in pond communities. Ecology 76:544–552.

Harrison, G. W. 1995. Comparing predator-prey models to Luckinbill's experiment with *Didinium* and *Paramecium*. Ecology 76:357–374.

Huffaker, C. B. 1958. Experimental studies on predation: dispersion factors and predator-prey oscillations. Hilgardia 27:343–383.

Hurlbert, S. H., J. Zedler, and D. Fairbanks. 1972. Ecosystem alteration by mosquitofish (*Gambusia affinis*) predation. Science 175:639–641.

Jaeger, R. G., and S. C. Walls. 1989. On salamander guilds and ecological methodology. Herpetologica 45:111–119.

Kareiva, P. 1989. Renewing the dialogue between theory and experiments in ecology. Pages 68–88 in J. Roughgarden, R. M. May, and S. A. Levin (eds.), Perspectives in Theoretical Ecology. Princeton University Press, Princeton, New Jersey.

Kondrashov, A. S. 1994. Sex and deleterious mutations. Nature 369:99–100.

Kurzava, L. M. 1994. The structure of prey communities: effects of predator identity and geographic variation in predators. Ph.D. dissertation, Rutgers University, New Brunswick, New Jersey.

Kurzava, L. M., and Morin, P. J. 1994. Consequences and causes of geographic variation in the body size of a keystone predator, *Notophthalmus viridescens*. Oecologia 99:271–280.

Lawler, S. P. 1989. Behavioral responses to predators and predation risk in four species of larval anurans. Animal Behavior 38:1039–1047.

———. 1993a. Direct and indirect effects in microcosm communities of protists. Oecologia 93:184–190.

———. 1993b. Species richness, species composition and population dynamics of protists in experimental microcosm. Journal of Animal Ecology 62:711–719.

Lawler, S. P., and P. J. Morin. 1993. Food web architecture and population dynamics in laboratory microcosms of protists. American Naturalist 141:675–686.

Lawton, J. H., S. Naeem, R. M. Woodfin, V. K. Brown, A. Gange, H. J. C. Godfray, P. A. Heads, S. Lawler, D. Magda, C. D. Thomas, L. J. Thompson, and S. Young. 1993. The ecotron: a controlled environmental facility for the investigation of population and ecosystem processes. Philosophical Transactions of the Royal Society of London B 341:181–194.

Leibold, M., and H. M. Wilbur. 1992. Interactions between food-web structure and nutrients on pond organisms. Nature 360:341–343.

Levins, R. 1968. Evolution in Changing Environments. Princeton University Press, Princeton, New Jersey.

Lewin, R. 1992. Complexity: Life at the Edge of Chaos. Macmillan, New York.

Lively, C. M., C. Craddock, and R. C. Vrijenhoek. 1990. Red Queen Hypothesis supported by parasitism in sexual and clonal fish. Nature 344:864–866.

McCauley, E. J., and W. W. Murdoch. 1990. Predator-prey dynamics in environments rich and poor in nutrients. Nature 343:455–457.

McGrady-Steed, J., and P. J. Morin. 1996. Disturbance and the species composition of rain pool microbial communities. Oikos, 76:93–102.

McMillan, P. A. 1995. Ecological interactions and the maintenance of sex: studies with experimental populations of the ciliate *Tetrahymena thermophila*. Ph. D. dissertation, Rutgers University, New Brunswick, New Jersey.

Morin, P. J. 1981. Predatory salamanders reverse the outcome of competition among three species of anuran tadpoles. Science 212:1284–1286.

———. 1983. Predation, competition, and the composition of larval anuran guilds. Ecological Monographs 53:119–138.

———. 1984. The impact of fish exclusion on the abundance and species composition of larval odonates: results of short-term experiments in a North Carolina farm pond. Ecology 65:53–60.

———. 1986. Interactions between intraspecific competition and predation in an amphibian predator-prey system. Ecology 67:713–720.

———. 1989. New directions in amphibian community ecology. Herpetologica 45:124–128.

Morin, P. J., and S. P. Lawler. 1995. Food web architecture and population dynamics: theory and empirical evidence. Annual Review of Ecology and Systematics 26:505–529.

————. 1996. Effects of food chain length and omnivory on population dynamics in experimental microcosms. Pages 218–230 in G. A. Polis and K. O. Winemiller (eds.), Food Webs: Integration of Patterns and Dynamics. Chapman and Hall, New York.

Murdoch, W. W., and A. Sih. 1978. Age-dependent interference in a predatory insect. Journal of Animal Ecology 47:581–592.

Murdoch, W. W., and E. McCauley. 1985. Three distinct types of dynamic behavior shown by a single planktonic system. Nature 316:628–630.

Naeem, S., L. J. Thompson, S. P. Lawler, J. H. Lawton, and R. M. Woodfin. 1994. Declining biodiversity can alter the performance of ecosystems. Nature 368:734–737.

Paine, R. T. 1966. Food web complexity and species diversity. American Naturalist 100:65–75.

Park, T. 1962. Beetles, competition, and populations. Science 138:1369–1375.

Pickett, S. T. A., and P. H. White (eds.). 1985. The Ecology of Natural Disturbance and Patch Dynamics. Academic Press, New York.

Pimm, S. L., and J. H. Lawton. 1977. Number of trophic levels in ecological communities. Nature 268:329–331.

Resetarits, W. J., Jr., and H. M. Wilbur. 1989. Oviposition site choice in *Hyla chrysoscelis*: role of predators and competitors. Ecology 70:220–228.

Sih, A., P. Crowley, M. McPeek, J. Petranka, and K. Strohmeier. 1985. Predation, competition, and prey communities: a review of field experiments. Annual Review of Ecology and Systematics 16:269–311.

Silvertown, J. S. 1987. Ecological stability: a test case. American Naturalist 130:807–810.

Smith, D. C. 1983. Factors controlling tadpole populations of the chorus frog (*Pseudacris triseriata*) on Isle Royale, Michigan. Ecology 64:501–510.

Sokal, R. R., and F. J. Rohlf. 1981. Biometry, 2nd ed., W. H. Freeman, San Francisco, California.

Sousa, W. P. 1979. Experimental investigations of disturbance and ecological succession in a rocky intertidal algal community. Ecological Monographs 49:227–254.

Wilbur, H. M. 1989. In defense of tanks. Herpetologica 45:122–123.

Wilbur, H. M., and R. A. Alford. 1985. Priority effects in experimental pond communities: responses of *Hyla* to *Bufo* and *Rana*. Ecology 66:1106–1114.

4

The Desert Granivory Experiments at Portal

JAMES H. BROWN

In 1977 D. W. Davidson, O. J. Reichman, and I began experimental manipulations in the Chihuahuan Desert near the little town of Portal in southeastern Arizona to study interactions among seed-eating rodents and ants and between these consumers and the annual plants that supply their primary food resources. The study site comprises about 20 ha of relatively homogeneous desert shrub habitat and contains 24 plots, each 0.25 ha in area (50 × 50 m), that were randomly assigned to different treatments: removal of some or all rodent or ant species or addition of supplementary seeds. While the plots initially assigned to seed additions and some of the other treatments were changed to additional replicates of rodent and ant removal treatments in 1988, several plots have had all rodents, just kangaroo rats, or all ants removed continuously and other plots have served as unmanipulated controls for the entire 19-year period. In the nearly two decades that we have been manipulating and monitoring a selected group of organisms on a small patch of ground, we have learned a great deal about how they affect each other and how they respond to natural variation—chiefly in climate—in their environment. Numerous papers describing the responses of the desert community to these long-term "press" experiments (see Bender et al. 1984) have been published, but no major review and synthesis of the experimental research program has been written since Brown, et al.'s chapter in Diamond and Case's *Community Ecology* (1986).

In the present chapter, I shall begin with some background information on the study site and the relevant groups of animals and plants. This will be followed by a brief description of the experimental design, sampling methods, and history of the treatments. Then I shall present an overview of some of the most interesting and important results. In this section I will present very little in the way of new data but instead summarize and synthesize information that has already been published in many separate papers. In the final section, I shall discuss some more philosophical and less tangible issues:

what these experiments have taught me about the advantages and pitfalls of doing long-term experimental ecology.

Background

Before Portal

In the early 1970s, stimulated in large part by the theoretical work of Robert Mac-Arthur, there was a great deal of interest in the role of interspecific interactions, especially competition, in the organization of communities of coexisting species. Nonexperimental studies, many involving comparisons of multiple and geographically separated assemblages of desert rodents and ants (e.g., Rosenzweig and Winakur 1969; Brown 1973, 1975; Brown and Lieberman 1973; Rosenzweig 1973; M'Closkey 1976, 1978; Davidson 1977a,b) produced results consistent with the hypothesis that competition between closely related, ecologically similar species played a major role in determining the composition of communities. At that time I was working on desert rodents and Dinah Davidson was my graduate student studying desert ants. Comparing and discussing our data for the two taxa, we came up with the idea that these distantly related organisms, which are the predominant consumers of seeds in most North American deserts, might also compete.

In 1973, with support of the International Biological Program (IBP), Davidson and I set up experiments to test the hypothesis that rodents and ants compete for the seeds of desert plants. We set up eight 0.1-ha circular plots, giving two replicates of a simple 2 × 2 factorial design: rodents were removed by fencing and trapping, ants were removed by poisoning, both rodents and ants were removed by a combination of the preceding procedures, and there were two unmanipulated control plots. These manipulations were conducted at the IBP's site in the Sonoran Desert northwest of Tucson, Arizona. They produced three important results. First, ants increased approximately twofold on plots where rodents had been removed and rodents appeared to have increased in the absence of ants. This result supported the hypothesis that the two taxa compete (Brown and Davidson 1977, Brown et al. 1979). Second, the removal of rodents led to major changes in the annual plants: principally an increase in the population densities of large-seeded species. This suggested that rodents forage selectively for large seeds and, in so doing, suppress populations of their preferred prey (Inouye et al. 1980, Davidson et al. 1984, Brown et al. 1986). Third, the increase of large-seeded annual plants on rodent-removal plots was accompanied by a decrease in the abundance of small-seeded species and, after a lag of at least two years, by a decrease in the ants that had initially increased in response to rodent removal. This suggested the existence of important indirect interactions: predation by rodents suppresses the populations of large-seeded annual plants, competition from large-seeded species suppresses small-seeded annuals, and the availability of small seeds limits the abundance of the ants (Davidson et al. 1984, Brown et al. 1986). Inouye (1981) also found an interesting indirect interaction between rodents and a parasitic plant fungus mediated by their shared food resource, a common large-seeded annual.

As informative as these results were, the experiments suffered from several important problems. The plots were too small to support good numbers of rodent individuals and ant colonies. Replication, with only two of each treatment, was minimal, and our first statistical analyses were flawed by pseudoreplication (Hurlbert 1984, Galindo 1986, Brown and Davidson 1986). And, finally, only the rodent removal plots were fenced to exclude small mammals, creating potentially serious artifacts, such as the differential exclusion of folivorous rabbits. Given the importance of the results, it was clearly desirable to repeat the experiments in a similar desert habitat, but with larger fenced plots and more replication.

Starting the Experiments at Portal

So, in the summer of 1976 Davidson and I decided to write a grant proposal to conduct a second set of experiments. In addition to repeating the study of competition between rodents and ants, we also wanted to test for interspecific competition within each taxon (i.e., between species of rodents and between species of ants) and to focus more closely on the relationships between the granivores and their food resources, in part by manipulating experimentally the availability of seeds. We elected to do the work in the Chihuahuan Desert near Portal, Arizona, where both of us had worked previously. Although only about 200 km east of the original Sonoran Desert site and similar in having shrubby vegetation growing on a sandy *bajada* (an alluvial outwash plain), the desert near Portal differed in several respects: (1) more precipitation, especially in the summer; (2) higher species diversity of rodents and ants, about twice as many common species in each group; and (3) different species composition, sharing some species with the Sonoran Desert site but adding distinctive Chihuahuan Desert forms.

We convened in Flagstaff to write the proposal, joined by O. J. Reichman, who became a third collaborator, and Zvika Abramsky, who shortly returned to Israel to take an attractive job. The brainstorming and writing of the proposal was one of those rare collaborative occasions where the chemistry was just right. I don't think any of us remembers who came up with the elegantly simple idea that we could use ''semipermeable membranes,'' fences with different-sized holes, to differentially exclude particular rodent species based on their body sizes. We contemplated treatments that would exclude seed-eating birds, which are the third major group of desert granivores, but this was rejected as impractical. The number of plots required for a reasonable factorial design would have been too large, and we calculated that covering just one plot with netting would cost as much (about $14,000) as installing 24 plots with potentially rodent-proof fences. (And, in retrospect, the nylon fish netting that we had planned to use to exclude birds would have deteriorated within a few years in the desert sun.)

We received a grant of $200,000 from the Ecosystem Studies Program of the National Science Foundation (NSF), the only ecological program at the NSF that funded large, multi-investigator projects. The experimental site was set up between July and October 1977 (Fig. 4-1). Using a combination of graduate and undergraduate students and local workers, we installed a barbed-wire fence to exclude livestock from the entire 20-ha site and then set up 24 experimental plots, each 0.25 ha in area (50 × 50 m). The plots were surrounded by fencing: ¼-inch (6.25 mm) wire mesh, buried 20 cm in the ground, extending 60 cm above ground, topped on the outside with a 15-cm strip

Figure 4-1. Aerial photograph of the study site taken in December 1979, showing the 24 experimental plots, each 50 × 50 m. Experimental treatments assigned to each plot are given in Table 4-1.

of aluminum flashing, and supported by ½-inch (1.25 cm) metal posts at intervals of approximately 2 m. This involved putting in a total of 4.8 km of fencing, using a Ditchwitch machine to dig the trench but doing all of the other work by hand.

Before starting the manipulations, we censused the plots to obtain baseline, pretreatment data. In July we counted the ants and then began poisoning to remove either all species or just the two largest and dominant granivores (*Pogonomyrmex rugosus* and *Po. barbatus*) from designated plots. For three months, July, August, and September, we trapped rodents with large holes in the fences so that all species had free access. Then in October we reduced the sizes of the holes and began removing either all rodents or the selected kangaroo rat species (*Dipodomys* spp.) from designated plots.

Experimental Design

The initial design and the changes in treatments made in 1986 and 1988 are given in Table 4-1. The most important manipulations involved removing selected species or all species of rodents and ants. Four categories of rodent removal were accomplished by cutting 16 holes of a specific size at ground level in the fencing surrounding a plot and then selectively trapping out the designated species: (1) all rodents removed, no holes; (2) all three kangaroo rat species (*D. merriami, D. ordii*, and *D. spectabilis*) removed, 1.9 × 1.9 cm holes; (3) banner-tailed kangaroo rats (*D. spectabilis*, the largest granivorous rodent species) only removed, 2.6 × 3.0 cm holes; and (4) no rodents removed (called equal-access plots), 3.7 × 5.7 cm holes. Ants were removed by applying minimal amounts of commercial baited poison (Myrex initially and AMDRO after 1980) either broadcast across an entire plot to remove all ants or applied just to the conspicuous mounds of the two large *Pogonomyrmex* species to remove them selectively. Treatments were assigned to plots at random in a partial factorial design. Altogether, 14 of the 24 plots were initially subjected to some combination of rodent or ant removal or served as unmanipulated controls.

The remaining 10 plots were initially assigned at random to two replicates of five seed addition treatments. Ninety-six kg of millet seed, ground to various sizes and applied in different temporal patterns, was added per year (Table 4-1). The seed addition produced results of only limited interest: banner-tailed kangaroo rats increased; the other two kangaroo rats decreased; other rodents and the ants were seemingly unaffected; and bird utilization increased on all of the seed addition plots, regardless of the specific treatment (Brown and Munger 1985, Thompson et al. 1991). Given this disappointing response to added food, we decided to reduce food supply by removing annual plants. Therefore, in 1986 we reassigned the seed addition plots to treatments in which we tried to remove seasonal assemblages of annuals by applying the herbicide Roundup (Table 4-1). These manipulations were not effective because we were unable to kill all seedlings and compensatory growth of surviving individuals resulted in seed crops nearly as large as on unmanipulated plots.

By the summer of 1987 it was clear that the annual plant removal experiments were a failure, and we decided to reassign these plots to "replicates," initiated 10 years later, of the original rodent and ant removal treatments. These manipulations were begun in January 1988, according to the design in Table 4-1. The original rodent and ant removal treatments were continued, so that 14 plots have been subjected to continuous press removal of particular combinations of rodents and ants since 1977 and the remaining 10 have been subjected to similar manipulations, but only since 1988. There are two exceptions. First, the two large *Pogonomyrmex* ant species and the similarly large but somewhat less granivorous ant *Aphaenogaster cockerellii* (see Chew 1995) had decreased rapidly to near-extinction by the mid-1980s, making the "removal" of these ants a useless manipulation. Therefore, the original *Pogonomyrmex* removal plots were reassigned in 1988 to have all ants removed, and no new *Pogonomyrmex* removal treatments were initiated. Second, the population of bannertail kangaroo rats also declined precipitously in the mid-1980s (Valone et al. 1995), making the continued study of the effects of *D. spectabilis* removal problematic. We decided to convert the original

Table 4-1. Experimental design and history of the treatments applied to each plot

Plot	1977–1985	1985–1987	1988–present
1	mixed-sized seeds added in pulse	all annuals removed	banner-tailed kangaroo rats removed
2	small seeds added	summer annuals removed	unmanipulated control
3	kangaroo rats and *Po. rugosus* removed	kangaroo rats and *Po. rugosus* removed	kangaroo rats and ants removed
4	ants removed	ants removed	ants removed
5	banner-tailed kangaroo rats removed	banner-tailed kangaroo rats removed	rodents removed
6	large seeds added	winter annuals removed	kangaroo rats removed
7	rodents removed	rodents removed	rodents removed
8	*Po. rugosus* removed	*Po. rugosus* removed	ants removed
9	mixed-sized seeds added	biseasonal annuals removed	banner-tailed kangaroo rats removed
10	rodents and ants removed	rodents and ants removed	rodents and ants removed
11	unmanipulated control	unmanipulated control	unmanipulated control
12	*Po. rugosus* removed	*Po. rugosus* removed	ants removed
13	large seeds added	winter annuals removed	kangaroo rats and ants removed
14	unmanipulated control	unmanipulated control	unmanipulated control
15	kangaroo rats removed	kangaroo rats removed	kangaroo rats removed
16	rodents removed	rodents removed	rodents removed
17	ants removed	ants removed	ants removed
18	mixed-sized seeds added in pulse	all annuals removed	kangaroo rats removed
19	kangaroo rats and *Po. rugosus* removed	kangaroo rats and *Po. rugosus* removed	kangaroo rats and ants removed
20	mixed-sized seeds added	biseasonal annuals removed	kangaroo rats and ants removed
21	kangaroo rats removed	kangaroo rats removed	kangaroo rats removed
22	small seeds added	summer annuals removed	unmanipulated control
23	ants removed	ants removed	ants removed
24	banner-tailed kangaroo rats removed	banner-tailed kangaroo rats removed	rodents removed

For additional details, see Brown and Munger (1985), Davidson et al. (1984), and Samson et al. (1992).

bannertail removal plots to rodent removal and assigned the two seed addition plots that had the highest numbers of bannertails to *D. spectabilis* removal. This manipulation was ineffective, however, because the bannertail populations never recovered and they eventually went locally extinct on the entire study site in 1994.

Data Collection Methods

We have carefully standardized the ways that we census rodent, ant, plant, and bird populations and take other data, so that changes in methodology do not diminish the value of our long-term data sets. Since July 1977 each plot has been laid out in a 7 × 7 grid, with 6.25 m separating each of the conspicuous and permanently placed metal grid stakes. Rodents are trapped during a 2- or 3-night period each month, as near as possible to the new moon, by setting a single Sherman live trap at each grid stake for one night. Captured individuals are marked, formerly with ear tags or toe clips and now with PIT (permanently implanted transponder) tags, and then released. Ants are censused during the period of maximal activity each summer by counting colonies of the large *Pogonomyrmex* species on each plot and colony entrances of the other species within a circle of 2-m radius around each grid stake. Annual plants are censused each spring and fall, at the end of the winter and summer growing seasons, respectively, by counting the individuals on 16 quadrats per plot, each 0.25 m² in area and located 1 m south of alternate grid stakes. Similarly standardized and spatially referenced methods have been used at less frequent intervals to census birds, to quantify seed removal by rodents, birds, and ants, to measure grass and perennial shrub cover, and to take other kinds of data.

In addition to standardizing data collection protocols, we make every effort to minimize the impact of human activities on the study site and its organisms and to ensure that the unavoidable impacts are equal across all plots and experimental treatments. Thus, for example, we try to confine foot traffic to east–west trails along each row of grid stakes and to ensure that all plots receive equal amounts of rodent trap bait and foot traffic (see Chew 1995).

Important Results

Competition

The competitive interactions revealed by our experiments at Portal are shown diagrammatically in Figure 4-2. Unlike the earlier study in the Sonoran Desert, the experiments in the Chihuahuan Desert provided little evidence of strong competition between rodents and ants. There were some changes in the abundances of certain ant species associated with the removal of all rodents or just kangaroo rats (Valone et al. 1994), but only one species, *Pheidole rugulosa*, showed a consistent, statistically significant increase in the absence of rodents. Since this ant is found predominantly in arid grasslands, it is uncertain whether its increase reflects release from competition with rodents or an indirect response to the changes in vegetation caused by excluding kangaroo rats (see following discussion). There was no consistent evidence that any rodents increased significantly in response to removal of ants.

Figure 4-2. Known and suspected direct competitive and predator–prey interactions. Lines illustrate reciprocal interactions, with points indicating negative effects but crosses indicating positive effect. The number of points and crosses provides a qualitative indication of the relative strength of the interaction. Some of the important indirect interactions are indicated by sequences of interaction. For example, through selective predation on large seeds kangaroo rats reduce large-seeded annuals, thereby releasing small-seeded species from competition.

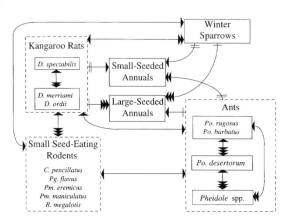

Early in the study, there was evidence of both direct and diffuse competition among the ant species (Fig. 4-2; Davidson 1980, 1985). However, the decline to extinction of the large *Pogonomyrmex* species by the mid-1980s made inoperative the only manipulation that would have enabled us to address interspecific interactions among ants over the long term.

There was abundant evidence of competition among the rodent species (Fig. 4-2). The removal of kangaroo rats resulted in large increases in the populations of the five most abundant species of smaller seed-eating rodents (*Chaetodipus penicillatus, Perognathus flavus, Reithrodontomys megalotis, Peromyscus maniculatus*, and *P. eremicus*; Munger and Brown 1981; Brown and Munger 1985; Brown et al. 1986; Heske et al. 1994; Valone and Brown 1995, 1996). Populations of all of these species have averaged at least two times higher on kangaroo rat removal plots than on control plots and other equal-access plots where kangaroo rats were present. None of these small rodent species have maintained permanent populations on plots or even on the entire study site, however. Instead, populations have gone locally extinct and recolonized, often repeatedly (Brown and Kurzius 1989, Brown and Heske 1990b, Valone and Brown 1996). For each of these five common species of small granivores, colonization rates were significantly higher, extinction rates were lower, or both for kangaroo rat removal plots than for control plots (Valone and Brown 1995). The result is that almost continually throughout the 19 years there have been consistently larger numbers of both individuals and species of other seed-eating rodents on plots from which kangaroo rats have been excluded.

This does not support Wiens's (1977) suggestion that interspecific competition should be episodic and confined to infrequent "ecological crunches." If such crunches occur, they should be pronounced in this desert ecosystem, where precipitation, seed supply, and rodent populations fluctuate widely (e.g., Brown and Heske 1990b, Valone and Brown 1996). Despite these fluctuations, however, areas without competing kangaroo rats are seemingly always used preferentially by the smaller seed-eating rodents.

While Wiens's hypothesis might seem logical, in retrospect it is not difficult to see why it does not apply in this case. Individual rodents should always benefit from settling and foraging on plots where kangaroo rats are absent. These plots will always tend to have more food available, because whenever kangaroo rats are present, they account for the vast majority of seeds consumed by rodents (Brown and Munger 1985). Kangaroo removal plots will also be refuges for the small rodents from any aggressive interference from the larger, behaviorally dominant kangaroo rats. While we have much to learn about the detailed mechanisms of competition among these rodent species, we do have good evidence that both exploitative preemption of food resources and aggressive interference play significant roles (Brown and Munger 1985, Bowers et al. 1987, Bowers and Brown 1992). Our results suggest that whenever there is substantial overlap in requirements for essential resources and/or direct behavioral interference there will be interspecific competition. While populations may be subjected to episodic ecological crunches caused by resource scarcity, interspecific competition will not be confined to such episodes (Heske et al. 1994; Valone and Brown 1995, 1996).

Predation and Disturbance

The experiments have shed considerable light on the roles of desert granivores, especially rodents, as predators on plants (Fig. 4-2). Most of the seeds consumed by desert granivores are produced by the annual (or ephemeral) plants. These plants spend most of their lives in the soil seed bank and germinate, grow, and reproduce only during brief periods when sufficient soil moisture is available. At Portal annual precipitation is distinctly bimodal, with peaks in winter (November to March) and summer (July and August). There are two corresponding assemblages of annual plants, winter and summer, which have virtually no overlap in phenology and species composition.

At Portal, as in the earlier experiments in the Sonoran Desert, large-seeded winter annual plants (seed mass > approximately 1 mg) increased on plots where rodents had been removed. By the winter season of 1983–1984, two species of *Erodium*—*E. cicutarium* and *E. texanum*—were several thousand times more abundant on rodent removal plots than on plots where kangaroo rats were present (Brown et al. 1986; see also Samson et al. 1992). Enormous differences in densities of *E. cicutarium* and also *Lesquerella gordonii* were apparent in the winter of 1994–1995 (Figs. 4-3, 4-4), but we have not yet completed analyses of the magnitude of the response. Complementing the increases in large-seeded species, small-seeded winter annuals have decreased on plots where rodents have been removed (Brown et al. 1986, Samson et al. 1992, Guo and Brown 1996).

We now know a good deal about the selective predation of rodents on large-seeded plant species. First, it is almost exclusively attributable to kangaroo rats. The increases of large-seeded species are just as large on plots where only kangaroo rats have been removed as on plots where all rodents have been excluded. From this we conclude that kangaroo rats are major, highly selective consumers of large seeds. The smaller seed-eating rodents either are not sufficiently abundant, even when they have more than doubled their densities on kangaroo rat removal plots, to have a detectable impact or do not feed so selectively on the large-seeded species.

Figure 4-3. Aerial photograph of the study site taken in March 1995. The dark squares are plots where either all rodents or just kangaroo rats have been removed since 1977 and the dense growth of large-seeded annual plants is clearly visible. Plots where all rodents or just kangaroo rats have been removed since 1988 have less dense vegetation, and plots where rodents have been present continuously since 1977 have the least annual cover. Compare with Figure 4–1 to see the vegetation change since 1979 and for numbers of plots, whose treatments are given in Table 4-1.

Second, large effects of selective predation by rodents are only apparent on winter annuals. Rodents have virtually no detectable impact on summer annuals even though the rodents are active and foraging when the summer annuals are in fruit and dispersing their seeds (Guo and Brown 1996). This initially surprising result seems to have a simple explanation: for some still unexplained reason, none of the summer annuals produce seeds as large (approximately 1 mg) as those of the winter annual species on

Figure 4-4. Top: Photograph across the fence of a kangaroo rat removal plot (*kangaroo rats are absent on the left-hand side*) taken in March 1995. Bottom: Photograph taken at the same time across the fence of a control plot; kangaroo rats have been present continuously since 1977. The distinctive appearance of the vegetation on the kangaroo rat removal plot is due to the very high density of the large-seeded, yellow-flowered annual *Lesquerella gordonii*.

which the kangaroo rats forage so selectively. In certain years, most spectacularly in 1983–1984 and 1994–1995, the winter annual assemblages was almost completely dominated by large-seeded species on plots where kangaroo rats had been removed. The large-seeded annuals germinate early in the winter season with the first rains, and they survive and enjoy a large competitive advantage over the smaller-seeded, generally later-germinating species in years when there is abundant winter precipitation that begins early and continues throughout the winter season. So the selective impact of kangaroo rats on annual plants can be understood largely as a consequence of the effects of seed size on the foraging of the rodents and on the phenology and competition of the plants.

While ants and birds have qualitatively and quantitatively smaller impacts as seed predators on annual plants than rodents, removal of all three groups of granivores caused significant changes in the species composition of annuals (Samson et al. 1992, Guo et al. 1995, Guo and Brown 1996). (We removed birds and rodents from 50-m² plots in a factorial experiment begun in 1982; see Guo et al. 1995 for details.) Again, these effects were conspicuous only in the winter annuals, where we observed a distinctly different plant assemblages on plots where either rodents, ants, or birds had been removed (Guo et al. 1995, Guo and Brown 1996).

Removal of kangaroo rats also had dramatic effects on the perennial vegetation, although as might be expected, these took longer to be observed than the impacts on the annuals. By 1988 there was much greater cover of perennial and annual grasses on plots where kangaroo rats had been removed than on plots where they were present. Again, there were no apparent differences between plots where just kangaroo rats or all rodents had been excluded, indicating that the impact could be attributed to kangaroo rats alone (Brown and Heske 1990a, Heske et al. 1993).

Three things are especially noteworthy about the effects of kangaroo rats on grass. First, grass cover increased significantly only on plots where kangaroo rats had been excluded since 1977, not on the new kangaroo rat removal plots established in 1988. This implied that a long time lag was required for the establishment of grass.

Second, although the differences in grass cover between the "old" kangaroo rat (and all rodent) removal plots and other plots were dramatic between 1988 and 1991, they subsequently diminished. In the late 1980s, it appeared that kangaroo removal plots were being transformed into a grassland habitat (Brown and Heske 1990a, Heske et al. 1993). Since the study area was historically arid grassland and has been degraded to desert shrubland since the 1870s (Chew 1995, Kelt and Valone 1995), it seemed that removal of kangaroo rats was facilitating reestablishment of the original vegetation. While this may be true to some extent, the factors that influenced the shrubland–grassland transition are complex and temporally variable (see following discussion). Grasses have decreased dramatically across the entire study site since the early 1990s, apparently due to a succession of severe summer droughts, but the remaining grasses are still conspicuously concentrated on those plots where kangaroo rats have been excluded since 1977.

Third, while the mechanism responsible for the increase of grass cover when kangaroo rats were removed requires further study, we believe that disturbance rather than selective seed predation by kangaroo rats was the primary cause. Three observations support this interpretation. First, several of the grass species that increased have tiny seeds, far smaller than the seeds of annual plants that increased in response to kangaroo rat removal. Second, kangaroo rats are major agents of small-scale disturbance of desert soils. They create networks of foraging trails through the vegetation, leave shallow scrapes in the soil where they have either collected or buried seeds, and dig extensive burrow systems. These disturbances are conspicuously absent from plots where kangaroo rats have been removed. Third, a dense litter layer accumulated on kangaroo rat and rodent removal plots. Together, these observations suggest that the disturbance by kangaroo rats tends to inhibit establishment of perennial grasses and to favor annual vegetation. Kangaroo rats may even be the first in the temporal sequence of important decomposers in desert ecosystems; their physical activities may be critical in breaking up and shallowly burying plant litter, thereby facilitating subsequent decomposition by arthropod and microbial detritivores. In 1994 W. H. Schlesinger sampled the nutrients in the soil and obtained preliminary results (personal communication) that suggest nitrogen compounds were clumped under shrubs on control plots where kangaroo rats were present and more evenly dispersed on the old, grassy kangaroo rat removal plots.

Indirect Interactions

We have documented a number of indirect interactions as a consequence of our experimental manipulations at Portal (Figs. 4-2, 4-5). There appear to be hierarchies of competitive interactions among both rodent and ant species, so that removal of the dominant species affects a second species, which in turn affects a third. In both rodents and ants, the cascading effects of such hierarchical competition were seen in shifts of small-scale spatial distributions of the subordinate species. Banner-tailed kangaroo rats, *D. spectabilis*, aggressively defend open microhabitats within their home ranges, especially the areas around their large, conspicuous burrows, against other rodents, especially other kangaroo rats (Frye 1983). Bowers et al. (1987) found that when *D. spectabilis* was removed, the smaller kangaroo rat (*D. merriami*) shifted its microhabitat use from under shrubs to more open areas, and several even smaller rodent species then shifted to forage under the shrub cover. Somewhat similarly, Davidson (1980) found that the large, behaviorally dominant harvester ant (*Po. rugosus*) excludes its smaller congener (*Po. desertorum*) from the vicinity of its nest entrances, and the even smaller *Pheidole* ants tend to be clustered around the *Po. rugosus* colonies, where they have a refuge from competition from *Po. desertorum*. When the dominant *P. rugosus* was removed, this spatial arrangement broke down, with *Po. desertorum* moving its nest entrances toward and *Pheidole* moving its nests away from the abandoned *Po. rugosus* colonies (Davidson 1985).

The most dramatic indirect interactions, however, were those caused by the changes in vegetation where either all rodents or just kangaroo rats had been removed (Fig. 4-5). As noted previously, small-seeded species of winter annual plants decreased in response to increased competition from large-seeded annuals. In the late 1980s, after the dramatic increase in grass cover on plots where kangaroo rats had been excluded since 1977, several species of rodents characteristic of arid grassland habitat colonized (Brown and Heske 1990a, Heske et al. 1994). Since several of these rodents (i.e., *Sigmodon* spp.) were folivorous and colonized differentially only the grassy plots where kangaroo rats

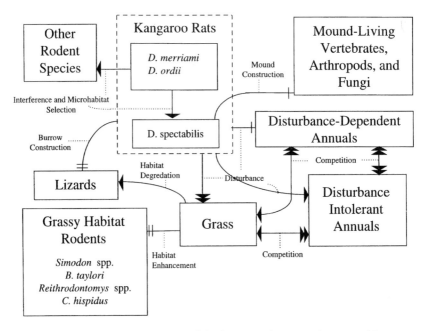

Figure 4-5. Diagram showing some of the important known and suspected keystone effects of kangaroo rats. Lines illustrate reciprocal interactions, with points indicating negative effects and crosses indicating positive effects. The number of points and crosses indicate the relative strength of the interactions. These may well be only a fraction of the direct and indirect effects of kangaroo rats in this ecosystem.

had been excluded since 1977 and not the shrubby plots where kangaroo rats had been removed only since 1988, we concluded that these species responded to the changes in vegetation and not to the absence of competition from kangaroo rats. However, the harvest mouse, *Reithrodontomys megalotis*, is granivorous and it also prefers grassy habitat; this species increased on all kangaroo rat removal plots, but from 1988 to 1991 it maintained higher densities on the older grassy ones than on the newer shrubby ones. Thus, the response of harvest mice to removal of kangaroo rats included both direct and indirect components: a release from competition that occurred in the short term and a response to habitat change that occurred only in the much longer term when grass had increased (Heske et al. 1994, Smith et al. 1997).

Birds also responded to the changes in habitat caused by removal of kangaroo rats (Fig. 4-5). Our first two National Science Foundation (NSF) grants did not contain sufficient funds to monitor the response of birds. We had predicted that seed-eating birds would show increased use of rodent and perhaps ant removal plots because of the increased food availability. In the first 2 or 3 years, anecdotal observations, including the small numbers of sparrows, quail, and doves caught in our rodent traps, suggested that such competitive release may have occurred. By the time we received funding to monitor the birds, however, just the opposite was occurring: mixed-species flocks of wintering sparrows were using rodent and kangaroo rat removal plots significantly less than plots where kangaroo rats were present (Thompson et al. 1991). Daniel Thompson's behavioral observations of the birds suggested an explanation for this unexpected result. The birds used the kangaroo rat trails to move through the living and standing dead vegetation and, being visual predators, were better able to detect seeds on the bare soil disturbed by rodents than on the litter-covered soil where kangaroo rats were absent. Thus, it appeared that although there should have been more seeds on plots where kangaroo rats had been excluded, the birds actually had higher success foraging where kangaroo rats were present.

These indirect interactions between birds and rodents were documented before the large increases in grass cover occurred on the ''old'' kangaroo rat removal plots in the late 1980s. Since then, we have not had sufficient funds to monitor birds or most other kinds of organisms that might have responded to the vegetation changes. However, some preliminary studies and anecdotal observations suggest that it would be interesting to do so. Donald Sias censused lizards on the plots in the summer of 1992 and found more species and individuals on plots where kangaroo rats were present (Fig. 4-5). He suggests (personal communication) that this reflects some combination of aversion to dense vegetation and utilization of rodent burrows by the lizards. Thomas Valone and I noted that in the late 1980s and early 1990s the grassy ''old'' kangaroo rat removal plots served as the centers of the territories of displaying male Cassin's sparrows, a species that, like the *Sigmodon* rodents, is usually restricted to arid grasslands. It would be interesting to assess the responses of other organisms, including grasshoppers, termites, and other arthropods, to the large changes in vegetation caused by excluding kangaroo rats.

Thus, removing kangaroo rats reveals them to be keystone organisms in this ecosystem (Fig. 4-5). They are not keystones in Paine's (1966, 1974) restricted sense that as selective predators on competitive dominants they facilitate the coexistence of subordinant species and thereby increase overall species diversity, although they are selective consumers of potentially dominant large-seeded annual plants. Kangaroo rats are keystones in the sense that they have strong interactions with many other organisms, so that community structure and ecosystem processes are drastically altered when they are removed (e.g., Paine 1980). The experimental design does not enable us to partition this keystone effect among the kangaroo rat species, but most of it is probably due to *D. merriami*, which has consistently been the dominant rodent species, in terms of both population density and biomass, on the site. It is the primary consumer of seeds and agent of widespread soil disturbance. The banner-tailed kangaroo rat, *D. spectabilis*, also acts as a keystone species—however, through a different mechanism. Its large, mounded burrow systems increase species diversity by creating distinctive microenvironments that are required or at least used preferentially by several species of vertebrates, arthropods, plants, and microbes (Kay and Whitford 1978, Reichman et al. 1985, Hawkins and Nicoletto 1992, Guo 1994). Two of these species, the burrowing owl and Mojave rattlesnake, are known to have decreased dramatically in the mid-1980s, coinciding with the catastrophic decline of *D. spectabilis* (see following discussion). The diverse impacts on the ecosystem of the two kangaroo rat species are especially impressive given the small quantities of these animals: *D. merriami* averaged 11.8 individuals and 505 g of biomass per hectare, *D. spectabilis* (before its extinction) 4.1 individuals and 505 g (Brown and Zeng 1989).

Long-Term Population Changes That Cannot Be Attributed to Experimental Manipulations

Some of the most interesting and important results of our study come from the long-term monitoring of rodent, ant, and plant populations. Major changes were documented, and since these occurred on unmanipulated control plots—and in a large area of habitat surrounding the study site—we can be confident that they were not caused by our manipulations. Presumably they reflect the influence of background environmental variation, especially in climate, on the animals and plants.

One of the most dramatic events was an approximately threefold increase in the cover of perennial shrubs between 1977 and 1995 (Brown et al. 1997). This increase occurred on our study site, where a fence has excluded grazing livestock, on the grazed land immediately adjacent to the study site, and on two other grazed sites of similar habitat 10 to 20 km away. We suspect that the cause was a change in regional climate to a regime of increased winter precipitation. Between 1978 and 1992 five weather stations surrounding the study site all reported significantly higher quantities of winter precipitation than the long-term average, and most stations received more than two standard deviations higher than the long-term average in 4 or 5 of the 15 years. This increase in woody vegetation associated with increased winter precipitation supports Neilson's (1986) suggestion that in this region of transition between Chihuahuan Desert shrubland and arid grassland cool-season rainfall favors winter-active C_3 shrubs, whereas summer precipitation favors summer-active C_4 grasses.

Associated with these changes in climate and vegetation, and very likely caused by some combination of them, was a catastrophic decline in the 1980s of the populations of several granivorous animals: the bannertail kangaroo rat (*D. spectabilis*) and four species of harvester ants (*Po. rugosus. P. barbatus, Po. desertorum*, and *A. cockerellii*). In 1977 *D. spectabilis* and *Po. rugosus* were dominant species with large, conspicuous burrow systems, but by the early 1990s they were locally extinct on the study site. Bannertails were the second most abundant rodent species until the winter of 1983–1984, when they suffered a precipitous decline (Valone et al. 1995). The population rebounded slightly in the next few years but then decreased again, and the last animal disappeared in 1994. The declines in the harvester ants were more gradual, but of the two species of *Pogonomyrmex* that had previously been common, *Po. rugosus* was locally extinct and the number of colonies of *Po. desertorum* had decreased by more than 75% by 1992. *A. cockerellii* went extinct on our study site and also on Robert Chew's long-term site 7 km away in somewhat different habitat (Chew 1995). A common feature of these species is that, as granivores, they store seeds in large granaries within their burrow systems. A likely cause of their decline was wetting and spoilage of their seed stores (Chew 1995, Valone et al. 1995). Horned lizards (*Phrynosoma cornutum* and *Py. modestum*), which are specialized ant predators, and burrowing owls (*Athene cunicularia*) and Mojave rattlesnakes (*Crotalus scutulatus*), which live in banner-tailed kangaroo rat burrows, also declined conspicuously during this same time period (personal observations and Mendelson and Jennings 1992, Taylor 1993).

By no means all of the animals decreased during the 1980s, however. Among the granivores, for example, overall rodent species diversity and populations of most rodent species and of the small harvester ant *Pheidole xerophila* fluctuated but showed no clear long-term trend (Valone and Brown 1995, 1996; Brown et al. 1997).

The patterns in the dynamics of the rodent populations highlight the special insights that can come from long-term studies. Between 1977 and 1988, total rodent population density and biomass showed five peaks. Three of these coincided with the El Niño years of 1977–1978, 1982–1983, and 1986–1987, which caused exceptionally heavy winter precipitation. This association appeared to support the long-standing interpretation that heavy rainfall results in large seed crops which in turn promote large increases in desert rodent populations (Brown and Heske 1990b). Imagine our surprise, therefore, when the El Niño events of 1991–1992 and 1992–1993 were associated with above-average winter rains but no subsequent increase in the rodent populations. In fact, total abundance and biomass of all rodents and populations of most rodent species remained near their all-time low throughout most of the early 1990s (Fig. 4-6; Valone and Brown 1996). This, together with the fact that the summer rains during this period were well below average, suggests that two successive seasons of high productivity may be required to produce peak rodent populations. While this hypothesis would require several decades of additional data for a definitive test, it is consistent with the observation that total rodent abundance and biomass attained an all-time high in the winter and spring

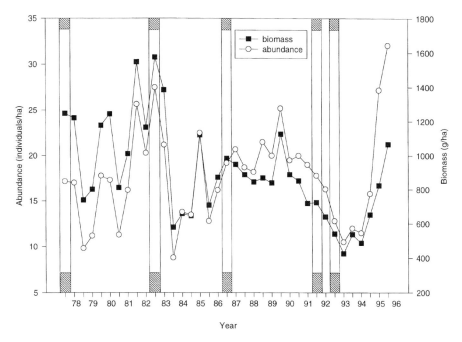

Figure 4-6. Fluctuations in the total abundance and biomass of all seed-eating rodents on the study site since 1977. Note that three of the first five peaks coincided with the heavy winter rains and El Niño events of 1977–1978, 1982–1983, and 1986–1987 (*vertical bars*; Brown and Heske 1990a), but the rodents declined after the El Niño events of 1991–1992 and 1992–1993 and then increased to an all-time high in 1995–1996.

of 1995–1996, following exceptionally high precipitation and primary production in the winter of 1994–1995 and above-average productivity in the summer of 1995 (Fig. 4-6).

A final insight from the long-term monitoring at the study site concerns species diversity. Our best data are for rodents, although anecdotal evidence suggests similar patterns in other groups, such as birds and certain arthropods. We have now recorded 24 species of desert rodents on the 20-ha study site. This is more than the number of native rodent species found in the entire states of Pennsylvania or Michigan! Aside from its sheer magnitude, two aspects of this diversity are noteworthy. First, only about half of these species were present on the site at any given time and the kangaroo rat (*D. merriami*) is the only species that was captured in every trapping session. The remaining species have colonized and gone extinct, often repeatedly in the 19-year duration of the study. Second, our experimental manipulations are responsible for some of these species having been recorded. These are species typical of arid grassland habitats which have differentially colonized the "old" grassy kangaroo rat removal plots (see preceding discussion). These species apparently disperse through the desert shrub habitat in very small numbers, but when they encounter the small patches of habitat that are perceived as favorable because of the absence of kangaroo rats and the substantial grass cover they establish residence and often breed. These observations point to the important effects of temporal and spatial environmental variation, even on scales of a few years and a few hectares, in promoting overall species diversity.

Summary

Issues of Data Quality, Experimental Design, and Statistical Analysis

Many chapters in this volume present the results of elegant experiments, with elaborate designs and excellent replication, intended to give convincing answers to specific questions about mechanisms. Nearly all of these experiments have been—at least by Portal standards—of short duration, a single season or at most a few years, and conducted in somewhat artificial settings, such as cattle tanks and artificial streams. By contrast, our experiments were very crude: a few targeted organisms have been removed in a simple design with minimal replication. Nevertheless, some of the results are very clear-cut and statistically significant, and they have held up well over the years as more data have accumulated. Our studies of competition among rodent species, selective predation of kangaroo rats on large seeds and the resulting effect on the composition of the annual plant assemblage, and indirect, keystone-type effects of kangaroo rats on both vegetation and animals provide some of the most thorough and convincing studies of these phenomena.

The success of the Portal project can be attributed largely to a few key factors, some of which were deliberately incorporated into the design from its beginning and others of which were encountered serendipitously and incorporated opportunistically. We have tried to maintain a balance between rigidity and flexibility.

We have rigidly adhered to parts of our original experimental design and to most of the standardized methods of data collection. Virtually all of our work is still centered on the original 24 fenced plots, which have been subjected to the same press treatments of rodent and ant removal since either 1977 or 1988. The standardized censuses of rodent, ant, and plant populations and other measurements that we take on these plots provide an invaluable record of the changes that have occurred.

While the value of maintaining the press experiments may seem obvious, we have been tempted to deviate from this design—to put in new plots or convert existing ones to new manipulation to test interesting hypothesis. Limited grant support made it impractical to install and monitor additional plots. We did change treatments on several plots in 1986, converting from seed addition to annual plant removal and, again in 1988, abandoning annual plant removal to establish additional replicates of the original rodent and ant removal experiment treatments. Given the spatial and temporal variation on the site and the long time required for some components of the system (e.g., and colonies and perennial plants) to respond to the manipulations, I want to keep this design for the foreseeable future. Nevertheless, there are always temptations to make changes. People frequently ask me what would happen if we allowed kangaroo rats access to plots from which they have long been excluded. This is a very interesting question. There was a time when I wanted to perform the experiment but delayed starting it because interesting changes were still occurring in response to the removal of kangaroo rats. Now the ecology of the original kangaroo rat removal plots has been so transformed by the absence of these keystone rodents that for the foreseeable future it will be more informative to continue the press experiment than to relax the treatment.

This is hardly to say that the original design is optimal. It is very imperfect, and if we were to start such a project anew, I would make many changes. In particular, I would increase replication, use a blocked design that better handles the underlying spatial variation in soils and vegetation, obtain more and better premanipulation baseline data, and perhaps not remove ants, although I would not be too surprised to see a major, long-delayed response to ant removal one of these years. But it would be counterproductive to try to incorporate such changes now.

Standardization of methods and insistence on a high level of data quality are essential for any empirical research program, but especially for a long-term study. Most of our interesting results come from documenting changes in response either to our experimental manipulations or to natural variation in the extrinsic environment, especially in climate. Our ability to document these changes, especially in a system characterized by enormous spatial and temporal variability on all scales, depends on taking carefully standardized censuses and other measurements. We can be sure that rodent, ant, and plant populations have changed because we have been counting them in exactly the same way, on the same patches of ground (delineated in relation to the permanent grid stakes), at regular intervals for the last 19 years. We take great pains to ensure that different field assistants use the same methods, make accurate identifications and measurements, and take responsibility for entering their data into the database. We have made a few changes, such as marking rodents with PIT tags rather than ear tags or toe clips, but only when we were confident that these changes did not compromise the comparability of the long-term data.

Conceptual Implications

Although we have been very conservative in adhering to the original experimental design and standardized sampling methods, we have also been very flexible. In particular, we have continually changed our questions, allowing them to be dictated by the long-term changes that we have documented. Thus, the original questions emphasized the direct effects of competition and predation by rodents and ants, while much of our subsequent work has focused on indirect interactions of these organisms and the keystone effects of kangaroo rats. Our experiments provide unique insights into these phenomena, because the effects of removing organisms with strong interactions cascade through the ecosystem for decades. Our long-term monitoring of plant and animal populations provides unique information on their population dynamics and responses to environmental change. The combination of experimental perturbation and monitoring enables us to examine how the interactions that we have manipulated play out in the context of varying extrinsic abiotic environment.

In a sense, the results of our studies at Portal have become an empirical model of how different kinds of organisms affect each other and how they are affected by temporal and spatial variation in their extrinsic environment. On the one hand, because there are few other studies of this kind, it is tempting to generalize from what we have learned at Portal and to suggest that the same patterns and processes will occur elsewhere.

On the other hand, our results show some of the pitfalls of trying to generalize prematurely from the results of short-term and small-scale studies, even experimental ones. We have not been able to repeat some results of our earlier experiments in the Sonoran desert, such as the apparently strong competition between rodents and ants, and some of our initial results at Portal, such as the increases in rodent populations following heavy precipitation during El Niño winters. In these cases, I think that the data obtained in the earlier studies were sound, but we misinterpreted the results because we did not appreciate the complexity of the system.

Our empirical results from Portal can be interpreted in the light of recent theoretical advances in nonlinear mathematics and complex adaptive systems (e.g., May 1974, 1987; Casdagli and Eubank 1992; Hastings et al. 1993; Brown 1994; Cowan et al. 1994). An ecosystem contains many individuals of many different species of organisms that interact with each other and with their abiotic environment to produce complex structures and dynamics. Such networks of interactions are likely to exhibit nonlinear behavior, in which slight differences in initial conditions or small perturbations can cause large changes in the system. Some of the most interesting results from Portal appear to be examples of such nonlinear behavior: both the experimental removal of kangaroo rats and the shift in winter precipitation regime caused extensive changes in vegetation and animal populations (Brown et al. 1997). It is extremely difficult—perhaps practically impossible—to predict how such complex systems will respond to perturbation. While in some cases the effects of small changes may be amplified and cause wholesale reorganization, in other cases they may be dampened and result in virtually no change. For example, we expect that removing kangaroo rats would not cause such dramatic changes in all arid ecosystems, and we know that the increased winter precipitation did not cause such large changes in vegetation in other habitats in the vicinity of our study site. It should be fruitful to continue to explore the relationships between the theory of complex, nonlinear systems and the empirical results of long-term and experimental ecological studies.

Reflections on the History of the Project

The Portal project began in 1977 as a collaborative enterprise by Dinah Davidson, O. J. Reichman, and myself. Within a few years, Reichman had moved from Arizona to Kansas, become interested in more behavioral questions, and greatly reduced his involvement. In the early 1980s, Davidson began shifting her fieldwork to Peru to pursue her growing interests in relationships between ants and plants in the tropics.

Although I have kept the project going, I could not have done so without the help of dedicated graduate students, postdocs, technicians, and undergraduate assistants. The list of collaborators and assistants must be approaching 100; I am reluctant to try to name all of them for fear of forgetting some important contributors. My reliance on assistants has increased as my time and stamina for fieldwork have diminished and as I have needed specialized expertise in data management, geographic information systems (for analyzing spatial data), ant and plant taxonomy, and other areas where my own knowledge and background are inadequate.

When we began the experiments, I expected that they would last only about 4 years. I have had a history of being involved in many diverse research projects—getting bored with or seeing diminishing progress on a current study and going on to something else that seemed more exciting and productive. For the first decade or so, the decision to continue the experiments was simply a matter of opportunism. Changes continued to occur in response to both our manipulations and natural environmental change, and I wanted to see what would happen if I waited a bit longer. It is only within the last few years that I have realized that interesting changes will always be occurring. They are almost inevitable in any broadly conceived, well-designed manipulative and monitoring study in ecology. They are what make the few very long-term experiments so valuable. Now I am committed to continuing the project and looking for someone to take over when I am no longer able to do so.

It has not always been easy. There have been struggles and frustrations. The most difficult time was the middle years, when funding was repeatedly interrupted. The continuity of the project required the skilled assistance of graduate students, postdocs, and technicians, some of whom worked for months without pay during periods between grants. After two initial grants from the NSF, in 1980 Davidson and I were informed that the project was not sufficiently concerned with biogeochemical processes for our third proposal to be funded by the Ecosystem Studies Program. Davidson became discouraged, but I kept the project going without funding for several months until I managed to obtain a grant from the NSF's Ecology Program. Again in 1985 our renewal proposal was rejected, and for the next 3 years I struggled to keep the project alive, sometimes without funding but most of the time with small, short-term grants from NSF's Ecology Program and with one grant from the Department of Energy (DOE). In 1987 a renewal proposal for a regular grant from NSF's Ecology Program was rejected, but I was encouraged to apply for a modest grant from the Long-Term Research in Environmental Biology (LTREB) Program. Funding has been tight, but constant since

1988, when we received the first of two successive 5-year LTREB grants. Altogether, we have had six grants from the NSF and one from the DOE for a total of $1,546,317 through 1997.

This may seem like a lot of money, but it has averaged less than $80,000 per year, not much for a major research program. The budget for 1996 was $76,662, which is barely enough to cover the expenses of a postdoc, two graduate research assistants, one or two summer undergraduate assistants (supported sometimes by supplements from the NSF's Research Experiences for Undergraduates [REU] Program), and some travel and incidental expenses (mostly software and hardware for data management). I get no summer salary and pay most of my own travel expenses. The current level of funding is essential to maintain the press manipulations, monitor the plant and animal populations, manage the gigantic long-term data base, and publish papers at regular intervals. The postdoc and two graduate assistants are especially indispensable. Among them they need to have the commitment to do arduous, accurate fieldwork; the taxonomic knowledge to work with the rodents, ants, and plants; the computer expertise to keep up and work with the database; and the scientific background to be collaborators and coauthors of the resulting publications.

I am not complaining about the level of funding. Support for basic ecological research is inadequate, and I feel privileged to have had an adequate share. I am happy to trade the relative security of a modest 5-year LTREB grant for the risk of trying to maintain higher levels of funding with shorter term but larger budget grants. And when one considers the influential publications, young scientists trained, and other impacts of the Portal project, I think both the scientific community and the American taxpayer have received good value for their grant money.

ACKNOWLEDGMENTS   I cannot adequately thank all of the graduate students, postdocs, undergraduates, area residents, and other friends who have worked on the Portal project. Without the collaboration of D. Davidson and O. J. Reichman the research would never have started; without the support of A. Kodric-Brown and the contributions of many others it would never have continued for 19 years. Most of the financial support has come from the National Science Foundation, most recently from Grant DEB-92-21238. E. Heske, P. Morin, J. Bernardo, and W. Resetarits made helpful suggestions on the manuscript. E. O'Brien prepared Figures 4-2 and 4-5.

Literature Cited

Bender, E. A., T. J. Case, and M. E. Gilpin. 1984. Perturbation experiments in community ecology: theory and practice. Ecology 65:1–13.

Bowers, M. A., and J. H. Brown. 1992. Structure in a desert rodent community: use of space around *Dipodomys spectabilis* mounds. Oecologia 92:242–249.

Bowers, M. A., D. B. Thompson, and J. H. Brown. 1987. Spatial organization of a desert rodent community: food addition and species removal. Oecologia 72:77–82.

Brown, J. H. 1973. Species diversity of seed-eating desert rodents in sand dune habitats. Ecology 54:775–787.

————. 1975. Geographical ecology of desert rodents. Pages 315–341 in M. L. Cody and J. M. Diamond (eds.), Ecology and Evolution of Communities. Harvard University Press, Cambridge, Massachusetts.

————. 1994. Organisms and species as complex adaptive systems: linking the biology of populations with the physics of ecosystems. Pages 16–24 in C. G. Jones and J. H. Lawton (eds.), Linking Species and Ecosystems. Chapman and Hall, London.

Brown, J. H., and D. W. Davidson. 1977. Competition between seed-eating rodents and ants in desert ecosystems. Science 196:880–882.

————. 1986. Reply to Galindo. Ecology 67:1423–1425.

————. 1990a. Control of a desert-grassland transition by a keystone rodent guild. Science 250:1705–1707.

————. 1990b. Temporal changes in a Chihuahuan Desert rodent community. Oikos 59: 290–302.

Brown, J. H., and M. Kurzius. 1989. Spatial and temporal variation in guilds of North American granivorous desert rodents. Pages 71–90 in D. W. Morris, Z. Abramsky, B. J. Fox, and M. R. Willig (eds.), Ecology of Small Mammal Communities. Special Publication no. 28, The Museum, Texas Tech University, Lubbock.

Brown, J. H., and G. A. Lieberman. 1973. Resource utilization and coexistence of seed-eating desert rodents in sand dune habitats. Ecology 54:788–797.

Brown, J. H., and J. C. Munger. 1985. Experimental manipulation of a desert rodent community: food addition and species removal. Ecology 66:1545–1563.

Brown, J. H., and Z. Zeng. 1989. Comparative population ecology of eleven species of Chihuahuan Desert rodents. Ecology 70:1507–1525.

Brown, J. H., D. W. Davidson, and O. J. Reichman. 1979. An experimental study of competition between seed-eating desert rodents and ants. American Zoologist 19: 1129–1143.

Brown, J. H., D. W. Davidson, J. C. Munger, and R. S. Inouye. 1986. Experimental community ecology: the desert granivore system. Pages 41–61 in J. Diamond and T. J. Case (eds.), Community Ecology. Harper and Row, New York.

Brown, J., T. Valone, and C. Curtin. 1997. Reorganization of an arid ecosystem in response to recent climate change. Proceedings of the National Academy of Sciences 94:9729–9733.

Casdagli, M., and S. Eubank (eds.). 1992. Santa Fe Institute Studies in the Science of Complexity, Proceedings: Vol. 12. Nonlinear Modeling and Forecasting. Addison-Wesley, New York.

Chew, R. M. 1995. Aspects of the ecology of three species of ants (*Myrmecocystus* spp. and *Aphaenogaster* sp.) in desertified grassland in southeastern Arizona, 1958–1993. American Midland Naturalist 134:75–83.

Cowan, G. A., D. Pines, and D. Melzer (eds.). 1994. Santa Fe Institute Studies in the Science of Complexity, Proceedings: Vol. 18. Complexity: Metaphors, Models, and Reality. Addison-Wesley, Reading, Massachusetts.

Davidson, D. W. 1977a. Species diversity and community organization in desert seed-eating ants. Ecology 58:711–724.

————. 1977b. Foraging ecology and community organization in desert seed-eating ants. Ecology 58:724–737.

————. 1980. Some consequences of diffuse competition in desert seed-eating ants. American Naturalist 116:92–105.

————. 1985. An experimental study of diffuse competition in harvester ants. American Naturalist 125:500–506.

Davidson, D. W., R. S. Inouye, and J. H. Brown. 1984. Experimental evidence for indirect facilitation of ants by rodents. Ecology 65:1780–1786.

Frye, R. J. 1983. Experimental field evidence of competition between two species of kangaroo rats (*Dipodomys*). Oecologia 59:74–78.

Galindo, C. 1986. Do desert rodent populations increase when ants are removed? Ecology 67:1422–1423.

Guo, Q. 1994. Dynamic desert plant community ecology: changes in space and time. Ph.D. dissertation, University of New Mexico, Albuquerque.

Guo, Q., and J. H. Brown. 1996. Temporal fluctuations and experimental effects in desert plant communities. Oecologia 107:568–577.

Guo, Q., D. B. Thompson, T. J. Valone, and J. H. Brown. 1995. The effects of vertebrate granivores and folivores on plant community structure in the Chihuahuan Desert. Oikos 73:251–259.

Hastings, A., C. L. Holm, S. Ellner, P. Turchin, and H. C. J. Godfray. 1993. Chaos in ecology: is mother nature a strange attractor? Annual Review of Ecology and Systematics 24:1–33.

Hawkins, L. K., and P. F. Nicoletto. 1992. Kangaroo rat burrow structure and the spatial organization of ground-dwelling animals in a semiarid grassland. Journal of Arid Environments 23:199–208.

Heske, E. J., J. H. Brown, and Q. Guo. 1993. Effects of kangaroo rat exclusion on vegetation structure and plant species diversity in the Chihuahuan desert. Oecologia 95:520–524.

Heske, E. J., J. H. Brown, and S. Mistry. 1994. Long-term experimental study of a desert rodent community: 13 years of competition. Ecology 75:438–445.

Hurlbert, S. H. 1984. Pseudoreplication and the design of ecological field experiments. Ecological Monographs 54:187–211.

Inouye, R. S. 1981. Interactions among unrelated species: granivorous rodents, a parasitic fungus, and a shared prey species. Oecologia 49:425–427.

Inouye, R. S., G. S. Byers, and J. H. Brown. 1980. Effects of predation and competition on survivorship, fecundity and community structure of desert annuals. Ecology 61:1344–1351.

Kay, F., and W. Whitford. 1978. The burrow environment of the banner-tailed kangaroo rat, *Dipodomys spectabilis*, in south-central New Mexico. American Midland Naturalist 92: 270–279.

Kelt, D. A., and T. J. Valone. 1995. Effects of grazing on the abundance and diversity of annual plants in Chihuahuan Desert scrub habitat. Oecologia 103:191–195.

May, R. M. 1974. Biological populations with non-overlapping generations: stable points, limit cycles and chaos. Science 186:645–647.

———. 1987. Chaos and the dynamics of biological populations. Proceedings of the Royal Society of London A 413:27–44.

M'Closkey, R. T. 1976. Community structure in sympatric rodents. Ecology 57:728–739.

———. 1978. Niche separation and assembly in four species of Sonoran Desert rodents. American Naturalist 112:683–694.

Mendelson, J. R., III, and W. B. Jennings. 1992. Shifts in the relative abundance of snakes in desert grassland. Journal of Herpetology 26:38–45.

Munger, J. C., and J. H. Brown. 1981. Competition in desert rodents: an experiment using semipermeable exclosures. Science 211:510–512.

Neilson, R. P. 1986. High-resolution climatic analysis and southwest biogeography. Science 232:27–34.

Paine, R. T. 1966. Food web complexity and species diversity. American Naturalist 100:65–75.

————. 1974. Intertidal community structure: experimental studies on the relationship between a dominant competitor and its principal predator. Oecologia 15:93–120.

————. 1980. Food webs: linkage, interaction strength and community structure. Journal of Animal Ecology 49:667–685.

Reichman, O. J., D. T. Wicklow, and C. Rebar. 1985. Ecological and mycological characteristics of caches in the mounds of *Dipodomys spectabilis*. Journal of Mammalogy 66: 643–651.

Rosenzweig, M. L. 1973. Habitat selection experiments with a pair of coexisting heteromyid rodent species. Ecology 54:111–117.

Rosenzweig, M. L., and J. Winakur. 1969. Population ecology of desert rodent communities: habitats and environmental complexity. Ecology 50:558–572.

Samson, D. A., T. E. Philippi, and D. W. Davidson. 1992. Granivory and competition as determinants of annual plant diversity in the Chihuahuan Desert. Oikos 65:61–80.

Smith, F. A., T. J. Valone, and J. H. Brown. 1997. Path analysis: a critical evaluation using long-term experimental data. American Naturalist 149:29–42.

Taylor, R. C. 1993. Location Checklist to the Birds of the Chiricahua Mountains. Borderlands, Tucson, Arizona.

Thompson, D. B., J. H. Brown, and W. D. Spencer. 1991. Indirect facilitation of granivorous birds by desert rodents and ants: experimental evidence from foraging patterns. Ecology 72:852–863.

Valone, T. J., and J. H. Brown. 1995. Effects of competition, colonization, and extinction on rodent species diversity. Science 267:880–883.

————. 1996. Desert rodents: long-term responses to natural changes and experimental manipulations. Pages 555–583 in M. L. Cody and J. A. Smallwood (eds.), Long-Term Studies of Vertebrate Communities. Academic Press, Orlando, Florida.

Valone, T. J., J. H. Brown, and E. J. Heske. 1994. Interactions between rodents and ants in the Chihuahuan Desert: an update. Ecology 75:252–255.

Valone, T. J., J. H. Brown, and C. L. Jacobi. 1995. Catastrophic decline of a keystone desert rodent, *Dipodomys spectabilis*: insights from a long-term study. Journal of Mammalogy 76:428–436.

Wiens, J. A. 1977. On competition and variable environments. American Scientist 65:590–597.

5

Experimental Compromise and Mechanistic Approaches to the Evolutionary Ecology of Interacting *Daphnia* Species

MATHEW A. LEIBOLD & ALAN J. TESSIER

To search for perfection is all very well, But to look for heaven is to live here in hell.

—Sting, 1985

The only rules of scientific method are honest observation and accurate logic.

—R. H. MacArthur, 1972

The experimental approach has been tremendously useful in community ecology, but it has led to a codified design for experiments, which has been characterized as phenomenological by some critics (Dunham, this volume; Petraitis, this volume). In the conventional approach, investigators select a particular system (i.e., an assemblage at a particular place and time) and conduct perturbation experiments to determine if certain general processes occur (e.g., species removals to detect the occurrence of predation or competition). In some cases, multiple factors are manipulated to determine the "relative importance" of these factors on aspects of the community. This tradition of experiments has rapidly escalated, and standard means have been developed and improved to deal with a wide array of methodological, statistical, and inferential issues (Hurlburt 1984, Hairston 1989, Manly 1992, Underwood 1995). Whereas early experiments often suffered from an array of such problems (Table 5-1), more recent work reflects a greater awareness of these important issues. Improvements on the experimental approach, however, continue to deal poorly with what are perhaps the most important concerns: *generality* and *extrapolation*. Advances in these areas come at great cost.

Consider, for example, the discussion by Underwood and Petraitis (1993) of a "perfect" experimental design to establish the importance of various factors that determine zonation patterns in intertidal communities. They do an excellent job of describing the array of issues that would need to be considered for such an experiment to be definitive and comparable. They identify the need to incorporate broad-scale spatial and temporal variability while maintaining adequate replication for powerful testing of effects. Be-

Table 5-1. Constraints on the application and interpretation of ecological experiments

Applying appropriate treatment levels: Use levels that clearly correspond to a reasonable view of fixed or random effects and avoid unrealistic manipulations.

Avoiding confounded manipulations: Ensure that treatments differ in no other way than the one manipulated factor level.

Choosing appropriate response parameters: Relate measured responses directly to hypothesis being tested and measure on appropriate spatial and temporal scales (e.g., accounting for dynamics).

Protecting from type I errors: Include issues of randomization to achieve independence, adjust for multiple testing of single hypotheses, and meet assumptions of statistical tests.

Protecting from type II errors: Include estimates of effect sizes for power analysis, provide adequate replication, minimize error variance, and meet assumptions of statistical tests.

Avoiding pseudoreplication: Carefully define experimental units in terms of independence and hypothesis scope.

Maximizing appropriate time scale and spatial scale: Ensure that extrapolation of results is only as broad as the range of conditions covered in the experimental design.

Employing appropriate controls: Avoid problems of intercorrelation of factors associated with treatments.

Addressing potential artifacts of manipulation: Use treatments that mimic natural conditions and contrast parameters between such treatments and the unmanipulated world.

cause of the large number of locations (random blocks or habitat strata) that would need to be encompassed, they conclude that the probability of such a comprehensive experiment ever being conducted is low due to limited resources.

One practical alternative is conducting experiments that are restricted in scope. Under this strategy, however, general conclusions about communities are convincing only after many such individual experiments are repeated (without necessarily having any other reason for doing them) and assessed via formal reviews. Further, despite a growing interest in metaanalysis of ecological experiments (see Gurevitch et al. 1992, Gurevitch and Hedges 1993, Arnqvist and Wooster 1995), comparability of results is often low except for the most qualitative of analyses. Further, a repetitive approach in general seems inefficient because it focuses only on parameters common to all these experiments and ignores the many individual clues that each system can provide about community processes as they are manifested in each particular system.

There is, however, a somewhat different approach to community ecology that is referred to as mechanistic (Price 1986, Schoener 1986). In this essay, we describe meaningful ways to combine experimental studies as part of an integrated, mechanistic research program. Our concern is the twofold problem of demonstrating causal processes and quantifying and evaluating the natural context for these processes.

The key component of the mechanistic approach is identification of a critical set of causal relations responsible for a particular result. While in principle a large number of factors could be involved in structuring a given community, a combination of deductive and inductive reasoning will often suggest ''fundamental'' causal relations that are critical to a given ecological pattern. The induction step we envision is basically the transition from pattern recognition to the postulation of general mechanisms. Prior observations and experimental results in the literature are synthesized to establish general hypotheses. Deduction is used to apply these general concepts to the particular system and in this sense is similar to "theory unpacking" (sensu Rosenzweig 1991). This general approach does not ignore the existence of complex interactions among the various processes but rather embraces "mechanistic holism" (sensu Wilson 1988) as a starting point. The primary goal, however, is to postulate clearly focused and operationally defined, albeit not comprehensive, models. Experiments are then employed sequentially to test the important assumptions and predictions from the model and separate studies conducted to extrapolate their application to a broader, more natural context. The approach allows each of the component studies to have different but complementary goals that together constitute an integrated research program.

This mechanistic experimental approach can be compared with a more descriptive approach that emphasizes detailed quantification of patterns and empirical model building and is used for prediction without necessarily resorting to basic mechanisms (Rigler 1975, Peters 1991). In contrast, a mechanistic approach retains the notion that the causal bases of patterns must themselves be understood before a hypothesis can be accepted as an explanation and presumes that an understanding of causal relations is an important element that allows extrapolation of results beyond the set of previously studied conditions. The mechanistic approach can also be contrasted with the more typical manner in which experiments are conducted in its focus on specific models. Frequently, the rationale for particular treatments used in more conventional experiments is poorly justified in terms of general or specific models. We agree with Quinn and Dunham (1983) that the analysis and interpretation of many experiments in ecology are more akin to statistical hypothesis testing than to strong inference but emphasize that experimental manipulation can also be employed to test clearly defined hypotheses that have been formulated with reference to patterns in nature and general theory.

Some Steps in a Mechanistic Experimental Approach

A list of some potentially important steps in a mechanistic approach is provided in Table 5-2. In contrast with much conventional work, an important first step is identifying a pattern of interest. Usually this will be motivated by observations from natural systems, but sometimes it could be motivated from theoretical expectations or from previous experimental results. This step is important in defining the scope of the research question and its context in the broader field of ecology.

A crucial next step is developing one or more explicit hypotheses to explain the pattern of interest. To the extent that the hypotheses are as explicit as possible about causal relations and assumptions, the empirical work that follows will be more convincing and rigorous. Thus, it is important to use mathematical models as much as possible and to avoid ambiguous or controversial simplifications. For example, if interspecific competition is invoked it will often be much better to specify whether it is via direct interference or whether resources or predators (via apparent competition) are involved and, if so, how. This modeling effort is useful by helping define crucial premises that underly the model and by providing a framework for generating explicit predictions from the model. An important reason for examining several models is to identify diagnostic predictions that contrast with those made by other reasonable alternative hypotheses.

Table 5-2. A mechanistic-experimental approach to a research program

Step	Methods
1. Identify a phenomenon	Description, exploratory data analysis
2. Hypothesize general causal models	Theory building Inductive reasoning from observations or prior experiments
3. Document major premises of the models	Experimental test of simple causes
4. Test diagnostic predictions of the models	Experimental test of complex causes
5. Quantify the mechanism	Calibration of key processes Nesting of mensurative experiments within manipulative experiments
6. Evaluate importance relative to other possibilities in nature	Random effects analysis of variance Incorporation of additional environmental factors/gradients

Experiments are useful for two different reasons in this approach. The first is to test the validity of important or potentially controversial premises made by the model. Often such experiments are reasonably simple because they are aimed at establishing the existence of direct causal links (rather than complex interactions). Furthermore, such experiments need not necessarily be conducted in the field setting, though of course they will be more convincing if they are.

Second and more important, experiments should be used as direct tests of model predictions. At this point, experiments are crucial because they are the primary way to identify causation in systems such as ecological communities, where almost all factors interact with each other via feedback loops. Further, such experiments are perhaps the most critical step in the development of the research program because they are the most likely point at which a qualitative decision will be made about the model. If the probability of type II errors (accepting the null hypothesis when it is false) is high, the model may be rejected inappropriately. If so, the consequences can be just as important as those of accepting a model when it is false (Shrader-Frechette and McCoy 1992). In general, the balance between type I and type II errors should depend upon how results will be used. However, in most "basic" research of the type we address here, psychological considerations (e.g., confirmation bias; Loehle 1987) favor an emphasis (perhaps misguided) on type I error.

Much of the "art" of being an experimental ecologist is manifested as the ability to minimize both types of errors, and it is difficult to do this when other constraints are involved. Hence, experiments whose purpose is to test the predictions of models typically emphasize the detection of qualitative results and postpone dealing with the more difficult evaluation of realism or quantificative predictions. These experiments may thus involve strategies such as imposing artificially large manipulations ("sledge-hammer" approach; Kitchell et al. 1988), using methods that have known artificial biases, etc. Although the context of these experiments should be as "representative" of the real world as possible, they are justifiably restricted in space and time. The results of such qualitative and experimental tests of a model are subject to concerns about their generality if treatment factors exceed the range of natural conditions. However, they can be invaluable in confirming or rejecting competing causal models, and in suggesting additional factors or causal mechanisms previously unappreciated (Wootton 1994).

Once a set of causal factors and a model of their interactions are identified and qualitatively tested, the research program objectives should focus on evaluating relevance to real systems, and generality in space and time. We advocate a pluralistic approach, recognizing that constraints differ with systems and patterns being explored. In particular, we discuss in the following sections three different approaches: random effects designs and related approaches, combining descriptive and manipulative techniques, and environmental gradients. One can view this stage of the research as a transition from the study of fixed effects to the incorporation of random effects and a shift from identifying important factors and causal mechanisms via the potential magnitude of their effects to partitioning variance within a representative context. While admirable, the transition is extremely difficult to accomplish in many systems. Often, there are no methods that are sufficiently free of artificial effects within a scale that allows adequate experimental replication. The problem of relating in vitro physiological studies, laboratory experiments, or "common garden" evolutionary studies with their expression in vivo or in the natural setting is completely analogous.

Random Effects Designs and Related Approaches

If there are manipulative methods that are largely free of artificial effects, more realistic experiments can be designed to quantify carefully the strength of particular factors over natural ranges of variation. Such experiments should seek to evaluate the predictive ability of the model relative to other possible sources of variance. This question is essentially one of trying to quantify how different potential factors affect the pattern of interest in an analysis of variance. In contrast with the designs for qualitative hypothesis testing described previously, artificial effects must be minimized because they could strongly alter the partitioning of variance. A number of other statistical issues become important in such mixed and random effects analyses of variance (Fry 1992, Bennington and Thayne 1994, Petraitis, this volume).

Combining Descriptive and Manipulative Techniques

Some types of artificiality can be partially circumvented by employing manipulations that mimic the actions of the mechanism without manipulating the causal factor itself. For example, if in a particular system it is difficult to manipulate predators within a range of natural densities, one might still be able to manipulate prey mortality rate or behavior by some comparable means in order to test particular models involving these mechanisms. As illustrated in the following discussion, such methods can be used to calibrate the relationship between the qualitative results obtained in experiments and the importance of mechanisms in a more natural setting (e.g., in the absence of artificial experimental effects).

An alternative, nonmanipulative approach to studying the relevance and generality of a model is refining descriptive techniques (often involving some degree of experimentation) that quantify the relationships among crucial factors identified by the model; Hurlburt (1984) has referred to such methods as "mensurative experiments." Combining manipulative experiments with these mensurative experiments is an effective solution to the common problems of artificiality and extrapolation. That is to say, the mensurative experiments can be conducted within both natural settings and manipulative experiments. The combination also aids greatly in interpretation of experiments where manipulation has immediate effects on more than one variable. Such intercorrelation of factors (Price 1986) can be best dissected by a combination of manipulation and detailed descriptive measurements.

The nesting of mensurative and manipulative experiments is, unfortunately, rare in most experimental approaches in community ecology. One application is in whole-system experiments (e.g., Carpenter and Kitchell 1993) where little or no replication demands that mechanisms be quantified at the same time that manipulations are conducted. We suggest that both press and pulse manipulations (Bender et al. 1984) employed in ecology would be more easily interpreted by such nesting of experimental approaches and strongly advocate their usage.

Environmental Gradients

The difference between random and fixed effects is central to statistical and conceptual interpretation of experimental manipulations. While there is no clear dichotomy, most experiments in ecology consider fixed factors, while many experiments in evolutionary biology consider random or mixed factors (e.g., random genotype and fixed environment; Fry 1992). Generality is best achieved when one draws samples or treatment levels at random from the natural range of a particular factor. It is the increasing concern with generality in space and time that leads to the incorporation of random factors in ecological experiments (Petraitis, this volume). However, many ecologists prefer to think of the world as composed of (fixed) categories or gradients of effects and design their experiments accordingly.

It is not immediately apparent to us that a focus on quantifying the magnitude of effects within fixed factor designs will be any less successful at achieving generality and extrapolation than a transition to random factor designs and variance partitioning. It seems clear, however, that simple presence/absence types of fixed effects designs are often inadequate; the goal should be increased refinement of the treatment levels based upon both qualitative and quantitative effects. Is it better to extrapolate by randomization in space or time or by recognizing and incorporating important factors that create spatial or temporal variation? An understanding of spatial or temporal effects seems eminently more pleasing to us than a simple partitioning of these effects. Typically, this understanding will involve the incorporation of important environmental gradients (e.g., productivity, seasonality, stress, disturbance, etc.) into causal models and experiments that test those models.

As should be apparent, it can be extremely difficult to design a single experiment that will simultaneously maximize the likelihood of success for all research program components. The approach we have outlined is sequential; as more is learned, experiments are designed to be more realistic and broader in scope. This differs from the more ''conventional'' experimental approach, which typically represents a large compromise among competing goals of treatment strength (type I error), realism, and generality. By conducting separate studies for each of these goals the design of each experiment can be more easily optimized with fewer constraints. This is because in the mechanistic approach advocated here (Table 5-2), sequential experiments focus on different priorities. Part of the advantage is that any given experiment need not attempt to simultaneously address all of the potential problems listed in Table 5-1. Given unlimited resources, it might be possible to design an experiment that simultaneously deals with all of the issues described above, but most working field ecologists have financial and temporal constraints that warrant a more piecemeal approach. There is not much point in allocating vast resources to study the relative importance of an effect without first knowing that it can exist.

An Example: Evolutionary Ecology of Interacting *Daphnia*

A number of studies could be used to illustrate these points (see Schoener 1986 for a review and chapters in this volume by Holbrook and Schmitt, Peckarsky, Power et al., and Werner). Here we focus on our work on the evolutionary ecology of competing *Daphnia* species to illustrate how such an approach can be developed in the context of a more complete research program. Though in the development of our research we did not consciously pursue the framework outlined above this review provides a useful example of the ways that different components of a research program can have complementary roles.

The Pattern

Figure 5-1 shows typical habitat distributions of two similar species of *Daphnia* that coexist in many stratified lakes in the Midwest United States. There are strong differences in habitat use suggestive of the notion that these species partition habitats. Further, the pattern of habitat use shows a seasonal shift that is concordant with changes in the relative densities of the two species (Threlkeld 1980, Leibold 1988, Leibold and Tessier 1991). In spring, *D. pulicaria* uses the two main habitats evenly and is more common than *D. galeata*, which is restricted to the shallow epilimnion. In summer, *D. pulicaria* is restricted to the hypolimnion of these stratified lakes and is less common than *D. galeata*, which is found more broadly in both habitats. Studies by Woltereck (1932), Tappa (1965), Haney and Hall (1975), Threlkeld (1979, 1980), and Kratz et al. (1987) indicate that this is a common pattern in lakes throughout the biogeographic range of these species. Among lakes, however, there is quantitative variation in the timing and degree of habitat restriction that occurs seasonally in these species (Leibold and Tessier 1991, unpublished data; Hu 1994).

The association between habitat use and population dynamics involving these two species (Leibold 1988) suggested that this system might be useful in investigating theories about how interspecific competition could result in habitat partitioning. Furthermore, there was evidence (Mittelbach 1981) that the seasonal shift in the relative abundances of species occurred synchronously with changes in the intensity of predation by bluegill sunfish (*Lepomis macrochirus*). This suggested that variance among and within lakes in patterns of relative abundance and habitat use might be regulated by the effects of predators on the outcome of competition (i.e., similar to the "keystone" predation effect; Paine 1966).

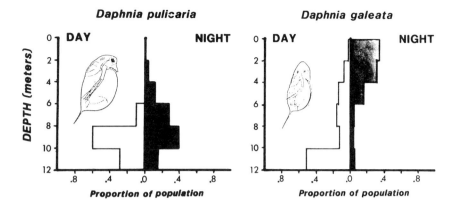

Midsummer vertical distributions of <u>Daphnia</u> in Lawrence Lake

Figure 5-1. Typical pattern of depth habitat use by the two dominant species of *Daphnia* in Lawrence Lake, Michigan. Histograms show the proportion of each species found at six different depth strata during the daytime (open bars) and nighttime (shade bars) using a 20-L Schindler trap. The thermocline is at 6 meters. *D. pulicaria* is almost completely hypolimnetic in distribution, whereas *D. galeata mendotae* undergoes strong diel vertical migration between the epilimnion during the nighttime and the hypolimnion during the daytime.

The Theory

Effects of predators on the outcome of competitive interactions have usually been modeled in the absence of habitat structure (e.g., Vance 1978; Leibold 1989, 1996). A few studies, however, have investigated how predation risk should affect habitat use (e.g., Werner and Gilliam 1984, Gilliam and Fraser 1988) and concluded that such effects could strongly modify the predictions of such models of predator-mediated competition among species. Modifying the model developed by Brown (1988, 1990; see Leibold 1988), we made several testable predictions about how predators and nutrients should affect *Daphnia* densities, resource densities, and habitat use in this system (Table 5-3). The derivation of these predictions (Leibold 1988, Leibold and Tessier 1997) follows from a synthesis of models that associate habitat selection under predation risk (Werner and Gilliam 1984; Gilliam and Fraser 1987, 1988) with models of trophic cascades and "bottom-up" effects in food webs (Leibold 1989). The model highlights the crucial importance that predation risk (due to planktivorous fish densities) plays in altering the outcome of competition between the two competing species of *Daphnia*.

Validating Model Premises

These models are based on several key assumptions. One of the most controversial is that consumers can alter patterns of habitat use in adaptive response to changes in

Table 5-3. Predictions from a mechanistic model of habitat use by competing *Daphnia* (Leibold 1988)

	Response by			
Effect of	*Daphnia* density	Epilimnetic algal density	Hypolimnetic algal density	Use of epilimnion by *D. pulicaria*
Higher fish density	Decrease	Increase	None	Decrease
Higher nutrient levels	Increase	None	None	None

predation risk and foraging gain. Field experiments in various types of enclosures demonstrated three different mechanisms for such responses: (1) *Daphnia* can detect variation in foraging gain and predation risk and respond facultatively (Dodson 1988; Leibold 1988, 1990); (2) *Daphnia* are genetically variable in habitat selection behavior (DeMeester 1993, 1994; Leibold et al. 1994) and can respond through rapid microevolutionary change; (3) *Daphnia* can acclimate to a specific habitat (Leibold et al. 1994), are genetically variable in other ecological traits that affect fitness in different habitats (Leibold and Tessier 1991, Tessier and Leibold 1997), and can change patterns of habitat use via indirect selection on such traits (see Hedrick 1990, Jaenike and Holt 1991 for similar examples in other arthropods).

In addition, the model assumes that *Daphnia* population growth is limited jointly by the effects of predators and of food resources. In experiments in which food levels were enhanced and fish presence was manipulated, both factors were shown to substantially affect *Daphnia* densities (Leibold 1989, 1991) as predicted by simple trophic cascade models (e.g., Rosenzweig 1971, Oksanen et al. 1981) in that resources were limiting both in the presence of artificially high fish predator densities and in the absence of any fish predators.

Finally, the model assumes that *D. pulicaria* and *D. galeata* show asymmetric patterns of limitation by predators and food. This appears to be closely related to differences in their genetically determined body sizes and pigmentation (Leibold and Tessier 1991): *D. galeata* is smaller, more transparent, and, hence, less vulnerable to fish predation than is *D. pulicaria* when both are found in the same habitat with fish. Attempts to document a simple difference in resource exploitative ability between the two species have not been as successful. *D. galeata* is not as sensitive to resource variance in the sense that it cannot exploit high resource conditions as well as can *D. pulicaria* (Tsao and Tessier, unpublished data). Which species is the superior competitor, however, depends on environmental conditions. By employing in situ incubation experiments within the context of both seasonal variation and experimental manipulations, we have concluded that temperature, food quality, and genetic variation within each taxon affect competitive interactions (Threlkeld 1979, Leibold 1991, Hu 1994, Hu and Tessier 1995, Tessier and Leibold, unpublished data). We return to this issue later.

Testing Model Predictions

Thus, several of the key assumptions of the model used to generate the predictions listed in Table 5-3 were validated. The predictions themselves were also examined in enclosure experiments and were for the most part supported (see Leibold 1991). Manipulations of nutrient levels and fish densities used in these experiments were not, however, representative of the natural range of variation of these two factors within or among lakes. Thus, the experiments serve as valid tests of the mechanistic hypotheses developed in the model but cannot evaluate the importance of these factors and mechanisms relative to other potential factors that affect the habitat use by *D. pulicaria* and *D. galeata*.

Quantitative Calibrations of Results

We took two different approaches in evaluating how useful the model was in explaining natural patterns in plankton assemblages involving these two *Daphnia* species. In the first approach, we attempted to calibrate the consequences of manipulations of *Daphnia* death rates in enclosures with the natural situation occurring in the lake. Large-scale enclosures (as in Leibold 1991) were deployed in Lawrence Lake and subjected to artificial harvesting using a 120 μm mesh net at six different levels. Harvested zooplankton (mostly *Daphnia*) were heat-killed and returned to their respective enclosures to prevent the export of nutrients. The experiment was allowed to run for a month, and final densities of zooplankton and estimates of phytoplankton concentration in each habitat were analyzed using regression methods. Results, shown in Figure 5-2, show a positive relationship between harvesting rate and phytoplankton concentration in both habitats. More important, phytoplankton levels in the lake at the same time corresponded well with the different harvesting rates chosen to mimic natural mortality in the epilimnion and hypolimnion. Epilimnetic lake phytoplankton levels corresponded to those expected for imposed mortality rates of about 20% per day, whereas hypolimnetic phytoplankton levels corresponded to those expected for imposed mortality rates of about 0%. Separate estimates of death rates that use conventional demographic methods (Rigler and Downing 1984) in the lake for *D. pulicaria* (the hypolimnetic species) and *D. galeata* (the migrating species that forages mostly in the epilimnion) are 5% and 20% per day, respectively. These correspond closely to the experimental results since 5% per day is easily attributable to factors other than predators (i.e., senescence and disease) and is not unusual for *Daphnia* cultured in the lab under good growth conditions.

However, the experimental results clearly differed from the results that occurred naturally in the lake in at least one important respect. Zooplankton densities that correspond to apparently natural levels of mortality were substantially lower in the enclosures than in the lake. The inference is that a lower number of *Daphnia* could successfully control phytoplankton in the enclosures than in the lake under the same mortality rate. Since productivity of plankton in enclosures is commonly lower than in the lake, due to the development of competing periphyton on the walls of the enclosures

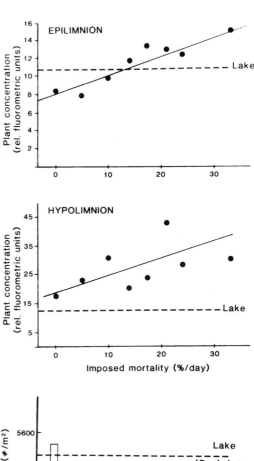

Figure 5-2. Results of manipulation of *Daphnia* mortality rate on phytoplankton and zooplankton levels in mesocosm experiments (10 m deep by 1 m diameter polyethylene "bags"). Phytoplankton and zooplankton were sampled at 1-meter intervals with a high-volume pump sampler and summed separately for the epilimnion (<4 m depth) and hypolimnion (>4 m depth). Phytoplankton levels were measured using in vivo fluorescence, and zooplankton densities were from subsample counts. For the zooplankton, counts of *Daphnia* are shown as shaded histograms and are overlaid by the total counts of microcrustaceans shown as open histograms. Ambient phytoplankton and zooplankton levels in the lake are shown as horizontal lines. Imposed mortality levels are significantly correlated (P <.05) with phytoplankton levels in both habitats (positive) and overall zooplankton levels (negative).

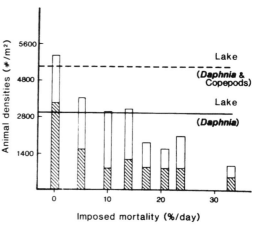

and possibly to higher sedimentation of particulates, this discrepancy is easily explained. However, it indicates that there are important enclosure effects in such experiments and that the method should not be used to evaluate the relative importance of nutrients and predators by the use of random effect designs because there might be significant statistical interactions between the enclosure effects and those of nutrients and predators.

Evaluating Natural Variance

Because solving the problems associated with the enclosure effects described previously seemed a difficult proposition (in terms of labor, materials, and opportunity to conduct whole-lake experiments or to partition lakes into large subdivided enclosures), we used a second approach and attempted to evaluate the consistency of the model with broader patterns of variation among different lakes. Our goal was to identify patterns that were expected on the basis of the hypotheses described previously but were incompatible with other possible explanations for habitat partitioning between *D. pulicaria* and *D. galeata*.

In a comparison of seven different lakes, we showed that variation in habitat use and relative abundance of *D. pulicaria* and *D. galeata*, similar to those caused by manipulations of fish in enclosures and predicted by the model described previously, was associated with natural variation in fish densities (Leibold and Tessier 1991). Though these lakes were not chosen randomly, the data indicate that the model can explain some of the variation that naturally occurs among lakes. In a separate study in which 30 study lakes were chosen almost exclusively on the basis of access and convenience (no presumed bias related to any of the factors potentially important to *Daphnia* populations), we found that patterns of abundance of these *Daphnia* species were strongly related to the availability of a deepwater refuge from fish predation. This pattern is consistent with the notion that interactions between these two species are regulated by predation risk as it alters competition for resources in alternate habitats (Tessier and Welser 1991). The survey also suggested an important environmental gradient (deepwater anoxia) that would need to be incorporated into the model before quantitative predictions were made.

One problem with the detailed study of seven lakes and our broader examination of 30 lakes is that they fail to reject more complex models. This is a typical shortcoming of purely descriptive work. However, these surveys did document natural variation in habitat use, within which we could retest some of the key mechanisms. When experiments were conducted in other seasons (Leibold and Tessier unpublished data), in other lakes (Hu 1994, Hu and Tessier 1995, Gonzalez and Tessier 1997), and with genetically divergent populations (Tessier and Leibold unpublished data) it became clear that other environmental and genetic (evolutionary) factors contribute to the broad pattern of habitat partitioning in these *Daphnia* species.

It is now apparent that both *D. pulicaria* and *D. galeata* populations in our local lakes contain substantial genetic variation. For example, most populations of *D. pulicaria* contain clonal morphs that are divergent in body size, behavior, and demographic traits (Leibold and Tessier 1991, Tessier and Leibold 1997). Large differences among lake populations in habitat use and depth partitioning are associated with shifts in the relative abundance of these morphs (Geedey et al. forthcoming). The *D. galeata* story is even more complex, since this taxon is now understood to contain at least two species, hybrids, and backcrosses (Taylor and Hebert 1993) throughout North America. Contrary to early assumptions (Hebert 1984), the ecological importance of this genetic variation is apparently large.

It is perhaps not surprising that mechanistic ecological investigations of community interactions such as competition and predation would ultimately develop an evolutionary perspective. The core assumptions in our model of habitat partitioning concern trade-offs in vulnerability to predators and in exploitative ability under different environmental conditions. If such trade-offs are of fundamental importance to the ecology of *Daphnia* today, then it is likely they also influenced the recent evolutionary history and distribution of each species. Although considering species as homogeneous with regards to ecological traits (and phenotype trade-offs) is a convenient starting point in community models, the real world may often harbor genetic variation within species that is of substantial ecological significance (Fauth, this volume).

Conclusions

Models of species interactions are often developed using fairly broad formulations in the hope that they will maintain some generality (Levins 1966). This level of generality involves describing any given mechanism in terms that focus on its most crucial components so that the model can be applied to systems that are potentially very different. This may be difficult, however, given the existence of numerous context-dependent processes that lead to complex interactions among components. Consequently, some ecologists feel that there may be few truly general theories likely to emerge in community ecology because different types of communities may be structured in very different ways for reasons that can be hard to determine (see Schoener 1986 for a discussion of this view). We believe that the mechanistic experimental approach we have outlined in this essay can be useful for comparing different types of communities by first identifying the key factors that influence broad patterns and then describing interactions among these factors within the context of general models.

So, for example, our results show some remarkable similarities to the results of mechanistic studies of habitat partitioning by gerbils in Israel (Abramsky et al. 1990, 1992, 1994; Kotler et al. 1991, 1993a, b, c; Brown et al. 1992, 1994). These studies show that two species of competing gerbils have different patterns of habitat use that are inter-and intraspecifically density-dependent. They also show that habitat use is affected by predation risk due to habitat-biased predators (owls that feed selectively on rodents found in the open habitat) and that predation risk alters the amount of habitat-specific resource depletion that gerbils impose on seeds in ways analogous to the effects of *Daphnia* on algae in lakes. There is even similarity between *Daphnia* species and gerbils in terms of an apparent trade-off in foraging exploitation at low versus high resource levels. Interactions among the gerbils, however, differ somewhat from those of *Daphnia* due to the effects of direct competitive interference (Ziv et al. 1993, Brown et al. 1994). Nevertheless, a relatively general mechanistic model seems to explain a number of natural patterns associated with assemblages as diverse as zooplankton and rodents.

Ecologists will continue to disagree and take different approaches to experimental design, but the results of any experiment should provide useful information for any reader. Almost any well–conceived and executed experiment provides valuable insights into the functioning of ecological processes and, thus results in important clues that can help enlarge our understanding. The tricky part is finding ways to organize the information from multiple, sometimes contradictory, experiments and studies into a cohesive research program. When experiments are dismissed as ''useless'' by critics, these dismissals are only valid if the logic behind the experiments is inappropriate for their stated goals or if the observations and their analyses are flawed. Of course, a major problem is that there are too many cases where the goals and limitations of a particular paper are not defined or these goals and limitations are subsequently ignored in the following citations (both supportive and critical). We believe philosophical discussions about the "best" ways to conduct experiments are often overly dogmatic and would like to advocate a more pluralistic approach, which recognizes that different experimenters have different goals in mind when they design particular studies. We particularly want to stress the distinct but complementary goals and constraints that distinguish experiments that attempt to test model predictions from those that attempt to quantify the relative importance of hypothesized causal factors.

ACKNOWLEDGMENTS   We acknowledge help in the field and lab research by N. Howe, G. Howe, E. Whitaker, M. McPeek, C. Osenberg, N. Consolatti, and P. Woodruff. The research described was supported by National Science Foundation grants BSR-90-07579, DEB-94-21539, and DEB-95-09004. This is Kellogg Biological Station contribution number 820.

Literature Cited

Abramsky, Z., M. L. Rosenzweig, J. S. Brown, B. P. Kotler, and W. A. Mitchell. 1990. Habitat selection: an experimental field test with two gerbil species. Ecology 71:2358–2369.

Abramsky, Z., M. L. Rosenzweig, and A. Subach. 1992. The shape of a gerbil isocline: an experimental field study. Oikos 63:193–199.

Abramsky, Z., O. Ovadia, and M. L. Rosenzweig. 1994. The shape of a *Gerbillus pyramidum* (Rodentia: Gerbillinae) isocline: an experimental field study. Oikos 69: 318–326.

Anrqvist, G., and D. Wooster. 1995. Meta-analysis in ecology and evolution. Trends in Ecology and Evolution 10:236–240.

Bender, E. A., T. J. Case, and M. E. Gilpin. 1984. Perturbation experiments in community ecology: theory and practice. Ecology 65:1–13.

Bennington, C. C., and W. V. Thayne. 1994. Use and misuse of mixed model analysis of variance in ecological studies. Ecology 75:717–722.

Brown, J. S. 1988. Patch use as an indicator of habitat preference, predation risk, and competition. Behavioral Ecology and Sociobiology 22:37–47.

———. 1990. Habitat selection as an evolutionary game. Evolution 44:732–746.

Brown, J. S., Y. Arel, Z. Abramsky, and B. P. Kotler. 1992. Patch use by gerbils (*Gerbillus allenbyi*) in sandy and rocky habitats. Journal of Mammalogy 73:821–829.

Brown, J. S., B. P. Kotler, and W. A. Mitchell. 1994. Foraging theory, patch use and the structure of a Negev Desert granivore community. Ecology 75:2286–2300.

Carpenter, S. R., and J. F. Kitchell (eds.). 1993. The Trophic Cascade in Lakes. Cambridge University Press, Cambridge.

De Meester, L. 1993. Genotype, fish-mediated chemicals, and phototactic behavior in *Daphnia magna*. Ecology 74:1467–1474.

———. 1994. Life histories and habitat selection in *Daphnia*: divergent life histories of *D. magna* clones differing in phototactic behaviour. Oecologia 97:333–341.

Fry, J. D. 1992. The mixed model analysis of variance applied to quantitative genetics: biological meaning of the parameters. Evolution 46:540–550.

Geedey, C. K., A. J. Tessier, and K. Machledt. 1996. Habitat heterogeneity, environmental change, and the clonal structure of *Daphnia* populations. Functional Ecology 10:613–621.

Gilliam, J. F., and D. F. Fraser. 1987. Habitat selection under predation hazard: a test of a model with stream-dwelling minnows. Ecology 68:1856–1862.

———. 1988. Resource depletion and habitat segregation by competitors under predation hazard. Pages 173–184 in L. Persson and B. Ebenman (eds.), Size-structured Populations: Ecology and Evolution. Springer-Verlag, Berlin.

Gonzalez, M. J., and A. J. Tessier. 1997. Habitat segregation and interactive effects of multiple predators on a prey assemblage. Freshwater Biology 38:179–191.

Gurevitch, J., and L. V. Hedges. 1993. Meta-analysis: combining the results of independent experiments. Pages 378–398 in S. M. Scheiner and J. Gurevitch (eds.), Design and Analysis of Ecological Experiments. Chapman and Hall, New York.

Gurevitch, J. L. Morrow, A. Wallace, and J. S. Walsh. 1992. A meta-analysis of field experiments on competition. American Naturalist 140:539–572.

Hairston, N. G., Sr. 1989. Ecological Experiments: Purpose, Design, and Execution. Cambridge University Press, Cambridge.

Haney, J. F., and D. J. Hall. 1975. Diel vertical migration and filter-feeding activities of *Daphnia*. Archiv für Hydrobiologie 75:87–132.

Hebert, P. D. N. 1984. Demographic implications of genetic variation in zooplankton populations. Pages 195–207 in K. Wöhrmann and V. Loeschcke (eds.), Population Biology and Evolution. Springer-Verlag, Berlin.

Hedrick, P. W. 1990. Genotypic-specific habitat selection: a new model and its application. Heredity 65:145–149.

Hu, S. 1994. Competition and seasonal succession of *Daphnia* in Gull Lake, Michigan. Ph.D. dissertation, Michigan State University, East Lansing.

Hu., S., and A. J. Tessier. 1995. Seasonal succession and the strength of intra- and interspecific competition in a *Daphnia* assemblage. Ecology 75:2278–2294.

Hurlburt, S. H. 1984. Pseudoreplication and the design of ecological field experiments. Ecological Monographs 54:187–211.

Jaenike, J., and R. D. Holt. 1991. Genetic variation for habitat preference: evidence and explanations. American Naturalist 137:S67–S90.

Kitchell, J. F., S. M. Bartell, S. R. Carpenter, S. R. Hall, D. J. McQueen, W. E. Neill, D. Scavia, and E. E. Werner. 1988. Epistemology, experiments, and pragmatism. Pages 263–280 in S. R. Carpenter (ed.), Complex Interactions in Lake Communities. Springer-Verlag, New York.

Kotler, B. P., J. S. Brown, and O. Hasson. 1991. Owl predation on gerbils: the role of body size, illumination, and habitat structure on rates of predation. Ecology 72:2249–2260.

Kotler, B. P., J. S. Brown, and W. A. Mitchell. 1993a. Environmental factors affecting patch use in two species of gerbilline rodents. Journal of Mammalogy 74:614–620.

Kotler, B. P., J. S. Brown, R. H. Slotwo, W. L. Goodfriend, and M. Strauss. 1993b. The influence of snakes on the foraging behavior of gerbils. Oikos 67:309–316.

Kotler, B. P., J. S. Brown, and A. Subach. 1993c. Mechanisms of species coexistence of optimal foragers: temporal partitioning by two species of sand dune gerbils. Oikos 67:548–556.

Kratz, T. K., T. M. Frost, and J. J. Magnuson. 1987. Inferences from spatial and temporal variability in ecosystems: long-term zooplankton data from lakes. American Naturalist 129:830–846.

Leibold, M. A. 1988. Habitat structure and species interactions in plankton communities of stratified lakes. Ph. D. dissertation, Michigan State University, East Lansing.

———. 1989. Resource edibility and consumer-resource interactions in predation and productivity gradients. American Naturalist 134:922–949.

———. 1990. Resource and predation can affect the vertical distribution of zooplankton. Limnology and Oceanography 35:938–944.

———. 1991. Trophic interactions and habitat segregation between competing *Daphnia* species. Oecologia 86:510–120.

———. 1996. A graphical model of keystone predators in food webs: trophic regulation of abundance, incidence, and diversity patterns in communities. American Naturalist 147: 784–812.

Leibold, M. A., and A. J. Tessier. 1991. Contrasting patterns of body size for coexisting *Daphnia* species that segregate by habitat. Oecologia 86:342–348.

———. 1997. Habitat partitioning by zooplankton and the structure of lake ecosystems. Pages 3–30 in B. Streit, T. Staedler, and C. M. Lively (eds.), Evolutionary Ecology of Freshwater Animals. Birkhouser Verlag, Basel.

Leibold, M. A., A. J. Tessier, and C. T. West. 1994. Genetic, acclimation and ontogenetic effects on habitat selection behavior in *Daphnia pulicaria*. Evolution 48:1324–1332.

Levins, R. 1966. The strategy of model building in population biology. American Scientist 54:421–431.

Loehle, C. 1987. Hypothesis testing in ecology: psychological aspects and the importance of theory maturation. Quarterly Review of Biology 62:397–409.

Manly, B. F. J. 1992. The Design and Analysis of Research Studies. Cambridge University Press, Cambridge.

Mittelbach, G. G. 1981. Foraging efficiency and body size: a study of optimal diet and habitat use by bluegills. Ecology 62:1370–1386.

Oksanen, L., S. Fretwell, J. Arruda, and P. Niemala. 1981. Exploitation ecosystems in gradients of primary productivity. American Naturalist 118:240–261.

Paine, R. T. 1966. Food web complexity and species diversity. American Naturalist 100:65–75.

Peters, R. H. 1991. A Critique for Ecology. Cambridge University Press, Cambridge.

Price, M. V. 1986. Structure of desert rodent communities: a critical review of questions and approaches. American Zoologist 26:39–49.

Quinn, J. F., and A. E. Dunham. 1983. On hypothesis testing in ecology and evolution. American Naturalist 122:602–617.

Rigler, F. H. 1975. Nutrient kinetics and the new typology. Verhandlungen Internationale Vereinigung für Theoretische und Angewandte Limnologie 19:197–210.

Rigler, F. H., and J. A. Downing. 1984. The calculation of secondary productivity. Pages 19–58 in J. A. Downing and F. H. Rigler (eds.), A Manual on Methods for the Assessment of Secondary Productivity in Freshwater. IBP handbook 17. Blackwell Scientific, Boston, Massachusetts.

Rosenzweig, M. L. 1971. Paradox of enrichment: destabilization of exploitation ecosystems in ecological time. Science 171:385–387.

———. 1991. Habitat selection and population interactions: the search for mechanism. American Naturalist 137:S5–S28.

Schoener, T. W. 1986. Mechanistic approaches to community ecology: a new reductionism? American Zoologist 26:81–106.

Shrader-Frechette, K. S., and E. D. McCoy. 1992. Statistics, costs and rationality in ecological inference. Trends in Ecology and Evolution 7:96–99.

Tappa, D. 1965. The dynamics of the association of six limnetic species of *Daphnia* in Aziscoos Lake, Maine. Ecological Monographs 35:395–423.

Taylor, D. J., and P. D. N. Hebert. 1993. Habitat-dependent hybrid parentage and differential introgression between neighboring sympatric *Daphnia* species. Proceedings of the National Academy of Science of the USA 90:7079–7083.

Tessier, A. J., and M. A. Leibold. In press. Habitat use and ecological specialization within lake *Daphnia* populations. Oecologia.

Tessier, A. J., and J. Welser. 1991. Cladoceran assemblages, seasonal succession and the importance of hypolimnetic refuge. Freshwater Biology 25:85–93.

Threlkeld, S. T. 1979. The midsummer dynamics of two *Daphnia* species in Wintergreen Lake, Michigan. Ecology 60:165–179.

———. 1980. Habitat selection and population growth of two cladocerans in seasonal environments. Pages 346–357 in W. C. Kerfoot (ed.), Evolution and Ecology of Zooplankton Communities. University Press of New England, Hanover, New Hampshire.

Tilman, D. 1986. A consumer-resource approach to community structure. American Zoologist 26:5–22.

Underwood, A. J. 1997. Ecological experiment & their logical design and interpretation using analysis of variance. Cambridge University Press, Cambridge.

Underwood, A. J., and P. S. Petraitis. 1993. Structure of intertidal assemblages in different locations: how can local processes be compared? Pages 39–51 in R. E. Ricklefs and D. Schluter (ed.), Species Diversity in Ecological Communities: Historical and Geographical Perspectives. University of Chicago Press, Chicago, Illinois.

Vance, R. R. 1978. Predation and resource partitioning in one predator-two prey model communities. American Naturalist 112:797–813.

Werner, E. E., and J. E. Gilliam. 1984. The ontogenetic niche and species interactions in size-structured populations. Annual Review of Ecology and Systematics 15:393–425.

Wilson, D. S. 1988. Holism and reductionism in evolutionary ecology. Oikos 53:269–273.

Woltereck, R. 1932. Races, associations and stratification of pelagic daphnids in some lakes of Wisconsin and other regions of the United States and Canada. Transactions of the Wisconsin Academy of Sciences 27:487–522.

Wootton, J. T. 1994. Predicting direct and indirect effects: an integrated approach using experiments and path analysis. Ecology 75:151–165.

Ziv, Y., Z. Abramsky, B. P. Kotler, and A. Subach. 1993. Interference competition and temporal and habitat partitioning in two gerbil species. Oikos 66:237–246.

6

Experimentation, Observation, and Inference in River and Watershed Investigations

MARY E. POWER, WILLIAM E. DIETRICH, & KATHLEEN O. SULLIVAN

Ecologists seek to understand the interactions between species and their heterogeneous, changing environments. This work is made more difficult by the fact that the underlying processes are often mediated through complex community- or ecosystem-level interactions. What tools do we have for unraveling this complexity?

Paine (1994) and Walters (1986, 1992) have reviewed the three fundamental methods available to field scientists: observation, modeling, and experimentation. All new information about nature is initially obtained through the first method: observation, which includes monitoring, mapping, and detecting correlations or other patterns. Observation alone, particularly when done in a hypothesis-free manner, has in the past disappointed ecologists by leaving them with messy data sets open to alternative interpretations. For this reason, ecologists in recent decades considered inferences from observations to be weak relative to inferences drawn from manipulative experiments (e.g., Connell 1974). We believe that field observations have been undervalued in contemporary community ecology, leaving ecologists poorly equipped to contribute to problems at large spatial scales at which manipulative experiments are infeasible. We will elaborate on this point later.

A second fundamental method for studying nature is modeling. We refer here to mechanistic modeling, either mathematical or qualitative, which attempts to portray key process that underlie phenomena of interest. The limitations of modeling are well known (e.g., Starfield and Bleloch 1986, Walters 1992). There is usually an assumption that the system's context is constant, and this is never true of real ecosystems. Also, modelers must assume that many (most) details are unimportant, but inevitably some omitted details will be important, probably more so than others chosen for representation in the model. Nevertheless, as Walters (1992) points out, modeling is unavoidable. If we have an idea about how our system works, we have a model of it. Therefore, he advises that we model openly, making assumptions explicit to ourselves and others.

The third method is experimentation, defined here in the narrow sense that community ecologists typically use. Experiments (sensu strictu, Paine 1994) involve study of replicated sample units which are subject to at least two treatments. One treatment is a control intended to represent the unmanipulated or background condition. In the other treatment(s), one or more factors are altered by the experimentalist, and their influence is evaluated by comparing responses of manipulated to control treatments. Replicated experiments cannot be performed in many situations, either because of logistical constraints (Matson and Carpenter 1990) or because adequate controls do not exist (see discussion of scale issues following).

Combined or Nested Approaches

While much has been written about the greater power or rigor of experimental over observational approaches (e.g., Paine 1977, Underwood 1990), it is usually more powerful to combine these approaches in a nested fashion (Fig. 6-1; see also Frost et al. 1988:252, Carpenter 1996). How experiments, observations, and modeling are combined depends on the scale of the study and the question addressed. A question that has served as an extremely productive opening gambit for community ecologists as they first explore a system has been: "What would happen if . . . ?" (Fig. 6-1a): "What would happen if I alter the density of species A or change factor B?" motivating what Art Dunham has called "kick it and see" experiments. This is the

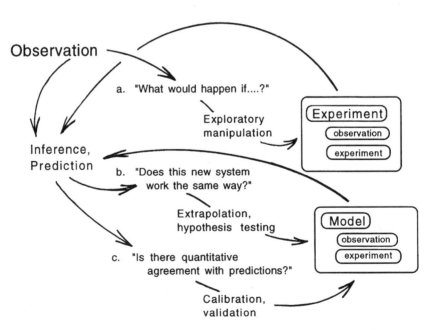

Figure 6-1. Nested experimental, observational, and modeling approaches to ask three types of questions.

approach that has revealed important surprises, such as keystone species (Paine 1966, 1969).

When experiments like these are done in intertidal systems on exposed rocky shores, as were the experimental removals of the starfish *Pisaster* that led to the original formulation of the keystone concept (Paine 1966, 1969), direct observations of underlying processes are difficult or impossible. Intertidal "action" (grazing, predation, growth, settlement, export, and reproduction) typically happens under conditions inconvenient for human observers (e.g., crashing surfs on moonless nights). Inferences must be drawn from periodic observations (typically at roughly monthly intervals) of changes in the states of assemblages. These interpretations are bolstered by knowledge of the local biology and the physical environment. When investigators are not able to observe the underlying processes in action, however, uncertainty may arise as to which components of the excluded biota were responsible for treatment effects (e.g., Edwards et al. 1980, Menge 1980, Underwood and Fairweather 1986) or whether alteration of consumer densities or behavior (Menge and Sutherland 1976, Menge 1980) or unintended habitat modifications (Virnstein 1978, Dayton and Oliver 1980, Hulberg and Oliver 1980) have caused or contributed to changes. Consequently, questions and controversies over the interpretations of experimental results persist.

Direct observations, when possible, can illuminate experimental black boxes, reducing the danger of misinterpreting experimental results. They are no panacea, given the problem of witnessing, let alone sampling, rare events with high impacts. But even in what would seem unlikely arenas, such as bottle experiments with microorganisms, direct observations have illuminated causality. Gause (1934) directly observed the spatial separation of two competing *Paramecium* species and their food resources (suspended bacteria vs. deposited yeast cells) and deduced his famous principle that this separation contributed to their coexistence. More recently, Balciūnas and Lawler (1995) used direct microscopic observation to detect an escape in size by a prey protozoan, *Colpidium*, from its predator *Euplotes*, which occurred when nutrient levels were increased in bottle experiments. Their observation uncovered the mechanism by which nutrient addition blocked top-down food chain control and shortened the length of the functionally important food chain, in contrast to previous predictions from simple food chain considerations (e.g., Lindeman 1942, Fretwell 1977, Oksanen et al. 1981). Clearly, direct observations can lead ecologists to both propose ecological generalizations and question them.

When we know more about a system, we can work within the framework of models, which are hypotheses about how the system works (Fig. 6-1b,c). For example, we can attempt to extrapolate. We might observe that a different system shares features with one that has been partially understood and ask whether the new system works the same way. We may instead be interested in whether the previously studied system will continue to work the same way under new conditions and whether our understanding is robust beyond the circumstances in which it was first attained. When we have a model of how the system works in mind, we can nest both observations and experiments within this model to test it (Fig. 6-1b). If a model has been developed to the point of making quantitative predictions, we can also use nested experiments and observations to calibrate it (Fig. 6-1c) and eventually to validate it (to test the match between predictions of a fully calibrated model and observations from nature). Nested experimental manipulations may be needed if parameters require calibration under a range of partially controlled conditions.

We will illustrate these three types of nested approaches with case histories drawn from river food web investigations and then discuss constraints on the application of these methods as the spatial extent of the system under study increases.

Nested Experimental and Observational Studies of River Food Webs

"What happens if . . . ?": The Eel River of Northern California

In the summer of 1989, Power experimentally manipulated fish in the South Fork Eel River to ask: "What happens if . . . the two most common species are excluded?" Enclosures and exclosures were distributed over a 1-km reach within the forested watershed of the Angelo Coast Range Reserve (formerly the Northern California Coast Range Preserve). Only two fishes—juvenile steelhead (*Oncorhynchus mykiss*) and California roach (*Hesperoleucas symmetricus*)—are common after winters with normal scouring floods. Surprisingly dramatic differences arose between fish enclosures and exclosures 5 to 6 weeks after the onset of the experiments. In the presence of fish, the dominant alga, *Cladophora*, which grew as 40 to 60–cm high turfs attached to boulder and bedrock at the start of the experiment, had collapsed down to a prostrate webbed mat no more than 1 to 2 cm high. The algae remained erect in the fish exclosures and became overgrown with nitrogen-fixing bluegreens and diatoms (Power 1990). Densities of benthic insects differed markedly between treatments as well. The fish enclosures were heavily infested with midges, *Pseudochironomus richardsoni*, that lived within the algae and wove it into tufts around its body. Heavy infestations of tuft-weaving midges collapsed the algal mats and produced a webbed and knotted architecture. This occurred several weeks later in the open river. Tuft-weaving midges were rare in the fish exclosures, where large numbers of small predators (lestid nymphs, sialid larvae, and young-of-the-year roach and stickleback) had recruited. These small predators were rare in the open channel and in fish enclosures but recruited in large numbers where larger fish were excluded and apparently suppressed the midge.

Power tested this last inference with a nested experiment and direct observations. She stocked 24 screened (1-mm mesh) buckets with ca. 7 g of cleaned *Cladophora* (picked free under 10× magnification of conspicuous macroinvertebrates). Six buckets received four roach fry, six received four stickleback fry, six received four lestid nymphs, and six were left as predator-free controls. After 20 days, the predator-free controls had been colonized by four times more midges than had any of the predator treatments (Power 1990). This nested experiment supported the interpretation that it was the guild of small predators that had, in fact, suppressed the recruitment of tuft-weaving midges to fish exclosures. Subsequent direct observations of feeding by larger fish and invertebrate predators revealed that the common predatory invertebrates (lestid damselflies, aeshnid dragonflies, and naucorid bugs) all are able to detect midges and extract them from their algal tufts. The odonates, after watching tufts for several minutes, shot their mouthparts in and extracted midges with a "surgical strike." The naucorid bugs probed cocoons with their beaks until they encountered the midge. In contrast, the larger fish in the Eel River did not seem able to detect chironomids within their algal tufts, although when midges were extracted and exposed these fish ate them readily (Power et al. 1992). These behavioral observations documented the predator-specific vulnerability of the prey, which was the mechanism that produced four functionally significant trophic levels in the Eel River. Herbivorous mayflies were the dominant prey in the guts of the larger fishes (Power et al. 1992). Observations of gut contents alone would have suggested that fish should exert control from the third trophic level and enhance, rather than suppress, plant biomass. Clearly, a combination of experimental and observational results produced a better understanding of food web interactions than either approach would alone, but at this early stage of investigating a poorly understood system experiments were particularly critical.

"Can we extrapolate?" The Eel River during Drought and Brier Creek, Oklahoma

Power and colleagues repeated these fish manipulations during the summers of 1990 and 1991, doubling the enclosure numbers from 12 to 24 and expanding the design to study the separate as well as the combined effects of roach and steelhead. In contrast to the 1989 results, however, fish had no functionally significant impacts on algae in either 1990 or 1991. In both the presence and the absence of fish, algae collapsed down to detritus within the first weeks of the experiment. A multiyear drought had begun in 1990, and in the absence of scouring floods large numbers of armored and sessile invertebrate grazers, invulnerable to most predators in the river, survived over the winter. When *Cladophora* began to grow in the late spring, these grazers quickly nibbled it back. These natural history observations finally motivated the definitive experiment, which was a 2 × 2 factorial manipulation of steelhead and the dominant armored caddisfly, *Dicosmoecus gilvipes*. The results showed that *Dicosmoecus*, not fish, controlled algal biomass during drought. Steelhead still had a statistically significant negative effect on algae (suggesting they were still at the fourth trophic level), but their effect was small in comparison to the two-level impact of the predator resistant grazers (Wootton and Power, unpublished data). Cross-watershed surveys of algae and invertebrates in two regulated channels with artificially stabilized flow and four unregulated rivers that all scoured in 1989 were also consistent with the inference that scouring floods reset primary consumers to earlier successional stages that are more vulnerable to predation and set the stage for trophic cascades that affect algal biomass (Power 1992).

The Eel River food web obtained under "normal" Mediterranean winter-flood, summer-drought conditions did not extrapolate to the same system during drought (Power 1995) or to regulated channels in the region that had been subject to anthropogenic "disturbance removal experiments." A food chain that had four functional trophic levels with respect to impacts of predators mediated through consumers on plants collapsed to a drought food chain with two functional trophic levels (Fig. 6-2), despite the fact that all the key species were still represented in the community.

Extrapolation is useful even when it fails, because expectations that one system may resemble another "prepare the mind" to make focused observations that either support

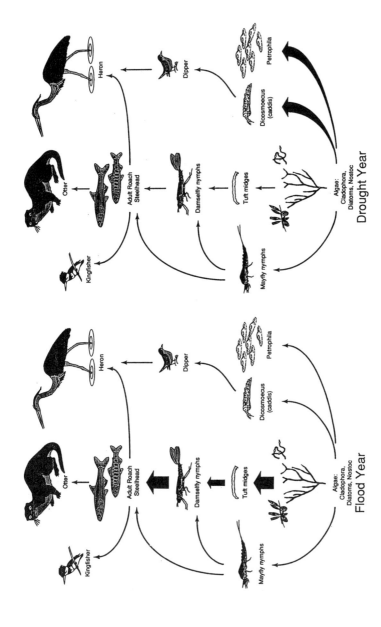

Figure 6-2. Functional food chain length in the Eel River during normal winter-flood, summer-drought regimes (*left*) and during year-round drought (*right*).

or refute the expectations. In contrast to the attempt to extrapolate from flood to drought years for the Eel River web, properties of a food web in an Oklahoma prairie stream were predictable by analogy with subtidal food webs in the northeastern Pacific, where sea otters suppress sea urchins and indirectly maintain kelp forests (e.g., Estes and Palmisano 1974). In Brier Creek, Oklahoma, some pools were filled with filamentous green algae, while adjacent pools were nearly barren. Observations of the distributions of predators and grazers quickly confirmed expectations from the sea otter–urchin–kelp model: piscivorous bass occurred in the green pools and were absent from the barren pools, where schools of grazing minnows (*Campostoma anomalum*) occurred (Power and Matthews 1983). Subsequent experimental transfers of bass and *Campostoma* among stream pools demonstrated that a trophic cascade did underlie the complementarity of bass, *Campostoma*, and algae. We electroshocked bass out of a green pool, split it down the middle, and added *Campostoma* to one side; within 5 weeks, algae on that side were grazed down to a barren state, while the control side without the grazing minnows remained green (Power et al. 1985). Clearly, herbivory accounted for the barren condition of pools with *Campostoma*. It was not obvious, however, whether the mechanism for the complementarity between bass and their minnow prey was predation or predator avoidance because, unlike algae, minnows have potential escape behavior.

We resolved this question about causality with direct observations. We added bass to a pool with a school of *Campostoma* but, before doing so, fenced off the upstream and downstream ends of adjacent pools, which were linked to the *Campostoma* pool by riffles which minnows could cross but which were too shallow for bass passage. These fenced areas served as potential ''escape ports'' for *Campostoma*. We also gridded the substrate of the *Campostoma* pool and, before bass addition, made behavioral observations of space use by adults and, during a spring repetition of the experiment, by young-of-the-year minnows. The adult fish tended to graze the deepest parts of the pool, with the young in slightly shallower water. After bass addition, both size classes moved into shallower water, which accounted for transient dynamics in the spatial distribution of the algae, which initially declined in these shallow areas (Power et al. 1985, Power 1987). Over the next 5 or 6 weeks, however, the entire pool became overgrown with green algae. Predator avoidance was an important contributing factor: in the spring experiment, we found 40 of the initial 74 adult minnows in the upstream escape port just 1 week after bass had been added. Whether bass convert *Campostoma* to bass meat or simply rearrange them spatially is of long-term significance to ecosystem dynamics. The partitioning of this causality required direct observation, within the context of a manipulative experiment.

Model Calibration and Validation—The Rio Frijoles of Central Panama

In the Rio Frijoles of central Panama, four species of armored catfish are the dominant algal grazers. Their algal foods renew faster in sunny stream pools than in dark pools. Power (1984) used an experimental manipulation to quantify this difference. She placed groups of unglazed clay tiles, with texture similar to that of the natural bedrock substrate, on pegs that elevated them above the streambed—in this position, they were not grazed, and visible standing crops of attached algae accrued over periods of 16 days. Harvesting these standing crops revealed that organic matter (mostly attached algae) accrued about seven times faster in two moderately sunny pools (25 to 50% open canopy) than in two dark pools (<10% open canopy). This difference corresponded quantitatively to the densities of armored catfish, which were about six to seven times denser in moderately sunny than in dark streams. Snorkeling censuses done of 16 stream pools over 12 consecutive months showed a consistent correlation of armored catfish densities (individuals per area grazeable substrate) with canopy and hence with primary productivity in the light-limited stream. These censuses suggested that fish tracked the productivity of their food, but the correspondence could have been misleading, as fish could have grazed outside pools where they were counted when not being observed. To resolve this issue, Power made direct behavioral observations (scan samples for density estimates and focal animal sampling for per capita rates [Altmann 1974]) of armored catfish feeding by day and night in two dark and two moderately sunny pools. These data were combined to compute an estimate of collective grazing pressure: the average return time, by any grazer, to a given small site on the substrate. These estimated return times varied from 9 to 10 hours in the two sunny pools and from 30 to 100 hours in the two dark pools. Multiplying return times by the algal accrual rates measured in each pool gave the standing crop of algae predicted to exist on a site at the time it was about to be regrazed. Estimates for this computed standing crop in the two dark pools bracketed the values estimated for all four pools, suggesting that collective grazing pressure was, as predicted, balanced with local algal growth rate so that standing crops of algae (food availability) were similar in dark, uncrowded pools and sunny, more crowded pools (Power 1984).

The Ideal Free Distribution model (Fretwell and Lucas 1970) predicts that if animals are ideal (able to evaluate the relative quality of habitats in their environment) and free (to settle in the best available habitat at any time), they should distribute themselves so that the fitnesses of individuals in crowded habitats of intrinsic high quality are similar to fitnesses of inhabitants of poorer but less crowded habitats. Data collected by following 1,308 individually marked armored catfish over 3 km of river during a 2.5-year period supported these predictions quantitatively. Growth rates of prereproductive *Ancistrus spinosus* (the most common species in stream pools) were similar among dark, half-shaded, and sunny pools in the dry and in the rainy seasons, and survivorship of all species was indistinguishable among these dark, sunny, and half-shaded pools (Power 1984).

These data on key components of fitness provided extremely strong quantitative support for Fretwell's Ideal Free Distribution model, for which field corroboration is still thought to be largely lacking (Kacelnik et al. 1992). Note that the bulk of this evidence was observational, with experiments nested within comparative observations to calibrate algal productivity in the absence of grazing. Along with other authors in this volume, we have had difficulty publishing observations, even when these provide evidence or contexts crucial to interpretations of results. In our opinion, there has been a bias in the culture of ecological publication that overvalues manipulative experiments and undervalues field observations. We consider this bias unfortunate, as it has left community ecologists relatively unprepared to study longer term, larger scale problems for which experimental approaches are not feasible.

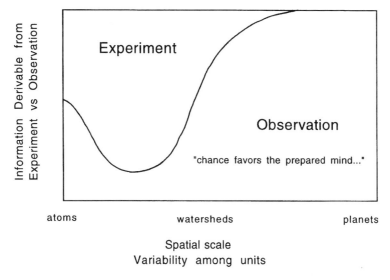

Figure 6-3. Information derivable from experiment versus observation as a function of the spatial scale, or variability among units, of the study system.

Utility of Experiments versus Observations

The amount of information that can be derived from experiment versus observation clearly depends on the spatial scale, as well as the variability among units of a study system (Fig. 6-3). As objects of study grow from single entities to systems that encompass many interacting components, the importance of experimentation grows. Experiments are the fastest way to learn about the workings of complex dynamic systems like ecological communities and, in some cases, may give us insights we cannot otherwise obtain. We should avoid the temptation, however, to do experiments that are too small or too short-term to manipulate the relevant processes. For example, consider large cattle exclosures in an arid landscape. If one were to observe grassland conversion to shrubland inside as well as outside these exclosures, one might infer that climate, not cattle, caused the conversion. This conclusion could be wrong if, for example, cattle trampling and destruction of vegetation had caused channel incision, which, in turn, lowered the water table and facilitated the invasion of xerophytic shrubs (Odion et al. 1988, Elmore and Kauffman 1994, Dudley et al. unpublished ms). Severe channel incision and water table lowering might not reverse inside cattle exclosures, even if these were several hectares in area and several decades in age. Historical study of geomorphic change is a crucial foundation for landscape-scale hypothesis testing.

In general, as systems get larger, the role of manipulative experiments must decrease, for two reasons. One is the well-known limitation by logistical constraints—resources are stretched thinner and thinner to study fewer replicates of larger units until these resources are exhausted. A more fundamental problem, however, is that as systems increase in scale it becomes hard and eventually impossible to find suitable replicates.

Valid controls simply become unavailable. It is clear that this is the case for planets: neither Venus nor Mars is an adequate control for Earth, so scientists who study global change are correct to confine their major efforts to observation and modeling (although experiments nested within observational studies may be useful for calibrating modeled process rates, e.g., effects of temperature or CO_2 on rates of photosynthesis or decomposition in particular biomes). As one scales down in size (moving from right to left across Fig. 6-3), where one crosses into the region where experiments are feasible as the primary, overarching approach to a problem (Fig. 6-1a) can be debated. We think this threshold occurs at or near the scale of natural watersheds. This is not to say that excellent, informative watershed-scale experiments do not exist—they clearly do (Likens et al. 1970, 1977). The reason that the Hubbard Brook experiments have been so valuable, however, is that they were preceded and followed by years of detailed observations. Hubbard Brook scientists know a great deal about chemical fluxes in runoff and groundwater following deforestation and during succession. They have studied changes in biological populations, as well as in physical and chemical processes that mediated energy flow and material cycling during the ecosystem's response and recovery period. The follow-up of the basic manipulative experiment with careful, detailed, well-planned, and prolonged observation underlies the great value of this large-scale, long-term project.

Watershed experiments have traditionally been done using a "paired basin" approach in which one or more basins do not receive treatment as others are manipulated. These paired watershed experiments suffer at least three types of problems (Reid et al. 1981). First, independent (treatment) variables are often only loosely characterized with qualitative designations (e.g., "managed vs. unmanaged" or "logged vs. unlogged") that are inadequate for assessing mechanisms. "Control" treatments are usually not "pristine" (areas may have been logged in the past; controls may have active roads). Even if control and treatment watersheds have similar general characteristics (aspect, slope, forest type, drainage density, and geological parent material), they may differ in subtle but important respects (e.g., structural orientation of bedrock, undetected ancient landslides, and disease history of vegetation). A second problem is that rather than making local, process-based observations, investigators have measured highly integrated response variables. Black-boxed signals, such as total sediment yield at the mouths of watersheds or changes in salmon escapement back to watersheds over the experimental period, are noisy and give little insight about causality, particularly when records are short (less than decades in duration).

A third problem in paired watershed studies mentioned by Reid et al. (1981) relates to spatial scale. Watersheds as study units are sufficiently large so that there is a reasonable probability that they will sample rare events with high impacts during an experiment. When, for example, major landslides occur in the "wrong" (control) basin, they can completely override signals from land use that experiments were set up to detect. In addition, lingering but undocumented effects of divergent landslide, fire, or land use histories may influence watershed responses to experimental manipulations in ways that may not be detected (e.g., via the amount of sediment available to be delivered to streams; Judy Meyer, personal communication).

A final problem is that it is now too late for this experimental design. In most parts of the world, comparable "unimpacted" control sites no longer exist for large-scale watershed studies. In the Pacific northwestern United States, for example, there are no large undisturbed forested watersheds left for comparison with harvested areas. One might as well search the solar system for a control for planet Earth.

Observations and the Prepared Mind: The Reference State

Given that there are important ecological problems that cannot be studied with manipulative experiments with controls, can we make observational studies more useful? One potentially useful approach has a long history in geology and is presently being promoted in ecology by Paine (1984, 1994). As background, recall that Connell (1974, 1975) pointed out that field experiments differ from laboratory experiments. In the laboratory, most factors are held constant and one or a few are manipulated in experimental treatments. In the field, a few factors (those being tested) are varied and their effects are then evaluated against a noisy natural background ("controls"). Therefore, only strong signals can generally be detected. Paine (1984, 1994, personal communication) has proposed evaluating noisy nature relative to an experimentally engineered, simplified reference state. This reference state would arise if only well-understood processes are operating. Therefore, it will occur rarely, if ever, in the real world. Once it is defined, however, the more poorly understood processes that complicate natural systems can be studied by evaluating the deviations they produce from the reference state.

This terminology leaves room for confusion, as the term *reference* has also been used by ecosystem scientists to refer to what community ecologists would call controls. For example, unmanipulated "reference lakes" (Schindler 1988, 1990; Carpenter and Kitchell 1993) or "reference streams" (Wallace et al. 1996) are followed to detect regional factors that, independent of the manipulation, may cause changes in response variables. In this sense, references are like Connell's controls in field experiments: their dynamics are not necessarily understood but reflect a noisy background against which we try to detect the impact of one or a few manipulated factors. We would like to distinguish these "background" references from two reference states that are simplified relative to nature: "manipulated" and "analytical" reference states. Manipulated reference states are portions of the real world that have been experimentally engineered so as to remove complicating factors. Analytical reference states are calculated expectations, derived from theoretical or empirical understanding of processes known to affect systems. Both manipulated and analytical reference states reflect conditions we would expect if only well-understood processes are operating. As illustrated in the following discussion, these reference states may be far from any natural (preimpact) background condition.

We offer three examples: (1) Paine's use of manipulated reference states to study coralline algal interactions on Tatoosh Island of the Washington State coast, (2) the use of an analytical reference state that predicts riverbed sediment size to judge effects of sediment supply changes and channel manipulations, and (3) ongoing efforts to understand the interaction of trophic dynamics and disturbance-succession regimes in rivers by examining, as manipulated reference states, trophic structures that develop in the absence of scouring floods, in channels with artificially regulated flow.

Coralline Algal Interactions on Tatoosh Island

On exposed rocky coasts of the northwestern United States, crustose coralline algae compete for space. Paine (1984, 1994) has experimentally studied this assemblage on Tatoosh Island for over 10 years. On an exposed rock bench from which he has continuously removed most grazers he has transplanted chips of various coralline species into competitive arenas made of nontoxic epoxy putty. Their growth and overgrowth have revealed a very deterministic competitive hierarchy. When the corallines grow into contact with each other, some species are better than others at lifting up their growing edges and overgrowing neighbors. These species tend to win in space competition. Grazing or physical disturbance can undermine the advantage of this trait, however, because herbivory or damage is often disproportionately high on these uplifted edges. In the presence of grazer or physical disturbance (e.g., log bashing), the outcome of competition among the algae is far less predictable. Therefore, the assemblage engineered by the elimination of grazers and disturbance, in which community structure results primarily from competition for space, becomes Paine's manipulated reference state, against which to evaluate the deviations produced by herbivory and other complicating biotic and abiotic factors at work in the natural world (Paine 1984, 1994).

The Threshold Channel Concept in a Watershed Context

In hilly and mountainous areas, riverbeds are typically gravel of mixed sizes, organized into fixed or slowly moving bars which, along with woody debris, create diverse habitat structure. Grain size influences the availability of spawning substrate for adult fish and of habitat and refuges for young fish and aquatic invertebrates (Brusven and Rose 1981, Minshall 1984, Kondolf et al. 1991). Bed movement influences food web structure (Power 1992, 1995) as well as spawning success of fish like salmonids (e.g., Kondolf and Wolman 1993). Habitats are generally degraded by land use that alters the supply to channels of coarse and fine sediment, alters the flow regime, removes the woody debris, or channelizes the river. Considerable effort is now under way to do "snapshot" analyses of river condition to infer effects of land use and possible benefits of land use prescriptions. The analytic reference state may prove useful in this context.

Field, laboratory, and theoretical studies show that for natural size mixtures of sediments found in gravel-bedded rivers significant bed mobility typically begins when stream forces exceed the resistance to motion of the median grain size of the bed (e.g., Leopold et al. 1964, Carling 1988, Parker 1990). When gravels of mixed sizes cooccur, the shear stresses needed to move small and large grains are, respectively, larger and smaller than on homogeneous beds, because the small grains lift large ones out of pockets into the flow, while large grains tend to shield the small grains from flow. Therefore, shear stresses that initiate motion of large grains and those that initiate motions of small grains both approach those needed to move particles of the median size. This suggests that gravel rivers, to a first approximation, are "threshold channels" (Henderson 1966), in which the threshold for initiating bed movement is crossed at some characteristic flow. Many studies have shown that this characteristic flow is typically close to bankfull (e.g., Jackson and Bestcha 1982; Andrews 1983, 1984; Carling 1988). The critical boundary shear stress that will initiate motion of the expected median grain size can be reasonably estimated based on the particle's diameter and specific gravity. Because significant bed motion in gravel-bedded rivers does not begin until flows approach or exceed bankfull, bankfull boundary shear stress can be used to calculate the median grain size of the riverbed. Over a long reach where we can consider the flow on average to be steady and uniform, the boundary shear stress at bankfull stage is just the product of the river slope, bankfull depth, fluid density, and gravity. Because of additional resistance due to drag over bars and woody debris, however, the boundary shear stress actually responsible for grain motion may be much less than the total available at bankfull, but these additional sources of resistance are difficult to estimate (Buffington 1995).

A simple analytical reference state emerges from this description. If we know the bankfull depth and river slope, we can calculate the expected median grain size of the bed. Deviations from this expected grain size arise from effects of woody debris and bar resistance (Buffington 1995) and from effects of sediment supply (Fig. 6-4; Dietrich et al. 1989, Buffington 1995). Form drag resistance from woody debris and bars will reduce the actual grain size from that expected from the depth–slope product alone. High gravel supply will cause the median grain size of the bed to become smaller, reducing the critical boundary shear stress and increasing bed mobility. Large woody debris in channels is common in less disturbed forested basins. This example illustrates the contrast between background and analytical reference states that might be applied, for example, to study effects of timber harvest. The background reference state would be a wood-choked channel before humans removed snags; the analytical reference state would be a bare channel, in which sediment transport and supply could be more easily predicted.

This threshold channel as an analytical reference state can serve as a null hypothesis against which to compare field observations and as a guide to interpretations of channel condition. Deviations from the median sediment size predicted by the analytical reference state would point to hypotheses about the influence of either sediment supply or

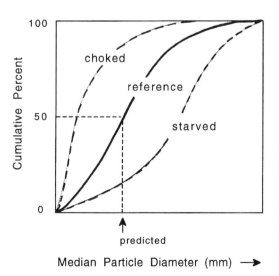

Figure 6-4. Use of an analytical reference state for evaluating sediment supplies to stream channels with mixed-sized gravel beds. Observations of median grain size diameter suggest that sediment supply to streams is higher (choked) or lower (starved) than would be expected for the reference state.

obstructions. Either hypothesis could be tested with additional field observations and nested experiments (e.g., small-scale removal or addition of wood and improvement of roads). For example, a reach of channel which is free of woody debris and which has a median grain size much less than that predicted should have high bed mobility. In order to maintain this high mobility, sediment supply to the channel from the watershed must be high; hence, deviation from the reference state may point upstream to land use effects (roads and forest harvest). Where supply has been cut off (e.g., below a dam), the bed grain size will coarsen until there is no movement at bankfull or higher stages. This may lead to elimination of spawning gravels and greatly reduce flood scour, which serves the role of resetting the stream to earlier successional biological stages in which prey are more vulnerable to predators (e.g., Power 1992, 1995). The addition of large woody debris on an otherwise naturally low-supply coarse gravel bed may cause bed material to reduce to a size more favorable for spawning (Buffington 1995), enhancing salmonid habitat in steep, bouldery channels, which generally lack both spawning gravel and well-formed pools (i.e., Montgomery and Buffington 1993).

Separating Disturbance and Succession from Trophic Dynamics in California Rivers

Our last example is a work in progress. We are attempting to apply the manipulated reference state approach to understanding the interactions of disturbance and succession with trophic dynamics in California rivers. As described in a previous section (on Eel River during drought), the length of functional food chains in the Eel River depends on whether or not scouring winter floods occurred during the previous high-flow season (Fig. 6-2). We are exploring the possibility of using artificially stabilized river channels as reference states against which to evaluate the effects of floods on river food webs.

In the winter of 1993, the South Fork Eel River finally experienced bed-scouring floods once again. In addition to the winter floods, however, the river received an anomalously late spring flood in June, which exported most of the algae, which was blooming at the time. Field experiments during the following summer revealed that effects of juvenile steelhead on algae were not negative but positive, as if steelhead were at the third rather than the fourth trophic level. With the removal of *Cladophora* by the June flood, the tuft-weaving midge did not recruit in large numbers during the experiments. Juvenile steelhead in enclosures consumed small predators, as in 1989, but also consumed all the remaining herbivores capable of suppressing algae. Therefore, in the presence of steelhead during 1993 blooms of diatoms overgrew enclosures, and algal standing crops were higher with these fish than without them. These results suggest that the interplay of disturbance, succession, and trophic dynamics in the Eel River is influenced by the timing as well as the annual occurrence of scouring floods.

Channels in which flow is artificially regulated may be useful for unraveling these interactions. In both channels downstream from dams and a regulated diversion that is not fed by upstream impounded water, preliminary observations suggest that biomass tends to be dominated by predator-resistant sessile or armored grazers. A fairly constant low-standing crop of attached algae is maintained, and biomass of vulnerable (mobile naked) grazers and predators is typically low (Power 1992, Parker and Power, unpublished data). We postulate that this state represents the late successional biomass distribution pattern of disturbance-free systems. This may prove useful as a manipulated reference state against which to evaluate the more indeterminate structures that arise when natural disturbance resets river communities.

This proposed application of a manipulated reference state is quite preliminary compared to the first two examples described here. Paine and Dietrich et al. both have solid empirical and experimental underpinnings for their reference states, based, respectively, on 10 years of field experimentation (Paine 1984, 1994) and on many decades of flume and field studies (Leopold et al. 1964, Henderson 1966, Parker 1978, Carling 1988). In addition, the first two reference states are simpler than the third proposed here. The threshold channel is the product of physical processes that are far simpler than ecological processes. Paine's reference state is ecological, but interactions are restricted to competition for an easily measured resource (space) among organisms that are sessile for most of their life histories (coralline algae). Our proposed third reference state (trophic-level biomass distributions in channels that do not experience scour) includes interactions of mobile higher trophic levels and, in fact, partially corresponds to Paine's less deterministic "natural state" in that consumers, but not disturbance, have been factored back into the system. Clearly, real food webs in streams, even in the absence of flood scour, do not maintain static distributions of trophic-level biomass as portrayed in Figure 6-5. These patterns can be disrupted by a variety of factors. Synchronized pupation or emergence of grazing insects may temporarily release algae. Formerly invulnerable primary consumers may come under attack if new types of predators invade or if epidemics break out (Kohler and Wiley 1992). Changes in physical and chemical factors other than scour will have effects. Would we be quicker to recognize the influence of these other factors if we searched for deviations from an expected state? The value of a reference state does not lie in how likely it is to occur in nature but in whether it prepares the mind to be surprised, triggering the pursuit of profitable new leads. There does remain an issue of how much uncertainty must be removed by a model like a reference state before it proves useful. The decision about how "well understood" processes need to be before they are used to formulate a reference state expectation is the investigator's choice, based on knowledge of the system, practical considerations, and the degree of uncertainty that can be tolerated when addressing particular issues or questions. We argue that manipulated or analytical reference states can be useful even at the onset of investigations of poorly understood systems, as they force fieldworkers to make, as Darwin recommended, observations that are "for or against some view" (Darwin [1861] in a letter to Henry Fawcett, cited in Gould 1995: 148). When systems are poorly understood and preliminary reference states are far from the mark, mental suppleness (rapid feedback between observations, model testing and revision, and new observations) is particularly important.

Ecologists have come to emphasize a posteriori interpretations of manipulative experiments because these have been more successful than a priori predictions in dealing with our dauntingly complex subject. We may never be able to predict ecological phenomena, but our postdictions will certainly be more timely for the effort we invest in trying to do so. We need to build up our a priori skills, like developing reference state expectations for field observations, if we are to contribute to the urgent problems

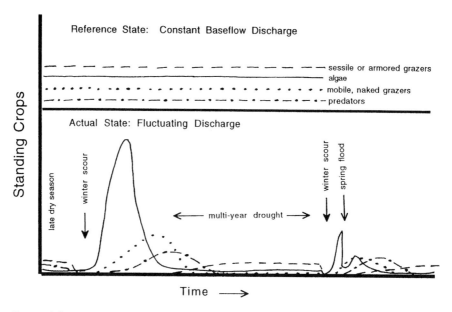

Figure 6-5. Manipulated reference state proposed for evaluating the effects of floods and flood timing on trophic structure and dynamics in western U.S. rivers. Patterns shown here are qualitative simplifications, partially derived from field observations, partially hypothesized. In the upper panel, the reference state expectation in channels that are not subject to periodic bed scour is shown. Trophic-level biomass is largely made up of predator-resistant sessile or armored grazers, which by persistent grazing maintain a constant low-standing crop of algae. Biomasses of vulnerable grazers and predators are relatively low in such channels (Power 1992, Parker and Power, unpublished data). The lower panel suggests trophic biomass changes following floods in a natural channel, represented by the South Fork Eel River. Late in the dry season, a biomass pattern similar to the reference state develops. After scouring winter floods (November 1988) remove most benthic biomass (and dilute water column biomass), algae recovers first during the following spring. Over the following low-flow season and during drought years that follow, a biomass distribution approaching the reference state develops. Floods in January 1993 reset the community, releasing algae initially. This algae was exported during a June 1993 flood, and the consequent recruitment failure of tuft-weaving midges set the stage for a three-trophic-level system, in which recovered algal biomass was protected by steelhead predation on the remaining functionally significant algivores.

of environmental management and biodiversity conservation that arise over spatial scales too large and temporal scales too short to permit experimental study as the primary approach.

Summary

Experimental approaches have enjoyed justified popularity in community ecology, so much so that they have overshadowed direct observations. Well-designed field experiments will continue to humble ecologists by revealing surprises about how nature works. We argue here, however, that more direct observations should be planned, made, and reported in the literature. Direct observations nested within manipulative experiments illuminate black boxes and can resolve causality. In addition, ecologists should improve their skills at making hypothesis-based field observations that allow them to contribute to problems which cannot be addressed experimentally. These observations should be designed as carefully as field experiments. As the spatial scale of investigation increases (e.g., from pools to reaches to watersheds of rivers), the value of planned observation relative to manipulative experimentation increases, because of logistic constraints on large-scale experimentation and, more fundamentally, because of the lack of comparable control sites. Watersheds for which adequate controls are lacking must be studied by observing and characterizing causal mechanisms and linkages within the watershed. Small-scale manipulative experiments nested within these observational studies can contribute to these efforts by clarifying functional relationships of key variables. Field observations that are motivated by testable hypotheses such as expectations from analytical reference states can, like manipulative experiments, surprise investigators, leading to new insights.

ACKNOWLEDGMENTS   We thank Steve Carpenter, Tom Dudley, Judy Meyer, Bob Paine, Michael Parker, Bill Resetarits, and Wayne Sousa for insightful comments and discussion, although they are not accused of agreeing with all of the points in this paper. MEP acknowledges support from the National Science Foundation (Grant DEB-93-19924 [Ecology]) and the Water Resources Center of California (Grant WRC-825).

Literature Cited

Altmann, J. 1974. Observational study of behavior: sampling methods. Behaviour 49:227–267.

Andrews, E. D. 1983. Entrainment of gravel from naturally sorted riverbed material. Geological Society of America Bulletin 94:1225–1231.

———. 1984. Bed-material entrainment and hydraulic geometry of gravel-bed rivers in Colorado. Geological Society of America Bulletin 95:371–378.

Balčiūnas, D., and S. P. Lawler. 1995. Effects of basal prey resources, predation, and alternative prey in microcosm food chains. Ecology 76:1327–1336.

Brusven, M. A., and S. T. Rose. 1981. Influence of substrate composition and suspended sediment on insect predation by the torrent sculpin, *Cottus rhotheus*. Canadian Journal of Fisheries and Aquatic Sciences 38:1444–1448.

Buffington, J. M. 1995. Effect of hydraulic roughness and sediment supply on surface textures in gravel-bedded rivers. M. S. thesis, University of Washington, Seattle.

Carling, P. A. 1988. The concept of dominant discharge applied to two gravel-bed streams in relation to channel stability thresholds. Earth Surface Processes and Landforms 13:355–367.

Carpenter, S. R. 1996. Microcosm experiments have limited relevance for community and ecosystem ecology. Ecology 77:677–680.

Carpenter, S. R., and J. F. Kitchell (eds.). 1993. The Trophic Cascade in Lakes. Cambridge University Press, Cambridge.

Connell, J. 1974. Ecology: field experiments in marine ecology. Pages 21–54 in Experimental Marine Biology. Academic Press, New York.

———. 1975. Some mechanisms producing structure in natural communities. Pages 460–490 in M. L. Cody and J. M. Diamond (eds.), Ecology and Evolution of Communities. Belknap, Cambridge, Massachusetts.

Dayton, P. K., and J. S. Oliver. 1980. An evaluation of experimental analyses of population and community patterns in benthic marine environments. Pages 93–120 in K. R. Tenore and B. C. Coull (eds.), Marine Benthic Dynamics. University of South Carolina Press, Columbia.

Dietrich, W. E., J. W. Kirchner, H. Ikeda, and F. Iseya. 1989. Sediment supply and the development of the coarse surface layer in gravel-bedded rivers. Nature 340:215–217.

Dudley, T. L., D. C. Odion, R. K. Knapp, K. R. Matthews, D. A. Sarr, and J. Owens. Unpublished ms. Livestock grazing impacts and the potential for riparian meadow recovery in the Golden Trout Wilderness Area, California.

Edwards, D. C., D. O. Conover, and F. Sutter. 1980. Mobile predators and the structure of marine intertidal communities. Ecology 63:1175–1180.

Elmore, W., and B. Kauffman. 1994. Riparian and watershed systems: degradation and restoration. Pages 212–231 in M. Vavra, W. A. Laycock, and R. D. Pieper (eds.), Ecological Implications of Livestock Herbivory in the West. Society for Range Management, Denver, Colorado.

Estes, J. A., and J. F. Palmisano. 1974. Sea otters: their role in structuring nearshore communities. Science 185:1058–1060.

Fisher, S. G. 1983. Succession in streams. Pages 7–27 in J. R. Barnes and G. W. Minshall (eds.), Stream Ecology: Application and Testing of General Ecological Theory. Plenum, New York.

Fisher, S. G., L. J. Gray, N. B. Grimm, and D. E. Busch. 1982. Temporal succession in a desert stream ecosystem following flash flooding. Ecological Monographs 52:93–110.

Fretwell, S. D. 1977. The regulation of plant communities by food chains exploiting them. Perspectives in Biology and Medicine 20:169–185.

Fretwell, S. D., and H. L. Lucas. 1970. On territorial behavior and other factors influencing habitat distribution in birds: I. Theoretical development. Acta Biotheoretica 19:16–36.

Frost, T. M., D. L. DeAngelis, S. M. Bartell, D. J. Hall, and S. H. Hurlbert. 1988. Scale in the design and interpretation of aquatic community research. Pages 229–258 in S. R. Carpenter (ed.), Complex Interactions in Lake Communities. Springer Verlag, New York.

Gause, G. F. 1934. The Struggle for Existence. Williams and Wilkins, Baltimore, Maryland.

Gould, S. J. 1995. Dinosaur in a Haystack. Harmony Books, New York.

Henderson, F. M. 1966. Open Channel Flow. Macmillan, New York.

Hulberg, L. W., and J. S. Oliver. 1980. Caging manipulations in marine soft-bottom communities: importance for animal interactions or sedimentary habitat modifications. Canadian Journal of Fisheries and Aquatic Sciences 37:1130–1139.

Jackson, W. L., and R. L. Beschta. 1982. A model of two-phase bedload transport in an Oregon coast range stream. Earth Surface Processes and Landforms 7:517–527.

Kacelnik, A., J. R. Krebs, and C. Bernstein. 1992. The ideal free distribution and predator-prey populations. Trends in Ecology and Evolution 7:50–55.

Kohler, S. L., and M. J. Wiley. 1992. Parasite-induced collapse of populations of a dominant grazer in Michigan streams. Oikos 65:443–449.

Kondolf, G. M., G. F. Cada, M. J. Sale, and T. Felando. 1991. Distribution and stability of potential salmonid spawning gravels in steep boulder-bed streams of the eastern Sierra Nevada. Transactions of the American Fisheries Society 120:177–186.

Kondolf, G. M., M. J. Sale, and M. G. Wolman. 1993. Modification of fluvial gravel size by spawning salmonids. Water Resources Research 29:2265–2274.

Lamberti, G. A., and V. H. Resh. 1983. Stream periphyton and insect herbivores: an experimental study of grazing by a caddisfly population. Ecology 64:1124–1135.

Leopold, L. B., M. G. Wolman, and J. P. Miller. 1964. Fluvial processes in geomorphology. W. H. Freeman, San Francisco.

Ligon, F. K., W. E. Dietrich, and W. J. Trush. 1995. Downstream ecological effects of dams. BioScience 45:183–192.

Likens, G. E., F. H. Bormann, N. M. Johnson, D. W. Fisher, and R. S. Pierce. 1970. Effects of forest cutting and herbicide treatment on nutrient budgets in the Hubbard Brook Watershed ecosystem. Ecological Monographs 40:23–47.

Likens, G. E., F. H. Bormann, R. S. Pierce, J. S. Eaton, and N. M. Johnson. 1977. Biogeochemistry of a Forested Ecosystem. Springer-Verlag, New York.

Lindemann, R. L. 1942. The trophic-dynamic aspect of ecology. Ecology 23:399–418.

Matson, P. A., and S. R. Carpenter (eds.). 1990. Statistical analysis of ecological response to large-scale perturbations. Ecology 71:2037–2068.

Menge, B. A. 1980. Reply to a comment by Edwards, Conover, and Sutter. Ecology 63:1181–1184.

Menge, B. A., and J. P. Sutherland. 1976. Species diversity gradients: synthesis of the roles of predation, competition, and temporal heterogeneity. American Naturalist 110:351–369.

Minshall, W. G. 1984. Aquatic insect-substratum relationships. Pages 358–400 in V. H. Resh and D. M. Rosenberg (eds.), The Ecology of Aquatic Insects. Praeger, New York.

Montgomery, D. R., and J. M. Buffington. 1993. Channel Classification, Prediction of Channels Response, and Assessment of Channel Condition. Washington State Department of Natural Resources Report TFW-SH10-93-002, Olympia.

Odion, D. C., T. L. Dudley, and C. M. D'Antonio. 1988. Cattle grazing in southeastern Sierran meadows: ecosystem change and prospects for recovery. Pages 277–292 in C. A. Hall and V. Doyle-Jones (eds.), Natural History of the White-Inyo Range, Symposium: Vol. 2. Plant Biology of Eastern California. Bishop, California.

Oksanen, L., S. D. Fretwell, J. Arruda, and P. Niemela. 1981. Exploitation ecosystems in gradients of primary productivity. American Naturalist 118:240–261.

Paine, R. T. 1966. Food web complexity and species diversity. American Naturalist 100:65–75.

———. 1969. A note on trophic complexity and community stability. American Naturalist 103:91–93.

———. 1977. Controlled manipulations in the marine intertidal zone, and their contributions to ecological theory. Pages 245–270 in Changing Scenes in Natural Sciences, 1776–1976. Special Publication no. 12, Academy of Natural Sciences of Philadelphia.

———. 1984. Ecological determinism in the competition for space. Ecology 65:1339–1348.

———. 1994. Marine Rocky Shores and Community Ecology: An Experimentalist's Perspective. Ecology Institute, Oldendorf/Luhe, Germany.

Parker, G. 1978. Self-formed straight rivers with equilibrium banks and mobile bed: Part 2. The gravel river. Journal of Fluid Mechanics 115:303–314.

———. 1990. Surface-based bedload transport relation for gravel rivers. Journal of Hydraulic Research 28:417–436.

Power, M. E. 1984. Habitat quality and the distribution of algae-grazing catfish in a Panamanian stream. Journal of Animal Ecology 53:357–374.

———. 1987. Predator avoidance by grazing fishes in temperate and tropical streams: importance of stream depth and prey size. Pages 333–351 in W. C. Kerfoot and A. Sih (eds.), Predation: Direct and Indirect Impacts on Aquatic Communities Hanover, New Hampshire, University Press of New England.

————. 1990. Effects of fish in river food webs. Science 250:811–814.

————. 1992. Hydrologic and trophic controls of seasonal algal blooms in northern California rivers. Archiv für Hydrobiologie 125:385–410.

————. 1995. Floods, food chains and ecosystem processes in rivers. Pages 52–60 in C. G. Jones and J. H. Lawton (eds.), Linking Species and Ecosystems. Chapman and Hall, New York.

Power, M. E., and W. J. Matthews 1983. Algae-grazing minnows (*Campostoma anomalum*), piscivorous bass (*Micropterus* spp.) and the distribution of attached algae in a prairie-margin stream. Oecologia 60:328–332.

Power, M. E., W. J. Matthews, and A. J. Stewart. 1985. Grazing minnows, piscivorous bass and stream algae: dynamics of a strong interaction. Ecology 66:448–456.

Power, M. E., J. C. Marks, and M. S. Parker. 1992. Community-level consequences of variation in prey vulnerability. Ecology 73:2218–2223.

Power, M. E., M. S. Parker, and J. T. Wootton. 1996. Disturbance and food chain length in rivers. Pages 286–297 in G. A. Polis and K. O. Winemiller (eds.), Food Webs: Integration of Patterns and Dynamics. Chapman and Hall, New York.

Reid, L. M., T. Dunne, and C. J. Cederholm. 1981. Application of sediment budget studies to the evaluation of logging road impact. Journal of Hydrology 20:49–62.

Schindler, D. W. 1988. Experimental studies of chemical stressors on whole lake ecosystems. Verhandlungen Internationale Vereinigung für Theoretische und Angewandte Limnologie 23:11–41.

————. 1990. Experimental perturbations of whole lakes as tests of hypotheses concerning ecosystem structure and function. Oikos 57:25–41.

Starfield, A. M., and A. L. Bleloch. 1986. Building Models for Conservation and Wildlife Management. Macmillan, New York.

Underwood, A. J. 1990. Experiments in ecology and management: their logics, functions and interpretations. Australian Journal of Ecology 15:365–389.

Underwood, A. J., and P. G. Fairweather. 1986. Intertidal communities: do they have different ecologies or different ecologists? Proceedings of the Ecological Society of Australia 14:7–16.

Virnstein, R. W. 1978. Predator caging experiments in soft sediments: caution advised. Pages 261–273 in L. A. J. Bulla and T. C. Cheng (eds.), Estuarine Interactions. Academic Press, New York.

Wallace, J. B., J. W. Grubaugh, and M. R. Whiles. 1996. Biotic indices and stream ecosystem processes: results from an experimental study. Ecological Applications 6:140–151.

Walters, C. J. 1986. Adaptive Management of Renewable Resources. Macmillan, New York.

————. 1992. Perspectives on adaptive policy design in fisheries management. Pages 249–260 in S. K. Jain and L. W. Botsford (eds.), Applied Population Biology. Kluwer Academic Publishers, Dordrecht.

7

From Cattle Tanks to Carolina Bays

The Utility of Model Systems for Understanding Natural Communities

WILLIAM J. RESETARITS JR. AND JOHN E. FAUTH

Experimental observations are only experience carefully planned in advance.

—Sir Ronald A. Fisher, 1935

You cannot step twice into the same river.

—Heraclitus, fl. 513 B.C.

A primary goal of ecological research is to identify generalities that can simplify the natural world from a jumbled collection of special cases to an ordered array of classifiable sets. This goal is shared by all ecologists and stems from the belief that complex ecological systems operate on a finite set of principles. Once these are understood, ecologists expect some level of predictive ability with regard to ecological phenomena. It has long been recognized that perhaps the most serious constraint in ecological research is the sheer number of factors affecting natural systems, coupled with the large number of unique natural communities we hope to understand. A parallel consideration is that sufficient resources will never be available to study every system on the planet. Thus, the search for generality and predictive power is not simply an abstract theoretical pursuit but an absolutely essential component of the ecological research paradigm. Ecologists use observation, experimentation, and deduction to generate predictive models that simplify the natural world and permit general statements about how it works. Only by obtaining a fundamental and general comprehension of the processes that shape natural systems can we hope to understand not only those systems that are intensively studied but also those innumerable systems that will never be studied. And only then can our basic understanding give rise to the informed and broadly applicable conservation decisions necessary to keep the world working.

As ecologists develop theories and models about how the world works, they also look for ways to rigorously test them. Experimentation is often referred to as a means of testing observed phenomena; often the "observed phenomena" are generated from theory rather than from observation or they are derived directly from previous experiments (Peckarsky, this volume). Fair tests of observed phenomena/theory require that null hypotheses can be rejected; this demands both adequate replication to obtain statistical power and sufficient control over experimental conditions to detect a signal (effect) against a background of noise (random variation). Achieving both these prerequisites is often difficult or impossible to achieve in nature (Lawton, this volume; Morin, this volume); we may be able to detect very strong signals in natural systems, but our ability to detect subtle yet potentially important effects in complex and noisy systems (Peckarsky, this volume) is often severely constrained by limits on replication and control. An alternative approach to manipulating natural communities is to re-create communities in artificial mesocosms that are under more complete control of the investigator. Model experimental systems, including mesocosms, allow a level of rigor and control either unobtainable or prohibitively expensive in natural systems; and because artificial mesocosms/microcosms are relatively inexpensive, they can be used to study a broad array of natural phenomena (Lawton, this volume; Morin, this volume).

While such model systems have proven their utility in testing ecological theory and exploring general ecological processes, their utility for understanding the ''real world'' of specific ecological systems has been questioned (e.g., Jaeger and Walls 1989, Petranka 1989, Carpenter 1996). The basic conundrum is that experiments that allow complete control of experimental conditions are inherently unnatural, while those conducted in nature vary in ways beyond the control of the experimenter (Diamond 1986; Hairston 1989a). The mesocosm approach combines rigorous control of experimental conditions with quasi-natural situations by using artificial environments (container communities) of intermediate size in which initial community composition and experimental conditions are controlled by the investigator but also may be exposed to natural photoperiods, temperatures, precipitation, and other factors. This approach is here exemplified by experiments in pond mesocosms designed to mimic natural temporary ponds (e.g., Morin 1981, 1983a; Wilbur 1984, 1987; Fauth and Resetarits 1991) and has been applied to a wide variety of systems (e.g., see literature cited in Hairston 1989a, Gurevitch et al. 1992, Lamberti and Steinman 1993, Rowe and Dunson 1995).

Temporary ponds themselves provide an ideal system for the study of ecological communities. Each time a pond refills, a new episode in community ecology begins: species colonize, a food web develops, trophic interactions become modified as predators and prey grow, and the community develops until the pond dries. This cycle can be repeated several times a year or just once in many years, depending on the morphology and hydrology of the pond and the vagaries of precipitation. Combined with the high productivity and diversity of ephemeral ponds, this temporal repetition provides a unique opportunity to explore the rules of community assembly, particularly for taxa amenable to experimentation, such as amphibians, microcrustaceans, and odonates.

Over the past 15 years, experiments in mesocosms have demonstrated the potential role of competition (e.g., Morin 1983b, Van Buskirk 1989, Fauth et al. 1990), predation (e.g., Morin 1983b, 1987; Morin et al. 1983; Wilbur et al. 1983; Semlitsch 1987; Semlitsch and Gibbons 1988; Van Buskirk 1988; Fauth 1990), and environmental stochasticity (in the form of pond drying, acting either alone or in concert with biotic interactions; e.g., Wilbur 1984 and references therein; Travis et al. 1985; Morin 1986; Semlitsch 1987; Wilbur 1987, 1988 and references therein; Wilbur and Fauth 1990) in the population dynamics and community structure of larval amphibians and insects that inhabit temporary ponds. Other experiments have illustrated the potential role of spatial scale (Pearman 1993), historical effects (in the form of breeding phenology: Alford and Wilbur 1985, Wilbur and Alford 1985, Morin 1987, Alford 1989, Morin et al. 1990), oviposition site choice (Resetarits and Wilbur 1989, 1991), and keystone predation (Morin 1981, 1983b; Fauth and Resetarits 1991; Kurzava and Morin 1994) in determining community structure in larval anuran assemblages. While these experiments were initiated to identify the processes that structure natural communities and to test theories related to that broad question, the experiments were carefully grounded in field data that remain largely unpublished (but see Harris et al. 1988). The experiments were thus designed so that their specific results would be applicable to the natural ponds they mimicked.

This assumption was challenged in the late 1980s by Petranka and others (e.g., Jaeger and Walls 1989, Petranka 1989 and references therein; also see Hairston 1989b, Morin 1989, and Wilbur 1989 for responses), who questioned the realism of experiments in both field enclosures and artificial ponds. They argued that experimental conditions, particularly high initial and final densities of larval anurans, favored biotic interactions over abiotic limiting factors that might be more important in nature. Subsequent work conducted in divided natural ponds (Petranka 1989) and large-scale field enclosures (Scott 1990) showed striking parallels to the results of the mesocosm experiments. Nonetheless, skeptics continue to question the utility of the artificial mesocosm/microcosm approach (e.g., Carpenter 1996).

Our research on the interactions between two salamanders, the lesser siren, *Siren intermedia*, and the broken-striped newt, *Notophthalmus viridescens dorsalis*, and their role in temporary ponds was largely conducted in such artificial communities (Fauth et al. 1990, Fauth and Resetarits 1991, Resetarits and Fauth unpublished ms., Fauth and Resetarits unpublished ms.). We were aware of criticisms leveled at mesocosm experiments and grew increasingly interested in whether the primary factors controlling community structure in natural communities were the same as in their artificial counterparts. One of us (WJR) used insights gleaned from the 20 years of mesocosm experiments that focused on temporary ponds in the southeastern United States (see following discussion) to develop a simple graphical model of the maintenance of anuran species richness in Carolina bays and other temporary ponds. In this paper we describe the model and test its qualitative predictions using field data collected independently as part of a baseline survey of Carolina bay amphibians (Fauth 1993). The model was developed before Fauth's data were collated and analyzed, thus obviating tautologized predictions.

Natural History of Carolina Bays and Their Diverse Faunas

Much of the biotic diversity within the Coastal Plain and nearby portions of the Sandhills Region of the southeastern United States is concentrated in and around temporary ponds. The principal type of natural temporary ponds in this region is the Carolina bays, shallow elliptical depressions of unknown geological origin that occur in a broad band on the Atlantic Coastal Plain from the Delmarva Peninsula to northern Florida and are most numerous in the Carolinas (Savage 1982, Sharitz and Gibbons 1982, Richardson and Gibbons 1993). Carolina bays range in size from <0.5 ha to thousands of hectares and in permanence from very ephemeral ponds that seldom allow metamorphosis of amphibians to large, permanent bay lakes such as Lake Waccamaw, which supports endemic molluscs and darters. Before European settlement Carolina bays numbered in the hundreds of thousands (Savage 1982), but today few remain unaltered by human impacts.

One of the most striking features of the ephemeral Carolina bays is their anuran diversity. Individual bays support up to 16 species of anurans out of a potential regional fauna of ~ 25 species (Morin 1983b, Wilbur and Travis 1984, Fauth 1993). Even greater numbers of anuran species may chorus from an individual pond on a single night (J. Fauth, personal observation; H. M. Wilbur, personal communication). Many tropical systems harbor greater regional diversity, but the anuran diversity supported by individual Carolina bays may be unsurpassed by that of any small body of water anywhere in the world. Smaller natural depressions and man-made temporary ponds (e.g., "borrow pits") support fewer species, but their faunas still are exceptionally diverse (Wilbur and Travis 1984). This anuran diversity provides a challenge (and a unique opportunity) to ecologists to explain its origin and maintenance.

A Model of Anuran Diversity in Carolina Bays and Other Temporary Ponds

In a series of papers (Fauth et al. 1990, Fauth and Resetarits 1991, Resetarits and Fauth unpublished ms., Fauth and Resetarits unpublished ms.), we explored the interactions between specific size classes of two competing predators, the broken-striped newt (*N. v. dorsalis*) and the lesser siren (*Siren intermedia intermedia*), and their effects on temporary pond communities. *N. v. dorsalis* and *S. intermedia* circumvent the predator avoidance strategy of many temporary pond species by persisting in the dry pond basin, which allows them to prey on the eggs and larvae of even the earliest colonists, including many species that are extremely vulnerable to predation. Both salamanders can influence species composition and community structure in temporary ponds. Because *N. v. dorsalis* and *S. intermedia* each have strong effects on prey populations, the dynamics of these two species and their interactions may be critical to the community ecology of Carolina bays and other temporary ponds.

These two species have markedly different impacts on anuran assemblages (Fig. 7-1). *N. v. dorsalis* is a keystone predator (Paine 1966) in assemblages that are characterized by strong competitive asymmetry (Morin 1983b, Fauth and Resetarits 1991), which is a common feature of anuran assemblages in temporary ponds (Morin 1981, 1983b; Wilbur and Alford 1985; Wilbur 1987; Morin and Johnson 1988; Van Buskirk 1988; Alford 1989; Fauth and Resetarits 1991). The keystone effect of *N. v. dorsalis* has been demonstrated repeatedly in artificial mesocosm experiments using anuran assemblages of varying species composition and density (Morin 1981, 1983b; Wilbur 1987; Fauth and Resetarits 1991) and is robust to the presence of a second predator, *S. intermedia* (Fauth and Resetarits 1991). The strength of the keystone effect can vary with newt density (Morin 1983b) and variation in body size among subspecies of *N. viridescens* (Morin 1986, 1987; Kurzava 1994). *N. v. dorsalis* has strong effects on the relative and absolute abundance of anurans in temporary ponds because newts prey selectively on the competitively superior species, allowing weaker competitors to persist (Morin 1981, 1983b; Fauth and Resetarits 1991). In our model, however, predation by

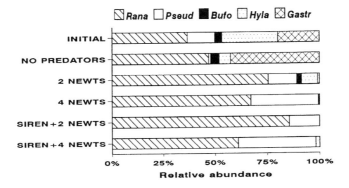

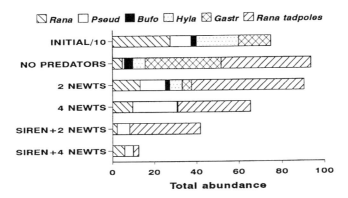

Figure 7-1. Differential effects of *Notophthalmus* and *Siren* on ensembles of larval anurans in artificial ponds. The anuran species are *Rana utricularia, Pseudacris crucifer, Bufo americanus, Hyla chrysoscelis,* and *Gastrophryne carolinensis.* The upper graph illustrates the effect of competitive asymmetries on species diversity in emerging anuran metamorphs and the highly significant effect of *Notophthalmus* on the relative abundance of species. The keystone effect of *Notophthalmus* is robust to the addition of *Siren,* but *Siren* has no effect on relative abundance. The lower graph illustrates the significant effect of *Siren* on the total abundance of larval anurans; *Notophthalmus* has no effect on total abundance. Modified from Fauth and Resetarits (1991).

Notophthalmus increases anuran species richness only across a relatively narrow range of potential anuran and newt densities (Fig. 7-2A). At low relative densities of *N. v. dorsalis* the keystone effect is overwhelmed (via predator satiation) and the species composition of metamorphs is largely determined by asymmetrical competitive interactions, resulting in lower species richness. At very high relative densities of *N. v. dorsalis,* anuran species richness decreases because newts eliminate some or all species (Fig. 7-2A; e.g., as in Morin 1983b).

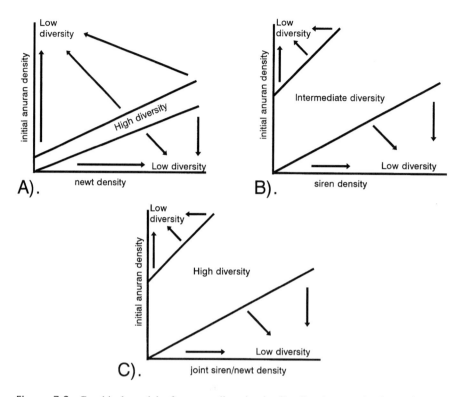

Figure 7-2. Graphical model of anuran diversity in Carolina bays and other ephemeral ponds. A, newt; B, siren; C, joint siren/newt density. Diversity is measured at the metamorph stage. The assumptions are: (1) Initial anuran species diversity (input diversity) is constant across the range of anuran densities. Increases in overall anuran densities result from increases (not necessarily equal) in the densities of component species. (2) The joint density of sirens and newts is at least 75% of their summed densities. This is a reasonable assumption based on the interactions between these two species (Fauth et al. 1990, Fauth and Resetarits 1991, Resetarits and Fauth unpublished manuscript).

S. intermedia is not a keystone predator but strongly affects the absolute numbers of larval and metamorphic anurans (Fauth and Resetarits 1991). *S. intermedia* preys on each species in direct proportion to its relative abundance (Fig. 7-1). Thus, the effects of *S. intermedia* on anuran assemblages are strikingly different from those of *N. v. dorsalis*. However, over much of the potential range of anuran and siren densities, species richness is determined by asymmetrical competitive interactions within the anuran assemblage rather than any direct impact of predation (Fig. 7-2C). Anuran diversity declines at high relative densities of *S. intermedia* because of the chance elimination of rare species and predation on weaker competitors whose populations are strongly impacted by interspecific competition. At high densities of both larval anurans and *S. intermedia*, diversity increases because predation reduces the overall density of larval anurans, thereby ameliorating the effects of interspecific competition. This is the classic case of predation being good for the survivors because it reduces the overall intensity of competition (Wilbur 1987, 1988). However, the potential gains in diversity achieved via the nonselective predation of sirens is necessarily smaller than that achieved via the keystone predation of newts. Anuran diversity is more severely reduced at high relative densities of *S. intermedia* than at high relative densities of *N. v. dorsalis* because *S. intermedia* has a much wider gape and is therefore an effective predator for a greater portion of most anuran larval periods.

The predicted joint effects of *S. intermedia* and *N. v. dorsalis* are additive and can be visualized by overlaying the two graphs (Fig. 7-2C); note especially that the region in which the keystone predator effect is expressed corresponds to a region in which *S. intermedia* has little impact on anuran diversity. By unselectively depressing prey abundance *S. intermedia* effectively extends the range of anuran densities over which *N. v. dorsalis* functions as a keystone predator. The two predators thus promote greater anuran diversity than is possible in ponds with either predator alone or in ponds lacking both predators. As illustrated in Figure 7-3, the mechanism underlying this joint effect depends on a relatively simple combination of characteristics of the anuran prey and the two predators. The characteristics of the prey assemblage are certainly not unique to this system, nor are the critical characteristics of the predators. Thus, we expect that these same or similar functional roles may be filled by different predators in other systems, and with different sets of prey.

The graphical model generates four testable predictions:

1. *At low levels of predation, increasing density of larval anurans will be negatively correlated with species richness.* Larvae of most anurans are generalist feeders (Altig and Johnston 1989, Duellman and Trueb 1990), so each additional larva raises the intensity of intra-and interspecific competition. Strong competitive asymmetries quickly lead to competitive exclusion and therefore to lower within-pond diversity (Morin 1981, 1983b; Wilbur and Alford 1985; Wilbur 1987; Morin and Johnson 1988; Van Buskirk 1988; Alford 1989; Fauth and Resetarits 1991).

2. *Ponds with the keystone predator* N. v. dorsalis *will have a richer anuran fauna only over a very restricted range of relative densities.* At low relative densities of newts, predator satiation overwhelms the keystone effect. At high relative densities of newts, species are directly eliminated by predation, thus negating the keystone effect (Morin 1983b).

3. *On average, ponds with* S. intermedia *will have a richer anuran fauna than ponds lacking* S. intermedia. Nonselective predation by *S. intermedia* over a broader range of larval anuran body sizes reduces the overall intensity of competition by reducing larval densities (Fauth and Resetarits 1991); this is the classic case of competitive release (Wilbur 1987, 1988).

4. *Ponds with both* S. intermedia *and* N. v. dorsalis *will have, on average, the richest anuran fauna.* Nonselective predation by *S. intermedia* extends the range of initial anuran densities over which the keystone effect of *N. v. dorsalis* can operate, thus allowing expression of the keystone effect over a greater number of ponds. Field data suggest that natural anuran densities typically fall within the range where the presence of *S. intermedia* is required to elicit the keystone effect of *N. v. dorsalis*.

We suggest that the anuran diversity that is such a striking feature of temporary ponds in the Sandhills and Coastal Plain of the Carolinas may be a direct result of the

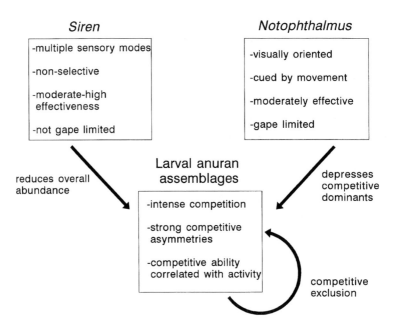

Figure 7-3. Diagram illustrating the critical features of the system. In the absence of predation, intense competition and strong competitive asymmetries reduce species diversity by means of competitive exclusion. The keystone predator effect is a simple consequence of the positive correlation between competitive ability and activity levels in larval anurans, combined with newts' reliance on prey movement as their primary mode of prey detection. Gape limitation enhances the effect on diversity by preventing complete elimination of competitive dominants. As a result, however, the keystone effect can be overwhelmed by predator satiation at high anuran densities. Sirens are consumate nonselective predators on larval anurans, using chemosensory, tactile, and visual cues to locate prey; being far less gape-limited also enhances their nonselective, generalist prey habits. In this system, *Siren* facilitates expression of the keystone effect simply by reducing overall abundances of larval anurans (continuing the architectural analogy, *Siren* is a voussior [a stone in the arch that supports the keystone] to *Notophthalmus*'s keystone).

dominance of these two predators. If they were not abundant, habitat diversity alone would be insufficient to maintain existing levels of anuran species richness. The intense competition and strong competitive asymmetries that are such reliable features of anuran assemblages (e.g., Morin 1983b, Wilbur 1987, Morin and Johnson 1988, Van Buskirk 1988, Alford 1989, Fauth and Resetarits 1991) would play themselves out, and many species would be eliminated from the region or at least restricted to the very few sites at which they held the competitive advantage. In a region dominated by ephemeral aquatic habitats that fill synchronously, the temporal axis along which species in more permanent aquatic habitats segregate is not available, and species are forced into greater overlap in both breeding sites and times. The potential for competition is greatly increased because water is itself a highly seasonal resource. Thus, mechanisms that offset competitive exclusion are essential in maintaining high species diversity.

Though the keystone predator effect is probably the critical biotic mechanism promoting increased anuran diversity, its predicted dependence on *Siren* predation for its expression over much of the typical range of anuran densities greatly elevates the importance of *Siren* in the system. If (in nature) initial species diversity scales with initial anuran density, then it is at the highest potential diversity levels that competition is also the most intense. Consequently, the expected final diversity of metamorphs is the lowest based on the effects of competition alone, and the keystone predator effect has the greatest capacity to affect diversity but the least potential to be evidenced. By having a strong suppressant effect on anuran densities *Siren* can prevent the keystone effect from being overwhelmed. Thus, both predators play a critical role in anuran species diversity in temporary ponds.

A Field Test of the Graphical Model of Anuran Diversity

We tested the qualitative predictions of the graphical model (Fig. 7-2) using observational field data from JEF's intensive faunal survey of seven Carolina bays in southern North Carolina (Fauth 1993). The data were analyzed by multiple regression of four factors, hydroperiod, anuran density, siren density, and newt density, on the response variable of anuran species diversity. Previous work had identified hydroperiod as a critical factor affecting community structure in temporary ponds, especially in interaction with competition and predation (e.g., Wilbur 1984, 1987; Werner and McPeek 1994; Rowe and Dunson 1995; Semlitsch et al. 1996). Thus, we expanded on the graphical model (Fig. 7-2) to include the effects of hydroperiod in the analysis of the field data.

The field survey began when the bays, which were spread across a 600-plus km^2 area, were synchronously refilled by torrential rains that ended a 2-year drought. The data consisted of minnow trap samples on 5–15 dates per bay, depending on the number of sampling trips on which each bay held water during one hydrologic cycle, from September 1988 through January 1990. On each sampling date 15 plastic minnow traps (opening 2.5 cm in diameter, maximum mesh gap 0.7 cm) were set in each bay and left overnight. Traps were set in two perpendicular lines, one with seven traps parallel to the shore in 12-cm-deep water and the other with eight traps extending from the pond's edge to the deepest water available. The shallow traps rested on the bottom, while the deepwater traps were suspended so their tops just breached the surface, allowing animals with lungs to breathe.

Hydroperiod was approximated by the number of sampling dates on which a pond held water. The maximum number of *S. intermedia*, *N. v. dorsalis*, and larval anurans caught on a single trap-night comprised the estimates of density. Anuran species richness was determined from minnow trap and dip net samples and from nighttime censuses of breeding frogs over the study period. The estimates of anuran species richness for each bay agreed with previous estimates made independently by personnel of the North Carolina Wildlife Resources Commission and the North Carolina State Museum of Natural History (A. Braswell, personal communication).

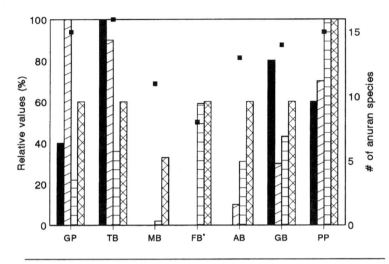

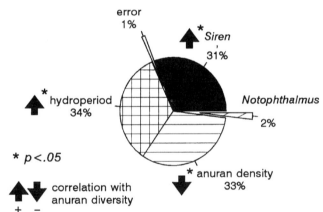

Figure 7-4. Anuran species richness in seven Carolina bays. (**Top**) Relative values (percentage of maximum observed value for each variable) of the four factors used in the multiple regression (*Siren, Notophthalmus,* and anuran densities, plus bay hydroperiod) measured in association with anuran diversity; solid squares = number of anuran species. Historically, *Notophthalmus* is recorded from all the bays but was not trapped in MB or FB during the study period; *S. intermedia* is not recorded from either of these bays. (**Bottom**) Results of the four-factor regression of anuran diversity, including partitioning of the variation and the direction of the effect. Regression equation: no. of anuran species = 1.06 + 1.99(hydroperiod) + 0.77(no. of sirens) − 0.02(no. of newts) − 0.20(no. of tadpoles). Total regression model R^2 = .997, $F_{4,2}$ = 70.27, P = .014.

Our four-factor multiple regression explained >99% of the variation in anuran species richness in the seven Carolina bays ($F_{4,2}$ = 70.27, R^2 = .997, P <.014; Fig. 7-4). The effects of hydroperiod, larval anuran density, and *S. intermedia* density were all significant, and each explained roughly 33% of the variation in anuran species richness. As predicted by the graphical model, *S. intermedia* had a strong positive effect, while the effect of increasing density of larval anurans was strongly negative. Hydroperiod also had a strong positive effect: temporary bays that held water longer permitted more anuran species to breed. Semlitsch et al. reported similar effects of hydroperiod on amphibian diversity in 15 years of data on a natural Carolina bay in South Carolina (Semlitsch et al. 1996).

Interestingly, however, the keystone predator *N. v. dorsalis* (Morin 1981, 1983b; Fauth and Resetarits 1991) had no significant effect on anuran species richness. We offer four testable (though not mutually exclusive) explanations for this last result:

1. Contrary to previous results (Morin 1981, 1983b; Fauth and Resetarits 1991), newts may not be keystone predators in natural ponds. This hypothesis seems improbable, given the repeated demonstrations of the keystone predator effect in mesocosm experiments that differed substantially in anuran species composition, initial density, and habitat heterogeneity (Morin 1981, 1983b, Wilbur 1987; Wilbur and Fauth 1990; Fauth and Resetarits 1991).
2. Although they may not have been present during the sampling period, newts have been recorded from all of the ponds studied. Contemporary estimates of their densities may not reveal their prior effects on anuran assemblage structure. Instead, the newt effect may simply be reflected as a higher (or lower) mean number of species being recorded across all sites (the y-intercept in the regression analysis). Detecting this "ghost of predation past" will require further experiments.
3. The model may be correct; the keystone predator effect of newts may be entirely dependent on sirens reducing overall anuran densities, at least across the range of densities found in the Carolina bays sampled.
4. The newt effect could be nonlinear; the field data support the notion (incorporated in the graphical model but not in the statistical analysis) that at high densities newts no longer function as keystone predators but instead eliminate species from the anuran assemblage. The anurans that typically metamorphose from ponds that have high densities of *N. v. dorsalis* scatter their eggs on the pond bottom or float them in small surface packets. Amphibians that lay their eggs in large clumps (e.g., the tiger salamander, *Ambystoma tigrinum*, and the southern leopard frog, *Rana utricularia*) often suffer complete prehatching mortality in GP and TB, the two Carolina bays with the highest newt densities (Fig. 7-4; see also Morin 1983a).

Alternative Models of Species Richness

If there is one rule in community ecology, it is to not blindly accept the first explanation for any hypothesis (see, for example, the exchange between proponents and detractors of species co-occurrence models in Strong et al. 1984). To test the effects of bay area, spatial heterogeneity, and other environmental influences on anuran species richness, JEF measured an additional 12 environmental variables: bay area, elevation, pH, tree density, distance to the nearest and second- and third-nearest bodies of water, minimum sandrim width, distance to the nearest road and building, and the number of buildings and percentage of forested cover within a 1-km radius. Species richness was regressed versus each variable, plus their first three principal components scores, with $\alpha = 0.05$. This procedure inflates the experimentwise Type I error but reduces the risk of Type II error and therefore provides a reasonable alternative to our anuran diversity model. Species richness and pond area were also regressed on a log by log plot to test the species-area relationship that predicts more species will be found in larger bays (MacArthur and Wilson 1967). JEF also constructed and compared UPGMA (unweighted pair-group method, arithmetic average; using PAUP) pheno-grams of the bays' anuran species composition and environmental/limnological char-acteristics to investigate possible concordance between these two data sets. Species were coded as binary data (present/absent), and environmental variables were coded accord-ing to a posteriori criteria established for each variable (e.g., acid versus neutral pH).

The alternative models tested failed to explain the observed pattern of anuran species richness in the Carolina bays. Bay area was not significantly correlated with species richness ($F_{1,5} = 3.00$, $R^2 = .37$, $P = .14$) and, in any case, was in a direction opposite the predictions of species-area relations (Fig. 7-5). There was no correlation between any of 13 standard environmental variables or their principal components scores and species richness (all $P \gg .10$; Fauth 1993). In addition, there was little concordance between UPGMA phenograms of the bays' environmental/limnological characteristics and anuran species composition. The only major congruence between the phenograms was identifying GP as a pond distinct from most or all of the others (Fig. 7-6). This was expected because GP lies in a different physiographic province (Sandhills Region) than the other six bays (Atlantic Coastal Plain) and thus has unique physiognomic and biotic features.

Testing these alternative explanations lends credence to our model of species diver-sity; our model not only explains most of the variation in species diversity but also does so far better than other, equally plausible, models based on potentially important environmental variables.

The Contribution of Mesocosms to Understanding Anuran Diversity in Carolina Bays

The field data strongly support the synthetic model developed from the larger body of mesocosm experiments, even while certain specific predictions from individual ex-periments are not upheld. As discussed previously, the lack of detectable effects of newts on species diversity is surprising based on the results of individual experiments but is entirely consistent with the synthetic model, which incorporates aspects not pres-ent in individual experiments. Thus, the simple graphical model has strong predictive power for species richness in Carolina bays, and that predictive power can be seen to derive not from a one-to-one match between the mesocosms and their natural analogue or from the identity of a particular experiment with the conditions seen in these specific bays, but from the detailed, emergent understanding of the processes potentially at work in temporary ponds. This understanding was achieved through thoughtful, detailed,

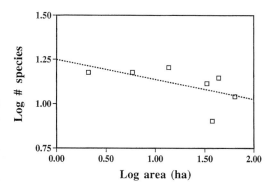

Figure 7-5. Species–area relations in the seven Carolina bays. Regression model: log (no. of species) = $1.25 - 0.12$ (log bay area [ha]). Model $R^2 = 0.37$, $F^{1,5} = 3.00$, $P > .10$.

rigorous experiments which required the control and statistical power afforded by a model system.

The field data strongly support the predicted roles of hydroperiod, anuran density, and *S. intermedia* density in determining anuran species diversity in Carolina bays. Hydroperiod has long been suggested as an important factor influencing anuran diversity in temporary ponds (Wilbur 1987), as has the role of interspecific competition and predation (e.g., Morin 1981, 1983b; Wilbur et al. 1983; Wilbur 1987; Alford 1989; Fauth and Resetarits 1991). *N. v. dorsalis* has long been suspected to play an important

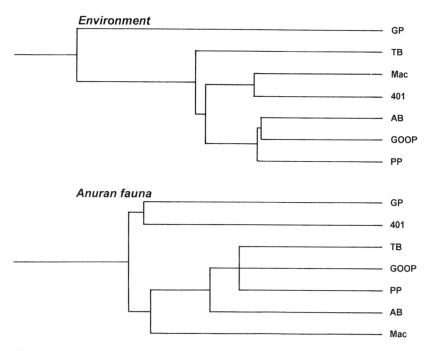

Figure 7-6. UPGMA phenograms of (*top*) environmental variables of seven Carolina bays and (*bottom*) species composition. Figure modified from Wilbur (1984).

role in temporary ponds; however, little was previously known regarding *S. intermedia*. The field data support the view that the keystone effect of *N. v. dorsalis* may largely depend on the complementary effects of *S. intermedia*. Because of its overland dispersal capabilities (and the lack of such capabilities in *S. intermedia*), *N. v. dorsalis* alone may be most important in smaller, less "historically long-lived" temporary ponds, man-made borrow pits, and similar habitats that lack *S. intermedia*. These habitats are common, especially in the Sandhills, and contain *N. v. dorsalis* almost without exception (often at very high densities; e.g., Morin 1983a) but individually do not support the level of diversity seen in the Carolina bays.

Our field test of the model of anuran diversity suggests that the remarkably high and apparently stable anuran diversity of Carolina bays and other ephemeral ponds is a product of internal biotic forces (intraspecific competition and predation) coupled with the environmental stochasticity imposed by pond drying. Biotic factors accounted for two-thirds of the variation in anuran diversity, with hydroperiod accounting for the remainder. The fragile nature of diversity in temporary aquatic habitats should not be underestimated. Simply preserving the physical features of these and other similar habitats is not sufficient to protect the diversity they contain; the balance between predation, competition, and hydroperiod that specifically maintain it must be preserved. One conservation lesson that arises from our research is that diversity is not a characteristic of place but a characteristic of process. Places with high species diversity can be identified, but unless the processes that maintain that diversity are understood and preserved, their unique characteristics cannot be protected.

The Dual Role of Mesocosm Experiments

Community ecologists began using experiments relatively early in the history of the discipline to reveal the mechanisms that affect community structure (e.g., Gause 1934). The experimental approach varied according to the interests and training of the observer and the limitations imposed by particular systems, but it soon became apparent that a fundamental trade-off existed between complete control of environmental conditions and realism (sensu Diamond 1986; see Morin, this volume). Artificial mesocosms were introduced as a hybrid approach that retained much of the interesting variation of nature while allowing the replication and rigorous control of initial conditions required for sophisticated experimental designs; this approach has been adapted to a wide variety of aquatic systems (e.g., Maguire et al. 1968, Vandermeer 1969, Hall et al. 1970, Addicott 1974, Neill 1975, Morin 1981, Hay 1986, Resetarits 1991, Bernardo 1994). Proponents of this approach have long argued that their experiments shed light on fundamental ecological principles, but little empirical evidence has been brought to bear on the question of predictive power in natural systems.

Our field test of a model derived from the results of artificial mesocosm experiments accomplishes precisely what critics rightly argued was lacking (e.g., Petranka 1989 and references therein, Jaeger and Walls 1989): it makes predictions about nature based on the lessons learned from mesocosm experiments and determines the accuracy and utility of those predictions in a natural system. In this case, we explicitly tested a model generated from the results of mesocosm experiments in the natural system that inspired them. Our results suggest that mesocosm experiments designed primarily to test general ecological theories could, because they also were properly grounded in natural history, simultaneously describe in detail and make specific predictions about the structure and dynamics of natural communities. Mesocosm-based experiments have been recognized for their valuable contributions to our understanding of general ecological processes. This heightened understanding undoubtedly contributes to our understanding of natural systems, even if only indirectly. Here we suggest that our reluctance to accept a more tangible, direct contribution to understanding specific natural systems or types of natural systems bears reconsideration.

We do not suggest that the level of agreement between our model and the field data will be common (in fact, we suspect we will never, ever, explain $> 99\%$ of the variation in anything again!). Much will depend on the care taken in designing the model system in the image of its natural counterpart and on the specific system being studied. Nevertheless, we are equally convinced that mesocosms can reveal much about the processes at work in natural systems. Both of us have used mesocosms to address a variety of questions in a relatively broad range of systems and organisms (e.g., Resetarits and Wilbur 1989; Fauth 1990, this volume; Fauth and Resetarits 1991; Resetarits 1991, 1995). In each case, mesocosms allowed us to look at the responses of organisms to predators or competitors (or both) at a level of detail that would be virtually impossible in field studies or field experiments on highly mobile organisms. In each case, we gained insight into both ecological theory and the dynamics of the system our mesocosms mimicked. And in each case, we were surprised by unexpected results or cryptic processes that were only revealed by the detailed responses we could assay in controlled mesocosm experiments. Although individual experiments may not reveal precisely what occurs in natural systems, we can use the cumulative knowledge from multiple experiments to ''build'' an emergent understanding of processes, just as we have done for the Carolina bays. Though the cost might seem rather high in terms of the number of person-years and experiments devoted to the temporary pond work, one must keep in mind the original goals of the research. Had the intent been specifically to understand the determinants of species diversity in temporary ponds or, in particular, Carolina bays, we could have done specific experiments to answer that question more directly. The critical lesson for ecology is that the trade-offs so often cited between rigor and ''realism'' may be more an artifact of how we think about natural systems and our own lack of creativity than a characteristic of the relationship between model systems and natural systems.

From the point of view of this article and this volume, our most important message is that we have forged a link between the results obtained in ''experimental abstractions'' and the dynamics of natural systems. Considering the amount of work that has been done in mesocosms and the growing use of model systems of all sorts, this result is a positive prescription for understanding ecological systems on a broad scale. The criticism that experiments in model systems cannot tell us what happens in natural systems is a hypothesis to be tested, just as any other scientific hypothesis is evaluated against the weight of evidence. We have already begun an experimental test of that hypothesis (and our model of anuran diversity) in natural temporary ponds. Here we have taken an important first step in demonstrating that mesocosm-based research, even that specifically designed to test general ecological principles, can, when properly grounded in the ecology of a specific natural system, provide a both rigorous and economical tool for understanding that system. Clearly, the greatest gains in generality and specificity can be made when rigorous experiments in model systems are combined with solid fieldwork and strong theoretical underpinnings (Werner, this volume), as has been the case with the temporary pond work. In such cases, the potential contributions of model system–based research to understanding natural communities may not be so limited as critics would have us believe (e.g., Carpenter 1996). At the very least, well-reasoned, well-conducted experiments in model systems should enable us to ask the right questions in natural systems (Lawton, this volume); at their very best, these experiments may indeed provide the answers to those questions.

ACKNOWLEDGMENTS   The body of experimental work on temporary ponds in North Carolina was largely supported by National Science Foundation (NSF) grants to H. M. Wilbur. Field sampling of Carolina bays was supported by grants to JEF from the North Carolina Nature Conservancy, the North Carolina Nongame Wildlife Commission, the Explorers' Club, the Theodore Roosevelt Fund of the American Museum of Natural History, and Sigma Xi. S. McCollum, P Harris, and M. Leibold helped sample the Carolina bays, and C. Gill collated the field data. T. Schultz provided access to PAUP. J. Bernardo, C. Cáceres, S. Kohler, R. Marquis, E. Marschall, C. Rogers, R. Semlitsch, and T. Smith provided useful comments on various iterations of this manuscript. We especially thank several generations of our academic siblings at Duke and our colleagues at the Savannah River Ecology Laboratory whose work has provided valuable insights into the ecology of Carolina bays and their diverse amphibians. Our continuing work on temporary ponds, now in South Carolina, is generously supported by grants from the Environmental Protection Agency and NSF.

Literature Cited

Addicott, J. F. 1974. Predation and prey community structure: an experimental study of the effect of mosquito larvae on the protozoan community of pitcher plants. Ecology 55: 475–492.

Alford, R. A. 1989. Variation in predator phenology affects predator performance and prey community composition. Ecology 70:206–219.

Alford, R. A., and H. M. Wilbur. 1985. Priority effects in experimental communities: competition between *Bufo* and *Rana*. Ecology 66:1097–1105.

Altig, R. A., and G. F. Johnston. 1989. Guilds of anuran larvae: relationships among developmental modes, morphologies, and habitats. Herpetological Monographs 3:81–109.

Bernardo, J. 1994. Experimental analysis of allocation in two divergent, natural salamander populations. American Naturalist 143:14–38.

Carpenter, S. R. 1996. Microcosm experiments have limited relevance for community and ecosystem ecology. Ecology 77:677–680.

Diamond, J. M. 1986. Overview: laboratory experiments, field experiments, and natural experiments. Pages 3–22 in J. Diamond and T. J. Case (eds.), Community Ecology. Harper and Row, New York.

Duellman, W. E., and L. Trueb. 1990. Biology of Amphibians. Harper and Row, New York.

Fauth, J. E. 1990. Interactive effects of predators and early larval dynamics of the treefrog *Hyla chrysoscelis*. Ecology 71:1609–1616.

————. 1993. Factors Affecting the Abundance and Distribution of Amphibians in Carolina Bays. Final report to the North Carolina Nature Conservancy and the North Carolina Nongame Wildlife Commission, Raleigh, North Carolina.

Fauth, J. E., and W. J. Resetarits Jr. 1991. Interactions between the salamander *Siren intermedia* and the keystone predator *Notophthalmus viridescens*. Ecology 72:827–838.

————. Unpublished ms. Detecting higher order interactions in biological systems: identifying appropriate ecological and statistical models.

Fauth, J. E., W. J. Resetarits Jr., and H. M. Wilbur. 1990. Interactions between larval salamanders: a case of competitive equality. Oikos 58:91–99.

Gause, G. F. 1934. The Struggle for Existence. Williams and Wilkins, Baltimore, Maryland.

Gurevitch, J., L. L. Morrow, A. Wallace, and J. S. Walsh. 1992. A meta-analysis of competition in field experiments. American Naturalist 140:539–572.

Hairston, N. G., Sr. 1989a. Ecological Experiments: Purpose, Design, and Execution. Cambridge University Press, Cambridge.

————. 1989b. Hard choices in ecological experimentation. Herpetologica 45:119–122.

Hall, D. J., W. E. Cooper, and E. E. Werner. 1970. An experimental approach to the production dynamics and structure of freshwater animal communities. Limnology and Oceanography 15:839–928.

Harris, R. N., R. A. Alford, and H. M. Wilbur. 1988. Density and phenology of *Notophthalmus viridescens dorsalis* in a natural pond. Herpetologica 44:234–242.

Hay, M. E. 1986. Associational plant defenses and the maintenance of species diversity: turning competitors into accomplices. American Naturalist 128:617–641.

Jaeger, R. G., and S. C. Walls. 1989. On salamander guilds and ecological methodology. Herpetologica 45:111–119.

Kurzava, L. M. 1994. The structure of prey communities: effects of predator identity and geographic variation in predators. Ph.D. dissertation, Rutgers University, New Brunswick, New Jersey.

Kurzava, L. M., and P. J. Morin. 1994. Consequences and causes of geographic variation in the body size of a keystone predator, *Notophthalmus viridescens*. Oecologia 99:271–280.

Lamberti, G. L., and A. D. Steinman (eds.). 1993. Research in artificial streams: applications, uses, and abuses. Journal of the North American Benthological Society 12:313–384.

MacArthur, R. H., and E. O. Wilson. 1967. The Theory of Island Biogeography. Princeton University Press, Princeton, New Jersey.

Maguire, B., Jr., D. Belk, and G. Wells. 1968. Control of community structure by mosquito larvae. Ecology 49:207–210.

Morin, P. J. 1981. Predatory salamanders reverse the outcome of competition among three species of anuran tadpoles. Science 212:1284–1286.

————. 1983a. Competitive and predatory interactions in natural and experimental populations of *Notophthalmus viridescens dorsalis* and *Ambystoma tigrinum*. Copeia 1983: 628–639.

————. 1983b. Predation, competition and the composition of larval anuran guilds. Ecological Monographs 53:119–138.

————. 1986. Interactions between intraspecific competition and predation in an amphibian predator-prey system. Ecology 67:713–720.

————. 1987. Predation, breeding asynchrony, and the outcome of competition among treefrog tadpoles. Ecology 68:675–683.

————. 1989. New directions in amphibian community ecology. Herpetologica 45:124–128.

Morin, P. J., and E. A. Johnson. 1988. Experimental studies of asymmetric competition among anurans. Oikos 53:398–407.

Morin, P. J., H. M. Wilbur, and R. N. Harris. 1983. Salamander predation and the structure of experimental communities: responses of *Notophthalmus* and microcrustacea. Ecology 64:1430 436.

Morin, P. J., S. P. Lawler, and E. A. Johnson. 1990. Ecology and breeding phenology of larval *Hyla andersonii*: the disadvantages of breeding late. Ecology 71:1590–1598.

Neill, W. E. 1975. Experimental studies of microcrustacean competition, community composition and efficiency of resource utilization. Ecology 56:809–826.

Paine, R. T. 1966. Food web complexity and species diversity. American Naturalist 100:65–75.

Pearman, P. B. 1993. Effects of habitat size on tadpole populations. Ecology 74:1982–1991.

Petranka, J. W. 1989. Density-dependent growth and survival of larval *Ambystoma*: evidence from whole-pond manipulations. Ecology 70:1752–1767.

Resetarits, W. J., Jr. 1991. Ecological interactions among predators in experimental stream communities. Ecology 72:1782–1793.

———. 1995. Limiting similarity and the intensity of competitive effects on the mottled sculpin, *Cottus bairdi*, in experimental stream communities. Oecologia 104:31–38.

Resetarits, W. J., Jr., and J. E. Fauth. Unpublished ms. Competition between juvenile sirens and adult newts: interspecific competition from hatching to maturity in a pair of generalist predators.

Resetarits, W. J., Jr., and H. M. Wilbur. 1989. Oviposition site choice in *Hyla chrysoscelis*: role of predators and competitors. Ecology 70:220–228.

———. 1991. Calling site choice by *Hyla chrysoscelis*: effect of predators, competitors, and oviposition sites. Ecology 72:778–786.

Richardson, C. J., and J. W. Gibbons. 1993. Pocosins, Carolina bays and mountain bogs. Pages 257–310 in W. H. Martin, S. G. Boyce, and A. C. Echternacht (eds.), Biodiversity of the Southeastern United States: Lowland Terrestrial Communities. Wiley, New York.

Rowe, C. L., and W. A. Dunson. 1995. Impacts of hydroperiod on growth and survival of larval amphibians in temporary ponds of central Pennsylvania, USA. Oecologia 102:397–403.

Savage, H. 1982. The Mysterious Carolina Bays. University of South Carolina Press, Columbia.

Scott, D. E. 1990. Effects of larval density in *Ambystoma opacum*: an experiment in large-scale field enclosures. Ecology 71:296–306.

Semlitsch, R. D. 1987. Paedomorphosis in *Ambystoma talpoideum*: effects of density, food, and pond drying. Ecology 68:994–1002.

Semlitsch, R. D., and J. W. Gibbons. 1988. Fish predation in size-structured populations of treefrog tadpoles. Oecologia 75:321–326.

Semlitsch, R. D., D. E. Scott, J. H. K. Pechmann, and J. W. Gibbons. 1996. Structure and dynamics of an amphibian community. Pages 217–248 in M. L. Cody and J. R. Smallwood (eds.), Long-Term Studies of Vertebrate Communities. Academic Press, New York.

Sharitz, R. R., and J. W. Gibbons. 1982. The Ecology of Southeastern Shrub Bogs (Pocosins) and Carolina Bays: A Community Profile. U.S. Fish and Wildlife Service FWS/OBS-82/04, Washington, D.C.

Strong, D. R., Jr., D. Simberloff, L. G. Abele, and A. B. Thistle (eds.). 1984. Ecological Communities: Conceptual Issues and the Evidence. Princeton University Press, Princeton, New Jersey.

Travis, J. T., W. H. Keen, and J. Julianna. 1985. The effects of multiple factors on viability selection in *Hyla gratiosa* tadpoles. Evolution 39:1087–1099.

Van Buskirk, J. 1988. Interactive effects of dragonfly predation in experimental pond communities. Ecology 69:857–867.

———. 1989. Density-dependent cannibalism in larval dragonflies. Ecology 70:1442–1449.

Vandermeer, J. H. 1969. The competitive structure of communities: an experimental approach with protozoa. Ecology 50:362–372.

Werner, E. E., and M. A. McPeek. 1994. Direct and indirect effects of predators on two anuran species along an environmental gradient. Ecology 75:197–213.

Wilbur, H. M. 1984. Complex life cycles and community organization in amphibians. Pages 195–224 in P. W. Price, C. N. Slobodchikoff, and W. S. Gaud (eds.), A New Ecology: Novel Approaches to Interactive Systems. Wiley, New York.

———. 1987. Regulation of structure in complex systems: experimental temporary pond communities. Ecology 68:1437–1452.

———. 1988. Interactions between growing predators and growing prey. Pages 157–172 in B. Ebenman and L. Persson (eds.), Size-structured Populations: Ecology and Evolution. Springer-Verlag, Berlin.

———. 1989. In defense of tanks. Herpetologica 45:122–123.

Wilbur, H. M., and R. A. Alford. 1985. Priority effects in experimental pond communities: responses of *Hyla* to *Bufo* and *Rana*. Ecology 66: 1106–1114.

Wilbur, H. M,. and J. E. Fauth. 1990. Experimental aquatic food webs: interactions between two predators and two prey. American Naturalist 135:176–204.

Wilbur, H. M., and J. T. Travis. 1984. An experimental approach to understanding pattern in natural communities. Pages 113–122 in D. R. Strong Jr., D. Simberloff, L. G. Abele, and A. B. Thistle (eds.), Ecological Communities: Conceptual Issues and the Evidence. Princeton University Press, Princeton, New Jersey.

Wilbur, H. M., P. J. Morin, and R. N. Harris. 1983. Salamander predation and the structure of experimental communities: anuran responses. Ecology 64:1423–1429.

8

Have Field Experiments Aided in the Understanding of Abundance and Dynamics of Temperate Reef Fishes?

SALLY J. HOLBROOK & RUSSELL J. SCHMITT

A central goal of ecology is to understand factors that determine the abundance of organisms and predict patterns of population size in space and time. Three different but related attributes of abundance have been identified and studied: the overall mean, spatial variation, and temporal variation. Although many different physical and biological processes potentially can impact these, it is often assumed that a process that affects one will automatically affect the other two. That is, a factor that has a strong effect on mean abundance will similarly affect spatial or temporal patterns of abundance. This assumption is often made in experimental studies of interspecific competition, because usually field experiments are designed to assess whether the interaction significantly reduces the average population size, with only limited evaluation of causes of spatial and temporal variation. In this essay, we explore and compare the mechanisms underlying the mean, spatial, and temporal abundance of two species of marine reef fish, striped surfperch (*Embiotoca lateralis*) and black surfperch (*Embiotoca jacksoni*), as well as consider the effects of interspecific competition.

Our investigations of surfperch have involved a combination of field experiments and observations spanning a period of about 13 years (Holbrook and Schmitt 1984, 1986, 1988a,b, 1989, 1992, 1995, 1996; Schmitt and Holbrook 1984a,b, 1985, 1986, 1990a,b; Holbrook et al. in press). Field experiments to explore interspecific competition between the two surfperch species have played a central role in this research (Holbrook and Schmitt 1989, 1992, 1995; Schmitt and Holbrook 1990b). However, the use of competition experiments in studies of patterns of abundance poses certain problems, one of the most important of which relates to scales at which experiments are conducted and the results interpreted. Due to logistic constraints, field experiments are often short in duration and circumscribed in space, yet the goal is usually to place their results in a larger ecological context. For example, one issue regarding experiments is whether their results can be "scaled up" to predict longer-term responses. Can a short-term behavioral competition experiment be used to accurately predict the population-level (abundance) consequences of the interaction? If it could, logistically demanding experiments to measure population dynamical responses of interacting populations might not be necessary.

A second scaling issue regards the time scales of experiments relative to longer term environmental processes. In marine systems, oceanographic processes that result in storms, altered productivity regimes, or changes in physical conditions can have dramatic effects on population sizes (Holbrook et al. 1994, Holbrook and Schmitt 1995), at time scales ranging from among seasons within a single year up to decades. While many field experiments span the seasons of the year or even a few years and thus encompass a certain level of environmental variation, some important sources of environmental variation occur less frequently and are thus less likely to be "captured" during an experimental period. For instance, in the Southern California Bight, El Niño Southern Oscillation (ENSO) phenomena result in above-average sea surface temperatures, reduced dissolved nutrients, elevated sea levels, changes in current regimes, and an increase in the frequency and intensity of winter storms (McGowan 1984, 1985; Mysak 1986; Murray and Horn 1989; Hayward 1993), all of which can substantially affect populations of nearshore and offshore fishes and their resources (Cowen 1985, Fiedler et al. 1986, Tegner and Dayton 1987, Cowen and Bodkin 1993). The average interval between moderate or strong El Niño events is about 5.4 years (Quinn et al. 1978). Clearly, the ENSO phenomenon is an example of an oceanographic process that occurs intermittently over time spans that are much longer than experimental studies of species interactions, yet the events are frequent enough that relating their effects to experimental results may help derive a more complete understanding of temporal patterns of abundance. It is also known that fish populations can be influenced by long-term environmental trends (Ware and Thomson 1991). These trends may be unidirectional, such as could result from global warming, or they may be fluctuating. Incorporating effects of such environmental variation on species populations obviously poses a challenge to experimentalists, yet this issue is not usually addressed in the discussion of field experiments on competition.

In this essay, we examine first the problem of scaling up, using examples from our studies of the effects of interspecific competition on mean abundance of surfperch. Then we discuss causes of spatial and temporal variation in these populations of fish in light of what we have learned about interspecific competition and environmental variation.

The Issue of Scaling Up and effects of Competition on Mean Abundance of Surfperch

Despite the central tenet of competition theory that the process depresses abundances of the participating species, most competition experiments that involve vertebrates have not measured density responses to altered competitive regimes (Table 8.1; also see Connell 1983). For instance, in a set of about 50 studies of competition involving vertebrate species reported by Schoener (1983), more than half employed behavioral (e.g., habitat shifts or changes in feeding rates) or demographic (e.g., changes in clutch size, growth rate, or survivorship) attributes as response variables. In general, these experiments were much shorter in duration than those that measured density effects.

Table 8-1. Response variables used in studies of competition among vertebrate species (reviewed in Schoener 1983, Table 1)

Aspects studied	No. of studies	Duration (mos.)
Behavior	5	6.8
Demography	19	8.2
Demography + Behavior	6	4.8
	30	7.3
Density	7	31.5
Density + Demography	5	20.6
Density + Behavior	4	23.6
Density + Demography + Behavior	4	27.3
	20	26.3

Given are the number of studies (out of a total of 50 involving vertebrates) and average duration (in months) that employed various types of response variables. Behavioral responses included foraging, habitat use and activity levels. Demographic aspects included growth, survival, reproductive output or condition of individuals, as well as rates of recruitment and emigration or immigration. Density responses assessed numbers of individuals.

Of course, when short-term responses are used in experiments to gauge strength of competition it is assumed that these "scale up"—that is, they reflect ultimate changes in population size that eventually would result under the altered competitive regime. However, there is very little information available to evaluate this assumption, since few experiments have combined behavioral and demographic studies with measurements of density responses (Table 8-1) and there has been virtually no discussion about whether the various responses are congruent.

In our explorations of the effect of interspecific competition on mean abundance of black surfperch and striped surfperch, we conducted a 4-year-long press (sensu Bender et al. 1984) experiment to measure changes in densities of each species on reefs where densities of the competitor were greatly reduced. We also examined two short-term behavioral responses after the first 2 months—shifts in habitat and changes in feeding behavior. The key question was whether knowledge of the magnitude of either one, or both, of the behavioral responses correctly predicted the magnitude of the density response after 4 years.

Both species of surfperch co-occur on shallow (<10 m water depth) rocky reefs at Santa Cruz Island, California (34° 05' N, 119°, 45' W). They are microcarnivores, diurnally harvesting small crustaceans from the surfaces of foliose algae and from turf, a low-growing carpet of plants, small animals, and debris. Availability of crustaceans differs among the various types of bottom substrates; the red alga *Gelidium robustum* contains extremely high levels of crustaceans during most periods of the year. Outside the winter period (when crustacean levels on it are low), *Gelidium* is the best microhabitat for feeding by surfperch. As such, it is the focus in the competitive interaction (Holbrook and Schmitt 1989, 1992, 1995; Schmitt and Holbrook 1990b). *Gelidium* is the most highly preferred feeding substrate for striped surfperch. Black surfperch also show a preference for *Gelidium* but overall show the highest preference for turf, from which they are able to extract prey items by winnowing, which striped surfperch cannot do. The ability to use turf enables black surfperch to thrive in areas that lack *Gelidium* and other foliose algae, and this can be a decided advantage because many reef zones lack well-developed stands of algae (see following discussion). Furthermore, black surfperch use turf extensively during the winter period when crustacean levels on *Gelidium* are low. Striped surfperch and *Gelidium* are most abundant in the shallow zones of the reef, while black surfperch and turf are more common deeper (Fig. 8-1). On the whole, the abundance of crustacean prey is highest in the shallow zones (Fig. 8-1).

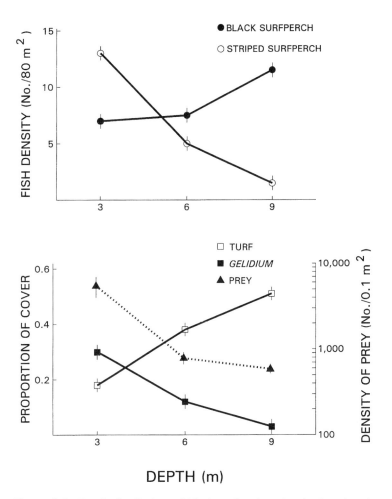

DEPTH (m)

Figure 8-1. Depth distribution of black surfperch and striped surfperch (top) and their feeding resources (bottom) at sites on Santa Cruz Island, California. Given are mean ($\pm$ 1 S.E.) of fish along 80 m²–band transects, abundances of crustacean prey items, and cover of turf and *Gelidium* along fixed 40-m transects at two control reefs. Prey densities are based on benthic samples ($N = 36$ per depth) taken during the prey-rich season of the year.

The reciprocal removal competition experiment was conducted at six sites on the north shore of Santa Cruz Island during 1982–1986. There were two unmanipulated control sites and two sites each where densities of striped surfperch or black surfperch were reduced by about 90% for a period of 4 years (fully described in Schmitt and Holbrook 1986, 1990b; Holbrook and Schmitt 1989). At each reef 40 × 2–m belt transects were established at the 3-, 6-, and 9-m depth contours prior to any manipulations, and sampling of fish densities, foraging behavior, and availability of food resources (crustacean prey and the substrates that contain them) was carried out in each of these depth zones. Since our experiments involved reductions in densities of the competitor species, we interpret changes in features such as densities or distribution of fish or in feeding behavior (relative to control sites) as evidence of competitive release.

During the initial 2-month period that followed reduction in densities of competitors on the reefs, the behavior of each target species showed clear signs of competitive release, although their overall abundances on the experimental sites did not change. (The experiment was initiated outside the time of year when young are born and enter adult populations, so there was no recruitment in the initial 2-month period. These surfperch reproduce for the first time at age 2 or 3; the average life span in the Santa Cruz Island populations is about 3 years, with a maximum of about 7 years.) The behavioral responses were not the same for each species (when the congener was removed), and the responses in habitat use suggested a fundamentally different scenario about the competitive interaction than did those of feeding behavior. Regarding habitat use, black surfperch individuals immediately moved into shallower zones of the reef upon removal of striped surfperch, whereas striped surfperch did not shift habitats when black surfperch were removed (Fig. 8-2). By contrast, upon competitive release both of the species showed marked changes in feeding behavior. In the absence of the competitor, each species greatly increased use of *Gelidium* as a foraging location, but this effect was seen in the shallow zones of the reef only (Fig. 8-3) where *Gelidium* is common (Fig. 8-1). In the deeper zones of the reef where *Gelidium* is very rare, neither species altered use of foraging microhabitats upon competitive release.

Overall, although both of the short-term behavioral responses indicated that interspecific competition was occurring, they yielded two very different predictions about the strength of the interaction. The asymmetry in shifts in depth distribution (black surfperch responding only) suggested that the interaction was asymmetrical, with striped surfperch the superior competitor. Extrapolation of future effects on abundance would imply a much stronger density response of black surfperch than striped surfperch (Table 8-2). By contrast, the foraging data indicated a strong, symmetrical interaction that ultimately should lead to similar increases in densities of each competitor. Further, in this case the response should be strong where *Gelidium* is abundant (shallow habitats) and weak where *Gelidium* is rare (deep zones) (Table 8-2).

Abundances of both species of surfperch increased markedly on the removal reefs by the end of the experiment. These responses resulted from both increased local production of young and some immigration of older individuals to the reefs (Schmitt and Holbrook 1990b). Overall, abundance increased on the order of 40% for each species on these reefs, with most of the increases occurring in the shallower zones of the reefs (Fig. 8-4). This result is contrary to the density predictions based on the asymmetrical habitat shifts but consistent with changes in feeding behavior during competitive release.

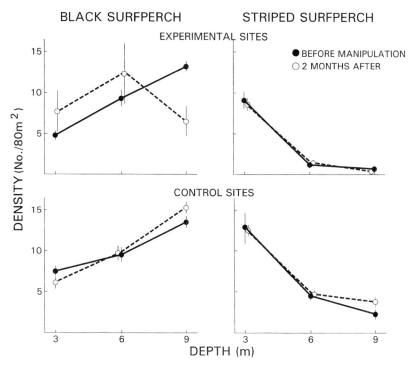

Figure 8-2. Change in density of black surfperch (*left panels*) and striped surfperch (*right panels*) within each depth zone following release from interspecific competition (*top graphs*) and at unmanipulated control sites (bottom graphs). Given are mean ($\pm$ 1 S.E.) number of fish along 80 m²–transects ($N = 8$ daily counts per depth per site during each time period).

The density responses indicated that interspecific competition was strong and more or less symmetrical.

The short-term changes in depth distribution occurred because the contested resource (*Gelidium*) is only found in shallow water. By removing striped surfperch, which are most abundant in the shallow zones, resources were freed that black surfperch could use, and they shifted into this zone (Fig. 8-2). But removing black surfperch, which are more common deeper, did not free *Gelidium* in those areas. So striped surfperch did not change in depth distribution (Fig. 8-2). Had *Gelidium* been uniformly distributed in space, our removals would not have produced any changes in depth distribution, and we might have drawn from these behavioral data the false conclusion that no competition was occurring. By contrast, changes in use of feeding microhabitats were much better predictors of the symmetry of the eventual numerical response, as well as of the existence of spatial variation in the strength of the interaction (Table 8-2). The reason for this is that feeding behavior is likely a much better proxy for demographic responses such as changes in growth or fecundity, which ultimately drive changes in population size.

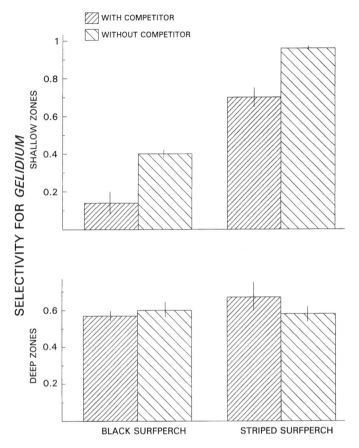

Figure 8-3. Selectivity (Manly's $\propto$; see Chesson 1983) for *Gelidium* of black surfperch and striped surfperch in the shallow zones and deep zones of experimental (without competitor) and control (with competitor) reefs. Given are means ($\pm$ 1 S.E., $N = 2$ reefs each treatment).

What do our results indicate about the appropriateness of the spatial scale of our experiments? We chose to work throughout the depth range for the two species at Santa Cruz Island, and this was possible because the reefs are relatively steeply sloping, so the various depths are in close proximity (a matter of tens of meters or less of linear distance). However, our findings and conclusions would have been quite different had we conducted experiments only in the shallow or in the deep zone. This would have been possible to do since the fish are quite residential (although not strictly territorial) and tagging studies indicated that individuals tend to remain within relatively small ranges of depth throughout their lives. If the manipulations had been confined to the deepest (9-m) areas, it is unlikely that much evidence for interspecific competition would have been obtained. The immediate exit of black surfperch from the deep zone following removal of striped surfperch (Fig. 8-2) resulted from competitive release in

Table 8-2. Scaling-up the results of short-term competitive release to predict long-term abundance responses

Short-term competitive release	Prediction	Reality
Depth shifts	Asymmetrical response Abundance of black sp. ↑ Abundance of striped sp. ↑	Wrong
Changes in microhabitat use	Symmetrical response Abundance of both ↑	Correct
	Spatial variation Response in shallow zone > deeper zone	Correct

Predictions regarding the symmetry and strength of density responses that would be expected based on patterns of competitive release observed for two behavioral variables (depth shifts and changes in microhabitat use) are given. "Reality" indicates how well the predictions matched the density responses that were ultimately observed.

the shallower zones. That this exodus did not reflect competition in the deep zone is evidenced by the failure of densities to increase in this area after four years of sustained removal of striped surfperch (Fig. 8-4). Striped surfperch densities never changed in the deep zone, regardless of whether black surfperch were present (Figs. 8-2 and 8-4). The weakness of interspecific competition in the deepest habitats is corroborated by the lack of changes in feeding behavior of each species upon competitive release (Fig. 8-3). By contrast, since the competitive interaction is very strong in the shallow reef habitats, if experiments had been conducted only in those zones our conclusions about the strength and symmetry of the interaction would have been similar to those obtained in the larger scale experiments that we actually did. These findings are not surprising given that the surfperch occupy a habitat gradient where their contested resources vary strongly over short spatial scales. However, they do point out the value of obtaining as much information as possible about patterns of resource distribution among habitats and explicitly considering such knowledge when deciding on the spatial scale of a competition experiment.

Our results on scaling up illustrate the difficulty in using responses at one spatial or temporal scale to infer responses at others (Wiens et al. 1986, Hughes et al. 1987, Roughgarden et al. 1988, Steele 1989). Like our study, several others have suggested that results of short-term field experiments must be scaled up with caution, if at all. For example, Hughes et al. (1987) found profound changes in the algal coral community in the Caribbean following mass mortality of an herbivorous sea urchin that were not anticipated by previous small-scale removal experiments (Sammarco 1980, 1982a,b). With respect to interspecific competition, there need not be obligatory coupling between effects at the individual and population levels (MacNally 1983). Manifestations of interspecific competition on individuals may be decoupled from effects on density if some other factor limits population size. For instance, competition for food between a hermit crab and a gastropod affected foraging behavior and patch use of the crab but did not impact abundance because that was set by shell limitation (Raimondi and Lively 1986). And despite a wrasse's strong behavioral interaction with a tripterygiid fish that greatly

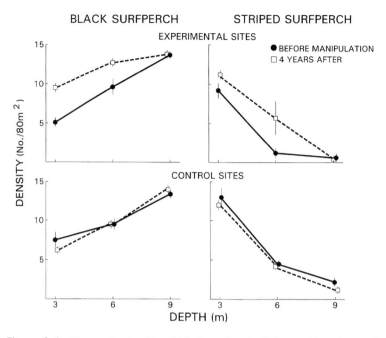

Figure 8-4. Change in density of black surfperch (*left panels*) and striped surfperch (*right panels*) within each depth zone after 4 years of sustained removal of the competitor (*top graphs*) and at unmanipulated control sites (*bottom graphs*). Given are mean (± 1 S.E.) number of fish along 80 m² transects, based on eight daily counts per depth per site during each time period.

influenced its patterns of foraging, there was little evidence that the tripterygiid had much effect on broader scale patterns of distribution and abundance of the wrasse (Thompson and Jones 1983). Clearly, although changes in demographic rates upon competitive release may provide the strongest inference regarding the ultimate density effects of competition, if other information about the system that would allow competition to be placed into a broader context is lacking, scaling up is best avoided.

Does Interspecific Competition Account for Variation in Abundance of Surfperches Among Reefs?

Despite the finding from our experiments that interspecific competition has a strong impact on the average abundance of black surfperch and striped surfperch, we found little evidence that it accounted for patterns of spatial variation at the among-reef scale. Rather, differences in the amount of food resources among reefs determined spatial patterns of abundance at Santa Cruz Island. Data from spatial surveys revealed that the availability of most-favored feeding substrates (turf for black surfperch and *Gelidium* for striped surfperch) accounted for a large amount (> 70%) of spatial variation in abundance (Schmitt and Holbrook 1986, 1990b); each of these substrates is highly correlated with surfperch abundance on a time-averaged basis (Fig. 8-5).

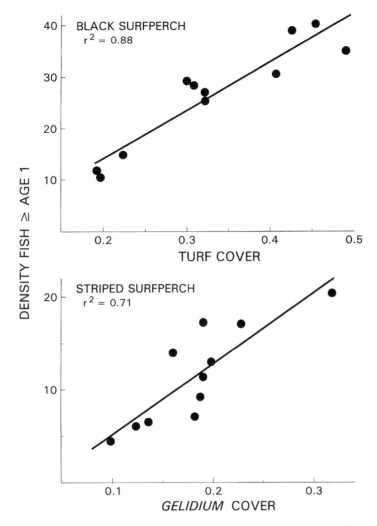

Figure 8-5. The relationship between density of black surfperch and striped surfperch and benthic cover (turf and *Gelidium*) on reefs ($N = 11$) at Santa Cruz Island. Each point represents a single reef. All data are time-averaged (1982–1990).

We tested for a causal link between availability of reef resources and abundance of surfperch. On six reefs at Santa Cruz Island we reduced the cover of *Gelidium* by 5–90% by removing the plants, and we measured subsequent changes in abundance of striped surfperch. The amount of decline in striped surfperch was strongly correlated with the amount by which the feeding resource (*Gelidium*) had been experimentally decreased (Fig. 8-6). So although interspecific competition can substantially depress abundance of the surfperch at particular places and times, availability of feeding resources for each species, not their competitive interaction, actually explains much of

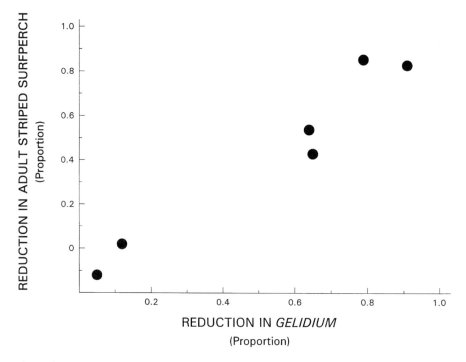

Figure 8-6. Reduction in adult striped surfperch following experimental reduction of cover of *Gelidium* on six reefs at Santa Cruz Island. Reduction at each site occurred along 60 m of reef from the intertidal zone to a depth of 12 m. Data are the proportional decline in adult fish 2 months following the manipulation.

the variation in abundance among different reefs. Interspecific competition will be most easy to detect at sites where there is ample supply of turf and *Gelidium*, thus affording an opportunity for a large numerical response following competitive release. It would have been difficult to detect the interaction had we conducted our experiments at sites where one species was extremely abundant and the other very rare, because the abundant species would experience relatively little increase in density upon removal of the rare competitor, and the potential numerical response of the rare competitor on competitive release could be constrained by its own (rare) resource supply. This emphasizes the importance of placing results of competition experiments in as broad a context as possible when examining spatial patterns of abundance.

Temporal Variation in Abundance and the Importance of Environmental Factors

Local surfperch populations can be quite constant through time or quite variable, like other species of temperate reef fish that have been examined (Ebeling et al. 1980, 1990; Stephens and Zerba 1981; Stephens 1983; Stephens et al. 1984, 1986, 1994; Cowen and Bodkin 1993; Holbrook et al. 1993, 1994). For surfperch, much of this temporal variation can be attributed to fluctuations in availability of key feeding substrates (*Gelidium* and turf), because they determine the amount of crustacean food actually available for consumption (Holbrook et al. 1990a,b; Schmitt and Holbrook 1990b). These feeding substrates, like other components of algal assemblages in Southern California, tend to be quite variable. Over a period of years, biotic factors such as urchin grazing, as well as abiotic events that include storms and upwelling, can cause many changes in cover. Species such as the canopy-forming giant kelp (*Macrocystis pyrifera*) can disappear from a reef only to recolonize it in subsequent years, resulting in dramatic effects on the understory algal cover. The supply of turf can also change markedly over time as amount of foliose algae varies; on our study sites, increases in giant kelp result in a decline in understory *Gelidium* and a larger supply of turf (Schmitt and Holbrook 1990a). There is relatively little information on the rates of interannual change of algal assemblages, but several long-term studies (Dayton et al. 1984, 1992; Dayton 1985), and our own studies and data indicate that substantial variation can occur on time scales of a few years. If algae are crucial to surfperch populations because they meditate the food supply, predictable changes in surfperch should occur as supply of algae or other reef substrates varies. In fact, the surfperch do track reef resources. This tracking is reflected both in the general amounts of year-to-year variability observed in fish populations and reef resources and in demographic responses of particular populations as their environment changes over time.

On reefs where the cover of *Gelidium* or turf is highly variable from year to year, numbers of adult surfperches also vary greatly, and reefs where the resource base fluctuates little similarly have fish populations that fluctuate little (Fig. 8-7). When individual reefs are examined in detail, there is clear evidence that key demographic features of the fish respond to the supply of food resources. For instance, as the amount of turf on a reef increases, black surfperch experience a higher birthrate and better early survival, and the population rises. Conversely, local populations decline (presumably due to a combination of emigration and mortality) as the resource supply dwindles (Schmitt and Holbrook 1990a).

During the early and middle 1980s, reefs at Santa Cruz Island tended to fluctuate asynchronously with respect to resource supply. That is, nearby reefs did not gain and lose cover of algae in synchrony, and the algal cover (and, as a result, surfperch populations) on some reefs tended to be more variable through time than on others (Fig. 8-7). However, despite asynchrony in numbers of fish per reef among local groups of reefs, the aggregate abundance of fish at our 11 study sites began to drift downward. We have no evidence that the downward trend started as a result of the very strong 1982 El Niño event, although the signal for the 1982 El Niño has been detected in other populations of reef fishes (Cowen and Bodkin 1993). Further, the trend downward of the surfperch was not strong during the 4-year competition experiment (Fig. 8-8), although we noted (Schmitt and Holbrook 1990b) in our discussion of the results of the experiments that populations of striped surfperch had declined somewhat, and we took this into account in our statistical analyses of the density responses. Even at the end of the decade, during 1989–1990, when we conducted our *Gelidium* manipulation experiment, abundances of fish were lower, but still a precipitous downward trend was not apparent (Fig. 8-8). Further, there was a more than adequate supply of fish on the reefs to permit population-level experiments. However, during the past several years

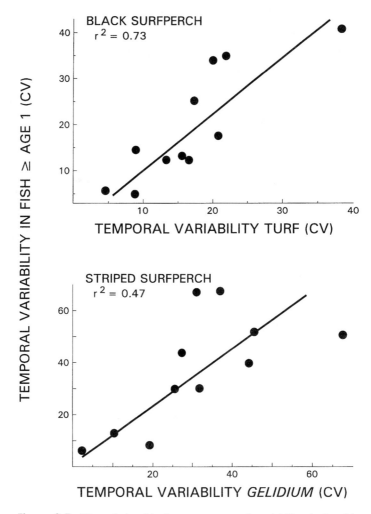

Figure 8-7. The relationship between temporal variability in benthic cover (turf and *Gelidium*) and in densities of black surfperch and striped surfperch at reefs ($N = 11$) at Santa Cruz Island. Each point represents a single reef over a 9-year period (1982–1990). Temporal variation for each variable is expressed as the coefficient of variation (CV).

dramatic changes have taken place. Populations of both surfperch have dropped tremendously at Santa Cruz Island and at other locations in Southern California (Holbrook and Schmitt 1996, Holbrook et al. in press). As we would have predicted based on earlier findings, a decline in food supply has been accompanied by a decline in the abundance of surfperch (Fig. 8-8).

These findings suggest that the 1990s have been a period of very low primary and secondary productivity; we estimate the magnitude of the decline in the two surfperches at our sites on Santa Cruz Island to be about 80–90% and that of their food base to

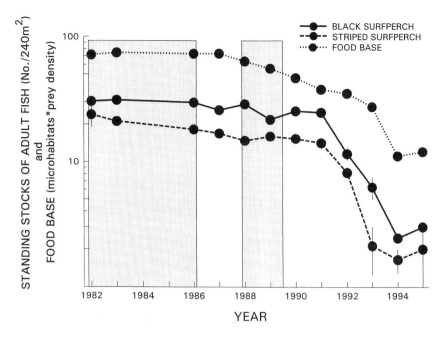

Figure 8-8. Temporal trends in surfperches and their food base (averaged across depth zones) at Santa Cruz Island. Shown are mean ($\pm$ 1 S.E.) number of adults per site (N = 2 sites) of black surfperch and striped surfperch and the mean value of the food base. The food base was estimated during the prey-rich season of each year (fall). Shaded areas in rectangles indicate time periods of competition and *Gelidium* reduction experiments. Some S.E. bars for densities of fish are obscured by the symbols.

be on the order of 60–70%. The temporal trends in population densities strongly suggest that the fish are tracking a declining food resource base, a finding that corroborates the idea that productivity regimes in the California Current have changed markedly in recent years. It is unclear whether these changes are related somehow to defined El Niño events (several years since 1990 have been classified as El Niño years based on oceanographic conditions) or the product of longer term trends in ocean conditions that might be reflective of global warming. Our analyses and those of others do suggest that changes in upwelling patterns that have resulted in decreased supply of nutrient-rich cooler waters to nearshore areas could be a contributing factor to the trends (Norton and Crooke 1994, Roemmich and McGowan 1995, Holbrook and Schmitt 1996). The declines of black surfperch and striped surfperch at Santa Cruz Island are representative of temporal patterns in abundance of a larger group of reef-associated fishes in the Southern California Bight. For instance, our analyses revealed declines of 75% in a group of 5 reef fishes at Santa Cruz Island (including the two surfperch) between 1982 and 1994 and of about 65% in a group of 33 fishes from King Harbor, California, between 1974 and 1993 (Holbrook et al. unpublished ms.). We predict that in the future numbers of these species will increase only as their resource supply improves.

Interpretation of Competition Experiments in Light of Long-Term Trends

The research in the surfperch system indicates that competition experiments must be conducted and interpreted cautiously. There are several reasons for this. First, scaling up results can present problems. We found that even very strong behavioral responses do not always reflect the density consequences of the interaction and that scaling up is likely to be most successful when behavioral responses reflect demographic responses that ultimately drive densities and when interspecific competition is of primary importance in limiting abundance. Clearly, these conditions will prevail in only a subset of natural systems. Further, the spatial scale at which an experiment is conducted can obviously influence the results that are obtained and, in the case of the surfperch system, experimentation on a small spatial scale that represents a subset of available habitats (e.g., the 9-m-depth zone) would lead to an incorrect conclusion regarding the existence and strength of interspecific competition.

Second, the interaction can fail to explain spatial or temporal patterns in variation even when it has a tremendous effect on mean abundance. For surfperch, strong variations in resource supply account for much of the observed variation in population densities. Our experimental design, because it explicitly included consideration of resource supply and species response across a spatial (depth) gradient, did encompass some of the spatial variation. But the bulk of our experiments were conducted during a time period when the system was rich in surfperch resources—and thus surfperch. The longer period temporal fluctuations that result from environmental forcing were not apparent in the early years of our work. However, the mechanistic understanding of the critical behavioral, demographic, and population linkages to the food supply that we gained by conducting our experiment along the habitat gradient gave us the foundation of knowledge we needed to explore and understand larger scale (among reef) spatial and temporal (decade and longer) patterns of variation. Importantly, our mechanistic understanding enables us to interpret the long-term environmental forcing of this resource-driven system. It is ironic that at the present time surfperch are so uncommon on reefs at Santa Cruz Island that competition experiments could not practically be conducted. In fact, interspecific competition would not be at the forefront of processes that would be thought to be impacting abundance, since neither species is abundant enough to be suspected of influencing the other's density. This does not necessarily imply that the species are not now competing for (currently) very limited resources; our resource-driven explanation of their dynamics would predict that they are still competing. Rather, a competition experiment, if it were conducted now, would likely fail to yield results simply because the species are collectively so uncommon that competitive release could not be detected easily simply due to sampling error.

ACKNOWLEDGMENTS   We thank the many divers who have helped us underwater at Santa Cruz Island, as well as the people who spent countless hours in the lab working on samples. The help of A. Breyer, J. Crisp, D. Canestro, and M. Angherra and the technical assistance of B. Williamson are especially acknowledged. M. Kingsford, W. Resetarits Jr., and T.

Behrents Hartney provided very helpful comments on the manuscript. We are deeply grateful for financial support from the National Science Foundation (Grant OCE-91–02191 and earlier awards), without which this research would not have been possible.

Literature Cited

Bender, E. A., T. J. Case, and M. E. Gilpin. 1984. Perturbation experiments in community ecology: theory and practice. Ecology 65:1–13.

Chesson, J. 1983. The estimation and analysis of preference and its relationship to foraging models. Ecology 64:1297–1304.

Connell, J. H. 1983. On the prevalence and relative importance of interspecific competition: evidence from field experiments. American Naturalist 122:661–696.

Cowen, R. K. 1985. Large scale patterns of recruitment by the labrid, *Semicossyphus pulcher*: causes and implications. Journal of Marine Research 43:719–742.

Cowen, R. K., and J. L. Bodkin. 1993. Annual and spatial variation of the kelp forest fish assemblage at San Nicolas Island, California. Pages 463–474 in F. G. Hochberg (ed.), Third California Islands Symposium: Recent Advances in Research on the California Islands. Santa Barbara Museum of Natural History, Santa Barbara, California.

Dayton, P. K. 1985. Ecology of kelp communities. Annual Review of Ecology and Systematics 16:215–245.

Dayton, P. K., V. Currie, T. Gerrodette, B. Keller, R. Rosenthal, and D. Van Tresca. 1984. Patch dynamics and stability of some southern California kelp communities. Ecological Monographs 54:253–289.

Dayton, P. K., M. J. Tegner, P. E. Parnell, and P. B. Edwards. 1992. Temporal and spatial patterns of disturbance and recovery in a kelp forest community. Ecological Monographs 62:421–445.

Ebeling, A. W., R. J. Larson, W. S. Alevizon, and R. N. Bray. 1980. Annual variability of reef-fish assemblages in kelp forests off Santa Barbara, California. Fisheries Bulletin U.S. 78:361–377.

Ebeling, A. W., S. J. Holbrook, and R. J. Schmitt. 1990. Temporally concordant structure of a fish assemblage: bound or determined? American Naturalist 135:63–73.

Fiedler, P. C., R. D. Methot, and R. P. Hewitt. 1986. Effects of California El Niño 1982–1984 on the northern anchovy. Journal of Marine Research 44:317–338.

Hayward, T. L. 1993. Preliminary observations of the 1991–1992 El Niño in the California Current. CalCOFI Reports 34:21–29.

Holbrook, S. J., and R. J. Schmitt. 1984. Experimental analyses of patch selection by foraging black surfperch (*Embiotoca jacksoni* Agassiz). Journal of Experimental Marine Biology and Ecology 79:39–64.

————. 1986. Food acquisition by competing surfperch on a patchy environmental gradient. Environmental Biology of Fishes 16:135–146.

————. 1988a. Effects of predation risk on foraging behavior: mechanisms altering patch choice. Journal of Experimental Marine Biology and Ecology 121:151–163.

————. 1988b. The combined effects of predation risk and food reward on patch selection. Ecology 69:125–134.

————. 1989. Resource overlap, prey dynamics and the strength of competition. Ecology 70:1943–1953.

————. 1992. Causes and consequences of dietary specialization in surfperches: patch choice and intraspecific competition. Ecology 73:402–412.

————. 1995. Compensation in resource use by foragers released from interspecific competition. Journal of Experimental Marine Biology and Ecology 185:219–233.

————. 1996. On the structure and dynamics of temperate reef fish assemblages: are resources tracked? Pages 19–48 in M. L. Cody and J. A. Smallwood (eds.), Long-Term Studies of Vertebrate Communities. Academic Press, San Diego.

Holbrook, S. J., R. J. Schmitt, and R. F. Ambrose. 1990a. Biogenic habitat structure and characteristics of temperate reef fish assemblages. Australian Journal of Ecology 15: 489–503.

Holbrook, S. J., M. H. Carr, R. J. Schmitt, and J. A. Coyer. 1990b. The effect of giant kelp on local abundance of demersal fishes: the importance of ontogenetic resource requirements. Bulletin of Marine Science 47:104–114.

Holbrook, S. J., S. L. Swarbrick, R. J. Schmitt, and R. F. Ambrose. 1993. Reef architecture and reef fish: correlates of population densities with attributes of subtidal rocky environments. Pages 99–106 in C. N Battershill, D. R. Schiel, G. P. Jones, R. G. Creese, and A. B. MacDiamid (eds.), Proceedings of the Second International Temperate Reef Ecology Symposium. National Institute of Water and Atmospheric Research, New Zealand Oceanographic Institute, Wellington.

Holbrook, S. J., M. J. Kingsford, R. J. Schmitt, and J. S. Stephens Jr. 1994. Spatial and temporal patterns in assemblages of temperate reef fish. American Zoologist 34:463–475.

Holbrook, S. J., R. J. Schmitt, and J. S. Stephens Jr. 1997. In press. Changes in an assemblage of temperate reef fishes associated with a climate shift. Ecological Applications.

Hughes, T. P., D. C. Reed, and M. Boyle. 1987. Herbivory on coral reefs: community structure following mass mortalities of sea urchins. Journal of Experimental Marine Biology and Ecology 113:39–59.

MacNally, R. C. 1983. On assessing the significance of interspecific competition to guild structure. Ecology 64:1646–1652.

McGowan, J. A. 1984. The California El Niño. Oceanus 27:48–51.

————. 1985. El Niño 1983 in the Southern California Bight. Pages 166–184 in W. S. Wooster and D. L. Fluharty (eds.), El Niño North: Niño Effects in the Eastern Subarctic Pacific Ocean. Washington Sea Grant Program, University of Washington, Seattle.

Murray, S. N., and M. H. Horn. 1989. Variations in standing stocks of central California macrophytes from a rocky intertidal habitat before and during the 1982–1983 El Niño. Marine Ecology Progress Series 58:113–122.

Mysak, L. 1986. El Niño, interannual variability and fisheries in the northeast Pacific Ocean. Canadian Journal of Fisheries and Aquatic Sciences 43:464–497.

Norton, J. G., and S. J. Crooke. 1994. Occasional availability of dolphin, *Coryphaena hippurus*, to Southern California commercial passenger fishing vessel anglers: observations and hypotheses. CalCOFI Reports 35:230–239.

Quinn, W. H., D. O. Zopf, K. S. Short, and R. T. W. K. Yang. 1978. Historical trends and statistics of the Southern Oscillation, El Niño, and Indonesian droughts. Fisheries Bulletin U.S. 76:663–678.

Raimondi, P. T., and C. M. Lively. 1986. Positive abundance and negative distribution effects of a gastropod on an intertidal hermit crab. Oecologia 69:213–216.

Roemmich, D., and J. McGowan. 1995. Climatic warming and the decline of zooplankton in the California Current. Science 267:1324–1326.

Roughgarden, J., S. Gaines, and H. Possingham. 1988. Recruitment dynamics in complex life cycles. Science 241:1460–66.

Sammarco, P. W. 1980. *Diadema* and its relationship to coral spat mortality: grazing, competition, and biological disturbance. Journal of Experimental Marine Biology and Ecology 45:245–272.

————. 1982a. Echinoid grazing as a structuring force in coral communities: whole reef manipulations. Journal of Experimental Marine Biology and Ecology 61:31–55.

————. 1982b. Effects of grazing by *Diadema antillarum* Philippi (Echinodermata: Echinoidea) on algal diversity and community structure. Journal of Experimental Marine Biology and Ecology 65:83–105.

Schmitt, R. J., and S. J. Holbrook. 1984a. Gape-limitation, foraging tactics and prey size selectivity of two microcarnivorous species of fish. Oecologia 63:6–12.

————. 1984b. Ontogeny of prey selection by black surfperch, *Embiotoca jacksoni* (Pisces: Embiotocidae): the roles of fish morphology, foraging behavior, and patch selection. Marine Ecology Progress Series 18:225–239.

————. 1985. Patch selection by juvenile black surfperch (Embiotocidae) under variable risk: interactive influence of food quality and structural complexity. Journal of Experimental Marine Biology and Ecology 85:269–285.

————. 1986. Seasonally fluctuating resources and temporal variability of interspecific competition. Oecologia 69:1–11.

————. 1990a. Contrasting effects of giant kelp on dynamics of surfperch populations. Oecologia 84:419–429.

————. 1990b. Population responses of surfperch released from competition. Ecology 71: 1653–1665.

Schoener, T. W. 1983. Field experiments on interspecific competition. American Naturalist 122:240–285.

Steele, J. H. 1989. Discussion: scale and coupling in ecological systems. Pages 177–180 in J. Roughgarden, R. M. May, and S. A. Levin (eds.), Perspectives in Ecological Theory. Princeton University Press, Princeton, New Jersey.

Stephens, J. S., Jr. 1983. The Fishes of King Harbor: A Nine Year Study of Fishes Occupying the Receiving Waters of a Coastal Steam Electric Generating Station. Research and Development Series 83-RD-1, Occidental College, Los Angeles, California.

Stephens, J. S., Jr., and K. E. Zerba. 1981. Factors affecting fish diversity on a temperate reef. Environmental Biology of Fishes 6:111–121.

Stephens, J. S., Jr., P. A. Morris, K. Zerba, and M. Love. 1984. Factors affecting fish diversity on a temperate reef: the fish assemblage of Palos Verdes Point, 1974–1981. Environmental Biology of Fishes 11:259–275.

Stephens, J. S., Jr., G. A. Jordan, P. A. Morris, M. M. Singer, and G. E. McGowen. 1986. Can we relate larval fish abundance to recruitment or population stability? A preliminary analysis of recruitment or a temperate rocky reef. CalCOFI Reports 27:65–83.

Stephens, J. S., Jr., P. A. Morris, D. J. Pondella, T. A. Koonce, and G. A. Jordan. 1994. Overview of the dynamics of an urban artificial reef fish assemblage at King Harbor, California, USA, 1974–1991. Bulletin of Marine Science 55:1224–1239.

Tegner, M., and P. K. Dayton. 1987. El Niño effects on southern California kelp forest communities. Advances in Ecological Research 47:243–279.

Thompson, S. M., and G. P. Jones. 1983. Interspecific territoriality and competition for food between the reef fishes *Forsterygion varium* and *Pseudolabrus celidotus*. Marine Biology 76:95–104.

Ware, D. N., and R. E. Thomson. 1991. Link between long-term variability in upwelling and fish production in the Northeast Pacific Ocean. Canadian Journal of Fisheries and Aquatic Sciences 48:2296–2306.

Wiens, J. A., J. F. Addicott, T. J. Case, and J. Diamond. 1986. Overview: the importance of spatial and temporal scale in ecological investigations. Pages 145–153 in J. Diamond and T. J. Case (eds.), Community Ecology. Harper and Row, New York.

9

Ecological Experiments with Model Systems

The Ecotron Facility in Context

JOHN H. LAWTON

Experimental ecology is not something you have to do in the field. Indeed, many experiments are both easier and less expensive to do in a laboratory. But by definition, laboratory experiments are less natural and more artificial than field experiments. There is therefore a trade-off between experimental tractability and ecological realism whenever ecologists do experiments in the laboratory. But as I hope to show in this chapter, that trade-off will often be worth making, not as a substitute for fieldwork, but as one of several powerful approaches to understanding and predicting the behavior of ecological systems, ranging from mathematical models, through analysis of large-scale data sets, to laboratory and field manipulation experiments. All have their place in an ecologist's armory.

The term *model systems* in ecology embraces many approaches. Here I restrict its usage to mean laboratory systems that involve two or more species set up in such a way that the populations persist and interact over several generations, without the repeated intervention of the experimenter, except, perhaps, to add food or water. We sometimes call the resulting assemblages microcosms if they are small and mesocosms if they are larger (with no agreed definition of what is "small" or "large"). I briefly review some of the ways in which ecologists have done experiments with model systems in the laboratory and some of the questions they have asked before I focus on a particularly large and sophisticated model system: the Ecotron facility at Silwood Park, Ascot, U.K. Throughout, my emphasis is on the advantages and disadvantages of model systems as experimental subjects for ecologists. Additional arguments and examples can be found in Kareiva (1989), Beyers and Odum (1993), Körner et al. (1993), and Lawton (1995, 1996).

Links Between Some Classic Laboratory Experiments and Current Ecological Thinking

Gause's Paramecium and Darwin's Finches

Ecologists have always used model systems to disentangle the complexities of nature. The classic laboratory experimental work by Gause (1934), for instance, on competitive exclusion and coexistence in three species of *Paramecium*, has inspired generations of field ecologists and spawned a vast subsequent literature. It was one of the works cited by MacArthur (1958) to provide the theoretical underpinning for his now equally famous paper on the coexistence of North American wood warblers. It also played a pivotal part in Lack's (1947) reinterpretation of data on Darwin's finches, forcing Lack to the conclusion that interspecific competition was the key to understanding the bill morphologies and interisland distributions of these famous little brown and black birds. Darwin's finches are arguably, now, *the* most thoroughly studied and most well-known field example of evolution in action and of the interplay between intra-and interspecific competition, evolution, and the distribution and abundance of animals (Grant 1986, Grant and Grant 1989, Weiner 1994). The pedigree of these wonderful field studies goes back directly to Gause.

Park's Tribolium and Shared Enemies

Equally well-known, but not as old, are Park's laboratory experiments on *Tribolium* flour beetles (Park 1948,1962), demonstrating the effects of chance, climate, and a shared sporozoan parasite on population persistence. This work is usually remembered because it shows particularly clearly how the outcome of two-species competition (for limited food and space, in which the two species both compete and eat one another) changes with climate; for instance, under hot-moist conditions, *T. castaneum* always wins, under cool-arid conditions the victor is invariably *T. confusum*, and in temperate-moist and temperate-arid environments the outcome is a lottery—sometimes *confusum* wins, sometimes *castaneum*. In his 1948 paper, Park also showed that the ''normal'' competitive outcome could be reversed by a shared sporozoan parasite, *Adelina tribolii*, which favors *confusum* over *castaneum*. Yet the consequences of and general principles that underpin this important result—how shared enemies, rather than shared resources, influence population persistence—remained oddly neglected by all but a few ecologists (Williamson 1957, Holt 1977) until recently.

Revived interest in the problem started with Holt (1977), who termed the phenomenon *apparent competition*, because formally and in general terms (but not in detail) the population dynamic effects of two species sharing an enemy (be it a predator, parasitoid, or, as in the case of Park's *Tribolium*, parasite) are identical to those in which two species compete exploitatively for a limiting resource. The phenomenon is now receiving increasing attention, both theoretically and in laboratory and field experiments (Holt and Lawton 1994). Apparent competition is probably at least as important as ''classical'' interspecific competition for limiting resources in determining the distributions and abundances of organisms.

Laboratory Predator–Prey Systems

Laboratory experiments that involve a single species of enemy and one species of victim (predator and prey or host and parasitoid) are reasonably common in the literature. Historically, these studies have been influential in at least two ways. First, although it is now well-known that different parameter values can give rise to markedly different population dynamics in structurally similar models (stable equilibria, persistent cycles, chaotic fluctuations, or extinction of one or both populations; e.g., Murdoch and McCauley 1985), Utida (1957) appears to have been among the first experimental ecologists to recognize this, linking differences in the dynamics of long-running laboratory populations of bean weevils (*Callosobruchus*) to differences in the fecundities and searching efficiencies of two species of parasitoids. To the pioneers of theoretical population biology (including Gause, Lotka, and Volterra, to name but three) the idea was obvious, and it was developed explicitly in their models; but the simplicity and pedagogic power of Utida's laboratory experiments undoubtedly helped to make the phenomenon more familiar to field biologists.

A second, even better known laboratory predator–prey experiment is Huffaker's work with *Eotetranychus* mites as prey (sustained by a supply of oranges as food, laid out in a grid—a strange but effective mesocosm!) and *Typhlodromus* mites as predators (Huffaker 1958). It remains the classic experimental demonstration of how environmental heterogeneity, generating spatial refuges, allows the persistence of an otherwise unstable predator–prey interaction. Huffaker also concluded that local dispersal from adjacent occupied patches was a key to the persistence of the system. Only relatively recently have sufficiently powerful computers made it easy for ecologists to model the spatial dynamics of coupled enemy–victim populations with local movement (e.g., Hassell et al. 1994). These models confirm that local movement in a patchy environment can stabilize an otherwise unstable enemy–victim interaction. They also show that deterministically generated spatial patterns in such populations can be exceedingly complex, as Huffaker's mites indeed demonstrate; there is clearly considerable potential in this area for further integration of theory with experimental work, using model laboratory systems. Indeed, probably because of the difficulty of working in the field at a sufficient number of sites and over a sufficiently large spatial scale, experimental tests of these ideas that use field experiments may take a long time to materialize, if they are ever attempted at all.

General Comments

These examples are by no means an exhaustive selection of laboratory experiments that have had a profound influence on the way in which modern ecologists view the natural world. I have not, for example, said anything about laboratory experiments with plants, partly for lack of space and partly because my own areas of interest and expertise lie in higher trophic levels! But zoologists do not have a monopoly of important and influential laboratory experiments, as Harper (1977) makes abundantly clear. The examples I have discussed serve simply to show that experiments do not have to be carried out in the field to be influential or to have relevance to ecological problems in the "real world." I doubt that more than a tiny fraction of the field experiments that currently fill the pages of ecological journals will either have as influential a pedigree as Gause, Park, Utida, Huffaker, or Harper or still be worth citing 30 years or more from now. Indeed, it has probably always been so in ecology. It is worth recalling the (undated) quotation attributed to Charles Manning Child in the dedication to Victor Shelford's masterpiece, *The Ecology of North America* (Shelford 1963): "Do experimental work but keep in mind that other investigators in the same field will consider your discoveries as less than one fourth as important as they seem to you." Of course, much the same comment can be made about laboratory experiments, and probably sometimes more so!

Modern Applications: Assembly Rules and the Dynamics of Complex Systems

The use of laboratory experiments has grown rapidly in the last few years. Again, it is neither possible nor necessary to provide a comprehensive review. Some examples will suffice. The approach has been particularly valuable for problems that are either too expensive, too difficult, or too time-consuming for field experiments to be a sensible alternative *at the moment*. These last three words are vital. In the longer run, once the potential importance of a phenomenon has been demonstrated in laboratory experiments, the goal must be to shift attention to the field. But we have to accept that for some ecological problems field experiments may be many years in the making, and others may have no realistic prospect of ever being attempted, because of cost and other logistical difficulties. Assembly rules and the dynamics of complex systems are good examples.

Assembly Rules

There can be no herbivores without plants, but are there less trivial assembly rules for ecological systems? For example, are some food web configurations more likely to persist than others (Pimm et al. 1991, Morin and Lawler 1995a,b), must species-rich communities be built up via particular subsets of species, and are there forbidden combinations and alternative stable states along that route (Drake et al. 1993); and how does the environment interact with the biota to determine community diversity and dynamics (Luckinbill and Fenton 1978; Weiher and Keddy 1995)? The first steps toward answering these questions are the small two-and three-species systems discussed in the previous section. These "modules" form the units from which larger communities are assembled; theory tells us that combining modules into more complex webs may fundamentally change the population dynamics of component species (e.g., Utida 1957, Hochberg et al. 1990, Holt and Lawton 1994).

Neill's Experiment

Using laboratory experiments to study community assembly is a relatively new field. In a pioneering study, Neill (1975) repeatedly inoculated freshwater microcosms with 12 species of crustaceans and more than 20 species of green and blue-green algae. After about 250 days, the 15 replicate communities each contained the same five species of algae, three cladocera, and one amphipod; all other taxa failed to establish. Neill's results show unequivocally that in a defined environment some combinations of organisms are unstable and fail to yield persistent sets of coexisting species; only some combinations of species "work," generating stable assemblages (exactly as theory predicts; e.g., May 1973). Lacking in Neill's work are theoretical predictions about which species would and would not be able to coexist in the system; indeed, sufficiently detailed models may still be some way off, although an appropriate framework now exists (Holt et al. 1994).

Food Webs

Recent laboratory experimental work summarized in, and much of it carried out by, Morin and Lawler (1995a,b) using aquatic bacteria and protists is starting to chip away at mechanisms of food web assembly. This work builds directly on the approach pioneered by Gause, and indeed, as Peter Morin has pointed out to me, Gause himself referred to some of his own experimental systems as food chains, for instance (Gause 1934:118) "the food chain: bacteria → *Paramecium* → *Didinium*." Gause observed that in "limited microcosms" this food chain is unstable.

More than 60 years later, a rich body of theory, much of it still to be tested experimentally, exists on the population dynamics and stability of food chains and webs (e.g., Pimm 1991). This theory predicts, for instance, that long food chains and those with abundant omnivory (species feeding on more than one trophic level) are less likely to be stable than shorter, simpler webs. Experimentally created food webs of bacteria, bacterivorous protists, and predatory protists support the first prediction but refute the second (Morin and Lawler 1995a). A second set of experiments by Lawler confirms model predictions about prey coexistence and exclusion with shared predators (Holt 1977; see previous discussion), specifically that the prey supporting the highest density of predators in paired prey-predator combinations should exclude alternative species of prey in more complex (two-prey, one-predator) food webs (Holt and Lawton 1994).

These laboratory experiments with aquatic microorganisms are now being extended to study the relative strengths of "top-down" (predation) versus "bottom-up" (nutrient) effects in food chains (Balčiūnas and Lawler 1995). The cost, time, and logistical ease of carrying out such experiments in small laboratory ecosystems are trivial compared with the investment needed to do similar experiments in the field, although a growing number of whole-lake manipulative experiments on "top-down" and "bottom-up" effects in food webs now exist (Carpenter and Kitchell 1993, Carpenter et al. 1995). So why continue with laboratory experiments? There is still a need for laboratory experiments because current field experiments do not test model predictions quantitatively. And "top-down" versus "bottom-up" effects aside, it has also proved to be extraordinarily difficult and time-consuming to test theoretical predictions about the role of population dynamics in the more general assembly and stability of food webs in the field (Pimm et al. 1991, Paine 1992) and, indeed, to discover whether population dynamics need be invoked at all (e.g., Warren 1995). Under these circumstances, laboratory experiments with microorganisms are a valuable "halfway house" between theory and the full complexity of field systems.

Alternative Persistent and Stable States?

Two current concerns (see Ricklefs and Schluter 1993 for numerous examples)—the roles of chance and history in determining contemporary community structure and the interplay of local and regional processes—have been addressed recently by Drake and colleagues (Drake et al. 1993). They used a unique laboratory "landscape" that consisted of interconnected 1-liter aquatic microcosms, through which an assemblage of four algal species and four crustacea invaded and spread. By the end of the experiment (after 80 days), species were distributed heterogeneously among microcosms (recall Huffaker's experiment, discussed previously) and had converged on one of several alternative states (defined by species' presence/absence and relative dominance), despite identical initial conditions.

How often have you observed different combinations of species coexisting in close proximity in a patchy habitat and mentally attributed the patterns to different environmental conditions in different patches? A naive investigator, observing Drake et al.'s system for the first time and ignorant of its history, would almost certainly invoke deterministic differences in the environments of the individual microcosms to explain the differences between them. This would, of course, be wrong. Rather, the patterns the investigator observe are the products of chance differences in the colonization and persistence of species in each microcosm; these differences may then remain for some time and be amplified or modified by species' interactions.

This is not to say that Drake et al.'s experiment is definitive. Because it was terminated after only 80 days (a relatively short time in terms of the generation times of the crustacea), it is a moot point whether the patch-to-patch differences would have survived indefinitely (i.e., as genuine, alternative stable, or at least persistent, states) or would simply have been transient (Grover and Lawton 1994). This matters if we are to obtain unequivocal experimental evidence for or against the existence of alternative persistent (Law and Morton 1993) or stable (Knowlton 1992) states in the species composition of communities under otherwise identical environmental conditions. There is clearly enormous potential (albeit involving a great deal of hard work) in extending and developing Drake et al.'s pioneering approach to community assembly in miniature laboratory landscapes.

Terrestrial Examples

Although the majority of recent laboratory experiments on community assembly use aquatic systems (probably because of the short generation times of the organisms involved—typically bacteria, protists, algae, and crustacea), there is no reason in principle why terrestrial communities cannot be similarly investigated. The Ecotron is a good example, on which I focus in "The Ecotron Facility at Silwood Park." But there are many others. For instance, Bazzaz (1995) has used experimental terrestrial communities to explore the importance of variation in early spring weather for the structure of annual plant assemblages in the eastern United States. He found that he was able to predict the responses of these model systems from a knowledge of the physiological ecology of seed germination. Using a similar approach with laboratory communities of British herbaceous plants, Grime et al. (1987) demonstrated that both mycorrhizal infection and grazing promoted plant species diversity, whereas soil heterogeneity did not. As they point out, "all three [mycorrhizae, grazing, and soil heterogeneity] have been implicated in diversity theories by earlier investigators but the effects of each are exceedingly difficult to quantify in natural vegetation." In contrast, Grime et al.'s laboratory experiments yielded unequivocal answers.

Looking for Chaos

The potential for studying processes in the laboratory that are simply too difficult, too time-consuming, or too expensive to do in the field is enormous. To take but one final example: The quickest, though certainly not the only, way to obtain a sufficiently long time series to discover whether complex ecological systems display chaotic dynamics (Kareiva 1989, Hastings et al. 1993) is by assembling and running communities of the desired complexity in a laboratory experiment with or without imposed environmental "noise" of the correct structure (Halley 1995). A similar test of theory in the field with birds, fish, or higher plants might have to be continued for hundreds of years to obtain a sufficiently long time series. Realistically, this is impossible.

Overview

It would be easy to continue with a much longer catalog of examples. But the point is, I hope, now well made. Laboratory experiments have a long and valuable pedigree in ecology, and they continue to provide provocative insights into the real world, as well as relatively quick, cheap, and powerful tests of theory. They are also becoming progressively more realistic, as the final section will demonstrate.

The Ecotron Facility at Silwood Park

The Ecotron facility is one of the most sophisticated laboratory experimental facilities currently available to ecologists anywhere in the world. It consists of 16 physically and electronically integrated environmental chambers, housed in the Centre for Population Biology at Silwood Park, and allows ecologists to construct and run replicate miniature (1 m^2 × 40 cm deep) terrestrial ecosystems for many months under controlled environmental conditions. Detailed descriptions of the facility are in Lawton et al. (1993) and Naeem et al. (1995).

The computer-controlled chambers simulate natural environments in diurnal light–dark cycles, temperature, humidity, and rainfall. Conditions within the chambers and ecosystems are monitored and recorded automatically. All the main inputs and outputs (e.g., rainfall and drainage water, macronutrients, and CO_2) are also known. In 1994 the facility was modified to allow CO_2 levels to be precisely controlled and elevated.

The ecosystems are established by first adding microbial assemblages to sterilized soil from a filtered soil wash taken from Silwood Park. Plants, insect herbivores, parasitoids, and soil fauna can then be added to the system as required (e.g., Thompson et al. 1993; Naeem et al. 1994, 1995). The specific identities of all plant and metazoan taxa (with the exception of nematodes) are known. We have chosen to work with small, "weedy," native annual plants that are self-compatible (eliminating the need for pollinators) and germinate without vernalization. Experiments are run long enough for the plants to have more than one generation; a typical experiment runs over three complete plant generations and lasts about 9 months, during which time most of the animals go through many more than three generations.

By ecological norms the Ecotron is an expensive and complicated piece of equipment. James H. Brown's classic long-term field manipulation experiments on desert plant and animal assemblages, for example, have been supported by an average of only U.S.$ 80,000 a year for 20 years from the National Science Foundation (Brown, this volume). It took three and a half years from conception to the first experimental run in the Ecotron, and it cost U.S.$ 1.5 million to build at early 1990s prices. Basic running costs (excluding the salaries of the team who run it and work in it, collaborators' salaries and costs, and all other associated administrative charges) are currently (1996) of the order of U.S.$ 150,000 per annum. The real cost of each experiment is at least double this. Nevertheless, these large sums of money are still tiny compared with the costs of the oceangoing ships, satellites, radio telescopes, and accelerators upon which oceanographers, astronomers, and particle physicists routinely depend. Why should ecology be cheap?

What Have We Learned?

Two major experiments have been carried out in the Ecotron, and a third has recently (May 1997) been completed. The first (Thompson et al. 1993) was designed as much to fully test the facility as to address a major ecological question. It examined the effects of snails and earthworms on plant population dynamics and produced two surprises. It revealed totally unexpected worm-by-snail interactions on plant dynamics and showed major effects of worms on the dynamics of one of the plant species (*Trifolium dubium*). Worms favored *Trifolium* by increasing nodulation of nitrogen-fixing *Rhizobium*, by creating safe microsites for the germination of *Trifolium* seeds (worm casts), and by differentially burying the seeds of competing plant species below the depth at which they could germinate successfully. In other words, this laboratory experiment points the way to some obvious field studies in a relatively neglected area of plant population dynamics and plant–animal interactions.

The second Ecotron experiment explored the role of species richness in ecosystem processes (Schulze and Mooney 1993) by creating ecosystems with three levels of higher plant and animal diversity (Naeem et al. 1994, 1995). The simplest ecosystems contained 9 species, the intermediate systems 15 species, and the most diverse 31 species, structured so that the lower diversity systems contained a subset of species from the higher diversity systems. All ecosystems contained plants, herbivores, at least one species of secondary consumer (parasitoids), and a soil fauna, and all trophic levels had their diversity manipulated. It is probably impossible with present technology to control plant, herbivore, and soil–animal diversity in the field with the precision that was achieved in this experiment.

The results showed clearly that most ecosystem processes varied significantly with species richness, but not in any systematic way. Uptake of CO_2 and plant productivity, however, both declined as species richness declined. Overall, the data support three of four possible theoretical relationships between diversity and ecosystem function (Schulze and Mooney 1993), depending upon the process. There results were the first to provide experimental evidence that species richness affects ecosystem processes (Ehrlich and Wilson 1991).

This second Ecotron experiment has since been followed by pioneering field experiments (Tilman et al. 1996, Tilman et al. 1997) in which plant species richness (but not animal species richness) has been manipulated; these field experiments yield similar conclusions (biodiversity matters for ecosystem processes) but are all the more powerful because they are field experiments. With a consortium of European partners I have recently established a similar set of field experiments (The BIODEPTH experiment) at eight sites across Europe; I doubt we could have contemplated or received funding for such a complex (and expensive!) field experiment without having first done the Ecotron experiment. At the same time, this whole research field has become more contentious theoretically, as we struggle to understand and interpret the new data (Huston 1997), and it remains an entirely open question whether field experiments on the impacts of reducing animal species richness can be carried out in a similar manner. In the present context, however, the point is a very simple one. Once again, we see model systems, if not triggering, then at least being in on the beginning of a major intellectual foray by ecological science.

The Ecotron is currently being used to explore the effects of enhanced CO_2, enhanced temperature, and their combination on population, community, and ecosystem dynamics (C. G. Jones, H. Jones, J. H. Lawton, L. J. Thompson et al., unpublished ms.). This is the first experiment, anywhere in the world, in which an attempt is being made to follow the response of an entire plant and animal assemblage to these two aspects of simulated global environmental change continuously for several generations. A similar experiment in the field would be prohibitively expensive and time-consuming—which is why nobody is doing it.

A Critique of Laboratory Experiments

All manner of criticisms can be, and often are, leveled at laboratory experiments, be they Gause's experiments with *Paramecium*, the Ecotron experiments, or anything in between. Most of these criticisms are misguided (Lawton 1996). Laboratory experiments, particularly the more complex kinds now being carried out by several research groups and briefly summarized in this chapter, have four main advantages. First, they offer a tractable yet biologically realistic bridge between the simplicity of mathematical models (which incorporate one or very few species and omit many essential linkages) and the full complexity of the real world. If we cannot understand simplified ecosystems like those in the Ecotron, we are unlikely to understand very complex ones. Second, the very act of trying to create and maintain simplified ecosystems in the laboratory tests ecological knowledge to the limits. Third, laboratory experiments speed up research. Fourth, these experiments give a degree of control and replication that is impossible in the field.

Critics argue, in contrast, that despite these (and other advantages), the artificial nature of laboratory systems, the restricted taxonomic nature of their inhabitants, the general absence of environmental disturbances, and the small scale mean that experiments carried out in them are at best a harmless game and at worst a waste of time and money. People who make these and other criticisms (see Lawton 1996 for a more detailed discussion) fail to recognize that most species assemblages we regard as "natural" in temperate regions generally appear to have had very short evolutionary histories. During glacial and interglacial periods, the responses of species to climate change have been highly idiosyncratic, generating assemblages with no modern equivalents and extant communities with no antecedents (Davis 1981, Coope 1994). Assembling a sample of species from the same geographic region and habitat type in an Ecotron is not dissimilar to what actually happens in nature.

I prefer to see these and similar criticisms as hypotheses, amenable to more sophisticated experimental testing in the laboratory and in the field. If you believe that environmental disturbance is the key to understanding an ecological process, design experiments that control and vary the intensity of disturbance; if you believe that scale matters, design the same experiments at different spatial scales; and so on. Many of these experiments will be most easily and quickly done in the laboratory. But of course, where possible, they should also be repeated in the field. However, as I have argued earlier, not all desirable ecological experiments can currently be done in the field, because we lack either the time, the money, or the ability to control the experimental conditions—and sometimes we lack all three!

There is, of course, no point in doing an experiment in the laboratory when it can be just as easily and quickly carried out in the field—or, indeed, if it can be done in the field with more time, effort, and money than in a laboratory but not prohibitively so. However, we often need to walk before we can run. Field experiments on interspecific competition became worthwhile after Gause. Knowing what we now know, costly and time-consuming attempts to manipulate biodiversity and to study concomitant changes in ecosystem function seem justified (and are under way; e.g., Tilman and Downing 1994, Tilman et al. 1996), and so on. It is equally obvious that experiments in controlled environment facilities like the Ecotron will never replace field observations and experiments, analyses of long-term databases, and the search for large-scale patterns and processes in ecology. We need a variety of approaches to understand nature and to grapple with the challenges of global change. Laboratory experiments are a tool; like all tools, they do some things well, some badly, and others not at all.

ACKNOWLEDGMENTS   I am grateful to the editors for asking me to write this chapter and to many colleagues for stimulating and thoughtful discussions on the use of laboratory experiments in ecology. Peter Morin, William Resetarits, and Hefin Jones made helpful and constructive comments on the manuscript.

Literature Cited

Balčiūnas, D., and S. P. Lawler. 1995. Effects of basal prey resources, predation, and alternative prey in microcosm food chains. Ecology 76:1327–1336.

Bazzaz, F. A. 1995. Plants in Rapidly Changing Environments: Linking Physiological, Population, and Community Ecology. Cambridge University Press, Cambridge.

Beyers, R. J., and H. T. Odum. 1993. Ecological Microcosms. Springer-Verlag, Berlin.

Carpenter, S. R., and J. F. Kitchell (eds.). 1993. The Trophic Cascade in Lakes. Cambridge University Press, Cambridge.

Carpenter, S. R., S. W. Chisholm, C. J. Krebs, D. W. Schindler, and R. F. Wright. 1995. Ecosystem experiments. Science 269:324–327.

Coope, G. R. 1994. The response of insect faunas to glacial-interglacial climatic fluctuations. Philosophical Transactions of the Royal Society of London B 344:19–26.

Davis, M. B. 1981. Quaternary history and the stability of forest communities. Pages 132–153 in D. C. West, H. H. Shugart, and D. B. Botkin (eds.), Forest Succession: Concepts and Application. Springer-Verlag, New York.

Drake, J. A., T. E. Flum, G. J. Witteman, T. Voskuil, A. M. Hoylman, C. Creson, D. A. Kenny, G. R. Huxel, C. S. Larue, and J. R. Duncan. 1993. The construction and assembly of an ecological landscape. Journal of Animal Ecology 62:117–130.

Ehrlich, P. R., and E. O. Wilson. 1991. Biodiversity studies: science and policy. Science 253:758–762.

Gause, G. F. 1934. The Struggle for Existence. Williams and Wilkins, Baltimore, Maryland.

Grant, B. R., and P. R. Grant. 1989. Evolutionary Dynamics of a Natural Population: The Large Cactus Finch of the Galápagos. University of Chicago Press, Chicago, Illinois.

Grant, P. R. 1986. Ecology and Evolution of Darwin's Finches. Princeton University Press, Princeton, New Jersey.

Grime, J. P., J. M. L. Mackey, S. H. Hillier, and D. J. Read. 1987. Floristic diversity in a model system using experimental microcosms. Nature 328:420–422.

Grover, J. P., and J. H. Lawton. 1994. Experimental studies on community convergence and alternative stable states: comments on a paper by Drake et al. Journal of Animal Ecology 63:484–487.

Halley, J. M. 1995. Ecology, evolution, and 1/f-noise. Trends in Ecology and Evolution 11: 33–37.

Harper, J. L. 1977. Population Biology of Plants. Academic Press, London.

Hassell, M. P., H. N. Comins, and R. M. May. 1994. Species coexistence and self-organising spatial dynamics. Nature 370:290–292.

Hastings, A., C. L. Hom, S. Ellner, P. Turchin, and H. C. J. Godfray. 1993. Chaos in ecology: is mother nature a strange attractor? Annual Review of Ecology and Systematics 24:1–33.

Hochberg, M. E., M. P. Hassell, and R. M. May. 1990. The dynamics of host-parasitoid-pathogen interactions. American Naturalist 135:74–94.

Holt, R. D. 1977. Predation, apparent competition, and the structure of prey communities. Theoretical Population Biology 12:197–229.

Holt, R. D., and J. H. Lawton. 1994. The ecological consequences of shared natural enemies. Annual Review of Ecology and Systematics 25:495–520.

Holt, R. D., J. Grover, and D. Tilman. 1994. Simple rules for interspecific dominance in systems with exploitative and apparent competition. American Naturalist 144:741–771.

Huffaker, C. B. 1958. Experimental studies on predation: dispersion factors and predator-prey oscillations. Hilgardia 27:343–383.

Huston, M. A. 1997. Hidden treatments in ecological experiments: re-evaluating the ecosystem function of biodiversity. Oecologia 110:449–460.

Kareiva, P. 1989. Renewing the dialogue between theory and experiments in population ecology. Pages 68–88 in J. Roughgarden, R. M. May, and S. A. Levin (eds.), Perspectives in Ecological Theory. Princeton University Press, Princeton, New Jersey.

Knowlton, N. 1992. Thresholds and multiple stable states in coral reef community dynamics. American Zoologist 32:674–682.

Körner, C., J. A. Arnone III, and W. Hilti. 1993. The utility of enclosed artificial ecosystems in CO_2 research. Pages 185–198 in E.-D. Schulze and H. A. Mooney (eds.), Design and Execution of Experiments in CO_2 Enrichment. Ecosystems Research Report 6. Commission of the European Communities, Brussels.

Lack, D. 1947. Darwin's Finches. Cambridge University Press, Cambridge.

Law, R., and R. D. Morton. 1993. Alternative permanent states of ecological communities. Ecology 74:1347–1361.

Lawton, J. H. 1995. Ecological experiments with model systems. Science 269:328–331.

———. 1996. The Ecotron facility at Silwood Park: the value of ''big bottle'' experiments. Ecology 77:665–669.

Lawton, J. H., S. Naeem, R. M. Woodfin, V. K. Brown, A. Gange, H. J. C. Godfray, P. A. Heads, S. Lawler, D. Magda, C. D. Thomas, L. J. Thompson, and S. Young. 1993. The Ecotron: a controlled environmental facility for the investigation of population and ecosystem processes. Philosophical Transactions of the Royal Society of London B 341:181–194.

Luckinbill, L. S., and M. M. Fenton. 1978. Regulation and environmental variability in experimental populations of protozoa. Ecology 59:1271–1276.

MacArthur, R. H. 1958. Population ecology of some warblers of northeastern coniferous forests. Ecology 39:599–619.

May, R. M. 1973. Stability and complexity in model ecosystems. Princeton University Press, Princeton, New Jersey.

Morin, P. J., and S. P. Lawler. 1995a. Effects of food chain length and omnivory on population dynamics in experimental food webs. Pages 218–230 in G. A. Polis and K. O. Winemiller (eds.), Food Webs: Integration of Patterns and Dynamics. Chapman and Hall, New York.

———. 1995b. Food web architecture and population dynamics: theory and empirical evidence. Annual Review of Ecology and Systematics 26:505–529.

Murdoch, W. W., and E. McCauley. 1985. Three distinct types of dynamic behavior shown by a single planktonic system. Nature 316:628–630.

Naeem, S., L. J. Thompson, S. P. Lawler, J. H. Lawton, and R. M. Woodfin. 1994. Declining biodiversity can alter the performance of ecosystems. Nature 368:734–737.

———. 1995. Empirical evidence that declining biodiversity may alter the performance of terrestrial ecosystems. Philosophical Transactions of the Royal Society of London B 347:249–262.

Neill, W. E. 1975. Experimental studies of microcrustacean competition, community composition and efficiency of resource utilization. Ecology 56:809–826.

Paine, R. T. 1992. Food-web analysis through field measurement of per capita interaction strength. Nature 355:73–75.

Park, T. 1948. Experimental studies on interspecific competition: I. Competition between populations of the flour beetles, *Tribolium confusum* Duval and *Tribolium castaneum* Herbst. Ecological Monographs 18:267–307.

———. 1962. Beetles, competition and populations. Science 138:1369–1375.

Pimm, S. L. 1991. The Balance of Nature? University of Chicago Press, Chicago, Illinois.

Pimm, S. L., J. H. Lawton, and J. E. Cohen. 1991. Food web patterns and their consequences. Nature 350:669–674.

Ricklefs, R. E., and D. Schluter (eds.). 1993. Species Diversity in Ecological Communities: Historical and Geographical Perspectives. University of Chicago Press, Chicago, Illinois.

Schulze, E.-D., and H. A. Mooney (eds.). 1993. Biodiversity and Ecosystem Function. Springer-Verlag, Berlin.

Shelford, V. E. 1963. The Ecology of North America. University of Illinois Press, Urbana.

Thompson, L., C. D. Thomas, J. M. A. Radley, S. Williamson, and J. H. Lawton. 1993. The effect of earthworms and snails in a simple plant community. Oecologia 95:171–178.

Tilman, D., and J. A. Downing. 1994. Biodiversity and stability in grasslands. Nature 367: 363–365.

Tilman, D., D. Wedin, and J. Knops. 1996. Productivity and sustainability influenced by biodiversity in grassland ecosystems. Nature 379:718–720.

Tilman, D., J. Knops, D. Wedin, P. Reich, M. Ritchie, and E. Siemann. 1997. The influence of functional diversity and composition on ecosystem processes. Science 277:1300–1305.

Utida, S. 1957. Population fluctuations, an experimental and theoretical approach. Cold Spring Harbor Symposium on Quantitative Biology 22:139–152.

Warren, P. H. 1995. Estimating morphologically determined connectance and structure for food webs of freshwater invertebrates. Freshwater Biology 33:213–221.

Weiher, E., and P. A. Keddy. 1995. The assembly of experimental wetland plant communities. Oikos 73:323–335.

Weiner, J. 1994. The Beak of the Finch. Vintage, London.

Williamson, M. H. 1957. An elementary theory of interspecific competition. Nature 180: 422–425.

10

How Can We Compare the Importance of Ecological Processes If We Never Ask, "Compared to what?"

PETER S. PETRAITIS

Shall I compare thee to a summer's day?

—Shakespeare, Sonnet XVIII

One of the most troublesome problems that faces modern ecology is how to place ecological processes into a meaningful context. The problem persists even though ecological processes such as competition and predation are reasonably well understood and the ground rules for experiments by which ecological processes are described are well established. Yet ecologists would like to generalize beyond a particular experiment, and this is where ecology runs into trouble because ecologists have had great difficulty in finding meaningful ways to evaluate the importance or relative effect of ecological processes. Often researchers simply assert that a given process is "important" or "large" without defining terms or providing a context. The problem in ecology is especially acute since we are being constantly pressed by environmental managers and policy makers to answer questions about the importance and magnitude of ecological processes. But to say that competition is an important determinant of community structure or that competition has a large effect begs the question: "Compared to what?"

The core of the issue, of course, is what we really mean when we use relative terms such as *strong* and *weak* and how we measure "strengths" and "importance" of effects observed experimentally. Some believe more thoughtful experimental designs, correct statistical analysis, and alternative methods of analysis will provide us with meaningful inferences about the magnitude of ecological effects and the importance of ecological phenomena. These issues do not have pat solutions, and in passing it is interesting to note that the field of psychology has been struggling with the same demons (Vaughan and Corballis 1969, Carroll and Nordholm 1975, Maxwell et al. 1981, Murray 1987, Strube 1988, Haase 1989).

One conventional approach for estimating the intensity or magnitude of an effect has been to use regression type approaches. For example, Welden and Slauson (1986) suggest that the regression coefficient, b, which is slope of the regression line, provides a good measure of what they call intensity. In contrast, importance is usually thought as the amount of variation that is explained by a treatment factor (e.g., Welden and Slauson 1986). The coefficient of determination, R^2, for regressions or a related measure such as ω^2 for analyses of variance (ANOVAs) is often used to estimate importance (see Maxwell et al. 1981, Underwood and Petraitis 1993). For both regression coefficients and coefficients of determination there is the feeling that these values allow us to say something about the effects beyond the context of a specific experiment.

Here I discuss why the use of regression coefficient and coefficients of determination in this fashion is fraught with danger. While Underwood and Petraitis (1993) have discussed many of the problems, I would like to continue this discussion and suggest some possible remedies. I will use data from one of my own published experiments (Petraitis 1989) to illustrate the problems of inferring magnitude of effects and importance. This experiment is not any better than many experiments done by others, but I hope that by using my own experiment I will help readers focus on particular problems and not specific experiments. For the same reason, I have avoided citing what I consider "good" and "bad" examples of these problems. Armed with the critique of my own work, I believe anyone can not only find ample examples in the current literature of the problems I raise here but also place his or her experiments in a more meaningful ecological context.

I will start my discussion of specific problems with a close look at regression coefficients and coefficients of determination because many inferences about relative importance are based on these measures. Throughout the discussion, I will use the results of my own experiment to highlight the difficulties and to demonstrate possible solutions.

For regression coefficients, I will examine how scaling, standardization, and co-variation can affect these coefficients and influence ecological inferences. In particular, I am concerned with how inferences are affected by the mismatch between the scale and covariation imposed by the experiment and the scale and covariation we see in nature. Clearly, the greater the disparity between experiment and nature, the bigger the leap between experimental and ecological inference. This point is not new (see Keppel's [1982:13] comments on the appropriateness of generalization), but this mismatch has unappreciated effects for various standardizations (such as using standardized regression coefficients) and for techniques (e.g., path analysis) that have been suggested as ways to broaden ecological inferences drawn from any particular experiment. I will show that attempts to standardize experimental results merely sweep the problems under the rug. I will also suggest that baseline data may be helpful for setting the size or magnitude of experimental effects into a meaningful ecological context, and I will point out that the very logic and design of experiments prevent us from making inferences about the magnitude of effects outside a specific experiment.

Another issue is the use of measures of the amount of explained variation (e.g., R^2) as a measure of relative importance. I will introduce the most commonly used measures for the proportion of explained variation. The similarities among these measures, which I call PEV (proportion of explained variation) measures, and their relationship to Cohen's (1977) measure of effect size and meta-analysis will be discussed. All PEV measures are affected by the number of replicates per treatment level and/or the number of levels per treatment factor. As a result, using PEV measures to compare the relative importance of ecological processes across different experiments may be flawed, and good remedies may not be possible (Underwood and Petraitis 1993). I do believe some limited progress can be made and will make several suggestions.

While my comments are very critical of what we can infer from experiments, I do not wish to imply that well-crafted experiments are not important. Results based on good designs and correct analyses provide invaluable information, but these results, by themselves, tell us little about the more general question of the relevance of ecological processes. The bridge between experiment and ecological processes depends on our rationales for the relationships between statistical outcomes and biological mechanisms. This bridge is built on two ideals. First, it requires an ecological inference based on our understanding of the various physical, chemical, and biological mechanisms at work (e.g., see Platt 1964, MacFadyen 1975, Underwood 1990, Yoccoz 1991, Dunham and Beaupre this volume). Second, it requires correct statistical inferences, which depend on proper designs and analyses. I suggest in the following discussion that most ecologists have been far too complacent about what one can and cannot infer from manipulative experiments.

An Example

In 1985, I undertook an experiment (Expt. 2 in Petraitis 1989) to examine inter- and intraspecific effects on growth of the herbivorous intertidal limpet *Tectura testudinalis* (note that *T. testudinalis* [Müller] was formerly placed in *Acaema* and *Notoacmea*; see Lindberg 1986 for recent revision). I held limpets in cages at densities equivalent to 26, 52, and 104 limpets per m^2. This treatment factor was crossed with a second factor in which I varied the density of a competitor, the periwinkle snail *Littorina littorea* (L.). Periwinkles were caged at densities equal to 0, 51, and 102 snails per m^2. There were three replicates of each of the nine treatment combinations. Lengths of limpets were measured at the start of the experiment, and data on lengths and dry weights were collected at 10 weeks.

I have reanalyzed a subset of the data here in order to simplify the presentation. From each cage I randomly chose one limpet that was between 13.5 and 16.5 mm in initial length. Final length was analyzed by analysis of covariance (ANCOVA) with initial length as the covariate. Slopes and variances were homogeneous. Analysis shows only that the density of periwinkles has a significant effect (Table 10-1, Fig. 10-1). The narrow ecological inference is that growth of limpets is depressed by periwinkles.

In my experiment, the magnitudes of the effects of periwinkles and limpets on the adjusted final length of a limpet (i.e., the regression coefficients) are $-.01$ mm per periwinkle per m^2 and $-.001$ mm per limpet per m^2 (see Table 10-2). This implies that growth of a 14.8-mm limpet will decline by 0.01 mm over a 10-week period with each addition of one periwinkle per m^2. In Welden and Slauson's (1986) terminology, the addition of one periwinkle is 10 times more "intense" than the addition of one limpet (Table 10-2). R^2 values are .365 and .004 for the effects of periwinkles and limpets, respectively (Table 10-1). In this particular example, the size of the regression coefficients suggests "strong" interspecific effects of periwinkles on limpets and "weak" intraspecific effects of limpets on themselves. Moreover, many would suggest that the effect of periwinkles is more relatively important since its coefficient of determination

Table 10-1. Analysis of variance (ANOVA) and measures of importance for competition experiment

| Source | df | Analysis based on density | | | | | Analysis based on S.D. units |
		SS	MS	R^2	ω^2	P	SS
Periwinkle density	2	6.505	3.252	0.365	0.296	0.01	
Linear regression	1	4.197	4.197				4.158
Deviation	1	2.308	2.308				4.158
Limpet density	2	0.078	0.039	0.004	0.000	0.93	
Linear regression	1	0.007	0.007				2.347
Deviation	1	0.071	0.071				0.071
Interaction	4	2.135	0.534	0.120	0.000	0.44	
Error	17	9.092	0.535	0.511	0.704		
Covariate	1	4.101	4.101				

Column headings use conventions of Sokal and Rohlf (1981). Computational formulae for R^2 and ω^2 are given in Table 10-3. Linear regressions were calculated as linear contrasts. For regressions in terms of density per m², the contrasts were -1, 0, $+1$ for periwinkles and -1.33, -0.33, $+1.66$ for limpets. For linear regressions based on standard deviation (S.D.) units, the contrasts were -0.4033, -0.0033, $+0.4066$ for periwinkles and -4.3333, -1.0833, $+5.4166$ for limpets.

accounts for 36.5% of the total variation while the coefficient for limpets accounts for 0.4% of the variation.

Limitations on Measures of Intensity

Interpretations of measures of intensity, such as regression coefficients, depend on at least three things. First is the range of variation of the independent variable. There is always a danger in extrapolating beyond the bounds of the data, but we should, at the very least, be certain that the range of the independent variable within an experiment matches the range of natural variation over which we wish to generalize. I will show some of the problems of extrapolation using my own experiment. Second are various standardizations that are often used, especially when multiple regressions involve independent variables of differing metrics (e.g., using weight, length, and caloric content in a single analysis) or when a path analysis is carried out. Standardizations do not, however, put variables on an equal footing, and I will show why this is so. Third is covariation. Covariation between independent variables is treated differently in regression and ANOVA. Since the covariance structure affects regression coefficients, estimates of intensity depend on the method of analysis. This problem is largely unappreciated in the ecological literature.

There are two ways that the range of experimental treatment levels may not match the range of natural variation (Fig. 10-2A). First, it is possible that one or more of the treatment levels are well outside the range of natural variation. Treatment level L1 in Figure 10-2A is an example of this. The inclusion of a "no predator" or "no competitor" treatment level is the most common example in ecological experiments. If individuals of the target species are never without predators or competitors, then use

Growth in the presence of periwinkles

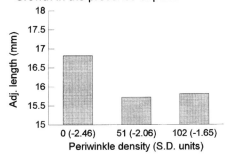

Periwinkle density (S.D. units)

Growth in the presence of limpets

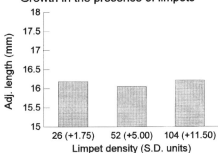

Limpet density (S.D. units)

Figure 10-1. Inter-and intraspecific effects on the adjusted final lengths of limpets. Numbers below each graph give the levels of treatment as the number of animals per m^2 and in terms of standard deviation (S.D.) units in parentheses. Unpublished field data from 1985 were used to calculate S.D. units. Field averages $\pm$ S.D.'s are 312 ± 127 per m^2 for periwinkles ($N = 47$) and 12 ± 8 per m^2 for limpets ($N = 47$). Correlations of average adjusted final length with periwinkle and limpet densities are -0.694 and -0.027 ($N = 9$), respectively.

Table 10-2. Regression coefficients (intensities) of the effects of periwinkles and limpets on the adjusted final length of limpets using different standardizations

Dependent variable	b	b'	b''	b'''
Periwinkles	-0.010	-0.694	-1.187	-0.768
Limpets	-0.001	-0.027	-0.004	-0.251
P/L	10	26	304	3

Row labeled P/L gives ratio of coefficients. Within ANOVA

$$b = \sqrt{SS_{LR} / \sum_{i=1}^{a} x_i^2}$$

where SS_{LR} equals the sums of squares of the linear regression and

$$\sum_{i=1}^{a} x_i^2 = cn \left[\sum_{i=1}^{a} X_i^2 - \left(\sum_{i=1}^{a} X_i \right)^2 / a \right].$$

X_i = the treatment levels (e.g., 0, 51, and 102 per m^2 for periwinkles) and c = number of levels in the crossed treatment (e.g., $c = 3$ levels for limpets); b = unstandardized partial regression coefficients; b' = standardized partial regression coefficients (i.e., simple correlations since ANCOVA forces the correlation between periwinkle and limpet densities to be zero); b'' = coefficients standardized by X but not Y; and b''' = standardized partial regression coefficients assuming the correlation between periwinkle and limpet densities equals -0.293.

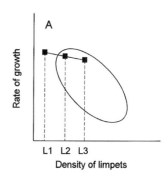

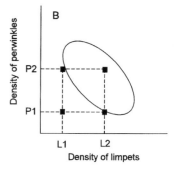

Figure 10-2. Potential mismatches between treatment levels and natural variation. Treatment levels are shown as shaded boxes. Natural variation and covariation are shown as 95% confidence ellipses. **A** is a plot of a single treatment (density of a competitor) versus the observed response (rate of growth). Note that one of the treatments (L1) is outside the range of natural variation and that the slope of the response does not match the natural covariation. **B** is a plot of two treatments in a factorial design. The ellipse shows the natural covariation of the two treatments. This factorial design not only breaks this covariation structure but also places treatment combination P1-L1 outside the ellipse.

of complete exclusion treatments will not be very informative of what happens in the field. Second, the range of experimental treatment levels may also fail to cover the complete range of natural variation. For example, treatment level L3 in Figure 10-2A is well below the upper limits of the natural variation. The combined effects of including treatment levels outside the natural range and failing to cover the complete range may alter the slope of the regression line. There are no a priori reasons why the regression line should match the natural covariation (see Fig. 10-2A) if the treatment levels do not cover a range of values that are ecologically relevant.

My experiment has both of these shortcomings. I chose what I thought were reasonable densities for periwinkles and limpets. Field data collected in 1985, the same year in which the experiment was done, suggest my optimism was unfounded. My treatment levels for both species failed to cover the complete range seen in the field (see Fig. 10-1). In addition, the upper level for limpets (104 limpets per m²) was 11.5 standard deviation (S.D.) units away from average field densities. Limpets were too common and periwinkles were too rare in the experiment. It is well-known that generalization beyond the range covered by the regression is foolhardy, and that appears to be the case here.

Standardization is the second issue. Standardizations are often employed to deal with problems of scaling. The unstandardized regression coefficients from my experiment suggest that one periwinkle has the same effect as 10 limpets (Table 10-2), but it remains an open question whether the one-to-one comparison of periwinkles and limpets is meaningful. What if periwinkles depress limpet growth by interference competition, but limpets affect growth via resource competition? Comparisons between more disparate treatments, such as the effects of a competitor versus the effects of resource quality, are even more problematic. The usual solution in multiple regression is standardization of the data. By extension, data from manipulative experiments are often standardized in a similar way (e.g., Wootton 1994). In the following discussion, I will show how fixed treatment levels from a manipulative experiment do not capture the natural variation in the field, and so such standardization is experiment-specific.

Standardization, as normally carried out in regression analysis, uses the S.D. of the data at hand. The standardized regression coefficient (b') is adjusted by the S.D.s of the independent (X) and dependent variables (Y), i.e.,

$$b' = b \frac{S_x}{S_y} \tag{1}$$

In a simple linear regression, the standardized regression coefficient (b') equals the correlation coefficient (r).

If data for a regression analysis are a random sample from the population of interest, then the S.D.s from the sample are unbiased estimates of the parametric values for the population. When this is true, inferences about changes in terms of S.D.s make sense. For example, suppose I wish to examine the relationship between weight and height in men. If my sample of men is randomly drawn from the whole population about which I wish to make inferences, then I could express that relationship in either the original units (i.e., a 10-cm change in height corresponds to a change in weight of so many grams) or in S.D. units (i.e., 1-S.D. unit change in height corresponds to . . .). Both statements are meaningful in the context of the natural population.

Standardization within an ANOVA is not so straightforward but is often based on the S.D.s of the levels of treatment and the average responses per cell. In ANOVAs, even with a factorial design, this sort of standardization is quite easy to do if the experimental design is orthogonal (i.e., the treatment factors are independent). When this is true, there is no correlation between treatment factors and the standardized regression coefficient equals the correlation coefficient. For the effect of periwinkles on limpet length in my experiment, $b' = -.694$ (calculated as the correlation between treatment levels 0, 51, and 102 periwinkles per m^2 and the adjusted average lengths for the nine treatment combinations). The effect of periwinkles is now 26 times greater than the effect of limpets.

In passing, I would note that regression coefficients for data from ANOVAs are usually calculated incorrectly by treating each replicate as a single observation in a linear regression (see Petraitis et al. 1997 for review of this issue in the context of path analysis). A regression done correctly within an ANOVA or ANCOVA (e.g., Table 10-1) is a linear contrast nested within the appropriate treatment (Sokal and Rohlf 1981: 488–490). Calculation of the regression coefficient should be based on the levels of treatment, which serve as the independent variables, and the average response to each treatment, which is the dependent variable. For example, a linear regression within a one-way ANOVA with six treatment levels and five replicates per level has 1 and 4 degrees of freedom, not 1 and 24 degrees of freedom.

Standardizations based on treatment levels, however, have no meaning outside the context of a particular experiment because the S.D.s are estimated from an experiment rather than from a random sample of the population. The standardized regression coefficient for the effect of periwinkles on the adjusted length of limpets is based on the S.D. for periwinkles from my experiment. Since I calculated b' based on the nine cell averages, this implies the S.D. for periwinkles is 41.17 (i.e., the three treatment levels— 0, 51, and 102 periwinkles per m^2—are each sampled three times). This S.D. not only tells us nothing about natural variation in field populations but also depends on how treatment levels are combined. I could have just as easily used the average adjusted lengths and ignored the effect of limpets. Then b' would be based on three pairs of observations: the average adjusted lengths at 0, 51, and 102 periwinkles per m^2. The S.D. differs (51.0), as does the correlation ($-.828$). To make matters worse, many experiments include treatment factors that are categorical. Categorical variables, such as the presence/absence of a predator, are usually scored as dummy variables (i.e., -1 for absence of predator and $+1$ for presence of predator). It is not at all clear how the S.D. of a dummy variable is ecologically relevant.

I would like to suggest one way out of this dilemma. Use field data to standardize the independent variables but not the dependent variable. Changes in limpet growth due to periwinkles and limpets would then reflect changes in growth in terms of the natural field variation of periwinkles and limpets. This would not only provide a means by which to standardize among treatments but would also capture the amount of variation that limpets "see" in the environment. Field data suggest that in my experiment the ranges covered by the treatment levels were too broad for limpets and too narrow for periwinkles (Fig. 10-1). In order to place the treatments on a more equitable footing, I recalculated the regressions using S.D. units based on abundance data collected in the field (see analysis based on S.D. units in Table 10-1 and column b'' in Table 10-2). The slope for the effect of periwinkles on limpets is now -1.187, which means the final adjusted length of a limpet declines by 1.187 mm with a change in periwinkle density of 1 S.D. Note that the effect of periwinkles is 304 times the effect of limpets, and so the interspecific effects are much stronger than previously suggested.

Covariation is the third and more troublesome issue because ANOVA and regression analysis treat covariation differently. On one hand, the very logic of experimental design is based on breaking the covariance among the independent variables. This is what allows us to test the hypotheses of main effects in factorial designs independently. So when orthogonal designs force the correlations among independent variables (see Fig. 10-2B) to be zero, it shouldn't be surprising that the regression coefficients based on experimental data reflect this fact. On the other hand, independent variables in nature do covary, and standard multiple regression analyses based on observational data incorporate this covariation, which affects the partial regression coefficients. Thus, estimates of "relative strengths" of effects that rely on regression coefficients will differ depending on the way in which the data were collected. What observational data preserve, experimental designs break.

This can be easily seen by examining how covariance affects standardized partial regression coefficients. Assuming that densities of periwinkles and limpets are X_1 and X_2, respectively, and growth rate of limpets is Y, then the standardized partial regression coefficient for the effect of periwinkles of limpet growth (b_1''') is

$$b_1''' = \frac{r_{1y} - r_{12}r_{2y}}{1 - r^2_{12}} \qquad (2)$$

Note that the magnitude of this effect depends on not only the correlation between periwinkles and limpet growth (r_{1y}) but also the correlations between limpets and limpet growth (r_{2y}) and between the densities of periwinkles and limpets (i.e., r_{12}). Now r_{1y} and r_{2y} can be obtained from the experimental results (e.g., see Fig. 10-1), but the correlation between periwinkle and limpet densities depends on how the data were collected. If the data are from a factorial experiment, then $r_{12} = 0$ and $b'_1 = r_{1y} = -.694$ (e.g., as in Table 10-2). Yet we might expect periwinkle and limpet densities to be inversely correlated because of interspecific interactions, and field data bear this out: $r_{12} = -.293$ ($N = 47$, $P = .045$; Petraitis, unpublished data 1985). Our estimates of intensity would now suggest that the effect of periwinkles is only three times the effect of limpets if we use $r_{12} = -.293$ to correct the regression coefficients obtained from the experiment (Table 10-2).

Which is the "right" measure of "relative strength"? The answer, I think, is none of them, because the notion of relative strength is context- or experiment-specific. The more relevant question is: What is the ecological inference of interest? This will determine not only the right measure but also the most appropriate experimental design.

Limitations on PEV Measures

There are a number of measures for the proportion of explained variation (what I call PEV measures). All PEV measures are designed to estimate the same quantity—that is, the proportion of explained variation—but they differ in how that estimation is done. For regression analysis, the most widely used index for importance is the coefficient of determination, r^2. For ANOVAs, several indices based on r^2 have been developed and widely used for estimates of the relative magnitude of treatment effects. The three most common measures are η^2, ε^2, and ω^2, which were developed as analogues to the coefficient of determination in regression. In fact, $\eta^2 = R^2$, and ε^2 is the adjusted or shrunken R^2 developed by Wherry (1931). Next I introduce r^2 in the context of regression analysis and give an overview of η^2, ε^2, and ω^2 in analysis of variance as analogues to r^2. The discussion of these measures is very brief since Maxwell et al. (1981) provide a very clear and complete review and Underwood and Petraitis (1993) discuss the difficulties of using these measures for comparisons among different experiments in ecology.

In regression analysis, the total variation of the dependent variable Y (S_y^2) can be partitioned into the variation explained by the independent variable X ($S_{y \cdot x}^2$) and the remaining unexplained variation is S_e^2. The variation explained by X is a function of the unstandardized regression coefficient, b, and thus

$$S_y^2 = Sy_{\cdot x}^2 + S_e^2 = b^2 S_x^2 + S_e^2 \qquad (3)$$

The relationship between the coefficient of determination, r^2, and b is $b^2 = (r^2 S_y^2)/S_x^2$. Thus,

$$r^2 = 1 - S_e^2/S_y^2 \qquad (4)$$

Table 10-3. Three most common proportion of explained variation (PEV) measures. The measures are very similar and are related to the effect size f. Their ranking is always the same: $\eta^2 \geq \varepsilon^2 \geq \omega^2$ (Glass and Hakstian 1969).

Name of measure	s_y^2	s_e^2	Measure in terms of f	Measure in terms of F	Computational formula
η^2 or R^2	$\dfrac{SS_T}{an}$	$\dfrac{SS_E}{an}$	$\dfrac{f^2 + \dfrac{a-1}{an}}{f^2 + \dfrac{an-1}{an}}$	$\dfrac{(a-1)F}{(a-1)(F-1) + an-1}$	$\dfrac{SS_A}{SS_T}$
ε^2 or R^2_{adjusted}	$\dfrac{SS_T}{an-1}$	MS_E	$\dfrac{f^2}{f^2 + \dfrac{an-1}{an}}$	$\dfrac{(a-1)(F-1)}{(a-1)(F-1) + an-1}$	$\dfrac{SS_A - (a-1)MS_E}{SS_T}$
ω^2	$\dfrac{SS_T + MS_E}{an}$	MS_E	$\dfrac{f^2}{f^2 + 1}$	$\dfrac{(a-1)(F-1)}{(a-1)(F-1) + an}$	$\dfrac{SS_A - (a-1)MS_E}{SS_T + MS_E}$

Columns s_y^2 and s_e^2 give the estimates of total and error variances used in the various measures. Table entries are based on a one-way ANOVA with a treatment levels with n replicates at each level. Abbreviations follow conventions of Sokal and Rohlf (1981). SS and MS refer to sums of squares and mean squares with subscripts T, A, and E denoting the total, treatment, and error contributions. The treatment expected mean square equals

$$\sigma^2 + \frac{n}{a-1} \Sigma \alpha^2,$$

which in terms of f^2 equals

$$\sigma^2 \left[1 + \frac{anf^2}{a-1} \right].$$

For regressions, it is well-known that the usual methods for calculating S_y^2 and S_e^2 provide good estimates of the parametric values σ_y^2 and σ_e^2, respectively, and thus provide a good estimate of the coefficient of determination.

For the random-effects model in ANOVAs, the usual calculations for r^2 also provide a good estimate and so the procedures used in regression analysis can be directly transferred. The reason for this are the meanings of "replicate" and "treatment levels" within the random-effects model. The random-effects model assumes that treatment groups (i.e., treatment levels) are randomly drawn from all possible groups in the population, and thus the collection of treatment groups is a random sample of all possible groups. By the same token, replicates, which are assumed to be randomly drawn, are a random sample of all possible replicates within a given group. For example, let us return to the height-versus-weight regression brought up in the previous section. Suppose I wanted to examine if men of different heights differed in weight. I could do the analysis as a regression, but I could just as easily define a set of height classes, assign men to the appropriate height classes, and carry out an ANOVA with weight as the variable of interest. Note that if men are chosen at random, the number of replicates per group will vary and capture height distribution of the whole population. As a result, the total variation of the dependent variable Y (weight in this ANOVA) is a good estimate of the total parametric variance. Membership in a group (i.e., membership in a height class) "explains" part of the total parametric variance. Since groups are randomly drawn from all possible groups and individuals are randomly chosen, the coefficient of determination for the random-effects model is an unbiased estimate of the proportion of explained variance.

The solution is not as clear for a fixed-effects model, where treatment levels are not randomly chosen and the number of replicates per treatment level is often set. Thus fixed-effects models present a problem of how to estimate the total and error parametric variances because the assignment to a level of treatment is not the same as membership in a group (Glass and Hakstian 1969, Dooling and Danks 1975).

The three measures—η^2, ε^2, and ω^2—differ in how the total and error parametric variances are estimated (see Table 10-3 for details). The oldest measure still commonly used is η^2, which is also called the correlation ratio. It is assumed $SS_{T/an}$ and SS_E/an are good estimates of σ_y^2 and σ_e^2, respectively. If this were true, then η^2 would equal R^2. In proposing ε^2, Kelley (1935) took a similar approach but assumed SS_T and SS_E must be adjusted by their respective degrees of freedom to provide unbiased estimates. Unfortunately, η^2 and ε^2 are the proportions of variance explained in a particular experiment and not populationwide estimates of the amount of variation explained by the treatment.

Hays (1963) was aware that assignment to treatment level was not the same as membership in a group and thus proposed ω^2 as an alternative to η^2 and ε^2. Hays suggested that the number of replicates per cell be used to define the probability of group membership in a fixed-effects model. For a one-way ANOVA, Hays defined m_j as the number of subjects found at the jth treatment level in the population and used $m_j/\Sigma m_j$ as an estimate of the populationwide probability of each treatment level. He then assumed that the number of replicates per treatment was proportional to each m_j. With equal cell sizes, $m_j/\Sigma m_j = 1/a$, and in this situation,

$$\sigma_y^2 = \sigma_e^2 + \Sigma\alpha_j^2/a \tag{5}$$

Thus,

$$\omega^2 = \frac{\Sigma\alpha_j^2/a}{\sigma_e^2 + \Sigma\alpha_j^2/a} \tag{6}$$

where $\Sigma\alpha_j^2/a$ equals the parametric variance due to the treatments and σ_e^2 equals the population error variance (see Table 10-3 for the computational formula). Hays viewed his measure for the fixed effects models to be analogous to the variance components in a random-effects model.

Which measure should ecologists choose for evaluating the proportion of explained variance? A good PEV measure would not only capture the ratio of the "signal" (i.e., $\Sigma\alpha_j^2/a$) to the "noise" (σ_e^2) but also be insensitive to changes in the number of treatment levels (a) and the number of replicates per treatment (n). This is exactly the notion of Cohen's (1977) effect size (f), and within the context of ANOVA, $f^2 = \Sigma\alpha_j^2/a\sigma_e^2$, Cohen gives the relationship between ω^2 and f^2 as

$$\omega^2 = \frac{f^2}{1 + f^2} \tag{7}$$

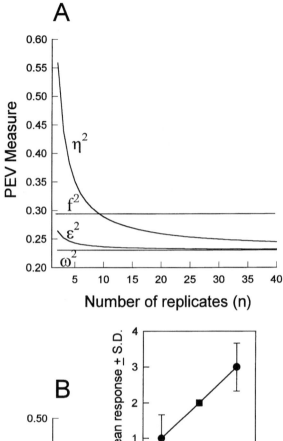

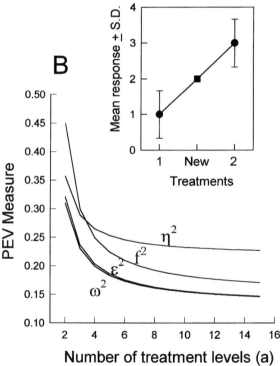

Figure 10-3. The effect of replication and treatment levels on PEV (proportion of explained variation) measures of importance and effect size (f^2; see Cohen 1977). (**A**) the effect of replication on the proportion of explained variance as estimated by different PEV measures when $a = 3$. I have chosen $f = 0.547$ so that the F test for a one-way ANOVA with three treatment levels and 10 replicates per level shows significance at the 1% level (i.e., $F_{(2,27)} = 5.49$). Cohen (1977) considers this a large effect size. Note that others (e.g., Winer 1971, Keppel 1982) use $\emptyset$ for effect size, and Cohen points out that $\emptyset = n'f^2$ where $n' = $ the required number of replicates. (**B**) the effect of number of treatment levels on PEV measures when $n = 10$. It is assumed that responses of additional treatment levels (*shaded box*) fall on the line defined by the first two treatments (*shaded circle*), and so the relationship between a and f^2 is defined by equation 8 in the text. $f = 0.547$ at $a = 3$ and $n = 10$.

Note that ω^2 is independent of the number of replicates, which is not the case for ε^2 and η^2 (see Table 10-3 and Fig. 10-3). It is not surprising that statisticians advise against using η^2 or ε^2 and favor using ω^2 for comparisons across experiments (e.g., Keppel 1982).

We must be careful, however, because changes in the number of treatment levels can alter f^2 and thus ω^2 even in situations where we might reasonably expect a good PEV measure not to decline. In order to see this, we must consider how the number of treatment levels can alter the added variance due to the treatment. Assume an experiment is carried out with two levels of treatments ($a = 2$), and assume the difference between the responses to these two levels is d (see Fig. 10-3B). Next assume we carry out a new experiment in which a third treatment level is added midway between the first two ($a = 3$), and the response to this new treatment lies on the line joining treatments 1 and 2 (Fig. 10-3B). Cohen (1977) has shown when this is true,

$$f^2 = \frac{d^2(a + 1)}{12(a - 1)} \tag{8}$$

Now d, which is the difference between the largest and smallest response, is the same in both experiments, yet f^2 will vary with the number of treatment levels (Fig. 10-3B). Thus, even ω^2 does not seem to be a very good measure of relative importance. This underscores the conclusions of Underwood and Petraitis (1993), who suggested that comparing the proportion of explained variation across experiments was an impossible task unless the experiments were done in identical fashion.

Some ecologists use P values instead of PEV measures, usually for one of two contradictory reasons! On one hand, many believe that the rank ordering of levels of significance will match the rank ordering of the proportions of explained variation. If this is true, why not use P values, which are provided by all statistical packages, rather than calculating PEV measures? P values, however, may not match PEV measures if the number of treatment levels per factor differs greatly among factors (see Table 10-4). A treatment factor with numerous levels may "explain" more of the total variance but be "less significant" than another factor. On the other hand, others believe significant levels capture the notion of importance better than PEV measures. Yet this notion assumes that the tests of different hypotheses are comparable (e.g., in Table 10-4, that the null hypotheses of treatment factor A and treatment factor B are, in some way, equivalent). For tests of very different hypotheses, there are no a priori reasons why ecologists should believe significance levels are any more informative than PEV measures.

Is there an acceptable alternative for comparisons across different experimental designs or within a single factorial experiment? I think so, but only in a very limited way. My suggested alternative relies on partitioning only the variance due to fixed treatment effects. The treatment variance, $\Sigma \alpha_j^2 / a$, can be divided into the effect of a linear trend among the means (i.e., $\beta^2 \sigma^2$) and deviations from the trend (σ^2_{Dev}). Following Hays's equation for total variance (see eq. 5), σ_y^2 can be written as

$$\sigma_y^2 = \sigma_e^2 + \sigma_{Dev}^2 + \beta^2 \sigma_x^2 \tag{9}$$

If $n = 1$, then the equation reduces to the formulation for simple regression, and thus the deviation variance is a measure of the regression error variance.

Table 10-4. Two-way ANOVA showing how the ordering of treatment factors by PEV measures (ω^2) may differ from ordering by significance level (P)

Source	df	SS	MS	F	P	ω^2
A	9	25.2	2.8	2.8	$0.025 < P < .050$	0.225
B	1	15.0	15.0	15.0	$P < .001$	0.194
AB	9	10.8	1.2	1.2	$P > .050$	0.025
Error	20	20.0	1.0			0.556

Treatment factors are A with 10 levels and B with 2 levels with two replicates per cell. The ω^2 column gives the partial ω^2, which is calculated as $\dfrac{df_i(F_i-1)}{\Sigma df_i(F_i-1)+N}$ where df_i = the degrees of freedom of the ith factor, F_i = the F ratio of the ith factor, and N = the total sample size ($N = 40$ in this example).

By analogy to the coefficient of determination, I believe a reasonable measure of importance would be

$$\rho_a^2 = \frac{\beta^2\sigma_x^2}{\beta^2\sigma_x^2 + \sigma_{Dev}^2} \tag{10}$$

For my experiment, using $(MS_{LR} - MS_{Dev})/(a - 1)$ as an estimate of $\beta^2\sigma_x^2$ and $(MS_{Dev} - MS_E)/n$ as an estimate of σ_{Dev}^2, the percentage of the variation explained by a linear response to periwinkles would be 61.5%, or roughly double ω^2.

This approach, however, still suffers from the major difficulties of inferring relative importance of ecological processes beyond the bounds of a particular experiment. PEV measures based on linear regression are still experiment-specific measures. They may provide a more reliable measure when comparing experiments with slightly different designs but they won't give us a better idea of how much of the variation we see in nature can be explained by a factor which we have experimentally manipulated. The reason for this is quite simple. Ecologists work very hard to minimize experimental error. Subjects for a particular experiment are rarely chosen at random, and the physical layouts of entire experiments are often placed in homogeneous areas. For example, if ecologists wish to test the effects of density on growth, they will often choose animals of the same size and place the entire experiment in a uniform area. There is nothing wrong with this approach as long as ecologists are only interested in testing the hypothesis that density affects growth. However, as soon as ecologists begin to ask questions about estimation that go beyond the bounds of the experiment (i.e., how much of the variation in growth in all animals within the population is explained by entire range of densities found in the field?), experiment-specific measures will provide little or no guidance. It is not possible to generalize in this way.

Meta-Analysis and PEV Measures

Meta-analysis has been suggested and debated as a way to rationalize comparisons across experiments of different designs (Rosenthal 1984, Wachter 1988, Thompson and Pocock 1991, Gurevitch et al. 1992, Wooster 1994). Meta-analysis of ecological experiments, however, typifies the problems of scaling response variables and the meaning of relative importance (e.g., Gurevitch et al. 1992, Wooster 1994). Meta-analysis takes the results from different experiments and standardizes the responses so that they can be compared on a single scale. Meta-analysis is usually used to examine difference between a single treatment ($Y\bar{A}$) and its control ($Y\bar{c}$), and unstandardized response is the difference between the two. The pooled S.D. or the S.D. of the control (S) and a sample size ($J(N)$) corrected for biases due to small sample sizes. Thus,

$$D = \frac{\overline{Y}_A - \overline{Y}_C}{S} J(N) \tag{11}$$

The collection of standardized responses is used to calculate an overall test statistic, akin to a t statistic. The expected value of the test statistic is zero under the null hypothesis of "no effect."

D is nothing more than the estimate of f, the effect size, in the situation where $a = 2$, and therefore any difficulties we have with measures based on d will also apply to test statistics based on D. Three problems are especially troubling. First, D depends on the difference in response under treatment versus control, and so the choice of treatment levels will alter D. For example, if we are looking at the effect of density on growth, then we might choose the average density in the field as the control density, but our choice of the treatment density is arbitrary. Meta-analysis, by itself, provides no guidance. We could just as easily choose 2 times or 10 times the average density, and it would be reasonable to assume that D for the 10-times treatment minus the control would be larger. Second, D will vary with the choice of organism and variable. The effect of density on growth, for example, depends on what we are measuring and comparing, and the D's for changes in shell height of gastropods and changes in biomass of annual plants are likely to very different. Of course, standardizing by s is supposed to correct for this, which brings us to the third problem. The S.D. used to calculate D is experiment-specific. Since in ecology we normally try to reduce the error variance in our experiments by using a nonrandom subset of all possible subjects (e.g., gastropods or plants of the same size) and by setting the entire experiment in a homogeneous area, s does not estimate the unexplained variance in the population as a whole. Thus, s, which is a measure of the variation within the experiment, is likely to be much smaller than the unexplained variance in the population at the level at which we hope to make inferences. Using s as an estimate of the unexplained variation across months, years, or geographic regions not included within the original experiment seems very foolhardy.

What Can Be Done?

First and foremost, we must accept that the conclusions drawn from an experiment depend on the experimental design itself. All PEV measures are affected by the numbers of replicates (n) and treatments (a). It seems reasonable that we would like a measure that did not decline with the number of treatment levels, yet even ω^2, which is the best measure, varies with the number of treatment levels (Fig. 10-3B). The sensitivity of PEV measures to changes in experimental design occurs because the numbers of replicates and treatment levels do not reflect the probability of membership. Probability of membership in a fixed-effects model is meaningless where the hypothesis under consideration sets the choice of treatment levels and the desired power of the test determines the number of replicates per treatment level. For example, if we are interested in whether a predator has an effect, we might use two treatment levels—one with and one without predators—and a sufficient number of replicates per level so that the test has enough power to detect an effect. In a typical experiment of this sort, neither the treatment levels nor the number of replicates reflects the natural variation in predator density. Our estimates of the error variance and the fixed effect never estimate the relative contributions of the error and treatment in the population as a whole (see Glass and Hakstian 1969).

Several tactics may improve the inferences based on the manipulation and analysis of a single factor.

1. A good start would be to choose levels of treatment that are relevant to baseline data. If the effect of a predator on the survivorship of a prey species is being examined, then collect baseline data on the predator density in order to set treatment levels. Rather than using an arbitrary number of predators per cage as treatment levels (e.g., cages with and without predators), use the average ($\bar{X}$) and S.D. of the number of predators in the field to set the treatment levels (e.g., $\bar{X} -$ S.D., $\bar{X}$, and $\bar{X} +$ S.D.).

2. Also, if you insist on calculating regression coefficients from experimental data, do it right. Use cell averages, not individual observations, because the number of treatment levels, not the number of replicates, sets the degrees of freedom for a regression done within an ANOVA (see Sokal and Rohlf 1981:447–491). This means that a regression with 1 and 4 degrees of freedom requires six treatment levels.

3. Finally, don't standardize independent variables unless you have a good reason to do so. For example, the magnitude of effect of periwinkles versus that of limpets can be interpreted as a threefold or three hundred–fold difference depending on the type of standardization (Table 10-2). It makes more sense to express effects in terms of the original variables. If standardization is done, I would suggest standardizing by using S.D.'s of the independent variables from baseline data rather than the experiment-specific treatment levels.

Often we wish to make inferences about relative strengths or effects of several factors. Unfortunately, in factorial designs there is a mismatch between the covariation imposed by the experimental design and the natural covariation found in the field. The mismatch occurs because balanced, orthogonal experimental designs break the correlational structure among the independent variables. Interaction tests in ANOVAs are tests of nonadditive effects of the independent variables and are not measures of the covariation among the independent variables. Since the interaction terms in a factorial design do not reflect the covariance among the independent variables, the tests of interaction tell us nothing about the effects of covariation. The latest incarnation of this problem is the use of path analysis to analyze experimental data, which has been recently advocated as a method for estimating relative strengths of interactions (e.g., Wootton 1994; also see Arnold 1972 and Power 1972 for earlier uses of path analysis in ecology and Petraitis et al. 1996 for a recent review of path analysis). While others have discussed the problems of matching experimental designs with natural variation (e.g., Underwood 1991), I would like to close with several specific suggestions of how ecologists might design experiments that capture the variance–covariance structure seen in the field.

It may be possible to use nonorthogonal designs and unequal cell sizes that reflect the probability of membership and capture the covariation structure of independent variables. Can we do this by using the natural covariation among the independent variables to set cell sizes? It seems reasonable that we could reverse Hays's logic and use field data to determine the number of replicates to be used in an experiment. For example, suppose I decided to redo my experiment and design the experiment so that treatment levels and the number of replicates per cell captured the variance–covariance structure in the field. The number of replicates per cell would then equal the probability of membership at that combination of treatment levels. In essence, the number of replicates per cell would be proportional to the probability of finding that combination of densities in the field. Table 10-5 shows the number of replicates per cell assuming that the average densities and S.D.s in the field are 312 ± 127 per m^2 for periwinkles ($N = 47$) and 12 ± 8 per m^2 for limpets (see Fig. 10-1) and that the correlation between periwinkle and limpet densities is $-.293$. Solving the membership problem, however, creates another. While cell sizes now capture the variance–covariance structure in the field, the design is now unbalanced and the resultant F ratios will not be exact.

Approaches of this sort that use unbalanced designs would be equivalent to a weighted regression. How far they may be extended remains to be explored. Even so, weighted regression approaches will still be of limited value for comparisons among experiments of very different designs. As pointed out by Underwood and Petraitis (1993), PEV measures of importance are experiment-specific, and it is likely the same will be true for importance measures based on weighted regressions done within ANOVAs.

While I have stressed how we might sharpen our experimental approaches and place our experiments in a more ecological context, ultimately what we must refine are our hypotheses. The more explicit the hypothesis, the clearer the choice of what is the critical experiment (Platt 1964; Underwood 1990, 1991). Explicit hypotheses may require models with very explicit mechanisms (e.g., see Dunham and Beaupre, this volume). Such detail works to the advantage of the experimenter. For example, if we have a very explicit mechanism in mind and good baseline data, we should be able to predict

Table 10-5. Replication needed to meet Hays' assumption

	Periwinkle densities		
Limpet densities	185/m^2 ($\overline{X}$ − S.D.)	312/m^2 ($\overline{X}$)	439/m^2 ($\overline{X}$ + S.D.)
4/m^2($\overline{X}$ − S.D.)	10	21	17
12/m^2($\overline{X}$)	21	33	21
20/m^2($\overline{X}$ + S.D.)	17	21	10

Hays assumed number of replicates is proportional to the probability of membership to a group. Table entries give the required number of replicates for the nine treatment combinations found in Petraitis's competition experiment. Entries were calculated by assuming that limpet and periwinkle densities follow a bivariate normal distribution and assuming the variance-covariance structure estimated from observational data (see Fig. 10-1).

the magnitude and shape of the response over an ecologically relevant scale. The predicted magnitude of the response from the model coupled with observations on the levels of natural variation would allow us to do a power analysis to set level of replications required to detect a significant difference. The shape of response will define number of treatment levels that is, if we predict a second-order function, we need at least three treatment levels. While this level of knowledge about ecological systems seems beyond the level of the science at the moment, it seems an appropriate goal if we wish to design experiments that will allow us to understand and measure the magnitude and importance of ecological processes.

ACKNOWLEDGMENTS   Art Dunham and I have had many heated discussions about experimental ecology during the last 15 years, and I thank him for his criticism and insight. I am especially indebted to Peter Fairweather, who, as always, provided many helpful comments during the development of this essay. I thank Mike Beck and Paul Nealen for suggesting several references. Joe Bernardo, Dave Hart, Stu Hurlbert, Bill Resetarits, Peter Stoll, Gerry Quinn, and Tony Underwood carefully read the final draft and offered many suggestions. Finally, over the last 5 years I have discussed the measurement of importance with many people, and although it is difficult to acknowledge them all, I thank Tony Underwood and members of the Institute of Marine Ecology at the University of Sydney, students and faculty in the Ecology Group at the University of Pennsylvania, and the scientific staff of the CSIRO Division of Water Resources in Griffith, New South Wales.

Literature Cited

Arnold, S. J. 1972. Species densities of predators and their prey. American Naturalist 106: 220–236.

Carroll, R. M., and L. A. Nordholm. 1975. Sampling characteristics of Kelly's ε^2 and Hays' $\hat{w}^2$. Educational and Psychological Measurement 35:541–554.

Cohen, J. 1977. Statistical Power Analysis for the Behavioral Sciences, rev. ed. Academic Press, New York.

Dooling, D. J., and J. H. Danks. 1975. Going beyond tests of significance: is psychology ready? Bulletin of the Psychonomic Society 5:15–17.

Glass, V. G., and A. R. Hakstian. 1969. Measures of association in comparative experiments: their development and interpretation. American Educational Research Journal 6:403–414.

Gurevitch, J., L. L. Morrow, A. Wallace, and J. S. Walsh. 1992. A meta-analysis of competition in field experiments. American Naturalist 140:539–572.

Haase, R. F. 1989. Multiple criteria for evaluating the magnitude of experimental effects. Journal of Counseling Psychology 36:511–516.

Hays, W. L. 1963. Statistics for Psychologists. Rinehart and Winston, New York.

Kelley, T. L. 1935. An unbiased correlation ratio measure. Proceedings of the National Academy of Science of the USA 21:554–559.

Keppel, G. 1982. Design and Analysis, 2nd ed. Prentice Hall, Englewood Cliffs, New Jersey.

Lindberg, D. R. 1986. Name changes in the "Acmaeidae." Veliger 29:142–148.

MacFadyen, A. 1975. Some thoughts on the behavior of ecologists. Journal of Animal Ecology 44:351–363.

Maxwell, S. E., C. J. Camp, and R. D. Arvey. 1981. Measures of strength of association: a comparative examination. Journal of Applied Psychology 66:525–534.

Murray, L. W. 1987. How significant is a significant difference? Problems with the measurement of magnitude of effect. Journal of Counseling Psychology 34:68–72.

Petraitis, P. S. 1989. Effects of the periwinkle *Littorina littorea* (L.) and of intraspecific competition on growth and survivorship of the limpet *Notoacmea testudinalis* (Müller). Journal of Experimental Marine Biology and Ecology 125:99–115.

Petraitis, P. S., A. E. Dunham, and P. H. Niewiarowski. 1996. Inferring multiple causality: the limitations of path analysis. Functional Ecology 10:421–431.

Platt, J. R. 1964. Strong inference. Science 146:347–353.

Power, D. M. 1972. Numbers of bird species on the California islands. Evolution 26:451–463.

Rosenthal, R. 1984. Applied Social Research Methods Series: Vol. 6. Meta-analytic Procedures for Social Research. Sage, Beverly Hills, California.

Sokal, R. R., and F. J. Rohlf. 1981. Biometry, 2nd ed. W. H. Freeman, San Francisco, California.

Strube, M. J. 1988. Some comments on the use of magnitude-of-effect estimates. Journal of Counseling Psychology 35:342–345.

Thompson, S. G., and S. J. Pocock. 1991. Can meta analyses be trusted? Lancet 338:1127–1130.

Underwood, A. J. 1990. Experiments in ecology and management: their logics, functions and interpretations. Australian Journal of Ecology 15:365–389.

———. 1991. The logic of ecological experiments: a case history from studies of the distribution of macro-algae on rocky intertidal shores. Journal of the Marine Biological Association of the United Kingdom 71:841–866.

Underwood, A. J., and P. S. Petraitis. 1993. Structure of intertidal assemblages in different locations: how can local processes be compared? Pages 39–51 in R. E. Ricklefs and D. Schluter (eds.), Species Diversity in Ecological Communities: Historical and Geographical Perspectives. University of Chicago Press, Chicago, Illinois.

Vaughan, G. M., and M. C. Corballis. 1969. Beyond test of significance: estimating strength of effects in selected ANOVA designs. Psychological Bulletin 72:204–213.

Wachter, K. W. 1988. Disturbed by meta-analysis? Science 241:1407–1408.

Welden, C. W., and W. L. Slauson. 1986. The intensity of competition versus its importance: an overlooked distinction and some implications. Quarterly Review of Biology 61:23–44.

Wherry, R. J. 1931. A new formula for predicting the shrinkage of the coefficient of multiple correlation. Annals of Mathematical Statistics 2:440–457.

Winer, B. J. 1971. Statistical Principles in Experimental design, 2nd ed. McGraw-Hill, New York.

Wooster, D. 1994. Predator impacts on stream benthic prey. Oecologia 99:7–15.

Wootton, J. T. 1994. Predicting direct and indirect effects: an integrated approach using experiments and path analysis. Ecology 75:151–165.

Yoccoz, N. G. 1991. Use, overuse, and misuse of significance test in evolutionary biology and ecology. Bulletin of the Ecological Society of America 72:106–111.

11

Insights and Application of Large-Scale, Long-Term Ecological Observations and Experiments

DAVID M. LODGE, STEVEN C. BLUMENSHINE,
& YVONNE VADEBONCOEUR

One of our goals as ecologists involved in basic research, should be to contribute knowledge that will enhance management of natural resources (Peterson 1993). As C. J. Krebs (1994:8) defines it, "ecology should be to environmental science as physics is to engineering." But the manner in which ecologists conduct research determines the extent of its usefulness to resource management and conservation efforts.

One frequent mismatch between basic research and resource management arises because the small spatial and temporal scale of much basic research does not provide information relevant to resource management (Kitchell et al. 1988, Cullen 1990, Hall et al. 1994, Naiman et al. 1995). Whereas basic research is often conducted at small temporal and spatial scales in simplified systems, management often operates on the whole-ecosystem scale and over many years. Even if only one species of economic interest is directly managed, the whole ecosystem is often affected because many indirect and feedback effects exist that cannot be adequately addressed in small-scale, short-term studies. Large-scale issues that require long-term research and management include acidic deposition, lake eutrophication and biomanipulation, fisheries management, forest management, role of large mammalian predators in affecting prey populations, invasion by exotic species, role of iron in controlling oceanic productivity, stratospheric ozone depletion, and global warming (Carpenter et al. 1995b). For many aspects of these topics, large-scale, long-term research (augmented by small-scale mechanistic experiments) can offer appropriate guidance to manage resources and mitigate anthropogenic effects (Fisher 1994, Brown 1995, Franklin 1995, Lubchenco 1995, Schindler 1995).

Conversely, resource management efforts often constitute very interesting large-scale and long-term manipulations of communities and ecosystems that could be exploited by research-oriented ecologists. The mutual benefits of cooperation between research ecologists and resource managers are clear from past and ongoing research relevant to anthropogenic stresses and resource management (Schindler 1974, Bormann and Likens 1979, Edmondson 1991, Magnuson 1991, Kitchell 1992, Likens 1992, Walters 1993, Carpenter et al. 1995b, Lewis et al. 1995, Naiman et al. 1995).

In this chapter, we contribute the following: a selective literature survey that documents the temporal and spatial scale of experiments used to test the effects of nutrient and food web manipulations in freshwater ecosystems; a vignette from our own lake research that documents the potential importance of benthic–pelagic links to the fate of nutrients added to lakes; and a simple test of the hypothesis that predicting the establishment of the exotic zebra mussel requires a cognizance of lake–stream links in the landscape. Both these case studies demonstrate the need for expanding the typical spatial scale of research. Because freshwater ecosystems are the focus of our research, our case studies also emphasize research relevant to the larger issues of biomanipulation, fisheries management, and eutrophication (Carpenter and Kitchell 1992) and the spread and impact of exotic freshwater organisms (Lodge 1993a,b). These are among the most common and important management issues involving freshwater resources, and they all operate at multiyear and whole-lake or landscape scales. We conclude this chapter with the suggestion that ecologists design their research more often to directly address scales and topics relevant to critical issues of environmental management.

At the outset, we acknowledge that the general themes of this chapter—that basic ecologists should add larger and longer term experiments and observational approaches to their methodological arsenal—are not new. They have been eloquently argued previously (e.g., Wiens 1986, Frost et al. 1988, Likens 1989, Mooney 1991, Dayton 1994, Hildrew and Giller 1994, Neill 1994). We reemphasize this theme, with new examples, in this chapter because the practice of ecology still has not responded adequately to this challenge. In the last 30 years, large-scale, long-term experiments have contributed substantially to our understanding of important applied problems—for example, eutrophication and acid deposition. However, we demonstrate that incorporating more habitats within lakes and taking into consideration a lake's position in the landscape substantially improve our ability to predict the response of freshwater ecosystems to anthropogenic stresses. For terrestrial ecologists, scaling ecological research to environmental issues will require an even greater change in practice because so great a proportion of experimental terrestrial ecology is small-scale (Karieva and Andersen 1988).

Literature Survey of Freshwater Nutrient and Food Web Experiments

We conducted a selective survey of the literature to assess changes over the last 30 years in the experimental methods used to test the roles of nutrients and food webs in determining freshwater community structure and ecosystem function (Table 11-1). Several authors have reviewed the *results* of terrestrial and/or aquatic experiments (e.g., Connell 1983, Schoener 1983, Sih et al. 1985, Leibold 1989, DeMelo et al. 1992), with limited attention to methods (Karieva and Andersen 1988, Tilman 1989). For the purposes of this chapter, we were more interested in the experimental *methods* per se. As our analyses demonstrate, the experimental design reveals a great deal about what topics and interactions ecologists think are important.

Table 11-1. Freshwater journal-years surveyed from 1965 to 1994

Journal	Years														Total
	1965	1966	1967	1968	1971	1974	1977	1980	1983	1986	1988	1990	1993	1994	
CJFAS	(0,0)/37	(0,0)/55	(0,0)/70	(1,0)/86	(3,0)/108	(1,0)/136	(0,0)/178	(4,0)/163	(2,0)/126	(6,0)/87	(6,0)/162	(6,4)/151	(4,1)/139	(3,0)/141	(34,6)/1639
Ecology	(1,0)/14	(1,0)/13	(0,0)/26	(2,0)/26	(1,0)/16	(3,0)/19	(4,0)/19	(3,2)/25	(4,1)/28	(7,2)/23	(8,2)/24	(13,3)/41	(8,5)/31	(6,3)/35	(61,18)/340
Freshw Biology							(4,0)/57	(0,0)/53	(0,2)/46	(5,0)/67	(4,3)/77	(13,7)/91	(3,2)/77	(2,3)/91	(31,17)/559
Hydrobiologia					(0,0)/51	(2,0)/58	(3,0)/52	(9,0)/182	(5,1)/159	(4,1)/377	(7,0)/207	(31,1)/280	(10,0)/288	(14,0)/347	(85,3)/2001
J-NABS										(1,6)/24	(2,1)/26	(1,3)/28	(0,0)/32	(4,3)/41	(8,13)/151
L & O	(2,0)/34	(1,0)/28	(0,1)/40	(0,0)/37	(0,0)/44	(4,0)/65	(1,0)/59	(2,0)/59	(4,1)/47	(4,1)/32	(8,1)/58	(10,0)/83	(5,0)/69	(7,0)/88	(48,4)/726
Total	(3,0)/85	(2,0)/96	(0,1)/136	(3,0)/149	(4,0)/219	(10,0)/278	(12,0)/348	(18,2)/482	(15,5)/406	(27,10)/610	(35,7)/554	(74,18)/674	(30,8)/636	(36,9)/743	(267,61)/5416

CJFAS = *Canadian Journal of Fisheries and Aquatic Sciences*; Freshw Biology = *Freshwater Biology* (University of Notre Dame library holdings started in 1977); J-NABS = *Journal of the North American Benthological Society* (publication started in 1986); L & O = *Limnology and Oceanography*. University of Notre Dame library holdings for *Hydrobiologia* started in 1971. In each cell, the following information is provided: (no. of lake papers meeting criteria, no. of stream papers meeting criteria)/total no. of freshwater papers published. Review papers were not included. Criteria included that the study be experimental and include a manipulation of a nutrient or a species (see text). Some papers that otherwise met our criteria were excluded because they did not provide the experimental unit area.

For six leading ecological journals that include or specialize in aquatic studies, we examined 5,416 papers from 14 haphazardly selected years (3 to 4 years per decade) between 1965 and 1994 (Table 11-1). Of those papers, 328 (Appendix A) met our criteria for closer examination: (1) the paper had to include an experimental manipulation of either a nutrient (usually N or P or both), the food web (by manipulating abundance of at least one species), or both, in either natural or laboratory lakes or streams, and (2) at least one response variable had to be biological (purely chemical studies were not included). In addition to traditional experiments, we included "natural" experiments and unreplicated whole-lake studies. We classified a study as a natural experiment if a clear perturbation was documented, the study included a before-after structure, and the authors themselves viewed their study as an experiment. Of the 327 papers (containing 415 experiments) that met our criteria, 266 papers (330 experiments) focused on the lentic environment and 61 papers (85 experiments) focused on the lotic system. Each datum in our analyses was an experiment (not a paper).

Prevalence of Experimentation

For all but 1 year surveyed, $< 9\%$ of the freshwater studies met our criteria (Fig. 11-1), primarily because most studies were descriptive; in addition, experiments that focused on behavioral questions, toxicology, acidification, impoundments, and other disturbances did not meet our criteria for food web or nutrient manipulation. Many of the earlier (before the mid-1980s) experiments that were included in our liberal definition of experiment would not meet current, post-Hurlbert (1984) standards of control and replication. Thus, the increase in rigorous experimentation has, if anything, been greater than indicated by the increase in the percentage of studies meeting our criteria (Fig. 11-1). The increase in the number of experiments that address the role of nutrients and food webs in freshwater probably reflects the increasingly experimental orientation of ecology in general. Hairston (1989:ix–x) noted a similar increase in ecological field

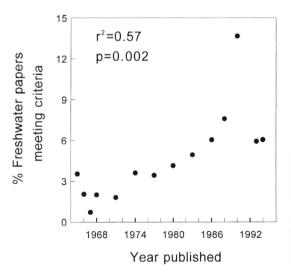

Figure 11-1. Percentage of freshwater papers by year for the journal-years indicated in Table 11-1 that met our criteria of examining a biological response to a manipulation of a nutrient or a species in the food web.

experiments during 1965–1987. This increase was in part a response to the epistemological debates that began in the late 1970s over how to detect competition in communities (Strong et al. 1984, McIntosh 1995). It remains to be seen whether the more recent vigorous promotion by limnologists (Peters 1986, 1991; Cole et al. 1991) and others (Brown 1995) of observational and comparative studies will reduce the percentage of experiments.

Duration and Size of Experiments

Over the last 30 years, the duration and size of experiments covaried (Fig. 11-2). Duration ranged from minutes (e.g., zooplankton feeding on phytoplankton) to 33 years (a resampling of Lake Michigan benthos after a period of eutrophication); size of experimental units ranged from 150-ml beakers to 6×10^4 km^2 Lake Michigan.

It is not surprising that the vast majority of experiments lasted less than one year (Fig. 11-2). Logistics, funding, and the pressure to publish quickly all dictate short experiments. Thus, most freshwater experiments include at best one to three generations of many benthic invertebrates and only a fraction of the multiyear life cycle of top

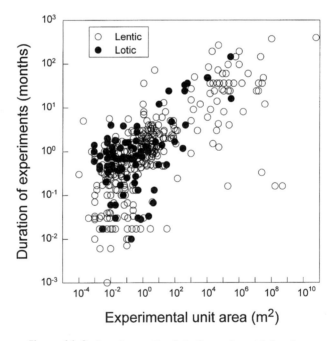

Figure 11-2. Log-log scatterplot of experimental duration versus area of an individual unit of each experiment. Each dot represents one experiment. For lotic studies, experimental unit area was estimated as mean stream width multiplied by the length of reach under study. For plotting, volumes (V) were converted to area (A) by $A = (\sqrt[3]{V})^2$.

consumers such as fish and waterfowl. Given the possible divergence between short- and long-term ecological responses (Tilman 1989, Grimm 1994, Schindler 1995) and typically high interannual variability in biological parameters (Carpenter et al. 1987), it is possible that results from many short-term studies are of limited use for management applications that have a much longer term context.

Long-term experiments (> 1 year) were not conducted in small containers, probably because experimental artifacts increase as experiment spatial scale declines and therefore conditions in small containers diverge more quickly from natural conditions (Carpenter and Kitchell 1988). Long-term experiments were conducted primarily in large (≥ 100 m^2) ponds/lakes or streams/rivers. The lack of experiments between about 100 m^2 and 1000 m^2 represents the shift from using in situ enclosures or small experimental ponds to using natural lakes and streams as experimental units.

The relationship between duration and size appeared to be similar for lakes ($r^2 = .473$, $P < .001$) and streams ($r^2 = .259$, $P < .001$), although the paucity of stream experiments (85) relative to lake experiments (330) makes the comparison weak (Fig. 11-2). It seems unlikely that the positive relationship between temporal and spatial scales of experiments evident in Figure 11-2 was an accidental result of independent increases in experiment duration and size. Over the last 30 years, there is no trend in experiment duration for either lakes ($r^2 = .001$) or streams ($r^2 = .037$) (Fig 11-3a). Similarly, no trend exists in experimental unit size for either lakes ($r^2 = .005$) or stream ($r^2 = .010$), whether examined as a scatterplot (Fig. 11-3b), as arithmetic or geometric means over time (not shown), or by examination of the prevalence of different unit area categories over time (not shown). The result for experiment size (Fig. 11-3b) is surprising because it shows that freshwater ecologists have for 30 years been conducting experiments on the large scales recently called for by both freshwater (Carpenter and Kitchell 1988) and terrestrial (Karieva and Andersen 1988) ecologists. However, it was clear from our examination of the literature that the quality of experimental designs (increased use of intentional manipulations, controls, and replication) and the sophistication of statistical analysis have increased.

Treatments and Responses of Interest

For each experiment in our literature survey, we determined the focal interests of investigators by categorizing treatment and response variables in two ways: (1) as either food web (manipulation of the abundance of at least one species), nutrients (manipulation of at least one nutrient), or both; and (2) as either benthic (e.g., manipulation of benthivorous fish, introduction of nutrient-diffusing substrates, and measurement of benthic invertebrate or periphyton response), pelagic (e.g., manipulation of planktivorous fish, *Chaoborus*, or water-column nutrients and measurement of zooplankton or phytoplankton response), or both. The first categorization describes treatments only, while the second categorization includes treatments and responses. The food web–nutrient categorization allowed us to quantify the relative emphasis by freshwater ecologists on testing top-down and bottom-up forces. The second categorization allowed us to quantify the long-standing separation between benthic-oriented and pelagic-oriented experiments.

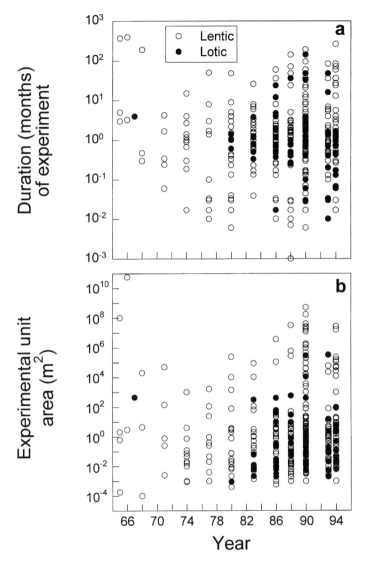

Figure 11-3. Scatterplots of (**a**) duration and (**b**) area of individual experimental unit versus year of publication. Each dot represents one experiment.

Streams   Because very few stream studies include a pelagic component, only the food web–nutrient treatment categorization is of interest for streams. Our selective literature survey did not pick up appreciable numbers of experimental stream studies until 1983 (Table 11-1). For the 79 stream experiments between 1983 and 1994, a clear trend exists: the focus gradually shifted from a majority of nutrient studies in 1983 to an overwhelming focus on food webs by 1994 (Fig. 11-4A). The shift in focus from bottom-up to top-down was even more pronounced than suggested by Figure 11-4a

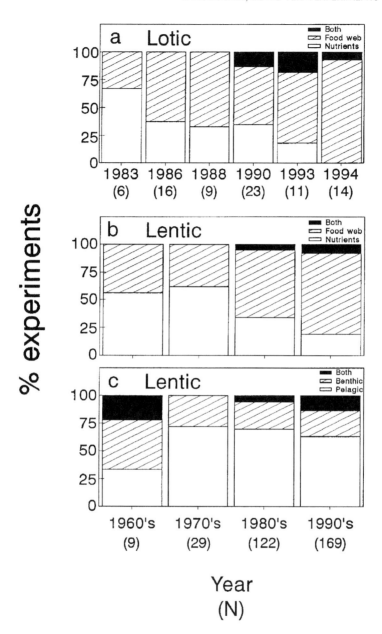

Figure 11-4. Percentage of (**a**) lotic and (**b**) lentic experiments within each year categorized by food web–nutrient treatment and (**c**) percentage of lentic experiments categorized by benthic–pelagic orientation. Sample number (number of experiments) is indicated below each year.

because many of the "food web" studies in the 1980s manipulated algal abundance (not predator abundance as was predominantly the case later). In the 1990s, a small proportion of studies included manipulations of both nutrients and food webs.

Lakes   For lakes, sample numbers over the entire 30 years are adequate for analysis (Table 11-1), and both the food web–nutrient and benthic-pelagic categorizations are of interest. As with lotic ecologists, the experimental focus of lentic ecologists has clearly shifted, from a slight preponderance of nutrient experiments in the 1960s to a clear preponderance of food web experiments in the 1980s and 1990s (Fig. 11-4b). Experiments that included manipulations of both nutrients and food webs also appeared in more recent years (Fig. 11-4b).

With respect to the benthic-versus-pelagic orientation of lentic experiments (Fig. 11-4c), any trend is less pronounced. The data and our qualitative assessment of the literature suggest that the intellectual disagreement in the late 1960s and 1970s about which macronutrient caused eutrophication led to the replacement of traditional limnological experiments (many of them European, in which fishes were manipulated and responses of both benthos and plankton were monitored) by more mechanistic experiments on pelagic nutrient–phytoplankton–zooplankton interactions (Fig. 11-4c). More recently, interest in benthos and especially in benthic–pelagic links has resurged (Fig. 11-4c).

These results demonstrate that researchers have only recently begun to explore the potentially critical interactions between nutrient status and food web configuration in lotic (Fig. 11-4a) and lentic (Fig. 11-4b) habitats. Similarly, the functional links between benthic and pelagic habitats in lakes have been understudied (Fig. 11-4c). Obviously, this has been a concession to feasibility. However, the resulting lack of attention to these complex interactions may lead researchers to erroneously attach a lack of importance to them.

For lentic habitats, an examination of the number of experiments in each cell of a cross-classification of our two categorizations reveals some important patterns (Fig. 11-5). First, because the recent focus has been on food webs (Fig. 11-4b) and the total number of papers published has increased with time (Fig. 11-3), far more food web than nutrient experiments have been published. Second, far more pelagic than benthic experiments have been published. The literature, therefore, is heavily biased toward pelagic food web studies; benthos are severely underrepresented (Fig. 11-5). Third, and perhaps most important, very few experiments in our lentic literature survey included an examination of both benthic and pelagic responses to both nutrient and food web manipulations (Fig. 11-5). Despite at least one early, now classic study (Hall et al. 1970) in the both-both category (Fig. 11-5), ecologists have been slow to address interactions between benthic–pelagic links and food web–nutrient effects.

In our first case study, in the next section, we provide a rationale and preliminary data on why understanding how benthic–pelagic links respond to food web–nutrient interactions may be critical to better predicting the response of lakes to nutrient loading. Given that eutrophication and manipulation of fish populations are probably the most common anthropogenic stressors of freshwater ecosystems, these issues are directly relevant to water resource management.

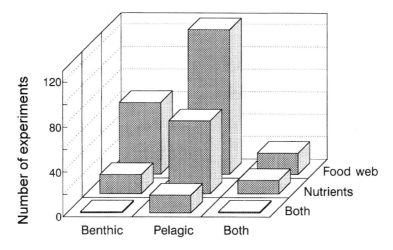

Figure 11-5. Number of lentic experiments in the nine combinations of treatments and responses created by the cross-classification of our food web–nutrient and benthic–pelagic categorizations.

Case Study of Benthic–Pelagic Links

In a series of ongoing whole-lake manipulations in the upper peninsula of Michigan, Carpenter, Kitchell, and colleagues (summarized in Carpenter and Kitchell 1993) have built on Arthur Hasler's legacy. Beginning in 1947, Hasler conducted the first whole-ecosystem experiment with both a manipulated and reference system, at what is now the University of Notre Dame Environmental Research Center. Hasler's focus in many of the early whole-lake experiments was on fish production, and the value of such large-scale experiments in both terrestrial (Bormann and Likens 1979) and aquatic (Schindler 1988, 1995; Schindler et al. 1990; Kitchell 1992; Carpenter and Kitchell 1993) ecosystems to test matters of basic and applied interest has become clear over time.

After focusing initially on the impact of food web manipulations on structuring lake communities (Carpenter et al. 1985, Carpenter and Kitchell 1993), Carpenter, Kitchell, and colleagues have more recently tested the interactions of food web and fertilization effects (Carpenter et al. 1995a, 1996). These and many other recent studies of lake communities (e.g., early experiments at the Canadian Experimental Lakes Area; Schindler 1995), however, have had an pelagic focus. Such phytoplankton-based studies, including one on which one of us (DML) was a coauthor (Carpenter et al. 1987), are often published as describing the controls on ''lake'' productivity; in reality, the role of the benthos in community structure and in ecosystem productivity has simply not been addressed. This is analogous to measuring the production of one vegetation layer in a multilayered forest canopy and ignoring the other layers in an extrapolation to the ''forest.''

In this section, we examine the potential role of benthic algae (periphyton) as a nutrient sink (Hansson 1990)—only one of the many potentially important benthic–

pelagic interactions (Lodge et al. 1988). We present preliminary evidence that including periphyton in whole-lake estimates of production may greatly enhance our understanding of how lakes respond to fertilization.

The abundance of periphyton in these lakes suggests that they could reduce phytoplankton response to water column fertilization. Two of our oligotrophic study basins, Central Long and East Long lakes (Fig. 11-6), were created in 1991 by partitioning Long Lake with a rubber curtain (Christensen et al. 1996). Because Central Long Lake is shallower than East Long Lake (Fig. 11-6), the ratio of periphyton to phytoplankton biomass was much higher in Central Long Lake than in East Long Lake (Fig. 11-7). Even in steep-sided East Long Lake, however, periphyton comprised at least 30% of the epilimnetic algae. In Central Long Lake, probably more representative of the world's predominantly small, shallow lakes, periphyton comprised at least 55% of the epilimnetic algae. In our Central Long Lake experimental mesocosms (2 m depth $\times 1$ m^2 cross-sectional circular area), periphyton, including that on the plastic walls, comprised about 85% of algae (Fig. 11-7).

Before beginning a whole-lake test of the hypothesis that periphyton might damp phytoplankton response to nutrients, we conducted a fertilization experiment in a set of mesocosms in Central Long Lake that contained abundant large *Daphnia*. With increased nutrient loading, phytoplankton increased about twofold (Cottingham et al. 1997; Figure 11-8), illustrating the well-known positive relationship between nutrient loading and phytoplankton abundance (Schindler 1978). Epipelic algae (algae growing on sediments) did not respond to increased nutrients because it was not nutrient-limited (Blumenshire et al. 1997). By far, the strongest response was the 3.5-fold increase in algae on the plastic mesocosm walls. This algal group is analogous to algae on coarse woody debris or on macrophytes in natural lakes because, unlike epipelic algae, it grew

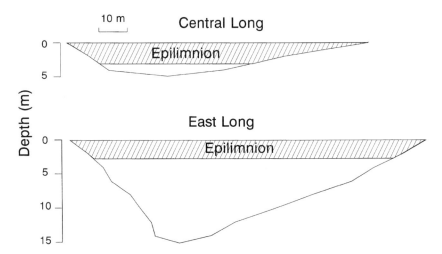

Figure 11-6. Bathymetry of Central Long and East Long lakes at the University of Notre Dame Environmental Research Center. Midsummer epilimnetic depth for 1992 is indicated.

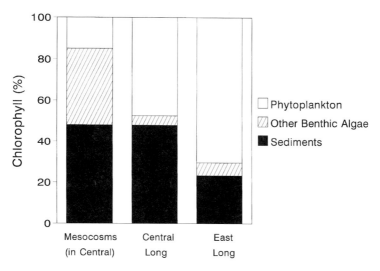

Figure 11-7. Percentage contribution to epilimnetic chlorophyll by different functional groups of algae: phytoplankton, algae growing on sediments, and algae growing on coarse woody debris (Central Long Lake and East Long Lake) or translucent polyethylene mesocosm walls. Epiphytic algae were not quantified because macrophytes are very sparse in these oligotrophic lakes. Data for phytoplankton (collected by K. L. Cottingham and S. R. Carpenter, University of Wisconsin, Madison) and periphyton in the lakes are from late July to mid-August 1992. When these data were collected, both basins had fish populations of about zero and neither had been fertilized. The mesocosm experiment was conducted in Central Long Lake from 12 June to 7 August 1992 (Blumenshine et al. 1996, Cottingham et al. 1996). Mesocosm chlorophyll data are means from unfertilized controls ($N = 3$ mesocosms) from 7 August.

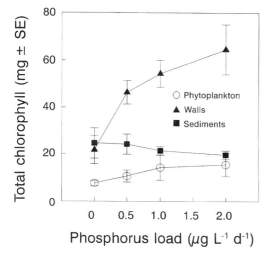

Figure 11-8. Standing stock of algae in different functional groups of algae at the conclusion of an 8-week (12 June–7 August 1992) mesocosm experiment in Central Long Lake. Replicate mesocosms ($N = 3$) were fertilized daily with N and P (at the ambient N:P ratio) at four loading rates above ambient. Adapted from Blumenshine et al. (1997).

on a non-nutrient-diffusing substratum. Clearly, a large proportion of added nutrients went into periphyton and was therefore unavailable to phytoplankton. Furthermore, logic and some preliminary evidence suggests that competition between phytoplankton and periphyton may be mediated by pelagic and benthic food web configuration.

As indicated in our literature survey (Fig. 11-5), few experiments have directly tested the important hypothesis that the configuration of the food web (e.g., abundance of *Daphnia*) might mediate the outcome of nutrient competition between phytoplankton and periphyton. However, in small artificial ponds high abundance of *Daphnia* enhances the flow of added nutrients to benthic algae relative to phytoplankton, while high abundance of a periphyton-grazing tadpole shifts the balance in favor of phytoplankton (Leibold and Wilbur 1992).

Thus, it is possible that much of the previously documented variation in phytoplankton response to nutrient additions (DeMelo et al. 1992) that is not attributable directly to pelagic food web configuration (Carpenter and Kitchell 1988) results from an interaction between the effect of the total food web (pelagic and benthic) and nutrient competition between phytoplankton and periphyton. This hypothesis complements other hypotheses about the importance of macrophytes and attached algae (Scheffer et al. 1993) and food web links between fish and benthic invertebrates (Threlkeld 1994, Schindler et al. 1996) in mediating phytoplankton response to nutrient enrichment. In whole-lake experiments, we are now testing the hypothesis that food web configuration determines the partitioning of added fertilizer between pelagic and benthic biota.

The potential importance of periphyton is not limited to artificial mesocosms (Fig. 11-9). Most of the world's lakes are small and shallow (Wetzel 1990), which makes the ratio of epilimnetic bottom area to epilimnion volume relatively high. In shallow lakes (≤ 6 m mean depth), benthic plants contribute 35 to 95% of annual lake primary production (Fig. 11-9). While some limnologists (e.g., Wetzel 1979) have appreciated the role of benthic processes including periphyton productivity, it has been largely

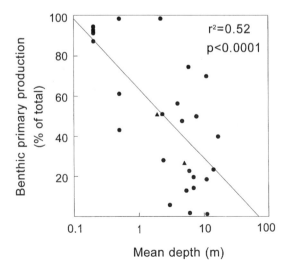

Figure 11-9. Percentage of total annual lake primary production contributed by benthic primary producers (macrophytes and algae). Triangles are for Central Long and East Long lakes; circles are for data gathered from the literature for all lakes for which we could locate benthic production (including macrophyte production where appropriate), phytoplankton production, and mean depth (lake volume/surface area). Data sources: Stanley (1976), Westlake (1980: Table 5.17), Wetzel (1983: Table 19-7).

ignored in the corpus of recent limnological work. The probable relationship between mean depth and the relative contribution of benthic production to whole-lake primary production has been noted (Stanley 1976, Westlake 1980), and much of the data in Figure 11-9 came from one of the most used limnology texts (Wetzel 1983). However, to our knowledge, this simple relationship has not been previously quantified.

We expect future nutrient abatement or biomanipulation plans designed to reach target phytoplankton standing crops will be made more accurate by incorporating the role of benthic communities in mediating phytoplankton response. Simple metrics like lake mean depth and lake size (Fee et al. 1994, Guildford et al. 1994) are suitable for incorporation into management models.

The scaling of benthic:pelagic production with lake depth also emphasizes why small-scale experiments may often lead to erroneous predictions of large-scale responses. If nutrient competition between benthic and pelagic algae is important, phytoplankton response to fertilization should be less in mesocosms than in whole-lake experiments. In fact, it is quite common for results of small-scale and large-scale results to differ, not only because of container "artifacts" but also because additional mechanisms operate at larger scales. For example, mesocosm experiments suggested that phytoplankton in an Argentine floodplain lake were not limited by nitrogen, whereas whole-lake fertilization demonstrated clearly that phytoplankton were N-limited (Carignan and Planas 1994). Whole-lake experiments often provide surprising results because unanticipated mechanisms and biotic interactions exist at the large scale that do not exist at the small scale (Carpenter and Kitchell 1993).

Case Study of Stream–Lake Landscape Links

Limnologists' understanding of lake productivity is advanced by expanding the spatial (see preceding example) and temporal (Carpenter and Kitchell 1993) scale of experiments. Yet a fuller understanding of aquatic community structure awaits more studies at an even larger spatial scale—the landscape scale—where the strong interactions between aquatic communities and terrestrial and groundwater communities can be recognized. Understanding aquatic responses to both anthropogenic (e.g., acidification and timber management) and natural (e.g., climate variability) disturbances often requires an incorporation of the watershed into the study (Hornung and Reynolds 1995). For a variety of freshwater, terrestrial, and estuarine communities, position in the landscape is strongly correlated with the variability of abiotic and biotic characteristics (Kratz et al. 1991). In northern Wisconsin, for example, lakes high in the groundwater table are more dependent on groundwater inputs of solutes than those low in elevation (Kratz et al. 1991). For high lakes, seasonal succession of phytoplankton cannot be understood without knowledge of groundwater sources of silica (Hurley et al. 1985).

Similarly, a lack of recognition of the important functional links between lakes and streams has hampered our understanding of what governs freshwater community structure. There are, of course, many reasons for the division in freshwater ecology between lake and stream studies, but it is partly maintained in the United States by the tradition of separate societies, each with its own journal, for stream and lake studies: the North American Benthological Society (*Journal of the North American Benthological Society*) and the American Society of Limnology and Oceanography (*Limnology and Ocean-*

ography), respectively. The segregation of lake and stream studies impedes progress in understanding the functional links between these two landscape elements.

Experiments at the scale of linked lakes and streams are very rare (Likens 1985, Swanson et al. 1988, Schlosser 1995), but observations at this scale should become a standard tool in studies on many anthropogenic effects on community and ecosystem function, including effects of acidic deposition, logging and deforestation, watershed liming, impact of chemical pollutants, resistance and resilience to disturbance (Tonn and Magnuson 1982), impact of climate warming (Lodge 1993a,b), and dispersal of exotic species (Lodge 1993a, Horvath et al. 1996). The dispersal of the exotic zebra mussel (*Dreissena polymorpha*) in North American inland waters is an instructive example of the need for a landscape perspective on freshwater habitats.

Zebra mussels were almost certainly transported to North America in the ballast water of ships from Europe (Nalepa and Schloesser 1993). In the sites of their earliest establishment in North America (around 1986)—lakes St. Clair and Erie—zebra mussels have reduced populations of native unionid clams, increased water clarity and abundance of macrophytes, and shifted much energy flow from the pelagic to the benthic habitat (Nalepa and Schloesser 1993). Zebra mussel control and removal from water intakes are already costing Great Lakes industries and municipalities millions of dollars annually. It is clear that the primary vectors of zebra mussel colonization for inland lakes are associated with boats (trailers, live wells, bait buckets, etc.; Carlton 1993). In order to appropriately target efforts to restrict further colonization or prepare for mitigation where establishment is inevitable, it is important to predict the future North American range of zebra mussels over areas ranging from continental to local scale.

On the continental scale, Strayer (1991) used the best information available from Europe (where, however, the species is still expanding its range) to predict cautiously that zebra mussels could ultimately occupy North America from southern Canada south to the central portions of California, Texas, Mississippi, Alabama, and Georgia (Fig. 3 in Strayer 1991). Already zebra mussels have colonized a number of inland lakes and rivers throughout the Midwest and the Tennessee River system and down the Mississippi River. Their recent establishment in Louisiana (Kraft 1995) already extends beyond the range predicted by Strayer (1991).

On a regional scale, zebra mussel distribution and abundance can be partly predicted by lake alkalinity, trophic status, human use, and substratum availability (Ramcharan et al. 1992, Koutnik and Padilla 1994, Mellina and Rasmussen 1994). Yet many regions of the midwestern and southeastern United States (excluding the coastal plain, where waters tend to be soft) have uncolonized lakes and streams that all fit within the range of physicochemical conditions necessary for self-sustaining zebra mussel populations (Horvath et al. 1996). Horvath et al. (forthcoming) think that links between lakes and streams will provide a key to predicting the trajectory of zebra mussel dispersal within such regions.

Strayer (1991) observed that in Europe streams colonized by zebra mussel were mostly large (mostly > 30 m wide); it seemed that stream size per se was important in determining habitat suitability. However, noting with Strayer (1991) the lack of obvious mechanisms for this pattern and the general difficulty of maintaining a stream population for a species with planktonic larvae (the veliger), Horvath et al. (1996) suspected that stream size was only correlated with an important unidentified factor.

They predicted that the hidden correlate was the presence or absence of a colonized upstream lake to provide a constant source of veligers.

A closer look at Strayer's European sources and our own U.S. midwestern studies confirmed that prediction (Horvath et al. 1996). Every stream site occupied by zebra mussels (five streams with adequate information available) had an upstream lake that was also occupied by zebra mussels. No stream without zebra mussels (59 streams with adequate information available) had a colonized upstream lake. Thus, there was a significant association (chi-square $P < .001$) between zebra mussel presence in lakes and zebra mussel presence in their outflowing streams. Streams that drain lakes tend to be larger than headwater streams, which explains the original stream size correlation. Despite our small sample size of colonized streams, the presence of an upstream lake with an established zebra mussel population appears to be a far better predictor of stream occupancy by zebra mussel than stream size (Horvath et al. 1996).

This more mechanistic understanding of zebra mussel dispersal will allow far better predictions of the spread of this exotic nuisance within regions and watersheds. This insight came only by expanding our perception of the landscape from only streams to the links between streams and lakes. Small-scale experiments in the stream could not have explained this pattern that required a larger scale context to be understood. This new understanding will allow better targeting of management of the zebra mussel. It will improve our ability to protect vulnerable native species like threatened and endangered unionid clams as well as our ability to mitigate the effects of the zebra mussel on industrial water intakes.

General Discussion and Recommendations

Over the last 30 years, the number of freshwater ecology papers published has increased dramatically, as has occurred in other scientific fields. Thirty years ago, ecological paradigms were based on far less evidence than the literature brings to bear on many current issues. Freshwater ecology has become more experimental in the last 30 years. While disagreements about scientific approach remain (Lehman 1986, Peters 1986), calls for pluralism are perhaps being heeded (McIntosh 1987, Ricklefs and Schluter 1993:2), as the value of both experimental and comparative studies is widely recognized.

Although freshwater ecologists have continued to conduct large-scale and long-term experiments, experimental designs have become much more rigorous. For both lotic and lentic habitats, goals of experiments have shifted from a concern with nutrients to a concern with food webs. This is clearly a manifestation of the top-down and bottom-up interplay that limnological research helped bring to the forefront of experimental ecology (Carpenter et al. 1985, Carpenter et al. 1991). Despite widespread agreement, if not consensus, that nutrients and food webs interact to shape communities and ecosystems (Matson and Hunter 1992), few published experiments were explicitly designed to quantify the direction and magnitudes of the interactive effects.

For lentic habitats, there have also been shifts in emphasis between pelagic and benthic studies; overall, benthic studies have been underrepresented. Joint studies of benthic and pelagic habitats seem to be increasing, but still few studies have devoted near equal attention to benthic and pelagic habitats. Surging interest in benthic–pelagic links among both freshwater (Lodge et al. 1988, Scheffer et al. 1993, Schindler et al. 1996) and marine (Anderson and Battarbee 1994, Denman 1994, Grassle and Grassle 1994, Rice and Lambshead 1994) researchers promises to reverse this soon.

Although we argue in the following discussion that within ecology freshwater ecology has been a leader in crossing subdisciplinary and habitat boundaries, it and the rest of ecology could benefit by moving more in that direction. The underrepresentation of studies that incorporate both nutrient and food web manipulations, the underrepresentation of studies that incorporate both benthic and pelagic components, and the absence of studies including all four are testaments to the conceptual and logistical difficulties inherent when multiple causes operate (Hilborn and Stearns 1982). However, the narrowness of much of our freshwater research and research in other areas of ecology limits the ability of the discipline to contribute to important scientific and societal issues (Brown 1995, Carpenter et al. 1995b).

We believe that for the most part, the temporal changes revealed by our literature survey reflect healthy development in freshwater ecology and ecology in general. Limnology and limnologists have provided some of the guiding concepts for community and ecosystem ecology for decades (Forbes 1887, Lindeman 1942, Hutchinson 1957, Brooks and Dodson 1965, Carpenter et al. 1985). One feature of aquatic systems that facilitates whole-ecosystem manipulations is the presence of more clear boundaries for aquatic communities relative to terrestrial communities. Freshwater ecology continues to lead in areas of both conceptual development and experimental methods.

For example, the recent increased appreciation for indirect effects and other complex interactions (Kerfoot and Sih 1987, Carpenter 1988), for the impact of acidification on communities (Likens and Bormann 1974, Schindler et al. 1985, Schindler 1988), and for the mutual importance to community structure and ecosystem productivity of nutrients and food webs (Carpenter et al. 1991) all had their genesis or substantial development in freshwater studies. From its inception, limnology has been a discipline that spans biology, geology, chemistry, and physics (Wetzel 1983, Lewis et al. 1995). As a consequence, both lake (Likens 1985, Kitchell 1992, Carpenter and Kitchell 1993, Frost et al. 1995, Schindler 1995, Sterner 1995) and stream (Grimm 1995, Power 1995) studies have been exemplars for integrating population, community, and ecosystem perspectives (Jones and Lawton 1995).

Freshwater ecologists have responded to the general ecological ferment about rigorous hypothesis testing, on the one hand (Strong et al. 1984), and the role of empiricism, on the other hand (Peters 1986, 1991), by making limnology a leader in both development of unreplicated experimental designs and statistical analysis (Matson and Carpenter 1990) and in comparative studies (Peters 1986, Carpenter et al. 1991), respectively. Additional contributions of appropriate unreplicated or poorly replicated experimental designs and analyses have come from a variety of ecologists—for example, before-after-control-impact designs (Stewart-Oaten et al. 1986, Green 1993), randomized intervention analysis (Carpenter 1989, Carpenter et al. 1989), and others (Underwood 1991, 1992, 1993; Lamont 1995). Use of both small-scale, mechanistic experiments and large-scale, unreplicated experiments, in addition to comparative studies, has contributed greatly to the success of limnology.

The experimental approach of freshwater ecology has been inescapably influenced by important applied interests like eutrophication (Edmondson 1991), acidification (Likens and Bormann 1974, Schindler et al. 1985, Schindler 1988), fish management (Magnuson 1991), and global change (Lodge 1993b). These have pushed the subdiscipline toward large-scale, long-term experiments that are directly relevant to the scales of human concerns with freshwater resources. We hope that the increased use of multifactorial experiments documented here is at the forefront of similar trends in other areas of ecology. As well as investigating more factors and habitats within a lake or stream, freshwater experiments need to expand the spatial scale to include links between lakes and streams and between aquatic and terrestrial communities.

The interplay of basic and applied research has increased the creativity of basic freshwater ecology and increased its usefulness to human society. When limnologists have had occasion to address an important applied question, they have not been paralyzed by statistical-design perfectionism. Rather, they have recognized that the test of some questions demanded a long temporal scale and large spatial scale. For example, questions that involve fish production or the role of fishes in structuring the community require at least an annual reproductive bout for top fish predators and the inclusion of both littoral and pelagic habitats for fish. When the logistics and expense of such critical scientific components require a poorly replicated design or observational approach, freshwater ecologists have been willing to forego statistically ideal designs. Some marine ecologists also advocate such a pragmatic approach (Dayton 1994).

We agree with this approach and suggest that if a research question is important, experimental methods should be tailored to the question and not solely dictated by the degrees of freedom available in an analysis of variance. Even if an experimental design renders statistical analysis impossible (because the goals cannot realistically be achieved any other way), a nonstatistical presentation is appropriate (Hurlbert 1984). For many such situations, however, statistical methods are now available (citations in previous discussion).

Increasing reliance on poorly replicated experiments also increases the importance of comparative approaches to ecological questions. Similar experiments conducted elsewhere become ''replicates,'' and meta-analysis (Arnqvist and Wooster 1995) and other comparative approaches become a critical logical link to infer causation. Comparative approaches become the test for generality (Carpenter et al. 1995b).

Large-scale, long-term experiments have also, of course, been conducted in terrestrial and marine habitats (Strayer et al. 1986; Carpenter et al. 1995b; Brown, this volume), but in general we recommend increased use of large-scale, long-term experiments in all subdisciplines of basic ecology. These should not replace small-scale, short-term experiments but should be complementary to them. Population and community phenomena are affected by large-scale, long-term ecosystem phenomena, and vice versa (Jones and Lawton 1995), and should be investigated at those scales. Incorporating these approaches in the research arsenal for population and community ecology (as well as ecosystem ecology) will allow researchers to better address important societal issues like acidic deposition, climate change, reductions in biodiversity, and ozone depletion. For many, if not all, of these issues, a large-scale, long-term perspective is necessary to separate small-scale versus large-scale variation and to separate short-term versus long-term effects and to judge anthropogenic effects against natural background variation.

ACKNOWLEDGMENTS   We thank Katie Wissing for a preliminary review of the literature and Laura Eidietis for substantial assistance with the final literature review. We also thank Steve Carpenter and Jim Kitchell for the opportunity to collaborate on whole-lake studies of lake productivity. Steve Carpenter and Kathy Cottingham kindly provided phytoplankton data for Figures 11-7 and 11-8. David Strayer made us aware of some of the data in Figure 11-9. Robert McIntosh provided several helpful references and a review of an early version. Reviews by Greg Cronin, Bill Perry, Peter Petraitis, Thomas Schlacher, and David Strayer greatly improved the manuscript. The National Science Foundation (Grant DEB-91-07569), the Environmental Protection Agency (CR820290-01-0), and the University of Notre Dame all provided substantial financial support for which we are grateful.

Literature Cited

Anderson, N. J., and R. W. Battarbee. 1994. Aquatic community persistence and variability: a paleolimnological perspective. Pages 233–260 in P. S. Giller, A. G. Hildrew, and D. G. Rafaelli (eds.), Aquatic Ecology: Scale, Pattern and Process. Blackwell Scientific, Oxford.

Arnqvist, G., and D. Wooster. 1995. Meta-analysis: synthesizing research findings in ecology and evolution. Trends in Ecology and Evolution 10:236–240.

Blumenshine, S. C., Y. Vadeboncoeur, D. M. Lodge, K. L. Cottingham, and S. E. Knight. 1997. Benthic-pelagic links: responses of benthos to water-column nutrient enrichment. Journal of the North American Benthological Society 16:466–479.

Bormann, F. H., and G. E. Likens. 1979. Pattern and Process in a Forested Ecosystem. Springer-Verlag, New York.

Brooks, J. L., and S. I. Dodson. 1965. Predation, body size, and composition of plankton. Science 150:28–35.

Brown, J. H. 1995. Macroecology. University of Chicago Press, Chicago, Illinois.

Carignan, R., and D. Planas. 1994. Recognition of nutrient and light limitation in turbid mixed layers: three approaches compared in the Paranà floodplain (Argentina). Limnology and Oceanography 39:580–596.

Carlton, J. T. 1993. Dispersal mechanisms of the zebra mussel (*Dreissena polymorpha*). Pages 677–697 in T. F. Nalepa and D. W. Schloesser (eds.), Zebra Mussels: Biology, Impacts, and Controls. Lewis, Ann Arbor, Michigan.

Carpenter, S. R. (ed.). 1988. Complex Interactions in Lake Communities. Springer-Verlag, New York.

———. 1989. Replication and treatment strength in whole-lake experiments. Ecology 70: 453–463.

Carpenter, S. R., and J. F. Kitchell. 1988. Consumer control of lake productivity. BioScience 38:764–769.

———. 1992. Trophic cascade and biomanipulation: interface of research and management—a reply to the comment by DeMelo et al. Limnology and Oceanography 37: 208–213.

——— (eds.). 1993. The Trophic Cascade in Lakes. Cambridge University Press, Cambridge.

Carpenter, S. R., J. F. Kitchell, and J. R. Hodgson. 1985. Cascading trophic interactions and lake productivity. BioScience 35:634–639.

Carpenter, S. R., J. F. Kitchell, J. R. Hodgson, P. A. Cochran, J. J. Elser, M. M. Elser, D. M. Lodge, D. Kretchmer, X. He, and C. N. von Ende. 1987. Regulation of lake primary productivity by food web structure. Ecology 68:1863–1876.

Carpenter, S. R., J. A. Morrice, P. A. Soranno, J. J. Elser, N. A. MacKay, and A. L. St. Amand. 1993. Primary production and its interactions with nutrients and light transmission. Pages 225–251 in S. R. Carpenter and J. F. Kitchell (eds.), The Trophic Cascade in Lakes. Cambridge University Press, Cambridge.

Carpenter, S. R., T. M. Frost, D. Heisey, and T. K. Kratz. 1989. Randomized intervention analysis and the interpretation of whole ecosystem experiments. Ecology 70:1142–1152.

Carpenter, S. R., T. M. Frost, J. F. Kitchell, T. K. Kratz, D. W. Schindler, J. Shearer, W. G. Sprules, M. J. Vanni, and A. P. Zimmerman. 1991. Patterns of primary production and herbivory in 25 North American lake ecosystems. Pages 67–96 in J. Cole, G. Lovett, and S. Findlay (eds.), Comparative Analyses of Ecosystems: Patterns, Mechanisms, and Theories. Springer-Verlag, New York.

Carpenter, S. R., J. F. Kitchell, K. L. Cottingham, D. E. Schindler, D. L. Christensen, D. M. Post, and N. Voichick. 1996. Chlorophyll variability, nutrient input and grazing: evidence from whole-lake experiments. Ecology 77:725–735.

Carpenter, S. R., S. W. Chisholm, C. J. Krebs, D. W. Schindler, and R. F. Wright. 1995b. Ecosystem experiments. Science 269:324–327.

Carpenter, S. R., D. L. Christensen, J. J. Cole, K. L. Cottingham, X. He, J. R. Hodgson, J. F. Kitchell, S. E. Knight, M. L. Pace, D. M. Post, D. E. Schindler, and N. Voichick. 1995a Biological control of eutrophication in lakes. Environmental Science and Technology 29:784–786.

Christensen, D. L., S. R. Carpenter, K. L. Cottingham, S. E. Knight, J. P. LeBouton, D. E. Schindler, N. Voichick, J. J. Cole, and M. L. Pace. 1996. Pelagic responses to changes in dissolved organic carbon following division of a seepage lake. Limnology and Oceanography 41:553–559.

Cole, J., G. Lovett, and S. Findlay (eds.). 1991. Comparative Analyses of Ecosystems: Patterns, Mechanisms, and Theories. Springer-Verlag, New York.

Connell, J. H. 1983. On the prevalence and relative importance of interspecific competition: evidence from field experiments. American Naturalist 122:66–696.

Cottingham, K. L., S. E. Knight, S. R. Carpenter, J. J. Cole, M. L. Pace, and A. E. Wagner. 1997. Response of phytoplankton and bacteria to nutrients and zooplankton: a mesocosm experiment. Journal of Plankton Research 19:995–1010.

Cullen, P. 1990. The turbulent boundary between water science and water management. Freshwater Biology 24:201–209.

Dayton, P. K. 1994. Community landscape: scale and stability in hard bottom marine communities. Pages 289–332 in P. S. Giller, A. G. Hildrew, and D. G. Rafaelli (eds.), Aquatic Ecology: Scale, Pattern and Process. Blackwell Scientific, Oxford.

DeMelo, R., R. France, and D. J. McQueen. 1992. Biomanipulation: hit or myth? Limnology and Oceanography 37:192–207.

Denman, K. L. 1994. Scale-determining biological-physical interactions in oceanic food webs. Pages 377–402 in P. S. Giller, A. G. Hildrew, and D. G. Rafaelli (eds.), Aquatic Ecology: Scale, Pattern and Process. Blackwell Scientific, Oxford.

Edmondson, W. T. 1991. The Uses of Ecology: Lake Washington and Beyond. University of Washington Press, Seattle.

Fee, E. J., R. E. Hecky, G. W. Regehr, L. L. Hendzel, and P. Wilkinson. 1994. Effects of lake size on nutrient availability in the mixed layer during summer stratification. Canadian Journal of Fisheries and Aquatic Sciences 51:2756–2768.

Fisher, S. G. 1994. Pattern, process and scale in freshwater systems: some unifying thoughts. Pages 575–592 in P. S. Giller, A. G. Hildrew, and D. G. Rafaelli (eds.), Aquatic Ecology: Scale, Pattern and Process. Blackwell Scientific, Oxford.

Forbes, S. A. 1887. The lake as a microcosm. Bulletin of the Peoria Scientific Association 77–87. Reprinted as pages 14–27 in L. A. Real and J. H. Brown (eds.), 1991, Foundations of Ecology: Classic Papers with Commentaries. University of Chicago Press, Chicago, Illinois.

Franklin, J. F. 1995. Why link species conservation, environmental protection, and resource management? Pages 326–335 in C. G. Jones and J. H. Lawton (eds.), Linking Species and Ecosystems. Chapman and Hall, New York.

Frost, T. M., D. L. DeAngelis, S. M. Bartell, D. J. Hall, and S. H. Hurlbert. 1988. Scale in the design and interpretation of aquatic community research. Pages 229–260 in S. R. Carpenter (eds.), Complex Interactions in Lake Communities. Springer-Verlag, New York.

Frost, T. M., S. R. Carpenter, A. R. Ives, and T. K. Kratz. 1995. Species compensation and complementarity in ecosystem function. Pages 224–239 in C. G. Jones and J. H. Lawton (eds.), Linking Species and Ecosystems. Chapman and Hall, New York.

Grassle, J. F., and J. P. Grassle. 1994. Notes from the abyss: the effects of a patchy supply of organic material and larvae on soft-sediment benthic communities. Pages 499–516 in P. S. Giller, A. G. Hildrew, and D. G. Rafaelli (eds.) Aquatic Ecology: Scale, Pattern and Process. Blackwell Scientific, Oxford.

Green, R. 1993. Application of repeated measures designs in environmental impact and monitoring studies. Australian Journal of Ecology 18:81–98.

Grimm, N. B. 1994. Disturbance, succession and ecosystem processes in streams: a case study from the desert. Pages 93–112 in P. S. Giller, A. G. Hildrew, and D. G. Rafaelli (eds.), Aquatic Ecology: Scale, Pattern and Process. Blackwell Scientific, Oxford.

———. 1995. Why link species and ecosystems? A perspective from ecosystem ecology. Pages 5–15 in C. G. Jones and J. H. Lawton (eds.), Linking Species and Ecosystems. Chapman and Hall, New York.

Guildford, S. J., L. L. Hendzel, H. J. Kling, E. J. Fee, G. G. C. Robinson, R. E. Hecky, and S. E. M. Kasian. 1994. Effects of lake size on phytoplankton nutrient status. Canadian Journal of Fisheries and Aquatic Sciences 51:2769–2783.

Hairston, N. G., Sr. 1989. Ecological Experiments: Purpose, Design, and Execution. Cambridge University Press, Cambridge.

Hall, D. J., W. E. Cooper, and E. E. Werner. 1970. An experimental approach to the production dynamics and structure of freshwater animal communities. Limnology and Oceanography 15:839–928.

Hall, S. J., D. Raffaelli, and S. F. Thrush. 1994. Patchiness and disturbance in shallow water benthic assemblages. Pages 333–376 in P. S. Giller, A. G. Hildrew, and D. G. Rafaelli (eds.), Aquatic Ecology: Scale, Pattern and Process. Blackwell Scientific, Oxford.

Hansson, L-A. 1988. Chlorophyll determination of periphyton on sediments: identification of problems and recommendation of method. Freshwater Biology 20:347–352.

———. 1990. Quantifying the impact of periphytic algae on nutrient availability for phytoplankton. Freshwater Biology 24:265–273.

Hilborn, R., and S. C. Stearns. 1982. On inference and evolutionary biology: the problem of multiple causes. Biotheoretica 31:145–164.

Hildrew, A. G., and P. S. Giller. 1994. Patchiness, species interactions and disturbance in the stream benthos. Pages 21–62 in P. S. Giller, A. G. Hildrew, and D. G. Rafaelli (eds.), Aquatic Ecology: Scale, Pattern and Process. Blackwell Scientific, Oxford.

Hornung, M., and B. Reynolds. 1995. The effects of natural and anthropogenic environmental changes on ecosystem processes at the catchment scale. Trends in Ecology and Evolution 10:443–449.

Horvath, T. G., G. A. Lamberti, D. M. Lodge, and W. L. Perry. 1996. Zebra mussel dispersal in lake-stream systems: source-sink dynamics? Journal of the North American Benthological Society 15:564–575.

Hurlbert, S. H. 1984. Pseudoreplication and the design of ecological field experiments. Ecological Monographs 54:187–211.

Hurley, J. P., D. E. Armstrong, G. J. Kenoyer, and C. J. Bowser. 1985. Ground water as a silica source for diatom production in a precipitation-dominated lake. Science 227: 1576–1578.

Hutchinson, G. E. 1957. Concluding remarks. Cold Spring Harbor Symposium on Quantitative Biology 22:415–427.

Jones, C. G., and J. H. Lawton (eds.). 1995. Linking Species and Ecosystems. Chapman and Hall, New York.

Karieva, P. M., and M. Andersen. 1988. Spatial aspects of species interactions: the wedding of models and experiments. Pages 38–54 in A. Hastings (ed.), Community Ecology. Springer-Verlag, New York.

Kerfoot, W. C., and A. Sih (eds.). 1987. Predation: direct and indirect impacts on aquatic communities. University Press of New England, Hanover, New Hampshire.

Kitchell, J. F. (ed.). 1992. Food web management: a case study of Lake Mendota. Springer-Verlag, New York.

Kitchell, J. F., and S. R. Carpenter. 1993. Synthesis and new directions. Pages 332–350 in S. R. Carpenter and J. F. Kitchell (eds.), The Trophic Cascade in Lakes. Cambridge University Press, Cambridge.

Kitchell, J. F., S. M. Bartell, S. R. Carpenter, S. R. Hall, D. J. McQueen, W. E. Neill, D. Scavia, and E. E. Werner. 1988. Epistemology, experiments, and pragmatism. Pages 263–280 in S. R. Carpenter (ed.), Complex Interactions in Lake Communities. Springer-Verlag, New York.

Koutnik, M. A., and D. K. Padilla. 1994. Predicting the spatial distribution of *Dreissena polymorpha* (zebra mussel) among inland lakes in Wisconsin: modelling with a GIS. Canadian Journal of Fisheries and Aquatic Sciences 51:1189–1196.

Kraft, C. 1995. Zebra mussels invade the south. University of Wisconsin Sea Grant Institute Zebra Mussel Update 24:3.

Kratz, T. K., B. J. Benson, E. R. Blood, G. L. Cunninham, and R. A. Dahlgren. 1991. The influence of landscape position on temporal variability in four North American ecosystems. American Naturalist 138:335–378.

Krebs, C. J. 1994. Ecology. 4th ed. Harper-Collins, New York.

Lamont, B. B. 1995. Testing the effect of ecosystem composition/structure on its functioning. Oikos 74:283–295.

Lehman, J. T. 1986. The goal of understanding in limnology. Limnology and Oceanography 31:1160–1166.

Leibold, M. A. 1989. Resource edibility and the effects of predators and productivity on the outcome of trophic interactions. American Naturalist 134:922–949.

Leibold, M. A., and H. M. Wilbur. 1992. Interactions between food-web structure and nutrients on pond organisms. Nature 360:341–343.

Lewis, W. M., Jr., S. W. Chisholm, C. F. D'Elia, E. J. Fee, N. G. Hairston Jr., J. E. Hobbie, G. E. Likens, S. T. Threlkeld, and R. G. Wetzel. 1995. Challenges for limnology in the United States and Canada: an assessment of the discipline in the 1990's. American Society of Limnology and Oceanography Bulletin (Special Issue 2) 4:1–20.

Likens, G. E. (ed.) 1985. An Ecosystem Approach to Aquatic Ecology: Mirror Lake and Its Environment. Springer-Verlag, New York.

———— (ed.) 1989. Long-Term Studies in Ecology. Springer-Verlag, New York.

———— 1992. The Ecosystem Approach: Its Use and Abuse. Ecology Institute, Oldendorf/ Luhe, Germany.

Likens, G. E., and F. H. Bormann. 1974. Acid rain : a serious regional environmental problem. Science 184:1176–1179.

Lindeman, R. L. 1942. The trophic-dynmaic aspect of ecology. Ecology 23:399–418.

Lodge, D. M. 1993a. Biological invasions: lessons for ecology. Trends in Ecology and Evolution 8:133–137.

————. 1993b. Species invasions and deletions: community effects and responses to climate and habitat change. Pages 367–387 in P. M. Karieva, J. G. Kingsolver, and R. B. Huey (eds.), Biotic Interactions and Global Change. Sinauer, Sunderland, Massachusetts.

Lodge, D. M., J. W. Barko, D. Strayer, J. M. Melack, G. G. Mittelbach, R. W. Howarth, B. Menge, and J. E. Titus. 1988. Spatial heterogeneity and habitat interactions in lake communities. Pages 181–208 in S. R. Carpenter (ed.), Complex Interactions in Lake Communities. Springer-Verlag, New York.

Lubchenco, J. 1995. The relevance of ecology: the societal context and disciplinary implications of linkages across levels of ecological organization. Pages 297–305 in C. G. Jones and J. H. Lawton (eds.), Linking Species and Ecosystems. Chapman and Hall, New York.

Magnuson, J. J. 1991. Fish and fisheries ecology. Ecological Applications 1:13–26.

Matson, P. A., and S. R. Carpenter (eds.). 1990. Statistical analysis of ecological response to large-scale perturbations. Ecology 71:2037–2068.

Matson, P. A., and M. D. Hunter (eds.). 1992. The relative contributions of top-down and bottom-up forces in population and community ecology. Ecology 73:723–765.

McIntosh, R. P. 1987. Pluralism in ecology. Annual Review of Ecology and Systematics 18:321–341.

————. 1995. H. A. Gleason's ''individualistic concept'' and theory of animal communities: a continuing controversy. Biological Review 70:317–357.

Mellina, E., and J. B. Rasmussen. 1994. Patterns in the distribution and abundance of zebra mussel (*Dreissena polymorpha*) in rivers and lakes in relation to substrate and other physicochemical factors. Canadian Journal of Fisheries and Aquatic Sciences 51:1024–1036.

Mooney, H. A. (ed.). 1991. Ecosystem Experiments. Wiley, New York.

Naiman, R. J., J. J. Magnuson, D. M. McKnight, and J. A. Stanford (eds.). 1995. The Freshwater Imperative. Island Press, Washington, D.C.

Nalepa, T. F., and D. W. Schloesser (eds.). 1993. Zebra Mussels: Biology, Impacts, and Controls. Lewis, Ann Arbor, Michigan.

Neill, W. E. 1994. Spatial and temporal scaling and the organization of limnetic communities. Pages 189–232 in P. S. Giller, A. G. Hildrew, and D. G. Rafaelli (eds.), Aquatic Ecology: Scale, Pattern and Process. Blackwell Scientific, Oxford.

Peters, R. H. 1986. The role of prediction in limnology. Limnology and Oceanography 31:1143–1159.

————. 1991. A Critique for Ecology. Cambridge University Press, Cambridge.

Peterson, C. H. 1993. Improvement of environmental impact analysis by application of principles derived from manipulative ecology: lessons from coastal marine case histories. Australian Journal of Ecology 18:21–52.

Power, M. E. 1995. Floods, food chains, and ecosystem processes in rivers. Pages 52–60 in C. G. Jones and J. H. Lawton, (eds.), Linking Species and Ecosystems. Chapman and Hall, New York.

Ramcharan, C. W., D. K. Padilla, and S. I. Dodson. 1992. A multivariate model for predicting population fluctuations of *Dreissena polymorpha* in North American lakes. Canadian Journal of Fisheries and Aquatic Sciences 49:150–158.

Real, L. A., and J. H. Brown (eds.). 1991. Foundations of Ecology: Classic Papers with Commentaries. University of Chicago Press, Chicago, Illinois.

Rice, A. L., and P. J. D. Lambshead. 1994. Patch dynamics in the deep-sea benthos: the role of a heterogenous supply of organic matter. Pages 469–498 in P. S. Giller, A. G. Hildrew, and D. G. Rafaelli (eds.), Aquatic Ecology: Scale, Pattern and Process. Blackwell Scientific, Oxford.

Ricklefs, R. E., and D. Schluter (eds.). 1993. Species Diversity in Ecological Communities: Historical and Geographical Perspectives. University of Chicago Press, Chicago, Illinois.

Scheffer, M., S. H. Hosper, M-L. Meijer, B. Moss, and E. Jeppesen. 1993. Alternative equilibria in shallow lakes. Trends in Ecology and Evolution 8:275–279.

Schindler, D. E., S. R. Carpenter, K. L. Cottingham, X. He, J. R. Hodgson, J. F. Kitchell, and P. A. Soranno. 1996. Food web structure and littoral zone coupling to pelagic trophic cascades. Pages 96–105 in G. A. Polis and K. O. Winemiller (eds.), Food Webs: Integration of Patterns and Dynamics. Chapman and Hall, New York.

Schindler, D. W. 1974. Eutrophication and recovery in the experimental lakes: implications for lake management. Science 184:897–899.

———. 1978. Factors regulating phytoplankton production and standing crop in the world's lakes. Limnology and Oceanography 23:478–486.

———. 1988. Effects of acid rain on freshwater ecosystems. Science 239:149–157.

———. 1995. Linking species and communities to ecosystem management: a perspective from the experimental lakes experience. Pages 313–325 in C. G. Jones and J. H. Lawton (eds.), Linking Species and Ecosystems. Chapman and Hall, New York.

Schindler, D. W., K. H. Mills, D. F. Mallery, D. L. Findlay, J. A. Shearer, I. J. Davies, M. A. Turner, G. A. Linsley, and D. R. Cruikshank. 1985. Long-term ecosystem stress: the effects of years of experimental acidification on a small lake. Science 228:1395–1401.

Schindler, D. W., K. G. Beatty, E. J. Fee, D. R. Cruikshank, E. R. DeBruyn, D. L. Findlay, G. A. Lindsey, J. A. Shearer, M. P. Stainton, and M. A. Turner. 1990. Effects of climate warming on lakes of the central boreal forest. Science 250:967–970.

Schlosser, I. J. 1995. Dispersal, boundary processes, and trophic-level interactions in streams adjacent to beaver ponds. Ecology 76:908–925.

Schoener, T. W. 1983. Field experiments on interspecific competition. American Naturalist 122:240–285.

Sih, A., P. Crowley, M. McPeek, J. Petranka, and K. Strohmeier. 1985. Predation, competition, and prey communities: a review of field experiments. Annual Review of Ecology and Systematics 16:269–311.

Stanley, D. W. 1976. Productivity of epipelic algae in tundra ponds and a lake near Barrow, Alaska. Ecology 57:1015–1024.

Sterner, R. W. 1995. Elemental stoichiometry of species in ecosystems. Pages 240–252 in C. G. Jones and J. H. Lawton (eds.), Linking Species and Ecosystems. Chapman and Hall, New York.

Stewart-Oaten, A., W. M. Murdoch, and K. R. Parker. 1986. Environmental impact assessment: "pseudoreplication" in time? Ecology 67:929–940.

Strayer, D. L. 1991. Projected distribution of the zebra mussel, *Dreissena polymorpha*, in North America. Canadian Journal of Fisheries and Aquatic Sciences 48:1389–1395.

Strayer, D. L., J. S. Glitzenstein, C. G. Jones, J. Kolasa, G. E. Likens, M. J. McDonnell, G. G. Parker, and S. T. A. Pickett. 1986. Long-Term Ecological Studies: An Illustrated Account of Their Design, Operation, and Importance to Ecology. Occasional Publication no. 2. Institute of Ecosystem Studies, Millbrook, New York.

Strong, D. R., Jr., D. Simberloff, L. G. Abele, and A. B. Thistle (eds.). 1984. Ecological Communities: Conceptual Issues and the Evidence. Princeton University Press, Princeton, New Jersey.

Swanson, F. J., T. K. Kratz, N. Caine, and R. G. Woodmansee. 1988. Landform effects on ecosystem patterns and processes. BioScience 38:92–98.

Threlkeld, S. T. 1994. Benthic-pelagic interactions in shallow water columns: an experimentalist's perspective. Hydrobiologia 275/276:293–300.

Tilman, D. 1989. Ecological experimentation: strengths and conceptual problems. Pages 136–157 in G. E. Likens (ed.), Long-Term Studies in Ecology. Springer-Verlag, New York.

Tonn, W. M., and J. J. Magnuson. 1982. Patterns in the species composition and richness of fish assemblages in northern Wisconsin lakes. Ecology 63:1149–1166.

Underwood, A. J. 1991. Beyond BACI: experimental designs for detecting human environmental impacts on temporal variations in natural populations. Australian Journal of Marine and Freshwater Research 42:569–587.

———. 1992. Beyond BACI: the detection of environmental impacts on populations in the real, but variable, world. Journal of Experimental Marine Biology and Ecology 161: 145–178.

———. 1993. The mechanics of spatially replicated sampling programs to detect environmental impacts in a variable world. Australian Journal of Ecology 18:99–116.

Walters, C. J. 1993. Dynamic models and large scale field experiments in environmental assessment and management. Australian Journal of Ecology 18:53–61.

Westlake, D. F. 1980. Primary production. Pages 141–246 in E. D. Le Cren and R. H. Lowe-McConnell (eds.), The Functioning of Freshwater Ecosystems. Cambridge University Press, Cambridge.

Wetzel, R. G. 1979. The role of the littoral zone and detritus in lake metabolism. Ergebnisse der Limnologie 13:145–161.

———. 1983. Limnology, 2nd ed. Saunders College Publishing, New York.

———. 1990. Land-water interfaces: metabolic and limnological regulators. Verhandlungen der Internationale Vereinigung für theoretische und angewandte Limnologie 24:6–24.

Wiens, J. A. 1986. Spatial scale and temporal variation in studies of shrubsteppe birds. Pages 154–172 in J. Diamond and T. J. Case (eds.), Community Ecology. Harper and Row, New York.

Appendix

Published papers reviewed, listed chronologically (and alphabetically within years). Lotic and lentic papers are listed separately. Our categorization of each study is indicated in parentheses at the end of each citation: before the comma, the letter(s) indicate whether the experiment manipulated a nutrient (N), a food web component (F), or both (NF); after the comma, the letter(s) indicate the focus habitat, pelagic (P), benthic (B), or both (PB). Each experiment was treated as a datum in our analysis. When the number of experiments reported in a paper exceeded one, it is indicated in parentheses.

Lotic

King, D. K., and R. C. Ball. 1967. *Limnology and Oceanography* 12:27–33. (N,B)

Peckarsky, B. L., and S. I. Dodson. 1980. *Ecology* 61:1275–1282. (F,B; 3 expts.)

Peckarsky, B. L., and S. I. Dodson. 1980. *Ecology* 61:1283–1290. (F,B; 2 expts.)

Lamberti, G. A., and V. H. Resh. 1983. *Ecology* 64:1124–1135. (F,B; 2 expts.)

Meyer, J. L., and C. Johnson. 1983. *Freshwater Biology* 13:177–183. (N,B)

Newbold, J. D., et al. 1983. *Freshwater Biology* 13:193–204. (N,B)

Peterson, B. J., et al. 1983. *Limnology and Oceanography* 28:583–591. (N,B)

Shanz, F., and H. Juon. 1983. *Hydrobiologia* 102:187–195. (N,B)

Culp, J. M. 1986. *Journal of the North American Benthological Society* 5:140–149. (F,B)

Dudley, T. L., et al. 1986. *Journal of the North American Benthological Society* 5:93–106. (F,B; 3 expts.)

Fuller, R. L., et al. 1986. *Journal of the North American Benthological Society* 5:290–296. (F,B)

Grimm, N. B., and S. G. Fisher. 1986. *Journal of the North American Benthological Society* 5:2–15. (N,B; 4 expts.)

Hawkins, C. P. 1986. *Ecology* 67:1384–1395. (F,B)

Lowe, R. L., et al. 1986. *Journal of the North American Benthological Society* 5:221–229. (N,B)

Pringle, C. M., et al. 1986. *Hydrobiologia* 134:207–213. (N,B)

Sweeney, B. W., and R. L. Vannote. 1986. *Ecology* 67:1396–1410. (F,B; 2 expts.)

Wallace, J. B., et al. 1986. *Journal of the North American Benthological Society* 5:115–126. (F,B)

Williamson, C. E., and N. M. Butler. 1986. *Limnology and Oceanography* 31:393–402. (F,P)

Bothwell, M. L. 1988. *Canadian Journal of Fisheries and Aquatic Sciences* 45:261–270. (N,PB)

Hershey, A. E., et al. 1988. *Ecology* 69:1383–1392. (N,B; 2 expts.)

Hershey, A. E., and A. L. Hiltner. 1988. *Journal of the North American Benthological Society* 7:188–196. (F,B)

Hill, W. R., and A. W. Knight. 1988. *Limnology and Oceanography* 33:15–26. (F,B)

Lancaster, J., et al. 1988. *Freshwater Biology* 20:185–193. (F,B)

Power, M. E., et al. 1988. *Ecology* 69:1894–1898. (F,B)

Schofield, K., et al. 1988. *Freshwater Biology* 20:85–95. (F,B; 2 expts.)

van Donk, E., et al. 1988. *Freshwater Biology* 20:199–210. (N,P)

Armitage, M. J., and J. O. Young. 1990. *Freshwater Biology* 24:101–107. (F,B)

Bott, T. L., and L. A. Kaplan. 1990. *Journal of the North American Benthological Society* 9:336–345. (F,B)

Coleman, R. L., and C. N. Dahm. 1990. *Journal of the North American Benthological Society* 9:293–302. (N,B)

Cuffney, T. F., et al. 1990. *Freshwater Biology* 23:281–299. (F,B)

DeNicola, D. M., et al. 1990. *Freshwater Biology* 23:475–489. (F,B)

Hart, D. D., and C. T. Robinson. 1990. *Ecology* 71:1494–1502. (N,B; 2 expts.)

Hill, W. R., and B. C. Harvey. 1990. *Canadian Journal of Fisheries and Aquatic Sciences* 47:2307–2314. (F,PB)

Holomuzki, J. R., and J. D. Hoyle. 1990. *Freshwater Biology* 24:509–517. (F,B)

Horner, R. R., et al. 1990. *Freshwater Biology* 24:215–232. (N,B)

Huang, C., and A. Sih. 1990. *Ecology* 71:1515–1598. (F,PB)

Johnston, N. T., et al. 1990. *Canadian Journal of Fisheries and Aquatic Sciences* 47:862–872. (N,B)

Kreutweiser, D. P. 1990. *Canadian Journal of Fisheries and Aquatic Sciences* 47:1387–1401. (F,B)

Lancaster, J., et al. 1990. *Hydrobiologia* 203:177–190. (F,B)

McCormick, P. V. 1990. *Canadian Journal of Fisheries and Aquatic Sciences* 47:2057–2065. (NF,B)

Peckarsky, B. L., et al. 1990. *Freshwater Biology* 24:181–191. (F,B)

Pringle, C. M. 1990. *Ecology* 71:905–920. (N,B; 2 expts.)

Robinson, C. T., et al. 1990. *Journal of the North American Benthological Society* 9:240–248. (F,B; 2 expts.)

Winterbourn, M. J. 1990. *Freshwater Biology* 23:463–474. (NF,B; 2 expts.) (N,B)

Gilliam, J. F., et al. 1993. *Ecology* 74:1856–1870. (F,P)

Krystyna, P. 1993. *Freshwater Biology* 29:71–78. (F,B)

Mather, M. E., and R. A. Stein. 1993. *Canadian Journal of Fisheries and Aquatic Sciences* 50:1279–1288. (F,B)

Peckarsky, B. L. 1993. *Ecology* 74:1836–1846. (F,B; 2 expts.)

Peterson, B. J., et al. 1993. *Ecology* 74:653–672. (N,PB; 2 expts.)

Rosemond, A. D., et al. 1993. *Ecology* 74:1264–1280. (NF,B) (NF,P)

Soluk, D. A. 1993. *Ecology* 74:219–225. (F,B)

Wiseman, S. W., et al. 1993. *Freshwater Biology* 30:133–145. (F,B)

Bergey, E. A., and V. H. Resh. 1994. *Freshwater Biology* 31:153–163. (F,B) (NF,B)

Bergman, E., and L. A. Greenberg. 1994. *Ecology* 75:1233–1245. (F,PB)

Covich, A. P., and T. A. Crowl. 1994. *Journal of the North American Benthological Society* 13:291–298. (F,B; 2 expts.)

Creed, R. P. 1994. *Ecology* 75:2091–2103. (F,B; 2 expts.)

Feminella, J. W., and C. P. Hawkins. 1994. *Journal of the North American Benthological Society* 13:310–320. (F,P;2 expts.)

Forrester, G. E. 1994. *Ecology* 75:1208–1218. (F,B)

Greenberg, L. A. 1994. *Freshwater Biology* 32:1–11. (F,B)

McCormick, P. V., et al. 1994. *Freshwater Biology* 31:201–212 (F,B)

Scrimgeour, G. J., et al. 1994. *Journal of the North American Benthological Society* 13:368–378. (F,PB)

Lentic

Carr, J. F., and J. K. Hiltunen. 1965. *Limnology and Oceanography* 10:551–569. (N,B)

Hairston, N. G., and S. L. Kellermann. 1965. *Ecology* 46:134–139. (F,P)

McConnell, W. J. 1965. *Limnology and Oceanography* 10:539–543. (N, PB; 2 expts.)

Eisenberg, R. M. 1966. *Ecology* 47:889–406. (F,B)

Robertson, A., and W. P. Alley. 1966. *Limnology and Oceanography* 11:576–583. (N,B)

Dickman, M. 1968. *Ecology* 49:1188–1190. (F,B)

Maguire, B., Jr., et al. 1968. *Ecology* 49:207–210. (F,P)

Smith, M. W. 1968. *Canadian Journal of Fisheries and Aquatic Sciences* 25:2011–2036. (N,P)

Kalff, J. 1971. *Ecology* 52:655–659. (N,P)

Mauck, W. L., and D. W. Coble. 1971. *Canadian Journal of Fisheries and Aquatic Sciences* 28:957–969. (F,P)

Sakamoto, M. 1971. *Canadian Journal of Fisheries and Aquatic Sciences* 28:203–213. (N,P)

Schindler, D. W., et al. 1971. *Canadian Journal of Fisheries and Aquatic Sciences* 28:1763–1782. (N,P;2 expts.)

Addicott, J. F., 1974. *Ecology* 55:475–492. (F,B; 3 expts.)

deNoyelles, F., Jr., and W. J. O'Brien. 1974. *Limnology and Oceanography* 19:326–331. (N,P)

Dodson, S. I. 1974. *Ecology* 55:605–613. (F,P)

O'Brien, W. J., and F. DeNoyelles Jr. 1974. *Hydrobiologia* 44:105–125. (N,P)

Peterson, B. J., et al. 1974. *Limnology and Oceanography* 19:396–408. (N,P)

Schelske, C. L. 1974. *Limnology and Oceanography* 19:409–419. (N,P)

Schindler, D. W., et al. 1974. *Canadian Journal of Fisheries and Aquatic Sciences* 31:647–662. (N,P)

Starling, M. B., et al. 1974. *Hydrobiologia* 45:91–113. (N,B)

Tunzi, M. G., and D. B. Porcella. 1974. *Limnology and Oceanography* 19:420–428. (N,P)

Werner, E. E., and D. J. Hall. 1974. *Ecology* 55:1042–1052. (F,P)

Allen, H. T., and B. T. Ocerski. 1977. *Hydrobiologia* 53:49–54. (N,P)

Berman, T., et al. 1977. *Freshwater Biology* 7:495–502. (N,P)

Jones, J. G. 1977. *Freshwater Biology* 7:67–91. (N,P)

Kerfoot, W. C. 1977. *Limnology and Oceanography* 22:316–325. (F,P)

Liao, C. F. H. 1977. *Hydrobiologia* 56:273–279. (N,P)

Nduku, W. K., and R. D. Roberts. 1977. *Freshwater Biology* 7:19–30. (N,P)

Stevens, R. J., and M. P. Parr. 1977. *Freshwater Biology* 7:351–355. (N,P)

Wassersug, R. J., and D. G. Sperry. 1977. *Ecology* 58:830–839. (F,B)

Werner, E. E., and D. J. Hall. 1977. *Ecology* 58:869–876. (F,B)

Wilbur, H. M. 1977. *Ecology* 68:196–200. (F,B)

Wilbur, H. M. 1977. *Ecology* 68:206–209. (F,B)

Barica, J., et al. 1980. *Canadian Journal of Fisheries and Aquatic Sciences* 37:1175–1183. (N,P; 2 expts.)

Cooper, S. D., and C. R. Goldman. 1980. *Canadian Journal of Fisheries and Aquatic Sciences* 37:909–919. (F,P)

DeCosta, J., and C. Preston. 1980. *Hydrobiologia* 70:39–49. (N,P)

Dodson, S. I., and D. L. Egger. 1980. *Ecology* 61:755–763. (F,P)

Fry, D. L., and J. A. Osborne. 1980. *Hydrobiologia* 68:145–155. (F,P)

Henrickson, L., et al. 1980. *Hydrobiologia* 68:257–263. (F,P)

Hewett, S. W. 1980. *Ecology* 61:1075–1081. (F,P)

Hunter, R. D. 1980. *Hydrobiologia* 69:251–259. (F,B)

Jana, B. B., et al. 1980. *Hydrobiologia* 75:231–239. (N,PB)

Kerfoot, W. C., and C. Peterson. 1980. *Ecology* 61:417–431. (F,P)

Knoechel, R., and F. deNoyelles Jr. 1980. *Canadian Journal of Fisheries and Aquatic Sciences* 37:434–441. (N,P)

McDiffett, W. F. 1980. *Hydrobiologia* 71:137–145. (N,P)

Mitchell, S. F., and R. G. Wetzel. 1980. *Hydrobiologia* 68:235–241. (NF,P)

Pabst, M. H., and M. G. Boyer. 1980. *Hydrobiologia* 69:245–250. (F,P)

Redfield, G. W. 1980. *Hydrobiologia* 70:217–224. (F,P)

Schindler, D. W., et al. 1980. *Canadian Journal of Fisheries and Aquatic Sciences* 37: 320–327. (N,P)

Stross, R. G. 1980. *Limnology and Oceanography* 25:538–544. (N,P; 2 expts.)

Williams, J. D. H., et al. 1980. *Limnology and Oceanography* 25:1–11. (N,P)

Brambilla, D. J. 1983. *Hydrobiologia* 99:175–188. (F,P)

Cattaneo, A. 1983. *Limnology and Oceanography* 28:124–132. (F,B; 2 expts.)

Crivelli, A. J. 1983. *Hydrobiologia* 106:37–41. (F,B)

Cuker, B. E. 1983. *Ecology* 64:10–15. (F,B)

Cuker, B. E. 1983. *Limnology and Oceanography* 28:133–141. (F,B; 2 expts.), (N,B)

Dodson, S. I., and S. D. Cooper. 1983. *Limnology and Oceanography* 28:345–351. (F,P; 2 expts.)

Golterman, H. L. 1983. *Hydrobiologia* 100:59–64. (N,P)

McCoy, G. A. 1983. *Canadian Journal of Fisheries and Aquatic Sciences* 40:1195–1202. (N,P)

Morin, P. J., et al. 1983. *Ecology* 64:1430–1436. (F,B)

Mundie, J. H., et al. 1983. *Canadian Journal of Fisheries and Aquatic Sciences* 40: 1702–1712. (N,B)

Murphy, T. P., et al. 1983. *Limnology and Oceanography* 28:58–69. (N,P)

Spencer, D. F., et al. 1983. *Hydrobiologia* 107:123–130. (N,P)

Werner, E. E., et al. 1983. *Ecology* 64:1525–1539. (F,PB)

Wilbur, H. M., et al. 1983. *Ecology* 64:1423–1429. (F,B)

Williams, J. B. 1983. *Hydrobiologia* 107:131–139. (F,P) (F,PB)

Anderson, M. R., and J. Kalff. 1986. *Freshwater Biology* 16:735–743. (N,B)

Arumugam, P. T., and M. C. Geddes. 1986. *Hydrobiologia* 135:215–221. (F,P)

Barko, J. W., and R. M. Smart. 1986. *Ecology* 67:1328–1340. (N,B; 2 expts.)

Bergquist, A. M., and S. R. Carpenter. 1986. *Ecology* 67:1351–1360. (NF,P) (N,P)

Bjørnsen, P. K., et al. 1986. *Freshwater Biology* 16:245–253. (F,P)

Burns, C. W., and J. J. Gilbert. 1986. *Limnology and Oceanography* 31:848–858. (F,P)

Cryer, M., et al. 1986. *Limnology and Oceanography* 31:1022–1038. (F,P; 2 expts.)

Drenner, R. W., et al. 1986. *Canadian Journal of Fisheries and Aquatic Sciences* 43: 1935–1945. (F,P)

Hambright, K. D., et al. 1986. *Canadian Journal of Fisheries and Aquatic Sciences* 43: 1171–1176. (F,P)

Hanson, J. M., and W. C. Leggett. 1986. *Canadian Journal of Fisheries and Aquatic Sciences* 43:1363–1372. (F,PB)

Hardy, F. J., et al. 1986. *Canadian Journal of Fisheries and Aquatic Sciences* 43:1504–1514. (N,P)

Havens, K. E., and J. DeCosta. 1986. *Hydrobiologia* 137:211–222. (N,P)

Hershey, A. E. 1986. *Canadian Journal of Fisheries and Aquatic Sciences* 43:2523–2528. (F,B; 2 expts.)

Holomuzki, J. R. 1986. *Ecology* 67:737–748. (F,B) (F,PB)

Istvánovics, V., et al. 1986. *Limnology and Oceanography* 31:798–811. (N,P)

Milstead, B., and S. T. Threlkeld. 1986. *Journal of the North American Benthological Society* 5:311–318. (F,B)

Morin, P. J. 1986. *Ecology* 67:713–720. (F,B)

Persson, L. 1986. *Ecology* 67:355–364. (F,P;2 expts.)

Pitts-Diner, M. P., et al. 1986. *Hydrobiologia* 133:59–63. (F,P)

Raess, F., and E. J. Maly. 1986. *Hydrobiologia* 140:155–160. (F,P)

Reinertsen, H., et al. 1986. *Canadian Journal of Fisheries and Aquatic Sciences* 43:1135–1141. (F,P)

Tátri, I., and V. Istvánovics. 1986. *Freshwater Biology* 16:417–424. (F,P)

Tessier, A. J. 1986. *Ecology* 67:285–302. (F,P; 2 expts.)

Threlkeld, S. T. 1986. *Freshwater Biology* 16:673–683. (F,P)

Vanni, M. J. 1986. *Ecology* 67:337–354. (F,P;3 expts.)

Vanni, M. J. 1986. *Limnology and Oceanography* 31:1039–1056. (F,P)

Vaughn, C. C. 1986. *Freshwater Biology* 16:485–493. (F,B)

Brönmark, C. 1988. *Hydrobiologia* 169:363–370. (F,B)

Carney, H. J. 1988. *Ecology* 69:664–678. (F,P)

Carrick, H. J., and R. L. Lowe. 1988. *Canadian Journal of Fisheries and Aquatic Sciences* 45:271–279. (N,B)

Carrick, H. J., et al. 1988. *Journal of the North American Benthological Society* 7:117–128. (N,B)

Dodds, W. K., and R. W. Castenholz. 1988. *Hydrobiologia* 162:141–146. (N,P;2 expts.)

Elser, J. J., et al. 1988. *Limnology and Oceanography* 33:1–14. (F,P;2 expts.) (NF,P)

Fairchild, G. W., and A. C. Everett. 1988. *Freshwater Biology* 19:57–70. (N,B)

Gilbert, J. J. 1988 *Ecology* 69:1826–1838. (F,P)

Grover, J. P. 1988. *Ecology* 69:408–417. (N,P)

Güde, H. 1988. *Hydrobiologia* 159:63–73. (F,P)

Hansson, L-A. 1988. *Limnology and Oceanography* 33:121–128. (N,PB; 2 expts.)

Kerfoot, W. C., et al. 1988. *Ecology* 69:1806–1825. (F,P;2 expts.)

Lathrop, R. C. 1988. *Canadian Journal of Fisheries and Aquatic Sciences* 45:2061–2075. (N,P)

Leff, L. G., and M. D. Bachman. 1988. *Freshwater Biology* 19:87–94. (F,B)

Lowe, R. L., and R. D. Hunter. 1988. *Journal of the North American Benthological Society* 7:29–36. (F,B)

Mazumder, A., et al. 1988. *Limnology and Oceanography* 33:421–430. (NF,P)

McMahon, T. E. 1988. *Ecology* 69:1871–1883. (F,B)

McQueen, D. J., and J. R. Post. 1988. *Hydrobiologia* 159:277–296. (F,P)

Mittelbach, G. G. 1988. *Ecology* 69:614–623. (F,PB)

Moore, M. V. 1988. *Limnology and Oceanography* 33:256–268. (F,P)

Morin, P. J., et al. 1988. *Ecology* 69:1401–1409. (F,B)

Morris, D. P., and W. M. Lewis, Jr. 1988. *Freshwater Biology* 20:199–210. (N,P; 56 expts.)

Pollingher, U., et al. 1988. *Hydrobiologia* 166:65–75. (N,P)

Prepas, E. E., and A. M. Trimbee. 1988. *Hydrobiologia* 159:269–276. (N,P; 2 expts.)

Reynolds, C. S., and J. W. G. Lund. 1988. *Freshwater Biology* 19:379–404. (N,P)

Riessen, H. P., et al. 1988. *Canadian Journal of Fisheries and Aquatic Sciences* 45: 1912–1920. (F,P; 2 expts.)

Sommer, U. 1988. *Limnology and Oceanography* 33:1037–1054. (NF,P)

Søndergaard, M., et al. 1988. *Hydrobiologia* 164:271–286. (N,P)

Stockner, J. G., and K. S. Shortreed. 1988. *Limnology and Oceanography* 33:1348–1361. (N,P; 2 expts.)

Suttle, C. A., and P. J. Harrison. 1988. *Limnology and Oceanography* 33:186–202. (N,P; 2 expts.)

Threlkeld, S. T. 1988. *Limnology and Oceanography* 33:1362–1375. (F,P; 5 expts.)

Van Buskirk, J. 1988. *Ecology* 69:857–867. (F,B)

Vanni, M. J. 1988. *Canadian Journal of Fisheries and Aquatic Sciences* 45:1758–1770. (F,P)

Anderson, N. J., et al. 1990. *Freshwater Biology* 23:205–217. (N,P)

Anholt, B. R. 1990. *Ecology* 71:1483–1493. (F,B; 2 expts.)

Barnese, L. E., et al. 1990. *Journal of the North American Benthological Society* 9:35–44. (F,B)

Blaustein, L. 1990. *Hydrobiologia* 199:179–191. (F,PB)

Blaustein, L., and R. Karban. 1990. *Limnology and Oceanography* 35:767–771. (F,B; 3 expts.)

Carrillo, P., et al. 1990. *Hydrobiologia* 200/201:49–58. (F,P; 2 expts.)

Chambers, P. A., et al. 1990. *Freshwater Biology* 24:81–91. (F,B)

Christoffersen, K. 1990. *Hydrobiologia* 200/201:459–466. (F,P)

Chrzanowski, T. H., and K. Šimek. 1990. *Limnology and Oceanography* 35:1429–1436. (P)

Cuker, B. E., et al. 1990. *Limnology and Oceanography* 35:830–839. (N,P)

Dawidowicz, P. 1990. *Hydrobiologia* 191:265–268. (F,P)

Dawidowicz, P. 1990. *Hydrobiologia* 200/201:43–47. (F,P)

Dawidowicz, P., et al. 1990. *Limnology and Oceanography* 35:1631–1637. (F,P)

Dodds, W. K., and J. C. Priscu. 1990. *Canadian Journal of Fisheries and Aquatic Sciences* 47:2328–2338. (N,P)

Drenner, R. W., et al. 1990. *Hydrobiologia* 208:161–167. (NF,P)

Ehlinger, T. J. 1990. *Ecology* 71:886–896. (F,P)

Elser, J. J., et al. 1990. *Hydrobiologia* 200/201:69–82. (F,P; 3 expts.)

Faafeng, B. A., et al. 1990. *Hydrobiologia* 200/201:119–128. (NF,P)

Feminella, J. W., and V. H. Resh. 1990. *Ecology* 71:2083–2094. (F,B; 3 expts.)

George, D. G. et al. 1990. *Freshwater Biology* 23:55–70. (N,P)

Gilbert, J. J. 1990. *Ecology* 71:1727–1740. (F,P)

Gliwicz, Z. M., and W. Lampert. 1990. *Ecology* 71:691–702. (F,B)

Hanazato, T. 1990. *Hydrobiologia* 198:33–40. (F,P; 2 expts.)

Hanazato, T., et al. 1990. *Hydrobiologia* 200/201:129–140. (F,P)

Hanson, J. M., et al. 1990. *Freshwater Biology* 24:69–80. (F,B)

Hanson, M. A., and M. G. Butler. 1990. *Hydrobiologia* 200/201:317–327. (F,P)

Hanson, M. A., et al. 1990. *Freshwater Biology* 24:547–556. (F,P)

Hansson, L-A. 1990. *Freshwater Biology* 24:265–273. (N,PB; 3 expts.)

Havens, K. E. 1990. *Hydrobiologia* 198:215–226. (F,P; 2 expts.)

Horppila, J., and T. Kairesalo. 1990. *Hydrobiologia* 200/201:153–165. (F,P)

Iwakuma, T., et al. 1990. *Hydrobiologia* 200/201:141–152. (F,PB)

Jackson, L. J., et al. 1990. *Canadian Journal of Fisheries and Aquatic Sciences* 47: 128–135. (N,P)

Koenings, J. P., et al. 1990. *Ecology* 71:57–67. (N,P)

Lammens, E. H. R. R. 1990. *Hydrobiologia* 191:29–37. (F,P)

Lancaster, H. F., and R. W. Drenner. 1990. *Canadian Journal of Fisheries and Aquatic Sciences* 47:471–479. (NF,P)

Leibold, M. A. 1990. *Limnology and Oceanography* 35:938–944. (F,P)

Leucke, C., et al. 1990. *Canadian Journal of Fisheries and Aquatic Sciences* 47:524–532. (F,P)

Leventer, H., and B. Teltsch. 1990. *Hydrobiologia* 191:47–55. (F,P; 2 expts.)

Lundstedt, L., and M. T. Brett. 1990. *Limnology and Oceanography* 35:159–165. (F,P)

Lunte, C. C., and C. Luecke. 1990. *Limnology and Oceanography* 35:1091–1100. (F,P)

Matveev, V. F., and E. G. Balseiro. 1990. *Freshwater Biology* 23:197–204. (N,P; 2 expts.)

Meijer, M-L., et al. 1990. *Hydrobiologia* 191:275–284. (F,P)

Miura, T. 1990. *Hydrobiologia* 200/201:567–579. (F,P)

Morin, P. J., et al. 1990. *Ecology* 71:1590–1598. (F,B)

Northcote, T. G., et al. 1990. *Hydrobiologia* 194:31–45. (F,P)

Olsson, H. 1990. *Freshwater Biology* 23:353–362. (N,P)

Persson, L., and L. A. Greenberg. 1990. *Ecology* 71:44–56. (F,PB)

Persson, L., and L. A. Greenberg. 1990. *Ecology* 71:1699–1713. (F,B)

Reinertsen, H., et al. 1990. *Canadian Journal of Fisheries and Aquatic Sciences* 47: 166–173. (F,P)

Riemann, B., et al. 1990. *Hydrobiologia* 200/201:241–250. (F,P)

Rothhaupt, K. O. 1990. *Freshwater Biology* 23:561–570. (F,P)

Sanni, S., and S. B. Waervagen. 1990. *Hydrobiologia* 200/201:263–274. (F,P)

Schneider, D. W. 1990. *Limnology and Oceanography* 35:916–922. (NF,P)

Schoenberg, S. A. 1990. *Freshwater Biology* 23:395–410. (F,P)

Shortreed, K. S., and J. G. Stockner. 1990. *Canadian Journal of Fisheries and Aquatic Sciences* 47:262–273. (N,P)

Smith, C. K. 1990. *Ecology* 71:1777–1788. (F,B; 2 expts.)

Sondergaard, M., et al. 1990. *Hydrobiologia* 200/201:220–240. (F,P)

Starling, F. L. R. M., and A. J. A. Rocha. 1990. *Hydrobiologia* 200/201:581–591. (F,P)

Tatrai, I., et al. 1990. *Hydrobiologia* 191:307–313. (F,P)

Tatrai, I., et al. 1990. *Hydrobiologia* 200/201:167–175. (F,P)

Theiss, J., et al. 1990. *Hydrobiologia* 200/201:59–68. (F,P)

Turner, A. M., and G. G. Mittelbach. 1990. *Ecology* 71:2241–2254. (F,PB)

Underwood, G. J. C., and J. D. Thomas. 1990. *Freshwater Biology* 23:505–522. (F,B)

van Donk, E., et al. 1990. *Hydrobiologia* 200/201:275–289. (F,B) van Donk, E., et al. 1990. *Hydrobiologia* 200/201:291–301. (F,P)

van Donk, E., et al. 1990. *Hydrobiologia* 191:285–295. (F,PB)

Vanni, M. J., and D. L. Findlay. 1990. *Ecology* 71:921–937. (F,P; 2 expts.)

Vanni, M. J., and J. Temte. 1990. *Limnology and Oceanography* 35:697–709. (NF,P)

Vanni, M. J., et al. 1990. *Hydrobiologia* 200/201:329–336. (F,P)

Walls, S. C. 1990. *Ecology* 71:307–314. (F,B)

Willey, R. L. 1990. *Limmnology and Oceanography* 35:952–959. (F,P)

Wright, D., and J. Shapiro. 1990. *Freshwater Biology* 24:43–62. (F,P)

Zagarese, H. E. 1990. *Freshwater Biology* 24:557–562. (F,P)

Zielinski, K., et al. 1990. *Hydrobiologia* 200/201:59–68. (F,P; 3 expts.)

Arnott, S. E., and M. J. Vanni. 1993. *Ecology* 74:2361–2380. (F,P; 4 expts.)

Bell, R. T., et al. 1993. *Limnology and Oceanography* 38:1532–1538. (NF,P)

Botts, P. S. 1993. *Freshwater Biology* 30:25–33. (F,B)

Burns, C. W., and J. J. Gilbert. 1993. *Freshwater Biology* 30:377–393. (F,P)

Carrick, H. J., et al. 1993. *Canadian Journal of Fisheries and Aquatic Sciences* 50: 2208–2221. (N,P)

Christoffersen, K., et al. 1993. *Limnology and Oceanography* 38:561–573. (F,P)

DeMeester, L. 1993. *Ecology* 74:1467–1474. (F,P)

Dudgeon, D. 1993. *Freshwater Biology* 30:189–197. (F,B)

Havens, K. E. 1993. *Hydrobiologia* 254:73–80. (F,PB)

Hofman, W., and M. G. Höfle. 1993. *Hydrobiologia* 255/256:171–175. (N,P)

Jackson, M. E., and R. D. Semlitsch. 1993. *Ecology* 74:342–350. (F,B)

Lawler, S. P., and P. J. Morin. 1993. *Ecology* 74:174–182. (F,B)

Lehman, J. T., and C. E. Cáceres. 1993. *Limnology and Oceanography* 138:879–891. (F,P)

Markošová, R. S., and J. Ježek. 1993. *Hydrobiologia* 264:85–99. (NF,P)

Marks, J. C., and R. L. Lowe. 1993. *Canadian Journal of Fisheries and Aquatic Sciences* 50:1270–1278. (N,B)

Mittelbach, G. G., and C. W. Osenberg. 1993. *Ecology* 74:2381–2394. (F,P)

Niederhauser, P., and F. Schanz. 1993. *Hydrobiologia* 269/270:453–462. (N,B)

Novales-Flamarique, I., et al. 1993. *Limnology and Oceanography* 38:290–298. (F,PB)

Ozimck, T., et al. 1993. *Hydrobiologia* 251:13–18. (N,B)

Richardson, W. B., and S. T. Threlkeld. 1993. *Canadian Journal of Fisheries and Aquatic Sciences* 50:29–42. (F,PB)

Roche, K. F., et al. 1993. *Hydrobiologia* 254:7–20. (F,P)

Starling, F. L. R. M. 1993. *Hydrobiologia* 257:143–152. (F,P)

Sterner, R. W. 1993. *Ecology* 74:2351–2360. (NF,P)

Sterner, R. W., et al. 1993. *Limnology and Oceanography* 38:857–871. (NF,P)

Telesh, I. V. 1993. *Hydrobiologia* 255/256:289–296. (F,P; 2 expts.)

Thatcher, S. J., et al. 1993. *Hydrobiologia* 250:127–141. (F,P)

Wehr, J. D. 1993. *Canadian Journal of Fisheries and Aquatic Sciences* 50:936–945. (N,P)

Weider, C. J. 1993. *Ecology* 74:935–943. (NF,P)

Wissinger, S., and J. McGrady. 1993. *Ecology* 74:207–218. (F,B; 2 expts.)

Yozzo, D. J., and W. E. Oduin. 1993. *Hydrobiologia* 257:37–46. (F,P)

Barnese, L. E., and C. L. Scheleske. 1994. *Hydrobiologia* 277:159–170. (N,B)

Berg, S., et al. 1994. *Hydrobiologia* 275/276:71–79. (F,PB)

Brett, M. T., et al. 1994. *Ecology* 75:2243–2254. (F,P)

Brönmark, C. 1994. *Ecology* 75:1818–1829. (F,PB)

Campeau, S., et al. 1994. *Canadian Journal of Fisheries and Aquatic Sciences* 51:681–692. (N,PB)

Caravalho, L. 1994. *Hydrobiologia* 275/276:53–63. (F,P)

Carignan, R., et al. 1994. *Limnology and Oceanography* 39:439–443. (N,B)

Carignan, R., and D. Planas. 1994. *Limnology and Oceanography* 39:580–596. (N,P; 2 expts.) (N,P)

Cline, J. M., et al. 1994. *Hydrobiologia* 275/276:301–311. (F,PB)

Covich, A. P., et al. 1994. *Journal of the North American Benthological Society* 13:283–290. (F,B; 2 expts.)

Edgar, N. B., and J. D. Green. 1994. *Hydrobiologia* 273:147–161. (F,P)

Findlay, D. L., et al. 1994. *Canadian Journal of Fisheries and Aquatic Sciences* 51:2254–2266. (N,P)

Gabor, T. S., et al. 1994. *Hydrobiologia* 279/280:497–510. (N,PB)

Garvey, J. E., et al. 1994. *Ecology* 75:532–547. (F,PB)

Gresens, S. E., and R. L. Lowe. 1994. *Journal of the North American Benthological Society* 13:89–99. (NF,B)

Hambright, K. D. 1994. *Limnology and Oceanography* 39:897–912. (F,P)

Hamilton, D. J., et al. 1994. *Ecology* 75:521–531. (F,B)

Hanson, M. A., and M. G. Butler. 1994. *Hydrobiologia* 278/279:457–466. (F,PB)

He, X., et al. 1994. *Freshwater Biology* 32:61–72. (F,P)

Lauridsen, T. L., et al. 1994. *Hydrobiologia* 275/276:233–242. (F,PB)

Lehman, J. T., and D. K. Branstrator. 1994. *Limnology and Oceanography* 39:227–233. (N,P)

Lodge, D. M., et al. 1994. *Ecology* 75:1265–1281. (F,B)

MacIsaac, H. J. 1994. *Journal of the North American Benthological Society* 13:206–216. (F,B; 2 expts.)

Meijer, M-L., et al. 1994. *Hydrobiologia* 275/276:31–42. (F,P)

Murkin, H. R., et al. 1994. *Hydrobiologia* 279/280:483–495. (N,PB)

Pace, M. L., and D. Vagne. 1994. *Limnology and Oceanography* 39:985–996. (F,P; 3 expts.)

Paul, A. J., and D. W. Schindler. 1994. *Canadian Journal of Fisheries and Aquatic Sciences* 51:2520–2528. (NF,P)

Perez, E. A. A., et al. 1994. *Hydrobiologia* 291:93–103. (N,P; 2 expts.)

Perrow, M. R., et al. 1994. *Hydrobiologia* 275/276:43–52. (F,PB)

Prejs, A., et al. 1994. *Hydrobiologia* 275/276:65–70. (F,P)

Sterner, R. W. 1994. *Limnology and Oceanography* 39:535–550. (N,P)

van de Burd, W. J., et al. 1994. *Journal of the North American Benthological Society* 13:532–539. (F,B)

Verdonschot, P. F. M., and C. J. F. Terbraak. 1994. *Hydrobiologia* 278:251–266. (N,B)

Wang, L., and J. C. Priscu. 1994. *Freshwater Biology* 31:183–190. (NF,P)

Werner, E. E., and M. A. McPeek. 1994. *Ecology* 75:1368–1382. (F,B)

Wiackowski, K., et al. 1994. *Limnology and Oceanography* 39:486–492. (F,P)

12

Ecology in a Bottle

Using Microcosms to Test Theory

SHARON P. LAWLER

The science of ecology is brimming with models that portray how community and population dynamics might unfold through time (Kareiva 1989). Such models typically predict dynamics over hundreds of generations of the focal organisms. In contrast, most field experiments encompass just a few generations or less of the species studied. Because investigators seldom record long-term responses of populations to community structure, some key ecological theories have rarely been tested. This is a pervasive problem for models that predict population fluctuations and persistence through time (e.g., theories of predator–prey dynamics or metapopulation dynamics) or species composition and turnover in communities (e.g., island biogeography, assembly rules, succession, or alternative community states). Microcosms can be used as "biological accelerators" to test these theories because they contain organisms with short generation times (Lawton 1995). In addition, some microcosms are very large relative to the size and motility of their inhabitants. In this subset of microcosms, it becomes feasible to measure processes peculiar to large populations or those which require relatively large spatial scales.

Microcosm means "small world," and microcosms are small, usually artificially bounded habitats that contain one to many types of organism. Microcosms are meant to act as scale models of larger systems. Ideally, microcosms should share enough features with larger, more natural systems so that studying them can provide insight into processes acting at larger scales or, better yet, into general processes acting at most scales. Of course, some processes may operate only at large scales, and big, long-lived organisms may possess qualities that are distinct from those of small organisms (and vice versa). Because large and small organisms differ biologically, it will not be feasible to study some questions using microcosms. However, to the extent that some ecological principles transcend scale, microcosms can be a valuable investigative tool.

There has been some confusion about the terms *microcosm* and *mesocosm*. For either term to apply to an experimental unit, it should be large enough relative to its inhabitants to be considered a ''world'' in some sense. I prefer that the terms be restricted to units that are at least able to contain populations of the inhabitants that are within the range of natural population sizes. A more liberal definition might be one which stipulates that the unit is large enough to allow free expression of the process of interest (i.e., dispersal, schooling behavior, etc.). Whether the term *microcosm* or *mesocosm* applies should depend on how much the experimental unit is reduced in scale from the system(s) or processes it is meant to represent. A microcosm represents a scale reduction of several orders of magnitude, while a mesocosm represents a reduction of about two orders of magnitude or less (e.g., using cattle watering tanks to represent ponds). For example, a petri-dish experiment on competition among bacteria would be considered a micro-cosm study if it were intended to model interactions at the scale of vertebrate popula-tions. However, the same petri dish could be considered a mesocosm if the goal was to model bacterial interactions in a box turtle's gut. The experiment would not be a micro- or mesocosm study at all if it were meant to show how bacteria interact at the scale of a petri dish. This usage reflects the degree of extrapolation in size from the experimental unit to the other systems of interest. The distinction between terms is admittedly rough, but I hope it is preferable to an anthropocentric view where a mi-crocosm is anything small on a human scale (smaller than a breadbox?) and mesocosms are somewhat larger.

There are many naturally small systems that contain organisms with rapid genera-tions, such as the communities which form in mushrooms (e.g., Worthen 1993) or in ''phytotelmata,'' which are plant-held waters such as those in tree holes, bracts, and pitcher plants (review, Frank and Lounibos 1983). These fascinating communities are often studied for their own sake, but they can also enhance general ecological under-standing when treated as natural microcosms (e.g., Addicott 1974; Naeem 1988, 1990). The risk of artifactual results may be lower in natural versus constructed microcosms, since the organisms may be better adapted to each other and to abiotic conditions in the experimental unit. The investigator can also be more sure that the habitat is of the correct size relative to the organisms. However, I decided to focus on artificial micro-cosms for this review to show that even synthetic systems can provide ecological in-sight.

Microcosms are popular research tools in ecotoxicology, soil biology, genetic en-gineering, and ecosystem research (see the review in Beyers and Odum 1993) but are less commonly used to test ecological theory regarding population and community dynamics. During the past decade there has been an average of 80 publications per year on microcosms, of which only a few addressed such theory (as searched through Bio-logical Abstracts on compact disk). Many ecologists are wary of microcosm studies, fearing that the results of small-scale experiments could be misleading if they are ex-trapolated to systems that operate on vastly different spatial and temporal scales (Car-penter 1996). This fear is justified! Extrapolation is dangerous in many fields, and ecologists must be particularly careful because we deal with extraordinarily complex systems. However, I believe that, with cautious interpretation, microcosm studies can advance ecology (see also Drake et al. 1996, Ives et al. 1996, Verhoef 1996, Jaffee 1996, Moore et al. 1996).

Despite the low rate at which ecologists have performed microcosm studies, those that exist include exceptionally well designed, innovative tests of hypotheses pertaining to population and community dynamics (Kareiva 1989, Lawton 1995). I hope to convince more ecologists of the value of microcosms by reviewing several classic and recent studies, including some of my own work with collaborators. The examples are illustrative rather than exhaustive. I chose to highlight studies in which the primary intent of the investigator was to test or refine theory, rather than to learn about particular species. Because I find interspecific interactions particularly captivating, I omitted many excellent studies of single populations (e.g., Luckinbill and Fenton 1978). I have also focused on studies pertaining to population and community dynamics because dynamic theories are difficult to test outside of microcosm systems. The studies reviewed have shown that some of our most important theories have sound foundations.

Microcosms are used to test theory, but they should not be "analogue computers" of mathematical models, designed and fine-tuned so that a fit to a particular model is guaranteed (Hutchinson 1978, Koehl 1989). As systems of living organisms, microcosms can hold biological surprises. Some of these might be artifacts caused by the artificial features of the microcosm, while others may reflect general biological phenomena that were omitted from the theoretical model. Many of the studies reviewed in the following discussion uncovered general phenomena that indicated that a theory should be changed, rather than the microcosm. The role of microcosm work is not to *prove* theory, but to *improve* theory.

Studies of Population Dynamics and Pairwise Interactions

Population Dynamics and Competition

G. F. Gause (1934) performed a series of elegant microcosm experiments to test Lotka-Volterra competition and predation models (Lotka 1925, Volterra 1926). His well-known studies still stand as some of the best work unifying population-dynamic theory with empirical data and are well worth reading in the original. Gause studied population growth, interspecific competition, and predator–prey dynamics in microcosms, using bacteria, yeasts, and protozoans as study organisms. First, he showed that population growth of yeast and *Paramecium* fit the logistic growth curves developed by Verhulst and Pearl. Gause was then able to predict the outcome of competition between two different yeasts or two species of *Paramecium* by measuring their growth curves and using those data to estimate the parameters of Lotka-Volterra competition equations. He found that competition coefficients estimated from each species' growth alone adequately predicted which species would be the competitive dominant. When two similar species were competing for the same resource, the species that used the resource more efficiently always excluded the other species. As he reported his results, Gause noted some biological complications that made the mathematical theory inexact. For example, the competition coefficients changed slightly through time as competition proceeded. This occurred because the organisms changed their environment. Wastes accumulated, and the competitors had different tolerances for the waste.

Thomas Park also demonstrated competitive exclusion between two species that used the same resource (Park 1948). Park's microcosms were glass vials containing flour and yeast, and his study organisms were the flour beetles *Tribolium confusum* and *T. castaneum*. *T. confusum* was the competitive dominant in most cultures, but there were occasional reversals. Park found that a coccidean parasite, *Adelina* sp., rendered *T. castaneum* a weak competitor. If the parasite was removed, *T. castaneum* was often able to drive *T. confusum* to extinction. This was an early demonstration of "keystone predation," where the outcome of competition depends on whether or not a predator is present (e.g., Paine 1966, Morin 1983). This investigation showed that competition theory was valid for a system that met the assumptions of the model but also revealed how the presence of just one more species could render the model inadequate.

The original Lotka-Volterra competition models assumed that the intensity of competition between two species could be determined by relative efficiency with which they exploited a single limiting resource. This situation is fairly easy to duplicate in a microcosm, but is it realistic? Probably not in most habitats. Resource gradients exist in our spatially complex world, and later models showed that many species could coexist along a resource gradient, if each was limited by a different resource (see the review in Tilman 1977). Tilman (1977) modified resource-based, mechanistic models of competition (Monod 1950, Droop 1974) so that species would compete for two resources. Here competitive ability was not based on a static measure of overlap in resource use, as in Lotka-Volterra models, but was a function of each species' ability to acquire and utilize available resources. Tilman ran 76 long-term competition experiments between two species of freshwater algae where the algae were grown in different mixtures of two limiting nutrients. As the models predicted, the outcome of competition depended upon the mixture of the two nutrients. The Monod model could also account for 75% of the variation in the relative abundance of the two algae in Lake Michigan.

Predator–Prey Dynamics

In addition to his work on competition, Gause (1934) performed extensive experiments on predator–prey dynamics, using *Didinium nasutum* and *Paramecium caudatum* as predator and prey. Lotka-Volterra models predicted that predator and prey abundances should cycle, with the magnitude of the cycles depending on the initial abundances of the species. However, Gause found that *D. nasutum* rapidly overexploited its prey, regardless of starting densities. He was able to prolong the coexistence of the pair by adding a spatial refuge for *Paramecium* (oat sediment) or by adding one immigrant predator and prey every three days. Gause concluded that spatial refuges and immigration were important to the persistence of predators and prey.

Similarly, Takahashi (1959) found that a spatial refuge aided the persistence of a parasitoid–flour beetle system. As in Gause's work, the refuge was deep medium that protected a portion of the prey from the predators. The protected prey continued to reproduce, but because all the offspring could not fit in the refuge, a steady supply of vulnerable prey was available to the predator. Translated into modern theoretical terms, the refuge shifted the system toward more stable, "donor-controlled" dynamics, wherein predators have less effect on prey dynamics (e.g., DeAngelis 1975, Pimm 1982, Hawkins 1992).

Microcosm studies have also shown that predator–prey coexistence does not require absolute refuges for prey. Huffaker (1958) and colleagues (Huffaker et al. 1963) demonstrated that an unstable predator–prey pair could coexist in patchy habitat. The pair were predatory and herbaceous mites living on oranges. The researchers constructed arenas where the habitat was either continuous (single oranges or several oranges in close contact) or subdivided (oranges separated by rubber balls). Predators skyrocketed and decimated prey in continuous habitat because they were efficient in searching for prey. However, when the habitat was subdivided and spread over a larger area, the predators could not search the entire area before most of them starved. This slowed their numerical response to the prey and promoted coexistence. Huffaker's work has been criticized because replication was low within experiments, and the persistence of the subdivided system was at times so precarious that differences between continuous and divided habitats could have been fortuitous (Kareiva 1990, Taylor 1990). These are valid criticisms; however, Huffaker did perform several broadly comparable experiments through the years that yielded qualitatively similar results. Recent work by Holyoak and Lawler (1996 a, b) concurs with Huffaker's conclusions. We demonstrated that a protist predator–prey pair could persist much longer in subdivided habitat than in continuous habitat of the same volume, because subdivision provided spatial and temporal refuges for the prey.

Luckinbill (1973) took the study of predator–prey coexistence one step further. Following Hutchinson (1961), he noted that some natural predator–prey pairs, such as zooplankton and algae, do manage to coexist without obvious refuges or patchiness in their environment. He speculated that the small size of Gause's microcosms could have promoted overexploitation of prey, because predators could easily search the entire volume. In larger, natural systems, prey might survive at low densities, because many predators would starve while trying to find scarce prey dispersed over a large area. Luckinbill devised a clever microcosm test of this hypothesis. He used a predator–prey pair which Gause had found to be unstable in small microcosms—*D. nasutum* and *P. aurelia*—and simulated prey scarcity by reducing the encounter rate of predators and prey. He accomplished this by adding methyl cellulose, an inert thickener, to the aqueous medium. The pair persisted for 10 hours without the thickener, versus 10 days in the thickened medium. This mechanism required less modification of the Lotka-Volterra model than adding patchiness or refuges, because the theory already predicted that predator numbers would drop when prey became scarce.

These studies confirmed that refuges, patchiness, and dispersion could permit unstable interactions to persist. As a result, modern predator–prey dynamic models often incorporate refuges and patchiness as stabilizing influences (Taylor 1990, Harrison and Taylor 1997).

The Paradox of Enrichment

In addition to his work on volume effects, Luckinbill (1974, 1979) produced definitive demonstrations of Rosenzweig's (1971) "paradox of enrichment" theory in laboratory microcosms, although he did not mention the theory in his papers. Rosenzweig analyzed predator–prey isoclines derived from Lotka-Volterra type models, across a cline in the prey's intrinsic rate of increase (presumed to correspond to a nutrient or productivity gradient). If prey reproductive rates became too high, the tendency for predators to overshoot their prey increased, destabilizing the interaction. This was considered paradoxical because it seemed that a larger resource base should allow larger, hence more "stable," populations. Luckinbill effectively tested this theory in tritrophic food chains of bacteria, bacterivorous protists (hereafter "prey" either *Colpidium campylum* or *P. aurelia*), and a predatory protist (*D. nasutum*), by varying nutrient conditions in the microcosms. Under low-nutrient conditions, these predator–prey pairs persisted. Under high-nutrient conditions, prey condition and/or numbers increased rapidly, resulting in a surplus of well-fed predators. These overexploited the prey and went extinct in all microcosms. Harrison (1995) has reanalyzed the data from this experiment in an attempt to decide which of several alternative predator–prey models correspond best to empirical results.

Leonard and Anderson (1991) also found that additional resources destabilized a predator–prey interaction, in their case populations of a collembolan and a fungus cultured in bottles. If the fungus was supplied with 2 mg of nitrogen per liter of medium, the fungus and the collembolan were able to coexist for at least 147 days. This represented several generations of the collembolan. However, if the nitrogen level was raised to 200 mg/l, the collembolan populations rapidly increased in numbers, overgrazed the fungus, and became extinct. These investigators also added spatial heterogeneity (small ceramic cylinders) to other low-and-high-nutrient microcosms and found that the heterogeneity lengthened the persistence of the unstable, enriched predator–prey system.

In contrast to these studies, McCauley and Murdoch (1990) were unable to induce cycles via nutrient supplementation to an algae–zooplankton microcosm system. In their study, the prey were not a single species but a natural assemblage of various algae. McCauley and Murdoch hypothesized that enrichment had increased the proportion of inedible algae in their microcosms, so that nutrients didn't actually serve to improve the quantity and quality of algae available to zooplankton.

The preceding results represent most of the available experimental evidence on the paradox of enrichment. Microcosm studies confirmed that enrichment can destabilize single predator–prey pairs, but these studies also demonstrated that features of natural systems, such as spatial heterogeneity or species diversity, may ameliorate instabilities caused by enrichment.

Studies of Community Structure and Dynamics

Higher Order Interactions

In the 1960s, ecologists extended Lotka-Volterra competition models to whole assemblages of competitors in an attempt to predict the structure of communities (MacArthur and Levins 1967, Levins 1968). One assumption necessary to extend the models to multispecies systems is that the competitive effects of several species on each member of the community are additive—that is, the level of competition between two species is unaffected by the presence of other species. This simplifying assumption was suspect; it seemed likely that higher order interactions were important.

Vandermeer (1969) used microcosm experiments to test whether competition among species was additive or whether ecologists needed to consider higher-order interactions in predicting the outcome of competition among multiple species. He grew all single and pairwise combinations of four bacterivorous protists in 5-ml cultures and used data from these cultures to parameterize a Lotka-Volterra model that predicted the performance of all four species. When all four species were grown together, three out of four populations closely fit the additive model and the fourth showed qualitative agreement. Vandermeer was dismayed by this unusually fine match between theory and data. He had hoped to demonstrate the importance of higher-order interactions.

Neill (1974) continued the search for higher-order interactions, again using microcosms. Wilbur (1972) had by then demonstrated nonadditive competition among tadpoles in mesocosms; however, the long life cycles of his organisms did not allow him to demonstrate the dynamic effects of higher-order interactions. Neill constructed complex microcosm communities by adding a variety of algae and zooplankton to 1500-ml cultures over a period of 8 months, until the communities showed little change in species abundances. The steady-state community included four species of crustaceans (three cladocerans and an amphipod). He then conducted one-and two-species removals of the four crustaceans, to see whether the effects of two-species removals on the remaining crustaceans could have been predicted by additively combining the effects of single-species removals. In many cases, the effects were not additive. Neill's study confirmed what many ecologists had suspected but were unable to prove using short-term studies of long-lived organisms: higher order interactions can affect dynamics.

Island Biogeography

Island biogeography theory predicts that the species richness of islands will vary with the area of islands and the distance of the islands to mainland, or source, populations. MacArthur and Wilson (1963, 1967) predicted that the number of species on an island would reflect a balance between the rate at which immigrants arrive at the island and the rate at which species on the island become extinct. Larger islands were thought to support more species, both because larger islands could support larger populations (thereby decreasing extinction risk) and because greater habitat diversity might be available on larger islands. This fascinating theory has only been tested in a handful of experiments, most notably two long-term, labor-intensive experiments on mangrove and *Spartina* islands (Simberloff and Wilson 1969, 1970; Rey 1981) and two microcosm studies by Dickerson and Robinson (1985) and Have (1993). Space constraints dictate that I restrict discussion to the earlier study, leaving Have's creative work for the interested reader.

To search for effects of area and invasion rate on species richness, Dickerson and Robinson (1985) created "islands" which were 300- and 1800-ml volumes of soil-water medium and subjected these islands to two different invasion rates of algae and protozoa. Their results partially supported island biogeography theory, with some intriguing exceptions. Levels of species richness were fairly constant in microcosms during the 28-week experiment, and there was some species turnover. Thus, immigration did seem to balance extinction. Microcosms that received more frequent invasions reached their equilibrium number of species sooner and had higher species richness than those that received immigrants less frequently. These differences were largely due to the more frequent introductions of "transient" species, which persisted long enough to raise average species richness. Thus, the predictions of a balance between colonization and extinction and dependence of species richness on immigration were supported. However, in contrast to the expected positive species–area relationship, small-volume microcosms maintained greater species richness than large microcosms. Dickerson and Robinson hypothesized that habitat quality or diversity was somehow better in the smaller, shallower microcosms. This study also showed that two distinct community types tended to form in different microcosms, providing evidence that community structure has a historical component.

In later work, Robinson and Edgemon (1989) measured the relative importance of invasion rate versus predation to the species richness of a community. Predation had been shown to increase species richness by preventing competitive exclusion (e.g., Paine 1966), but no study had yet compared the relative importance of predation and invasion rate. As expected, both high invasion rates and predation increased species richness in microcosms, but invasion rate was four times more important than predation. More work is needed to discover whether this result is general.

Alternative State Theory

In the past, the existence of alternate community states has generated debate among ecologists (see reviews in Connell and Sousa 1983, Drake 1990). Alternate states occur when two or more distinct, persistent communities form from a single-species pool in a particular habitat, depending on historical processes such as invasion sequence or number of immigrants. It is almost impossible to prove that alternative states exist in natural communities because it is difficult to find several habitats that unequivocally share identical species pools and abiotic conditions and to monitor community composition for long enough to determine whether the community is relatively stable (Connell and Sousa 1983). These are problems of measurement, not evidence that alternative states do not exist.

While theory predicts that alternative states may occur (see review in Law and Morton 1993), there have been almost no examples in the literature (exceptions include Gilpin et al. 1976, Cole 1983, Barkai and McQuaid 1988). Dickerson and Robinson's (1985) work on assembling communities in microcosms showed some evidence that alternate community states could form from a single-species pool. However, their ''alternate states'' were somewhat confounded with different container sizes (see also Robinson and Edgemon 1989). One exceptionally clear example of alternate states is that of Drake (1991). He created identical habitats in laboratory microcosms and varied the order of their invasion by species out of a standard species pool. Different invasion sequences produced different communities, and the resulting communities showed divergent abilities to resist subsequent invaders. Drake monitored the microcosms for several generations of the species, long enough to show that the different community states were persistent. This experiment confirmed that community structure can have a substantial historical component, even in identical habitats subject to invasion by identical species. These results suggest that the outcome of species interactions is not entirely based on each species' per capita ability to compete or withstand predation. Instead, initial abundances and priority effects may help control community structure. However, priority effects are not inevitable. In similar work, Sommer (1991) varied the relative abundances of phytoplankton and zooplankton species assembled in microcosm communities and found that communities that were initially very different often converged through time.

Food Chain Theory

In 1977, Pimm and Lawton challenged the prevailing wisdom that food chain length depended on energetics. The traditional view was that food chains were short because energy attenuates as it is passed through successive trophic levels (e.g., Hutchinson 1959, Slobodkin 1961). Instead, Pimm and Lawton suggested that population dynamics might constrain food chain length if long food chains are particularly unstable. They tested their idea by constructing Lotka-Volterra models of a large set of food chains that varied in length between one and four trophic levels. Stability analysis indicated that long food chains would take longer to return to equilibrium than short food chains. Pimm and Lawton also explored the consequences of constructing food chains that included omnivores—that is, species that feed on more than one trophic level. Model food chains with omnivores were unstable more often than those lacking omnivores. Pimm and Lawton speculated that omnivores had a particularly destabilizing effect on intermediate prey because omnivores would compete with the intermediate prey as well as eating them. Sixteen years later, Peter Morin and I attempted to verify these intriguing results using microcosm experiments (Lawler and Morin 1993, Morin and Lawler 1995a).

We constructed food chains composed of bacteria, bacterivorous protists (hereafter, ''prey''), and predatory protists and monitored the population dynamics of bacterivores in chains of lengths two versus three. We studied 13 different combinations of predator and prey species (two species of prey and eight species of predators). Two of the predators were omnivorous, in the sense that they could divide by feeding on bacteria alone, as well as on bacteria plus prey. This enabled us to measure whether prey dynamics were less stable if the top predator was omnivorous. We compared the variation in abundance of prey populations over time in chains of different length. We assumed that higher variability corresponded to lower stability, as it does in some mathematical models (Taylor 1992). In 10 out of 13 predator–prey combinations prey were less abundant in the longer food chain, and in 8 of these cases prey abundance also fluctuated more widely in the longer food chain. We concluded that the lower, fluctuating abundances of prey in longer chains indicated that long food chains were less stable, which supports Pimm and Lawton's model. However, omnivores did not seem to destabilize prey any more than other predators (for further discussion, see Morin and Lawler 1995a).

Our studies were valuable as a direct comparison of the stability of chains of slightly different length. We have also attempted to construct longer food chains in the microcosms, but with limited success. It is possible that longer food chains are too unstable to persist in these simple microcosms. Because longer food chains occur in natural systems, we suspect that spatial or temporal heterogeneity, or perhaps species diversity, somehow stabilizes long food chains in more complex, natural food webs. Recently, Spencer and Warren (1996) have found evidence that habitat size may be important to food chain length. They used a standard group of species to assemble food webs in two different sizes of microcosms and observed that longer food chains formed in the larger microcosms. The reason for this difference in food chain lengths has not been elucidated. Discovering the factors that allow long food chains to persist is an exciting goal for future research.

Indirect Effects

Indirect effects are effects of one population on another that are mediated through a third species or a resource (see reviews, in Strauss 1991 and Wootton 1994). Ecologists have become increasingly aware that indirect effects can be strong enough to affect community structure. I will focus on two important indirect effects: cascading effects and apparent competition.

A trophic cascade is an indirect, positive effect of predators on their prey's resources that occurs when predators reduce prey abundance or prevent prey from foraging. Cascades were first discussed by Hairston et al. (1960) and Fretwell (1977), who hypothesized that top predators ultimately control the biomass of lower trophic levels, directly or indirectly (see also Oksanen et al. 1981, Hairston and Hairston 1993). In theory, effects of a top predator could reverberate through many links in a food chain if each trophic level is vulnerable to the one above. In a four-link chain, for example, top predators consume intermediate predators, thereby releasing consumers from predation; consumers then become abundant and have a strong effect on producers (e.g., Wootton and Power 1993). Other ecologists have pointed out that changes in the basal resources supplied to a food chain may cause effects that ripple upward, affecting the abundances of species at higher trophic levels (see review in Hunter and Price 1992). The relative importance of top-down versus bottom-up effects has been debated for years, and it is now evident that both processes can influence the biomass of trophic levels (see reviews in Hunter and Price 1992, Menge 1992, Power 1992, Strong 1992, Polis 1994, Morin and Lawler 1995b).

Balčiūnas and Lawler (1995) performed a microcosm experiment designed to measure the relative strengths of top-down versus bottom-up effects in a simple food chain composed of bacteria, a bacterivorous protist (*Colpidium*), and a predatory protist (*Euplotes*). We grew a two-link food chain (bacteria plus the bacterivore) and a three-link food chain (bacteria, bacterivore, and predator) in 30-ml cultures supplied with high or low levels of basal resources. Classic top-down theory predicts that the predator should become more abundant in the high-nutrient microcosms, because it would benefit from the increase in bacterivore productivity. Instead, the predator became extinct after an average of 27 days in high-nutrient microcosms but persisted for an average of 38 days in low-nutrient microcosms (Fig. 12-1). The extinction occurred because prey grew to a much larger size when resources were augmented, so large that the predator could no longer ingest them. In other words, the bottom-up effect obliterated the top-down effect. The lesson is a general one: resources can change the phenotype of organisms and in so doing affect interactions among species. This concept is certainly not original, but our microcosm work highlighted its omission from trophic cascade theory.

Apparent competition is an indirect effect of one species on another, mediated through a predator that feeds on both (Holt 1977). In the short term, prey that share a predator may have indirect, positive effects on each other through predator satiation or switching (see review in Holt and Lawton 1994). In the long term, however, the combination of prey may support a higher density of predators than either alone, so that the indirect effect is negative. The ultimate effect of adding alternate prey to a food chain thus hinges on the predators' numerical response. Apparent competition theory predicts that whichever prey supports a higher density of the predator will displace the

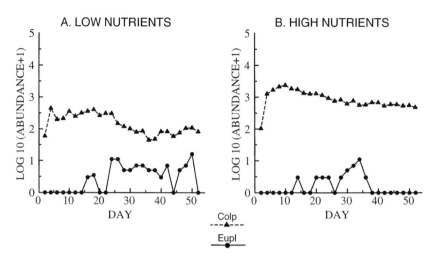

Figure 12-1. Representative population dynamics of a protist predator–prey pair in low-and high-nutrient microcosms. The predator was the hypotrichous ciliate *Euplotes patella* (abbreviated as "Eupl"), and the prey was the holotrichous ciliate *Colpidium striatum* (abbreviated as "Colp"). The pair were grown in 30-ml cultures of Carolina Biological Protist Pellet™ medium mixed at full strength plus three wheat seeds (high nutrients) or one-third strength plus one wheat seed (low nutrients). There were four replicates per treatment. Predators were introduced to cultures on day 6. Sample volume was 0.27 ml. The predator became extinct 10 days earlier in the high-nutrient microcosm because well-fed prey attained a size refuge. Redrawn from Balčiūnas and Lawler (1995).

other prey, assuming that both are equally edible and accessible to the predator (Holt 1977). This occurs because the density of each prey reflects a balance between its intrinsic growth rate, r, and the predator's attack rate, a. If alternate prey support a higher number of predators, losses to the predator will swamp r.

I demonstrated apparent competition in a microcosm experiment by growing a predator on either of two prey species and also on both species together (Lawler 1993). The predator was *Euplotes patella* and prey were *Tetrahymena pyriformis* and *Chilomonas paramecium*. Separately, each prey could coexist with the predator for at least 30 predator generations (Fig. 12-2b,c). The two prey could also coexist indefinitely in the absence of the predator (Fig. 12-2a). However, when both prey and the predator were together in microcosms, *C. paramecium* became extinct rapidly in all replicates (Fig. 12-2d). In keeping with apparent competition theory, the surviving prey supported four times more predators than the displaced prey when grown alone with the predator. This is one of the clearest demonstrations of apparent competition (Holt and Lawton 1994), largely because of the advantages afforded by microcosm work. The rapid generation times of the protists allowed me to measure equilibrium population sizes of the species, and the small scale of the system made it easy to replicate all species combinations.

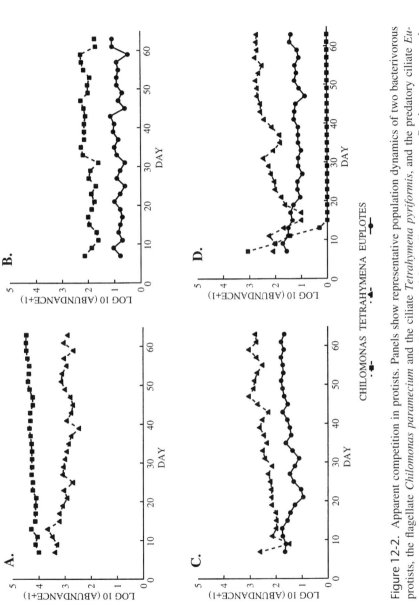

Figure 12-2. Apparent competition in protists. Panels show representative population dynamics of two bacterivorous protists, the flagellate *Chilomonas paramecium* and the ciliate *Tetrahymena pyriformis*, and the predatory ciliate *Euplotes patella*, in various combinations. Each panel shows one of five replicates per treatment. Protists were grown in 100-ml cultures containing Carolina Biological Protist Pellet℗ medium and bacteria. Panel **A** shows that the two bacterivores coexist without competitive exclusion. Panels **B** and **C** show that each bacterivore can coexist with the predator. Panel **D** shows that *C. paramecium* cannot withstand the high abundance of predators supported by *T. pyriformis*. Redrawn from Lawler (1993).

Discussion

Most ecologists prefer to study natural communities composed of relatively long-lived organisms, but tests of population dynamic theories in such systems require prohibitively long spans of time. In some cases (e.g., forest communities), the time required would far exceed the 30 to 40–odd productive years of a scientist's career. Consequently, much of the support for some theories is based on correlational evidence. Correlational evidence serves to show that a particular model may apply to natural systems, and so it justifiably heightens interest in the theory. However, because circumstantial evidence is notoriously misleading, scientific acceptance of a hypothesis can only arise after experimental verification. The preceding examples illustrate that microcosm experiments have been indispensable in testing and developing ecological theory.

I am not proposing that all ecologists troop indoors, deserting fields, lakes, oceans, and forests (Carpenter 1996). After all, the ultimate goal is to understand nature. We need to establish many more long-term studies in natural systems in order to know which theories work best. Microcosms are a tool to be used along the way—to check whether theories work at all when translated from numbers into organisms. There are many areas of undertested ecological theory. Metapopulation dynamics have rarely been demonstrated for single-species or predator–prey systems and never in more complex systems. Kareiva (1990) provides a comprehensive discussion of the sorts of experiments needed in this area. Chaos theory is practically impossible to test outside microcosm, because of the large number of generations required (Hastings et al. 1993). Because chaotic dynamics are possible in many population-dynamic models, empirical work is desperately needed to discover whether real systems are governed by initial conditions and transient dynamics or by equilibrium dynamics (e.g., Hastings and Higgins 1994). Community assembly processes are still poorly understood. Theories of community assembly suggest how resources and predation can interact to structure communities (e.g., Holt, Grover, and Tilman 1994); these theories seem amenable to microcosm tests. Ecologists also need to know how quickly population parameters might change through evolutionary time as a result of species interactions, since evolution might affect population dynamics. Pioneering microcosm work by Chao, Levin, and Stewart (1977; see also Levin et al. 1977, Lenski and Levin 1985) shows that it is experimentally tractable to study how evolution affects dynamics.

There are certainly limitations involved in choosing to work with microcosms. Organisms must be naturally suited to microcosm conditions, which usually means that they must be small and require only limited diversity in habitat or abiotic conditions. Some organisms with complex life cycles are difficult to study in microcosms, although there are exceptions (e.g., small insects such as fruit flies, blowflies, and parasitoids). Results from microcosms may not scale well to natural communities which include large organisms, because small organisms differ from large ones in many ways. For example, big organisms are more likely to be buffered from the environment by their larger size and, in some cases, greater mobility. Because microcosms are typically more "closed" systems than natural communities, inputs and outputs of materials often lack normal stochastic and seasonal variation unless special care is taken to mimic natural conditions. Finally, microcosm work usually traps the investigator in a laboratory, away from all the messy, wonderful complexity that makes a career in ecology attractive in the first place.

While microcosm work can often advance theory, it is still risky to generalize from theoretical or laboratory results to natural systems. Simple theories, even when substantiated by simple experiments, will often be woefully inadequate for predicting how a complex natural system will operate. Yet basic theory still needs to be tested, because it forms the backbone of the detailed, special-case models that can be more predictive. Ironically, ecologists sometimes seem willing to apply mathematical theory directly to observations of nature, without the intermediary step of testing the theory in some biological system (e.g., the search for chaotic behavior in population dynamics and the assumption that "dispersal corridors" will benefit fragmented populations). The preceding examples show that numbers are not organisms and that biology intrudes into even the simplest attempts to verify theory. Microcosms have been instrumental in revealing the biologically interesting ways in which theories fail, thus paving the way for improved, more predictive theory (Kareiva 1990).

ACKNOWLEDGMENTS   I thank the National Science Foundation's Experimental Program to Stimulate Competitive Research at the University of Kentucky for support while I was writing part of this chapter, and I thank Philip Crowley for providing encouragement. I am grateful to my collaborators Peter Morin, Marcel Holyoak, and Dalius Balčiūnas for their assistance in ensuring that our projects were successful and for stimulating discussions about microcosm work. Joseph Bernardo, Marcel Holyoak, Matthew Leibold, Shahid Naeem, and Philip Warren provided insightful comments that improved the manuscript.

Literature Cited

Addicott, J. E. 1974. Predation and prey community structure: an experimental study of the effects of mosquito larvae on the protozoan communities of pitcher plants. Ecology 55: 475–492.

Balčiūnas, D., and S. P. Lawler. 1995. Effects of basal resources, predation, and alternative prey in microcosm food chains. Ecology 76:1327–1336.

Barkai, A., and C. McQuaid. 1988. Predator-prey role reversal in a marine benthic ecosystem. Science 242:62–64.

Beyers, R. J., and H. T. Odum. 1993. Ecological Microcosms. Springer-Verlag, Berlin.

Carpenter, S. R. 1996. Microcosms have limited relevance for community and ecosystem ecology. Ecology 77:677–680.

Chao, L., B. R. Levin, and F. M. Stewart. 1977. A complex community in a simple habitat: an experimental study with bacteria and phage. Ecology 58:369–378.

Cole, B. J. 1983. Assembly of mangrove ant communities: patterns of geographic distribution. Journal of Animal Ecology 52:349–355.

Connell, J. H., and W. P. Sousa. 1983. On the evidence needed to judge ecological stability or persistence. American Naturalist 121:789–824.

DeAngelis, D. L. 1975. Stability and connectance in food web models. Ecology 56:238–243.

Dickerson, J. E., and J. V. Robinson. 1985. Microcosms as islands: a test of the MacArthur-Wilson equilibrium theory. Ecology 66:966–980.

Drake, J. A. 1990. Communities as assembled structures: do rules govern pattern? Trends in Ecology and Evolution 5:159–164.

———. 1991. Community-assembly mechanics and the structure of an experimental species ensemble. American Naturalist 137:1–26.

Drake, J. A., G. R. Huxel, and C. L. Hewitt. 1996. Microcosms as models for generating and testing community theory. Ecology 77:670–666.

Droop, M. R. 1974. The nutrient status of algal cells in continuous culture. Journal of the Marine Biological Association of the United Kingdom 54:825–855.

Frank, J. H., and L. P. Lounibos. 1983. Phytotelmata: Terrestrial Plants as Hosts for Aquatic Insect Communities. Plexus, Medford, New Jersey.

Fretwell, S. 1977. The regulation of plant communities by the food chains exploiting them. Perspectives in Biology and Medicine 20:169–185.

Gause, G. F. 1934. The Struggle for Existence. Reprinted 1971. Dover, New York.

Gilpin, M. E., M. P. Carpenter, and M. J. Pomerantz. 1976. The assembly of a laboratory community: multispecies competition in *Drosophila*. Pages 23–40 in J. Diamond and T. J. Case (eds.), Community Ecology. Harper and Row, New York.

Hairston, N. G., Sr., and N. G. Hairston, Jr. 1993. Cause-effect relationships in energy flow, trophic structure, and interspecific interactions. American Naturalist 142:379–411.

Hairston, N. G., F. E. Smith, and L. B. Slobodkin. 1960. Community structure, population control, and competition. American Naturalist 94:421–425.

Harrison, G. W. 1995. Comparing predator-prey models to Luckinbill's experiment with *Didinium* and *Paramecium*. Ecology 76:357–374.

Harrison, S., and A. D. Taylor. 1997. Empirical evidence for metapopulation dynamics: a critical review. Pages 27–42 in I. Hanski and M. E. Gilpin (eds.), Metapopulation Dynamics: Ecology, Genetics and Evolution. Academic Press, San Diego.

Hastings, A., and K. Higgins. 1994. Persistence of transients in spatially structured ecological models. Science 263:1133–1136.

Hastings, A., C. L. Hom, S. Ellner, P. Turchin, and H. C. J. Godfray. 1993. Chaos in ecology: is mother nature a strange attractor? Annual Review of Ecology and Systematics 24:1–33.

Have, A. 1993. Effects of area and patchiness on species richness: an experimental archipelago of ciliate microcosms. Oikos 66:493–500.

Hawkins, B. A. 1992. Parasitoid-host food webs and donor control. Oikos 65:159–162.

Holt, R. D. 1977. Predation, apparent competition, and the structure of prey communities. Theoretical Population Biology 12:197–229.

Holt, R. D., and J. H. Lawton. 1994. The ecological consequences of shared natural enemies. Annual Review of Ecology and Systematics 25:495–520.

Holt, R. D., J. Grover, and D. Tilman. 1994. Simple rules for interspecific dominance in systems with exploitative and apparent competition. American Naturalist 144:741–771.

Holyoak, M., and S. P. Lawler. 1996a. Persistence of an extinction-prone predator-prey interaction through metapopulation dynamics. Ecology 77:1867–1879.

———. 1996b. The role of dispersal in predator-prey metapopulation dynamics. Journal of Animal Ecology 65:640–652.

Huffaker, C. B. 1958. Experimental studies on predation: dispersion factors and predator-prey oscillations. Hilgardia 27:343–383.

Huffaker, C. B., K. P. Shea, and S. G. Herman. 1963. Experimental studies on predation: complex dispersion and levels of food in an acarine predator-prey interaction. Hilgardia 34:305–330.

Hunter, M. D., and P. W. Price. 1992. Playing chutes and ladders: heterogeneity and the relative roles of bottom-up and top-down forces in natural communities. Ecology 73: 724–732.

Hutchinson, G. E. 1959. Homage to Santa Rosalia, or why are there so many kinds of animals? American Naturalist 93:145–159.

———. 1961. The paradox of the plankton. American Naturalist 95:137–145.

———. 1978. Introduction to Population Ecology. Yale University Press, New Haven, Connecticut.

Ives, A. R., J. Fourfopoulos, E. D. Klopfer, J. L. Klug, and T. M. Palmer. 1996. Bottle or big-scale studies: how do we do ecology? Ecology 77:681–684.

Jaffee, B. A. 1996. Soil microcosms and the population biology of nematophagous fungi. Ecology 77:690–693.

Kareiva, P. 1989. Renewing the dialogue between theory and experiments in population ecology. Pages 68–88 in J. Roughgarden, R. M. May, and S. A. Levin (eds.), Perspectives in Ecological Theory. Princeton University Press, Princeton, New Jersey.

———. 1990. Population dynamics in spatially complex environments: theory and data. Philosophical Transactions of the Royal Society of London B 330:175–190.

Koehl, M. A. R. 1989. Discussion: from individuals to populations. Pages 39–53 in J. Roughgarden, R. M. May, and S. Levin (eds.), Perspectives in Ecological Theory. Princeton University Press, Princeton, New Jersey.

Law, R., and R. D. Morton. 1993. Alternative permanent states of ecological communities. Ecology 74:1347–1361.

Lawler, S. P. 1993. Direct and indirect effects in microcosm communities of protists. Oecologia 93:184–190.

Lawler, S. P., and P. J. Morin. 1993. Food web architecture and population dynamics in laboratory microcosms of protists. American Naturalist 141:675–686.

Lawton, J. H. 1995. Ecological experiments with model systems. Science 269:328–331.

Lenski, R. E., and B. R. Levin 1985. Constraints on the coevolution of bacteria and virulent phage: a model, some experiments, and predictions for natural communities. American Naturalist 125:585–602.

Leonard, M. A., and J. M. Anderson. 1991. Growth dynamics of Collembola (*Folsomia candida*) and a fungus (*Mucor plumbeus*) in relation to nitrogen availability in spatially simple and complex laboratory systems. Pedobiologia 35:163–173.

Levin, B. R., F. M. Stewart, and L. Chao. 1977. Resource-limited growth, competition, and predation: a model and experimental studies with bacteria and bacteriophage. American Naturalist 111:3–24.

Levins, R. 1968. Evolution in Changing Environments: Some Theoretical Explorations. Princeton University Press, Princeton, New Jersey.

Lotka, A. J. 1925. Elements of Physical Biology. Williams and Wilkins, Baltimore, Maryland.

Luckinbill, L. S. 1973. Coexistence in laboratory populations of *Paramecium aurelia* and its predator *Didinium nasutum*. Ecology 54:1320–1327.

———. 1974. The effects of space and enrichment on a predator-prey system. Ecology 55: 1142–1147.

———. 1979. Regulation, stability, and diversity in a model experimental microcosm. Ecology 60:1098–1102.

Luckinbill, L. S., and M. M. Fenton. 1978. Regulation and environmental variability in experimental populations of protozoa. Ecology 59:1271–1276.

MacArthur, R. H., and R. Levins. 1967. The limiting similarity, convergence, and divergence of coexisting species. American Naturalist 101:377–385.

MacArthur, R. H., and E. O. Wilson. 1963. An equilibrium theory of insular zoogeography. Evolution 17:373–387.

———. 1967. The Theory of Island Biogeography. Princeton University Press, Princeton, New Jersey.

McCauley, E., and W. W. Murdoch. 1990. Predator-prey dynamics in environments rich and poor in nutrients. Nature 343:455–457.

Menge, B. A. 1992. Community regulation: under what conditions are bottom-up factors important on rocky shores? Ecology 73:755–765.

Monod, J. 1950. La technique de culture continue: theorie et applications. Annales de l'Institute Pasteur Lille. 79:390–410.

Moore, J. C., P. C. De Ruiter, H. W. Hunt, D. C. Coleman, and D. W. Freckman. 1996. Microcosms and soil ecology: critical linkages between field studies and modelling food webs. Ecology 77:694–705.

Morin, P. J. 1983. Predation, competition, and the composition of larval anuran guilds. Ecological Monographs 53:119–138.

Morin, P. J., and S. P. Lawler. 1996. Effects of food chain length and omnivory on population dynamics in experimental food webs. Pages 218–230 in G. A. Polis and K. O. Winemiller (eds.), Food Webs: Integration of Patterns and Dynamics. Chapman and Hall, New York.

———. 1995b. Food web architecture and population dynamics: theory and empirical evidence. Annual Review of Ecology and Systematics 26:505–529.

Naeem, S. 1988. Resource heterogeneity fosters coexistence of a mite and a midge in pitcher plants. Ecological Monographs 58:215–227.

———. 1990. Resource heterogeneity and community structure: a case study in *Heliconia imbricata* phytotelmata. Oecologia 84:29–38.

Neill, W. E. 1974. The community matrix and interdependence of the competition coefficients. American Naturalist 108:399–408.

Oksanen, L., S. D Fretwell, J. Arruda, and P. Niemela 1981. Exploitation ecosystems in gradients of primary productivity. American Naturalist 118:240–261.

Paine, R. T. 1966. Food web complexity and species diversity. American Naturalist 100:65–75.

Park, T. 1948. Experimental studies of interspecific competition: 1. Competition between populations of the flour beetles, *Tribolium confusum* Duval and *Tribolium castaneum* Herbst. Ecological Monographs 18:265–307.

Pimm, S. L. 1982. Food Webs. Chapman and Hall, London.

Pimm, S. L., and J. H. Lawton. 1977. Number of trophic levels in ecological communities. Nature 268:329–331.

Polis, G. A. 1994. Food webs, trophic cascades and community structure. Australian Journal of Ecology 19:121–136.

Power, M. E. 1992. Top-down and bottom-up forces in food webs: do plants have primacy? Ecology 73:733–746.

Rey, J. R. 1981. Ecological biogeography of arthropods on *Spartina* islands in northwest Florida. Ecological Monographs 51:237–265.

Robinson, J. V., and M. A. Edgemon. 1989. The effect of predation on the structure and invasibility of assembled communities. Oecologia 79:150–157.

Rosensweig, M. L. 1971. The paradox of enrichment: destabilization of exploitation ecosystems in ecological time. Science 171:385–387.

Simberloff, D., and E. O. Wilson. 1969. Experimental zoogeography of islands: the colonization of empty islands. Ecology 50:278–296.

————. 1970. Experimental zoogeography of islands: a two-year record of colonization. Ecology 51:934–937.

Slobodkin, L. B. 1961. Growth and Regulation of Animal Populations. Holt, Rinehart, and Winston, New York.

Sommer, U. 1991. Convergent succession of phytoplankton in microcosms with different inoculum species composition. Oecologia 87:171–179.

Spencer, M., and P. H. Warren. 1996. The effects of habitat size and productivity on food web structure in small aquatic microcosms. Oikos 108:764–770.

Strauss, S. Y. 1991. Indirect effects in community ecology: their definition, study and importance. Trends in Ecology and Evolution 6:206–210.

Strong, D. R. 1992. Are trophic cascades all wet? Differentiation and donor-control in speciose ecosystems. Ecology 73:747–754.

Takahashi, F. 1959. An experimental study on the suppression and regulation of the host population by the action of the parasitic wasp. Japanese Journal of Ecology 19:225–232.

Taylor, A. D. 1990. Metapopulations, dispersal, and predator-prey dynamics: an overview. Ecology 71:429–433.

————. 1992. Deterministic stability analysis can predict the dynamics of some stochastic population models. Journal of Animal Ecology 61:241–248.

Tilman, D. 1977. Resource competition between planktonic algae: an experimental and theoretical approach. Ecology 58:338–348.

Vandermeer, J. H. 1969. The competitive structure of communities: an experimental approach with protozoa. Ecology 50:362–371.

Verhoef, H. A. 1996. The role of soil microcosms in the study of ecosystem processes. Ecology 77:685–689.

Volterra, V. 1926. Fluctuations in the abundance of a species considered mathematically. Nature 118:558–560.

Wilbur, H. M. 1972. Competition, predation, and the structure of the *Ambystoma-Rana sylvatica* community. Ecology 53:3–21.

Wootton, J. T. 1994. The nature and consequences of indirect effects in ecological communities. Annual Review of Ecology and Systematics 25:443–466.

Wootton, J. T., and M. E. Power. 1993. Productivity, consumers, and the structure of a river food chain. Proceedings of the National Academy of Science of the USA 90:1384–1387.

Worthen, W. B. 1993. Effects of ant predation and larval density on mycophagous fly communities. Oikos 66:526–532.

13

The Interplay Between Natural History and Field Experimentation

GARY A. POLIS, DAVID H. WISE, STEPHEN D. HURD,
FRANCISCO SANCHEZ-PIÑERO, JAMES D. WAGNER,
CHRISTOPHER TODD JACKSON, & JOSEPH D. BARNES

Modern ecology can be considered the "new natural history." It emerged from the fusion of elements from classical natural history and physiology (McIntosh 1985) and is the progeny of the marriage of two nineteenth-century traditions: the systematic, detailed cataloging of facts and a reductionist experimental approach to understanding function. As a result, experimentation and mathematical modeling emerged as useful tools to uncover the effects of discrete variables and processes on observed patterns, avoiding mere speculation. However, an extreme (or naive) acceptance of these new approaches could promote misinterpretation of ecological phenomena. McIntosh (1985) points out that Hutchinson "noted the need to temper mathematical abstraction with sound knowledge of natural history." Clearly, experimentalists as well as mathematically inclined theoreticians must not forget ecology's roots in basic natural history. As field experimentation becomes common and is no longer considered innovative, the danger increases that eagerness to conduct experiments will lead to field manipulations as irrelevant as some elegant sets of differential equations.

Natural history information is essential for both the design and interpretation of realistic ecological experiments. Spatial and temporal variation, indirect and higher order effects in complex trophic networks, effects of scale, and ignorance of initial conditions—all these make field experiments difficult to interpret. As Bender et al. (1984) argue, one aid to interpretation is "supplementing ecological experiments with the fullest amount of descriptive natural history." Ecologists generally recognize that the most effective experiments are those designed and interpreted in the context of a well-known system.

In this chapter, we examine the ways in which natural history data can strengthen manipulative experiments. With this goal, we delineate potential problems and make recommendations to lessen their impact. In an important sense, these recommendations could be summarized to indicate the need to acquire as deep an understanding as possible of the natural history of the focal species before planning, executing, or inter-preting a field experiment. Much of this supporting natural history information, how-ever, may be available only through experimental studies. We explain the interplay between the use and generation of natural history information in the design of ecological field experiments. To begin, we argue that although testing explicitly stated hypothesis is a major goal of field experiments, experiments also play other important roles.

The Value of Experiments

Field experiments play multiple roles. We emphasize that the most widely accepted function of experiments, to test explicitly stated hypotheses, is a major goal. Less widely recognized roles include revealing the relative magnitude of a process, identifying key factors in the dynamics of complex networks, and uncovering unsuspected phenomena. These latter functions can be categorized as the acquisition of basic natural history data about a species or a system.

Testing Hypotheses

We take for granted that well-designed field experiments can provide unambiguous tests of a priori hypotheses. Hairston (1989) provides a fine review of the rationale for field experiments, as well as a detailed critique of numerous experiments. Other reviews evaluate how field experiments test hypotheses on competition (Connell 1983, Schoener 1983, Underwood 1986), the effects of predation on prey (Sih et al. 1985), and how spiders have been used to test hypotheses about food limitation, competition, predation, habitat structure, and trophic dynamics (Wise 1993).

Quantitative Estimates of Processes

An underutilized and underappreciated feature of experiments is that by manipulat-ing a factor and observing the magnitude of change in response variables quantitative data on the relative contribution of that factor to the dynamics of the system can be obtained. For example, the relative impacts of predators A and B on prey species C could be evaluated by observing the responses of C to separate experimental reductions in densities of A and B. Combined manipulation of A and B could reveal the strength of indirect effects on densities of C (e.g., Spiller and Schoener 1994). Two tendencies discourage an emphasis on using field experiments to measure such quantitative effects. Each reflects the view that field experiments are solely a means to test hypotheses.

The first tendency is focusing too narrowly on whether or not the statistical null hypothesis is rejected. A narrow focus on hypothesis testing relegates the actual mag-nitude of the response to a position of secondary interest. Imagine a very well replicated field experiment that detects a statistically significant effect of excluding a predator with a low associated probability of $P < .001$. Such an outcome provides strong support for the hypothesis that excluded predators limit prey densities, but the change in prey numbers may be small. Did excluding predators produce a 5%, 10%, or 50% increase in prey? Previous information may make the hypothesis itself so plausible that uncov-ering the magnitude of the interaction becomes the experiment's major contribution.

Thus, a statistically significant result itself may reveal little new. However, by enabling us to know the magnitude of an effect the experiment strengthens understanding.

The second tendency that leads ecologists to underutilize the findings of experiments is disappointment with statistically nonsignificant results. Imagine that our field experiment fails to produce a statistically significant effect. Often erroneously termed a *negative* result, such an outcome does not fail to inform us about the system. Preoccupation with the null hypothesis, chagrin over the negative result, and failure to use to full advantage the concept of Type II error often prevent ecologists from using experimental results to full advantage (Wise 1984, 1993). The cause is our failure to calculate the statistical power of an experimental design as a regular feature of analyzing and interpreting field experiments. If excluding predators has no effect on prey densities, what is the minimum effect that could have been detected, with a specified probability, with the experimental design used? In our imaginary experiments, it might be quite valuable to determine that excluding a predator does not result in a detectable increase in prey density of at least 25% of the control value. Greater replication might uncover a significant effect under 25%, but it may not be necessary to measure the impact of predation with such precision to improve our understanding of a system. Toft and Shea (1983) and Rotenberry and Wiens (1985) previously emphasized the value of power analysis. Rotenberry and Wiens point to Cohen's (1977) "comparative detectable effect size" as a valuable statistic to interpret the results of field experiments.

Identification of Key Factors in the Dynamics of Complex Networks

Observational data alone may not yield data on the dynamics of a system—an experimental perturbation may be necessary to uncover the intensity and even presence of an interaction. The difficulty of inferring causation from patterns of correlation is a major rationale to use field experiments to test if a correlation between variables X and Y represents a causal connection—that is, to test an explicitly stated hypothesis and to choose among alternative explanations. But what if X and Y are not correlated? For example, if resource levels and consumer densities do not vary over years or among habitats, it does not necessarily follow that resource density does not affect consumer density, or vice versa. Or what if there is extensive variation in both resource and consumer but no correlation—how can we infer absence of a casual relationship in all years or habitats? An appropriately designed field experiment could answer these questions directly. Likewise, experiments are necessary to understand the dynamics of food webs. Data on diet and energetics may provide erroneous estimates of interaction coefficients. The actual influence of feeding links on the dynamics of complex webs may only reliably be uncovered by experimentally manipulating key elements of the web (Paine 1992, Polis 1994).

Unexpected Results

A field experiment may uncover entirely unexpected interactions. Of course, serendipity is not confined to the experimental approach, but its value in building basic understanding should not be overlooked. For example, a long-term removal experiment with darkling beetles, designed to test the hypothesis that interspecific competition among adults influences community structure, uncovered no evidence of competition but revealed indirect facilitation (Wise 1981). In the third year of the experiment, the combined density of the less dominant species in the community was lower in removal plots than in control plots. Small mammal activity was highest this year, leading to speculation that a mammalian predator had shifted from the removed species to the remaining beetles. Small mammals can influence densities of darkling beetles, as revealed by a field experiment that reduced densities of mammalian predators (Parmenter and MacMahon 1988). Vandermeer, Hazlett, and Rathcke (1985) point out that almost 20% of the experimental studies of competition reviewed by Connell (1983) revealed facilitation—a surprising result.

In summary, ecological field experiments are powerful tools to test hypotheses, to gather basic data needed to interpret the results of other manipulations, and to uncover patterns that suggest hypotheses to be tested by further experimentation. The value of field experiments makes it imperative that ecologists confront the plethora of potential problems associated with the design and interpretation of field experiments. Next we outline some major problems and offer recommendations to attack them.

Problems Inherent in Field Experiments

Ecology as a distinct discipline developed quite recently (Cittadino 1980, Ricklefs 1990). Only since the early 1970s have ecologists fully recognized the central importance of field experimentation (Hairston 1989). Thus, it is not surprising that much impetus for the primacy of experimentation came from comparisons with disciplines such as physics, chemistry, and physiology (Allen 1977, Coleman 1977, Cittadino 1980, Diamond 1986). With experimental considerations already long debated in these disciplines, much of the framework for experimental work was simply transplanted directly into field ecology via physiology (Allen 1977, Cittadino 1980), resulting in tremendous growth in the frequency and importance of field experiments (Hairston 1989, Wise 1993). However, this growth also spotlights many weaknesses that resulted from our failure to adapt this inherited experimental framework to the peculiarities of ecological research. These include the following.

Lack of Control over Experimental Conditions

The classic formulation of the experimental method includes minimizing variation of extrinsic factors which might otherwise confound the effects of the target manipulation. This formulation is well established in laboratory research, where investigators define most or all elements in the experimental universe, readily excluding complicating factors (e.g., the use of controlled-temperature rooms, controlled photoperiods, cell culture lines, defined growth media, sterile fields, age-and gender-specific samples, inbred experimental animal lines, and controlled diets). The remaining variation found in controlled laboratory environments is minuscule compared to both the tremendous variability seen in populations, species, and communities and to the stochasticity of the external environment (Strong et al. 1984). In the face of such variability, the investigator in a natural (nonagricultural) environment cannot always rely solely on Fisherian randomization. While such randomization may indeed yield a ''valid'' statistical test (R. A. Fisher, from Senn 1994), the underlying biological (and statistical) significance of the target variable may be completely submerged in a morass of random fluctuation (i.e., in the absence of some degree of control, the ''valid'' statistical test will often support a false null hypothesis because of the tremendous variation in outcomes). Large enough sample sizes would allow randomization to mitigate the effects of natural variation, but sample sizes in field ecology are often constrained to be small for a number of practical reasons (see following discussion). The combination of extremely high natural variability and low replication can make the lack of control of extrinsic factors a serious limitation on the ability of investigators to discern relevant effects that arise from experimental manipulations.

Experimental Artificiality

The very act of attempting to control extrinsic factors in the field makes such experiments artificial (Allen 1979, Cittadino 1980, Schaffer 1985). Ecologists seek to understand the operation of natural systems; given the complexity of these systems, the more controlled the experimental design, the greater the uncertainty as to whether target variables will operate similarly in the presence of a multitude of other factors (Gilpin et al. 1986, Hairston 1989, R. L. Smith 1990; see discussion of indirect effects following). Alternatively, a manipulation necessary to determine experimentally the effect of a particular factor may be so disruptive that it negates other important elements of the system, thus creating problems with artificiality and invasiveness. Obviously, experimental artificiality and lack of control are opposite sides of the same coin; if variation in confounding variables is handled successfully, problems associated with experimental artificiality increase in importance. In the worst case, field experiments fall squarely between the two poles: designs fail to control the variety of factors impinging a target variable but are sufficiently invasive (i.e., affect enough nontarget variables) that extension of results to intact systems may be questioned.

Inability to Discriminate among Alternative Hypotheses

Discussing the experimental method, Henri Poincare noted: ''If a phenomenon admits of a complete mechanical explanation it will admit of an infinity of others which will account equally well for all the peculiarities disclosed by the experiment'' (1946: n.p.). Although this problem afflicts all experimental sciences, the inability to predefine all factors in field experimentation makes it extremely difficult to validate any particular substantive hypothesis based on the rejection of a statistical null hypothesis. Manipulations aimed at a particular taxon or factor may influence, either directly or indirectly, other taxa or factors; it may be these other taxa which cause the experimentally observed effect and subsequent rejection of the null hypothesis (Brown et al. 1986, Diamond 1986).

Hidden Influences

Cursory (and sometimes even extended) observation of a system may fail to detect a key dynamic factor. Critical species are not necessarily the most abundant nor do they necessarily exhibit the greatest biomass (Polis 1994, Polis and Strong 1996). Critical processes may occur frequently but not continuously or predictably, or they may occur infrequently and be difficult to observe—for example, periodic physical disturbances (Armesto et al. 1992, Service and Feller 1992, Kennelly and Underwood 1993, Dayton et al. 1992) and climatic disturbances (Wiens 1977, Quinn and Neal 1983, Dayton and Tegner 1984, Tovar et al. 1987, Covich et al. 1991, Allison 1992, Pounds and Crump 1994, Polis et al. forthcoming)—and cycling populations may cause a minor consumptive pressure to become extreme at certain times (Oksanen and Oksanen 1981). Such a complex mosaic can cause experiments to be directed at unimportant factors, ignore hidden factors of paramount importance, or lend credence to a false alternative hypothesis.

Natural Variability

The importance of ecologically relevant factors varies naturally through time and space. Factors may be important continually or sporadically, concurrently or sequentially. Such temporal variability creates unique problems for field experiments. Philosophically, laboratory experiments also face natural temporal variability: chance or chaotic small changes could vastly affect the outcome of the same experiment on different days. However, underlying laboratory conditions generally change little, if at all, from day to day. In sharp contrast, field experiments must contend both with chance chaotic changes and with large random or directional variation in factors and forces over scales from hours to decades (e.g., Leigh 1975, Wiens 1977, H. Smith 1981, Gray 1991, Hawke 1992, Joern 1992, Karl and Tien 1992, Root and Cappuccino 1992).

Experimental results can also be affected via time lags in response variables. Depending on the specifics of focal processes or organisms, ''short-term'' experiments may not detect real changes—for example, perturbation susceptibility (Connell and Sousa 1983, Sousa 1984), community interactions (Brown et al. 1986), complex dynamics cycles (Brown and Heske 1990), or indirect effects (Bender et al. 1984, Yodzis 1995). Even in multiyear studies, time lags may obscure a central process such as density dependence (Turchin 1990). Finally, even if target variables could respond to manipulation, their response in their natural community may be constrained by a second variable with a longer response time (Lauenroth and Sala 1992).

Ecologically relevant forces also vary spatially. A sampling unit may be unable to demonstrate the effect of the target factor: it could be too small to manifest the effect—for example, stability (Connell and Sousa 1983), adult habitat preference (Sale and Dybdahl 1975), and habitat structure correlations (Wiens 1986)—or it could be so large that it incorporates variations in factor strength, for example density dependence (Hastings 1993). Even if factors are relatively constant, the scale over which they operate may vary. Population and community dynamics are affected by processes at many spatial scales (Addicott et al. 1987, Wiens 1989a, Holling 1992, Levin 1992, Mac-

Laughlin and Roughgarden 1993)—for example, competition and predation (Levin 1974, 1978; Hanski, 1983; Wiens 1986, 1989a, b; Holling 1992; Holt 1993; Goldwasser et al. 1994; Tilman 1994). Several scales are recognizably important: those of individual organisms (its ambit), of local populations, of groups of dynamically interlinked populations (mesoscale), and of regional and geographical distribution (species range; Wiens 1986, Menge and Olson 1990, Holt 1993, Ricklefs and Schluter 1993). An experiment that tests the effects of two factors on a population may not be able to manipulate each within the same-scaled experiment if scales over which each operates differ greatly—for example, predation by large, far-ranging mammals on movement-restricted prey (Wiens et al. 1986) and interspecific competition and conspecific migration (Schroder and Rosenzweig 1975). In general, patterns and processes at even adjacent scales may be quite different (Allen and Star 1982, Louda 1982, Holling 1992, Levin 1992), and studies conducted at inappropriate scales may provide erroneous perceptions of ecological processes (Schroder and Rosenzweig 1975, Wiens 1989a, b, Bennett 1990, Auerbach and Shmida 1987, Menge 1992, MacNally 1995). Lack of evidence for density dependence, for example, may often be attributed to use of the wrong spatial scales (Ray and Hastings in press).

Difficulty with Intraexperiment Replication in the Field

True replication, a critical component of traditional experimental design (Fisher 1973, Hurlbert 1984), is difficult for two reasons. First, the need for replicates to be as similar as possible may be hard to meet. Hurlbert (1984) notes that "we *know*, on first principles, that two experimental units *are* different in probably every measurable property." This fact often affects field experiments through pseudoreplication, in which "replicates" are not independent, and premanipulation differences between two independent sites (control and manipulated) remain throughout the analysis. Replicate similarity is also affected by hidden influences and by natural temporal and spatial variability. The only way to solve these problems is to increase the number of independent replicates in each treatment (Hulbert 1984). This solution runs into the second difficulty with replication: the difficulty of finding enough replicates. This is problematic when (a) the experiment focuses on large-scale phenomena, such as the dynamics of a watershed, lake, island, and so forth, or (b) multifactor experiments require large numbers of different treatments, each with numerous replicates. Attempts to simplify either of these situations by manipulating several "replicate" sample sites with a single action lead back to the problem of pseudoreplication.

Improper or Invalid Range of Manipulation

Experimental manipulations may be outside the natural range of conditions or changes by being either too large or too small. If manipulations are unrealistically large relative to the natural range of variation, the effects produced may be irrelevant to the question posed (Schindler 1987). If manipulations are too small, they may be insufficient to produce a biologically important effect; or more likely, even if the effect is biologically important, it will not be detected statistically. Given the possible misinterpretation of "negative" results as *supporting* the null hypothesis (rather than as *not* rejecting it; Oakes 1986), the danger of bolstering evidence against a hypothesis through improperly designed experiments is apparent.

Indirect, Higher order, and Interactive Effects

Ecologists now appreciate that many factors concurrently influence distribution and abundance and that single-factor analyses or experiments are insufficient to understand how populations, groups of species, and, ultimately, communities are structured (Dunson and Travis 1991; Polis 1991a, b, 1994; Hunter and Price 1992; Power 1992). A dependent variable (e.g., population size) is influenced by multiple factors: the abiotic environment (e.g., competition, predation, parasitism, disease, and mutualism). Further, these factors interact. For example, aquatic ecologists analyze the complex dynamics that occur from the interaction between variable ''top-down'' (predation) and ''bottom-up'' (productivity) factors on community regulation and structure (McQueen et al. 1989, Carpenter and Kitchell 1993). Thus, species are interconnected via multiple direct and indirect links that can reinforce or counter each other (Bender et al. 1984, Polis et al. 1989, Strauss 1991, Polis and Holt 1992, Diehl 1993, Abrams et al. 1995, Yodzis 1995, Polis and Strong 1996). In real webs with diversities of 102 to 104 species, a very large number of direct and indirect, weak and strong, trophic and nontrophic, and positive and negative links operate.

Finally, in many situations field experiments are either simply impossible (focal variables cannot be manipulated cleanly or in a timely fashion) or very time-consuming and difficult to establish. The use of experiments must be assessed against other uses of time, energy, and money for research.

Implications of Problems with Experiments

Ecological field experiments are beset with a myriad of theoretical, pragmatic, and logistic problems. But many of these problems also confront research in other disciplines without altering the fundamental role ascribed to experimentation in these fields. Furthermore, the most important theoretical problem, the inability to distinguish effectively among competing hypotheses, affects purely observational ecological studies to a far greater extent. There is certainly no shortage of elegant and successful field experiments in ecology; thus, despite their many problems, field experiments have been and certainly should remain the fundamental tool for understanding ecology.

However, we believe that ecologists must acknowledge the inherent trade-off between controlling confounding natural variability and risking artificiality/intrusiveness. We maintain that the distinction between field and laboratory experiments represents a significant dividing line between the experimental traditions and norms of much of physics, chemistry, engineering, and molecular and cellular biology versus field ecology. Excellent field experiments in ecology will not always abide by the expectations of accepted experimental design in these other disciplines.

Rather than argue against field experiments, we urge that they be placed in their proper context: as one element in an overall set of investigative tools used to assess any hypothesis. Such tools include purely observational techniques, mensurative ''experiments'' (Hurlbert 1984), unreplicated whole-system manipulations or observations, replicated manipulative field experiments, and controlled laboratory experiments on specific elements of the natural system. Together, these methods marshal much more reliable evidence for a particular alternative hypothesis than is possible with excessive reliance on controlled field experiments alone.

Recommendations for Field Experiments

We conclude with several recommendations as to how natural history information can be used to conduct field experiments more effectively:

1. Conduct long-term studies whenever feasible
2. Conduct experiments and observations at several spatial scales
3. Use tractable organisms and systems to elucidate processes
4. Use natural replicates where possible
5. Be conscious of repeatability
6. Embrace, do not shun, natural variation

These recommendations overlap to some extent but are presented separately to emphasize different facets of related challenges that confront field experimentation. Each is woven from a common philosophical thread—ecologists need to develop a deep appreciation and knowledge of the natural history of a system before initiating experiments. Such information is particularly important to diminish the potential significance of many of the problems discussed previously and to help frame precise questions, design protocols, and interpret results.

For example, natural history observation allows one to distinguish among the many potential processes that can explain an experimental result. To illustrate, intraguild predation (IGP) lowers densities of some species via either a direct (predation) or an indirect interaction (exploitation competition [EC] on resources used by predator and prey species) (Polis et al. 1989). Observational data allowed Polis and McCormick (1987) to distinguish these two possibilities. Ten years of natural history data combined with a two-and-one-half-year experiment were used to show that the predation component of intraguild predation rather than exploitation competition reduced spider and scorpion abundance. Without ancillary observation, only the phenomenological result of the experiment (here, changes in abundance after removal of the dominant scorpion species) can be known, with little insight for its mechanistic basis.

Recommendation 1: Conduct Long-term Studies Whenever Feasible

Processes operate on temporal scales from seconds to centuries, depending, in large part, on the focal species. These processes generate patterns that hopefully reflect underlying mechanisms. Since pattern is merely the variability of a response under a given set of conditions, the ecologist need only reproduce these conditions to design and execute experiments. This scheme may work well in the laboratory, where one can control tightly all but the dependent variable of choice. However, in field studies many (most?) initial conditions are beyond such control. Further, natural variability in conditions, even at a given location and season, can be significant over time (see earlier).

To alleviate these problems, long-term studies are crucial. First, they allow one to quantify the background ''natural'' variability in important factors (Brown and Heske 1990). Second, they may reveal ''hidden'' or infrequent processes that are key to dynamics. Third, they allow one to evaluate whether a preconceived temporal scale is appropriate. Fourth, one can document and reinforce the natural history of focal organisms.

The appropriate temporal scale of a study, either observational and/or experimental, must consider the duration and seasonality of the phenomenon. Obviously, duration is variable, depending on the focal organisms. The effect of resource competition on reproduction could be measured in weeks for *Drosophila* but months or years for trees. Interestingly, the temporal scale of the observed pattern may be different from that of the mechanism that produces the pattern (Levin 1992). This often produces a time lag between mechanism and its pattern. Recent analyses demonstrate that density-dependent population regulation may be common but is usually accompanied by a time lag (Turchin 1990, Hanski and Woiwood 1991, Godfray and Hassell 1992, Holyoak 1994, Ostfeld and Canham 1995). Similarly, an organism's ability to store resources may delay the manifestation of a process or manipulation; such storage effects are important to understand consumer-resource and food web dynamics (Polis et al. 1995). Natural history of the focal system and organisms should be the guide to determine the duration of important processes and the length of experiments.

Many ecological phenomena vary in their strength or even presence through time. Some occur unpredictably; others occur more regularly. For example, pompilid wasps significantly depress spider populations on islands in the Gulf of California, but only during infrequent wet years after unpredictable El Niño events (Polis et al. forthcoming). Otherwise, these wasps appear totally unimportant to spider dynamics. Predictable variability occurs on several temporal scales (annual, monthly, and diel). Organisms at the same place during the same period may interact or not according to their particular schedules. For example, Polis and McCormick (1987) show that scorpion species and age classes within species segregate activity times within the same night and among months.

Intensive long-term observations are necessary to identify processes that may otherwise be overlooked, particularly in terms of relatively infrequent factors that can exert great impact on dynamics. Pulsed or short-term phenomena (e.g., some predation events and rare recruitment) provide examples. *Notophthalmus* salamanders eat *Bufo* and *Rana* tadpoles; however, gape limitation constrains feeding to a short period when tadpoles are small (Wilbur et al. 1983). Nevertheless, such predation is a key determinant of tadpole densities. Other important events may be frequent but rarely observed. For example, in desert scorpions, cannibalism and intraguild predation are quite important and represent a major mortality factor (Polis 1980, 1991b; Polis and McCormick 1987). Nocturnal behavior, low feeding rates, and relative inactivity of these animals combine to make observations of IGP and cannibalism relatively rare.

Long-term studies are best suited to analyze temporally varying phenomena. Unfortunately, logistical constraints often make such studies difficult, and hence they are not common. The establishment of 15 Long-Term Ecological Research (LTER) sites at a variety of biomes in the United States provides such opportunities (Franklin 1987, Callahan 1991). The National Science Foundation (NSF) also funds long-term research that provides less money annually but over periods longer than those of other grants. Moreover, many field stations maintain a long-term data base. For example, the Desert Ecological Research Unit of Namibia has existed since 1954; such stability has facilitated many long-term projects (Seely 1991).

Franklin (1987) enumerates explicit ecological phenomena for which long-term studies are essential. We expand on his list of categories: (1) *slow processes* such as succession (Risser et. al. 1981, Peet and Christensen 1987, Tilman 1988); population dynamics of long-lived organisms (wolf/moose over a 29-year study [Peterson et. al. 1984, Peterson 1987], newts over a 10-year study [Gill et. al. 1983]; trees [Franklin and DeBell 1988]); and decomposition, soil formation, and nutrient cycling (Boul et. al. 1973, Harmon et al. 1986); (2) *rare and episodic phenomena* such as major disturbance (flood, drought, windstorms, ice storms, volcanic eruptions, earthquakes, unusual temperatures, and fires), episodic reproduction in long-lived species (e.g., masting in bamboo), or quasi-regular determinants of climate (El Niño and sunspots); (3) *processes with high inherent variability* such as desert productivity (MacMahon 1980, MacMahon and Wagner 1985, Polis 1991b) and litter fall in deciduous forests (Gosz et. al. 1972); (4) *processes that are subtle*, such as those that slowly change over time (e.g., paleoclimatic changes and global warming) or whose year-to-year variance is greater than the magnitude of the observed trend (e.g., acid rain [Likens 1983]); and (5) *complex phenomena*, such as food web dynamics (Polis 1994, Polis and Strong 1996) or multivariate system analyses (George and Harris 1985).

The key point is that ecologists need to maintain long-term studies to establish a background of normal variability, to document episodic events, and to discover subtle or hidden processes. Without such insights, it is difficult to interpret field experiments, which of necessity usually are of shorter duration than many ecological processes.

Recommendation 2: Conduct Experiments and Observations at Several Spatial Scales

Problems with spatial variability and scale mandate this recommendation. The scaling of experiments is a most difficult problem because different species use space in radically different ways, depending on their specific traits (size, mobility, and dispersal ability, Addicott et al. 1987, Wiens 1989a, Kotliar and Wiens 1990, Levin 1992, Robinson et al. 1992, Holt 1993). In addition, different scales are appropriate to approach different problems. It is at the scale of an animal's space requirements (i.e., its individual area; Wiens et al. 1986) that behavioral experiments should be conducted. Research on population dynamics should include information from many levels, including the individual area, the confines of the focal populations, and the area in which linked populations affect each other (the mesoscale [Holt 1993]; e.g., via migration, metapopulation, or source-sink dynamics). Multispecies research must consider spatial use and dynamics by all focal species and spatial scale of important abiotic processes. This can be difficult as organisms that vary in size, mobility, and dispersal often interact (e.g., a predatory guild of a bird, lizard, and spider; Hodar 1993). Moreover, processes outside the focal habitat may be quite significant. For example, food web dynamics is influenced to various degrees by the input of nutrients, detritus, and organisms from other systems, both near and far (Polis et al. 1995); most systems are relatively open to important external trophic influences, even those traditionally considered closed, such as islands (see Wiens et al. 1986, Holt 1993, Polis and Hurd 1995). Similarly, phenomena that occur in other habitats (e.g., mortality, recruitment, oceanic currents, or fronts) can exert immense ''supply-side'' effects (Roughgarden et al. 1987).

How can one establish appropriate spatial scales? Although spatial heterogeneity may vary continuously with scale (Mandelbrot 1983, O'Neill et al. 1991, Palmer and White 1994), a hierarchical approach seems the most practical way to scale (Morris 1987, Kotliar and Wiens 1990, Menge and Olson 1990, Holling 1992). In such an approach, the environment is composed of patches containing other patches at finer scales and themselves nested within other patches (Kotliar and Wiens 1990). Patch structure is influenced by lower levels and influences higher levels. Organisms are not restricted to one patch level; scales of organismal response are delineated by the lower and upper limits in which an interaction occurs—that is, grain and extent. The selection of grain is essential because it is the study unit that supports the entire hierarchy. Extent is likewise important, but selection of the largest scale is usually imposed by logistical considerations, not the ecology of the system. In general, one should select the largest scale possible in agreement with the traits of the organism, environment, and interaction. The next problem is establishing the identity of important scales between grain and extent. Because different species experience the environment differently, the number of levels will vary by case. A thorough natural history knowledge of the system, its organisms, and their interactions becomes an absolutely essential component during this process.

Finally, we should strive to conduct experiments that allow comparison across scales. Unfortunately, the ability to replicate decreases as scale increases, not only from logistical constraints but also because the number of possible natural replicates (e.g., lakes and islands) usually decreases. Carpenter (1990) suggests that this problem must not limit experimentation at large scales. Progress can be made by combining results of different studies and comparing experiments from different times or by using a hierarchical approach to compare dynamics across scales (Kotliar and Wiens 1990). Empirical Bayesian statistics can be a key. The comparative method has been used successfully with many taxa, environments, and interactions (e.g., Louda 1982, Ward and Saltz 1994; see Underwood and Petraitis 1993). The development of new mathematical approaches to analyze scale, such as fractals, could facilitate comparisons across multiple scales (Mandelbrot 1983, Levin 1992, Virkkala 1993).

Recommendation 3: Use Tractable Organisms and Systems to Elucidate Processes

Ecology is firmly entrenched within what may be called the naturalist tradition with regard to how systems are selected for study. Investigators are often drawn to one taxon or system by personal interest or training, and research programs often proceed from this basis. We suggest that such nondirected, idiosyncratic selection criteria do not efficiently elucidate the functional interrelationships that underlie communities and ecosystems (particularly with the realization of the diversity of and complex interconnections among the 10 to 50 million species on earth).

We believe strongly that organisms or systems should be chosen specifically with an eye toward *revealing process*. It is only through the study of processes that we can attempt to synthesize the peculiarities in the natural history of every species into a coherent, predictive framework. Ecologists should not approach each field situation with the sole question of what taxa are of interest but rather ask what processes are manifested and by which taxa. This feature contrasts pure ecologists with taxon-oriented researchers—for example, mammalogists or arachnologists—whose work is typically designed to elucidate the specifics of a particular taxon regardless of its suitability for a particular question. As ecologists, we study phenomena and describe processes with the goal of transcending the specifics of a particular taxon or system—we hope that some of our results describe general processes.

With process as the guiding principle, the important selection criterion is the amenability of a biological system to data collection and manipulation. We urge strongly the use of "model" organisms or systems. Three features are important. First, and most fundamental, a model organism must have some target variable—density, body size, fecundity, or fruit production—that exhibits a large range of response to variation in the strength of the independent variable. The frequency of Type II error (acceptance of a false null hypothesis) is a function of the response range of the dependent variable (the narrower the range, the greater the potential for Type II error). Second, a model organism must be sufficiently observable within the temporal and spatial constraints of the study. Use of organisms that offer few observations yields small data sets that are unstable predictors of effect strength and offer low resolution power in significance tests. Third, effective models should be manipulable in the least intrusive manner possible. The more intrusion required, the more likely that artificiality is introduced into experimental results.

Many taxa exhibit traits that facilitate research on particular phenomena; no one taxon is perfect. We have heard or read ecologists who preach model status for such diverse taxa as mantids, ants, fish, amphibians, lizards, birds, and protists; we believe that scorpions (Polis 1990, 1993; Brownell and Polis in press) and spiders (Wise 1993, Polis and Hurd 1995) are useful vehicles to advance behavioral, population, and community studies. Yet we stress that no taxon always conforms to model status; rather, a taxon may be a model *for the study of certain ecological processes.*

Recommendation 4: Use Natural Replicates Where Possible

Earlier (see the discussion of problems), we pointed out the paradox that as a field experiment seeks to control for variables extraneous to the design, it becomes more artificial and thus potentially less applicable to the question for which it was designed. The ability to control for all extraneous variables is beyond the reach of field experiments. Therefore, the real question is how to reduce intrusiveness and artificiality while not greatly decreasing the extent of control over various factors. One means to accomplish this is the use of natural replicates. Natural replicates are physically distinct and spatially separated objects which, *by nature*, are independent of one another with regard to the target variables, rather than being independent by intentional human structuring. Because of the separate and independent nature of these items, they can often serve as units of replication. Examples of such items and of their use in mensurative and manipulative experiments include isolated rock pools (Smith 1983; Van Buskirk 1992, 1993), stream pools (Fox 1975, Power et al. 1985), stream rocks (Fox 1977), whole lakes (Schindler 1977, Henrikson et al. 1980, Carpenter 1989a,b; Schindler 1990), whole islands (Simberloff and Wilson 1969, Boag and Grant 1981, Gibbs and Grant 1987, Schoener 1988, Losos et al. 1993, Polis, unpublished data), individual trees or shrubs not in a canopy (Southwood and Reader 1976, Jones 1987, Southwood et al. 1989, Lightfoot and Whitford 1991, Polis 1993), and phytotelmata (water-filled plant parts; e.g., tree holes [Kitching 1971, 1987; Frank and Lounibos 1983; Fincke 1992], pitcher plants [Clarke and Kitching 1993], bromeliads [Diesel 1992, Cotgreave et al. 1993, Diesel and Schuh 1993] flower bracts [Naeem 1990a, b], and fruit capsules [Caldwell 1993]).

Key characteristics of a good natural replicate are isolation and relative lack of dispersal by focal taxa among separate units. If this is the case, then the unit represents the ''universe'' available to these taxa in the ecological time frame of the experiment. To a lizard, the island it occupies is effectively its universe, just as the lake is to a fish and an isolated *Acacia* is to an arboreal scorpion. The natural replicates available that meet these requirements may be assigned to each treatment after statistical blocking based on area, volume, or any important character. To reduce interreplicate variability, certain sampling ground rules can be explicitly adopted a priori (e.g., islands < 1 km² and shrubs with volume 1 to 2 m³). Once assignments are made, manipulations of species (e.g., addition, removal, and augmentation) or the physical environment (e.g., change of the lake pH, augmentation of detrital matter, defoliation of trees, and removal of flowers) can be undertaken; in addition, large key interactants (e.g., consumers or competitors) can be excluded via fencing or cages.

The advantages of natural replicates are twofold. First, at the level of experimental validity, natural replicates, manipulated in such simple and nonintrusive ways, much more closely represent natural systems. In many cases where natural replicates can be used, the manipulation is so minor that it effectively interferes with no other local processes (e.g., manipulating resource availability [Naeem 1990a,b], population densities [Smith 1983, Van Buskirk 1993], and presence of predators/cannibals [Diesel 1992, Van Buskirk 1992, Polis 1993]). Therefore, the results can show with clarity the effect of altering that single variable. Second, at the level of experimental design and execution, natural replicates are easier to set up (the basic structure is already there), to manipulate (little physical disturbance is required), and to maintain (the isolation of each replicate relative to the taxa involved makes maintenance of additions and removals much easier). In fact, where numerous potential natural replicates are available, the smaller effort expended to prepare each replicate may make it possible to increase sample sizes.

Natural replicates have potential disadvantages. First, interreplicate differences do occur; even with a priori sampling rules and randomization, natural replicates will always differ to some degree. A related issue is that natural replicates may vary in some significant but unseen property that affects the focal phenomenon (e.g., interindividual differences in plant chemistry). Further, when using large and relatively rare natural replicates (e.g., whole lakes or islands), the matching of criteria to ensure close control of variables among replicates is likely to be ignored in favor of increasing sample size.

This point raises another problem specific to lakes, islands, or other large, potential natural replicates: probable small sample sizes and consequent low statistical power. Experimental designs that utilize such natural replicates often have small sample sizes—in the extreme, a sample size of one. Such small sample sizes yield little (or no) power under classical frequentist statistics in tests of significance among treatments. Loss of power is the price for such large-scale manipulations. It has been suggested that in such cases results should be presented, along with calculations of the power of the test given the sample sizes, even if P values do not reach the conventional $P < .05$ level of significance (Bakan 1967, Barnett 1973, Oakes 1986). Such results, if the effect is reasonably large and in the hypothesized direction, may still be used in conjunction with other evidence to support the substantive hypothesis.

In the case of samples with $N = 1$, Carpenter et al. (1989) presented a randomized intervention analysis (RIA), which analyzes changes in a variable in an unreplicated manipulated system relative to an unreplicated reference system. The authors make clear that this technique provides only statistical evidence that a change occurred at the time of the manipulation, not evidence of what caused that change: "To establish that the manipulation caused the response, one must show beyond reasonable doubt that no alternative causes could product the observed change . . . in unreplicated ecosystem experiments, determination of causality will rest on ecological rather than statistical arguments" (Carpenter et al. 1989: n.p.). As with the case of small sample sizes, even unreplicated "natural replicates" may be used as another tool to provide evidence for a particular hypothesis.

Recommendation 5: Be Conscious of Repeatability

Interexperiment replication, or "repeatability," is "the hallmark of a convincing scientific test" (Hairston 1989). Repeatability is a far more important measure of evidence for a hypothesis than is the P value (Meehl 1967, Goodman 1992), yet it is rarely acknowledged as a significant issue in experimental field ecology (but see Hairston 1989). The practical hurdles to repeating rigorously another investigator's results are large and often prohibitive. However, it is especially problematic if the same researcher at the same site is unable to "repeat" findings with some congruence from one year to the next. Two potential explanations exist. Given the impact of repeatability as an evidentiary tool, it is important for ecologists to explore the implications of these explanations.

First, noncongruent experimental findings may simply result from natural temporal variability. We argued that temporal changes can be large, and it would not be surprising if such changes led to markedly different results for the same field experiment from year to year. Our discussion of hidden influences gives an explicit example of how a periodic influence might cause experiments to vary (predictably or unpredictably) through time. In addition to such large-scale environmental variations, ecologists deal with extremely complex systems in which subtle, chance effects may change the overall development of the system (Hall et al. 1970, Wilbur et al. 1983). With both large systemic temporal changes and chance chaotic variation, the "failure" to repeat a result from year to year is anything but a failure; it reveals both the complexity of ecological systems and how factors and forces vary in importance through time.

Second, a repeated experiment may not produce similar results because statistical repeatability is generally much lower than thought. It is often assumed incorrectly that $P = .05$ implies a 95% probability of repeating a statistically significant result. Goodman (1992) analyzed the relationship between P values and "replication probability," the probability that given an initial experiment with $P = c$, a repeat will yield $P < .05$ (the conventional level of significance). He found that the replication probability for a given P value is markedly lower than one would expect: for an initial experiment result with $P = .01$, there is only a 73% chance that repeating the experiment would yield a result with $P < .05$. In fact, Goodman finds that "we do not achieve a 95 percent probability of replication until $P = 0.00032$, and a 99 percent probability at $P = 2 \times 10^{-5}$" (1992: n.p.).

It is not surprising, then, that some "statistically significant" results are not repeated. We should not weigh results from repeated experiments solely in terms of the correlation between P values, but rather on the correspondence of effect size and direction and the power of the test (Tversky and Kanheman 1971, Oakes 1986). In practice, however, there is a strong tendency to do exactly the opposite: to view the difference in P value (even with excellent agreement on effect size and direction) as an inconsistency that must be "explained" (Meehl 1967, Tversky and Kahneman 1971, Oakes 1986, Goodman 1992). This search for an explanation—when the most parsimonious one lies in the normal effects of random sampling or in the lack of statistical power—can lead to the use of ad hoc explanations for experimental findings that don't "pan out" (Meehl 1967).

Our recommendation to reduce the perceived problems associated with nonrepeatable results is again to urge the need for long-term studies. If the inability to repeat from year to year is due to simple vagaries of sampling, then not only will the results from each year yield important data about the size and direction of the effect, but also over time it will become evident that an occasional statistically nonsignificant result does not destroy the proposed substantive hypothesis. If the inability to repeat from year to year is due to true natural temporal variability, then the real question is not if a factor is important but how often, for how long, under what conditions, and to what degree that factor is important. These questions can only be answered by multiyear, repeated analyses of the same system. Indeed, short-term studies carry the danger that they will report as important some factor which is only occasionally or periodically significant.

Recommendation 6: Embrace, Do Not Shun, Natural Variation

Natural variation is what we seek to explain and is also what makes field experiments so difficult. Experimentalists must recognize the conflict between *controlling variation* and *studying phenomena in the context of natural variation*. One way to clarify this distinction is to clarify the concept of what constitutes a "control" in a field experiment.

We can recognize three types of "controls." First is the "control treatment"—the one with undisturbed levels of the variable being manipulated. Response variables in this treatment are compared to the "experimental treatment." Second is an open control or a series of partially caged controls. As we often need to control for unwanted effects of the manipulation—that is, for fence or cage effects—we need to establish this kind of control in an attempt to "control for" unwanted side effects of the artificial barrier. Controlling for this type of unwanted variation usually presents a particularly difficult challenge—one that makes most field experiments studies of "managed environments" (Hairston 1989). It is this difficulty that prompts our recommendation to use natural habitat units for experimental replicates whenever possible. Third is controlling for spatial and temporal variation, the final type of "control" that confronts the field experimentalist. Accounting for this type of variation depends on several critical components of experimental design: (a) spatial replication, (b) contemporaneous experimental and control (sensu types 1 and 2 above) treatments, (c) incorporation of knowledge of initial conditions and changes in potentially relevant variables throughout the experiment by statistical techniques such as a randomized block design, covariance analysis, and so forth, and (d) long-term experimentation and temporal replication.

The dilemma is knowing how much of the third type of control to exert. Excessive control can produce short-duration experiments with small, artificial managed environments in which the temporal and spatial scales may be inappropriate for the process being studied. One solution is to perform experiments at different scales and collect ancillary natural history data about the system throughout the experiment. Of course, shifting to larger scales decreases replication and statistical power. Thus, experimentalists may feel they have to choose between two equally unattractive options: gathering precise information about processes in a potentially irrelevant artificial environment and gathering imprecise and ambiguous data about processes in an almost natural setting.

We suspect that the solution is to tackle the problem by approaching the middle ground from both extremes: perform field experiments and supporting studies at different scales as much as possible, in order to address simultaneously the problems of replicability and natural variation. Improving replicability increases our ability to predict the behavior of ecological systems, once values of the relevant variables are known. Decreasing replicability, by both increasing the spatial heterogeneity of replicate experimental units and repeating experiments, sheds light on the extent of variation in the processes under investigation. Both goals are integral to a complete understanding of ecological systems.

Summary

Field experiments are difficult to perform and yield results that often are open to more than one interpretation. Nevertheless, our overall message is positive: the insight and potential payoffs from well-designed experiments make them well worth the challenge. We conclude by offering a few generalizations that may help crystallize our recommendations and points:

Ancillary descriptive data should always be used to design and interpret field experiments. Experimental ecologists will always benefit from data of careful observation of unmanipulated systems as aids to plan and interpret experiments. Successful programs will effectively synthesize diverse approaches and types of information.

"Model" systems in which experimentation can play a major role should be favored for study. However, because the definition of model systems is ambiguous and many

processes act at scales or involve variables that make experimentation impossible, we should not restrict our questions to those that only can be answered by field experiments.

Field experiments yield valuable natural history data about ecological systems. Although a variety of approaches have evolved—modeling, inferential statistics, engineering technology, and field experiments—we still seek to understand, and are still fascinated by, the same phenomena that early natural historians documented and attempted to explain. We encourage a broad interpretation of ''natural history'' for two reasons. First, some basic information about important processes (i.e., degree of food limitation, identity of crucial competitors and natural enemies, etc.) can often only be uncovered by controlled manipulations. The misleading concept that a complete set of ''natural history data'' awaits discovery by the patient observer and that data obtained by field experiments are somehow entirely different hampers the achievement of a pluralistic ecology. Second, the artificial dichotomy between data from natural history and those from experiments reinforces the misconception that the sole function of experiments is to test explicit a priori hypotheses. We argue that experiments can be an equally valid road to achieve new natural history insights.

Experiments can be used as a powerful means to build as well as test hypotheses. We must not focus solely on whether results of our field experiments are statistically significant. If we pay greater attention to statistical power in both the planning and interpretation of field experiments, we will move closer to making hypotheses that we build and test more quantitatively. Greater focus on the magnitude of effects that are statistically significant and the minimum effect size that an experiment is likely to detect will strengthen the contribution of field experiments to building more quantitative ecological theory.

We conclude that in a successful multifaceted, pluralistic research program there will be a productive interplay between field experimentation, the use and generation of natural history data, and the building and testing of theory.

ACKNOWLEDGMENTS   Financial support to GAP provided by NSF grants DEB-92-07855 and DEB-95-27888, the Natural Science Committee and University Research Council of Vanderbilt University, and Earthwatch Foundation. Support to DHW provided by NSF grants DEB-92-21786 and DEB-93-06692 and Kentucky Agricultural Experiment Station Hatch Project KY-00711. FSP was supported by grant PF94 52300837 of the Spanish Ministerio de Educación y Ciencia. We thank three anonymous reviewers and Joe Bernardo for very thorough and helpful reviews.

Literature Cited

Abrams P., B. A. Menge, G. G. Mittelbach, D. Spiller, and P. Yodzis. 1995. The role of indirect effects in food webs. Pages 371–395 in G. A. Polis and K. H. Winemiller (eds.), Food Webs: Integration of Patterns and Dynamics. Chapman and Hall, New York.
Addicott, J. F., J. M. Aho, M. F. Antolin, D. K. Padilla, J. S. Richardson, and D. A. Soluk. 1987. Ecological neighborhoods: scaling environmental patterns. Oikos 49:340–346.
Allen, G. E. 1977. Life Science in the Twentieth Century. Cambridge University Press, Cambridge.

————. 1979. Naturalists and experimentalists: the genotype and the phenotype. Pages 179–209 in W. Coleman and C. Limoges (eds.), Studies in History of Biology, Vol. 3. Johns Hopkins University Press, Baltimore, Maryland.

Allen, T. F. H., and T. B. Star. 1982. Hierarchy. University of Chicago Press, Chicago, Illinois.

Allison, S. K. 1992. The influence of rainfall variability on the species composition of a northern California salt marsh plant assemblage. Vegetation 101:145–160.

Armesto, J. J., I. Casassa, and O. Dollenz. 1992. Age structure and dynamics of Patagonian beech forests in Torres del Paine National Park, Chile. Vegetation 98:13–22.

Auerbach, M., and A. Shmida. 1987. Spatial scale and the determinants of plant species richness. Trends in Ecology and Evolution 2:238–242.

Bakan, D. 1967. On Method. Jossey-Bass, San Francisco, California.

Barnett, V. D. 1973. Comparative Statistical Inference. Wiley, London.

Bender, E. A., T. J. Case, and M. E. Gilpin. 1984. Perturbation experiments in community ecology: theory and practice. Ecology 65:1–13.

Bennett, W. A. 1990. Scale of investigation and the detection of competition: an example from the house sparrow and house finch introductions in North America. American Naturalist 135:725–747.

Boag, P. T., and P. R. Grant. 1981. Intense natural selection in a population of Darwin's finches (Geospizinae) in the Galápagos. Science 214:82–85.

Boul, S. W., F. D. Hole, and R. J. McCracken. 1973. Soil Genesis and Classification. Iowa State University Press, Ames.

Brown, J. H., and E. J Heske. 1990. Temporal changes in a Chihuahuan Desert rodent community. Oikos 59:290–302.

Brown, J. H., D. W. Davidson, J. C. Munger, and R. S. Inouye. 1986. Experimental community ecology: the desert granivore system. Pages 41–61 in J. Diamond and T. J. Case (eds.), Community Ecology. Harper and Row, New York.

Brownell, P. H., and G. A. Polis (eds.). In press. Scorpion Biology and Research. Oxford University Press, Oxford.

Caldwell, J. P. 1993. Brazil nut fruit capsules as phytotelmata: interactions among anuran and insect larvae. Canadian Journal of Zoology 71:1193–1201.

Callahan, J. T. 1991. Long-term ecological research in the United States: a federal perspective. Pages 9–21 in P. G. Risser (ed.), Long-Term Ecological Research: An International Perspective. Wiley, Chichester.

Carpenter, S. R. 1989a. Replication and treatment strength in whole-lake experiments. Ecology 70:453–463.

————. 1989b. Large-scale perturbations: opportunities for innovation. Ecology 71:2038–2043.

Carpenter, S. R. 1990. Large scale perturbations: opportunities for innovation. Ecology 71:2038–2043.

Carpenter, S. R., and J. F. Kitchell (eds.). 1993. The Trophic Cascade in Lakes. Cambridge University Press, Cambridge.

Carpenter, S. R., T. M. Frost, D. Heisey, and T. K. Kratz. 1989. Randomized intervention analysis and the interpretation of whole-ecosystem experiments. Ecology 70:1142–1152.

Cittadino, E. 1980. Ecology and the professionalization of botany in America, 1890–1905. Pages 171–198 in W. Coleman and C. Limoges (eds.), Studies in History of Biology, Vol. 4. Johns Hopkins University Press, Baltimore, Maryland.

Clarke, C. M., and R. L. Kitching. 1993. The metazoan food webs from six Bornean *Nepenthes* species. Ecological Entomology 18:7–16.

Cohen, J. 1977. Statistical Power Analysis for the Behavioral Sciences, rev. ed. Academic Press, New York.

Coleman, W. R. 1977. Biology in the Nineteenth Century: Problems of Form, Function and Transformation. Cambridge University Press, Cambridge.

Connell, J. H. 1983. On the prevalence and relative importance of interspecific competition: evidence from field experiments. American Naturalist 122:661–696.

Connell, J. H., and W. P. Sousa. 1983. On the evidence needed to judge ecological stability or persistence. American Naturalist 121:789–824.

Cotgreave, P., M. J. Hill, and D. A. J. Middleton. 1993. The relationship between body size and population size in bromeliad tank faunas. Biological Journal of the Linnean Society 49:367–380.

Covich, A. P., T. A. Crowl, S. L. Johnson, D. Varza, and D. L. Certain. 1991. Post–Hurricane Hugo increases in aytid shrimp abundances in a Puerto Rican montane stream. Biotropica 23:448–454.

Dayton, P. K., and M. J. Tegner. 1984. Catastrophic storms, El Niño and patch stability in a southern California kelp community. Science 224:283–285.

Dayton, P. K., M. J. Tegner, P. E. Parnell, and P. B. Edwards. 1992. Temporal and spatial patterns of disturbance and recovery in a kelp forest community. Ecological Monographs 62:421–445.

Diamond, J. M. 1986. Overview: laboratory experiments, field experiments, and natural experiments. Pages 3–22 in J. Diamond and T. J. Case (eds.), Community Ecology. Harper and Row, New York.

Diehl, S. 1993. Relative consumer sizes and the strengths of direct and indirect interactions in omnivorous feeding relationships. Oikos 68:151–157.

Diesel, R. 1992. Maternal care in the bromeliad crab, *Metopaulias depressus*: protection of larvae from predation by damselfly nymphs. Animal Behaviour 43:803–812.

Diesel, R., and M. Schuh. 1993. Maternal care in the bromeliad crab, *Metopaulias depressus* (Decapoda): maintaining oxygen, pH and calcium levels optimal for the larvae. Behavioral Ecology and Sociobiology 32:11–15.

Dunson, W. A., and J. Travis 1991. The role of abiotic factors in community organization. American Naturalist 138:1067–1091.

Fincke, O. 1992. Interspecific competition for trees holes: consequences for mating systems and coexistence in neotropical damselflies. American Naturalist 139:80–101.

Fisher, R. A. 1973. Statistical Methods and Scientific Inference, 3rd ed. Hafner Press, New York.

Fox, L. R. 1975. Factors influencing cannibalism, a mechanism of population limitation in the predator *Notonecta hoffmanni*. Ecology 56:933–941.

———. 1977. Species richness in streams—an alternative mechanism. American Naturalist 111:1017–1021.

Frank, J. H., and L. P. Lounibos (eds.). 1983. Phytotelmata: Terrestrial Plants as Hosts for Aquatic Insects. Plexus, Medford, New Jersey.

Franklin, J. F. 1987. Importance and justification in long term studies in ecology. Pages 3–19 in G. E. Likens (ed.), Long Term Studies in Ecology, Approaches and Alternatives. Springer-Verlag, New York.

Franklin, J. F., and D. S. DeBell. 1988. Thirty-six years of tree population change in an old-growth *Pseudotsuga-Tsuga* forest. Canadian Journal of Forest Research 18:633–639.

George, D. G., and G. P. Harris. 1985. The effect of climate on long-term changes in the crustacean zooplankton biomass of Lake Windermere, U.K. Nature 316:536–539.

Gibbs, H. L., and P. R. Grant. 1987. Ecological consequences of an exceptionally strong El Niño event on Darwin's finches. Ecology 68:1735–1746.

Gill, D. E., K. A Berven, and B. A. Mock. 1983. The environmental component of evolutionary biology. Pages 1–36 in C. E. King and P. S. Dawson (eds.), Population Biology Retrospect and Prospect. Columbia University Press, New York.

Gilpin, M. E., M. P. Carpenter, and M. J. Pomerantz. 1986. The assembly of a laboratory community: multispecies competition in *Drosophila*. Pages 23–40 in J. Diamond and T. J. Case (eds.), Community Ecology. Harper and Row, New York.

Godfray, H. C. J., and M. P. Hassell. 1992. Long time series reveal density dependence. Nature 359:673–674.

Goldwasser, L., J. Cook, and E. D. Silverman. 1994. The effects of variability on metapopulation dynamics and rates of invasion. Ecology 75:40–47.

Goodman, S. N. 1992. A comment on replication, *p*-values and evidence. Statistics in Medicine 11:875–879.

Gosz, J. R., G. E. Likens, and F. H. Bormann. 1972. Nutrient content of litter fall on the Hubbard Brook Experimental Forest, New Hampshire. Ecology 53:769–784.

Gray, C. A. 1991. Temporal variability in the demography of the palaemonid prawn *Macrobrachium intermedium* in two seagrasses. Marine Ecology Progress Series 75:227–237.

Hairston, N. G., Sr. 1989. Ecological Experiments: Purpose, Design, and Execution. Cambridge University Press, Cambridge.

Hall, D. J., W. E. Cooper, and E. E. Werner. 1970. An experimental approach to the production dynamics and structure of freshwater animal communities. Limnology and Oceanography 15:838–928.

Hanski, I. 1983. Coexistence of competitors in a patchy environment. Ecology 64:493–500.

Hanski, I., and I. Woiwood. 1991. Delayed density dependence. Nature 350:28.

Harmon, M. E., J. F. Franklin, F. J. Swanson, P. Sollins, S. V. Gregory, J. D. Lattin, N. H. Anderson, S. P. Cline, N. G. Aumen, J. R. Sedell, G. W. Lienkaemper, K. Cromack Jr. and K. W. Cummins. 1986. Ecology of coarse woody debris in temperate ecosystems. Environmental Conservation 11:11–18.

Hawke, D. J. 1992. Salinity variability in shelf waters near Otago Peninsula, New Zealand, on a time scale of hours. New Zealand Journal of Marine and Freshwater Research 26: 167–173.

Henrikson, L., H. G. Nyman, H. G. Oscarson, and J. A. E. Stenson. 1980. Trophic changes, without changes in the external nutrient loading. Hydrobiologia 68:257–263.

Hodar, J. A. 1993. Relaciones troficas entre los Passeriformes insectivoros de dos zonas semiaridas del sureste peninsular. Dissertation, Universidad de Granada, Granada, Spain.

Holling, C. S. 1992. Cross-scale morphology, geometry, and dynamics of ecosystems. Ecological Monographs 62:447–502.

Holt, R. D. 1993. Ecology at the mesoscale: influence of regional processes on local communities. Pages 77–88 in R. E. Ricklefs and D. Schluter (eds.), Species Diversity in Ecological Communities: Historical and Geographical Perspectives. University of Chicago Press, Chicago, Illinois.

Holyoak, M. 1994. Appropriate time scales for identifying lags in density dependent processes. Journal of Animal Ecology 63:479–483.

Hunter, M. D., and P. W. Price. 1992. Playing chutes and ladders: Bottom-up and top-down forces in natural communities. Ecology 73:724–732.

Hurlbert, S. H. 1984. Pseudoreplication and the design of ecological field experiments. Ecological Monographs 54:187–211.

Joern, A. 1992. Variable impact of avian predation on grasshopper assemblies in sandhills grassland. Oikos 64:458–463.

Jones, R. E. 1987. Ants, parasitoids, and the cabbage butterfly *Pieris rapae*. Journal of Animal Ecology 56:739–749.

Karl, D. M., and G. Tien. 1992. MAGIC: a sensitive and precise method for measuring dissolved phosphorus in aquatic environments. Limnology and Oceanography 37:105–116.

Kennelly, S. J., and A. J. Underwood. 1993. Geographic consistencies of effects of experimental physical disturbance on understory species in sublittoral kelp forests in central New South Wales. Journal of Experimental Marine Biology and Ecology 168:35–58.

Kitching, R. L. 1971. An ecological study of water-filled treeholes and their position in the woodland ecosystem. Journal of Animal Ecology 40:281–302.

———. 1987. Spatial and temporal variation in food webs in water-filled treeholes. Oikos 48:280–288.

Kotliar, N. B., and J. A. Wiens. 1990. Multiple scales of patchiness and patch structure: a hierarchical framework for the study of heterogeneity. Oikos 59:253–260.

Lauenroth, W. K., and O. E. Sala. 1992. Long-term forage production of North American shortgrass steppe. Ecological Applications 2:397–403.

Leigh, E. G. 1975. Population fluctuations, community stability and environmental variability. Pages 51–73 in M. L. Cody and J. M. Diamond (eds.), Ecology and Evolution of Communities. Harvard University Press, Cambridge, Massachusetts.

Levin, S. A. 1974. Dispersion and population interactions. American Naturalist 108:207–228.

———. 1978. Population models and community structure in heterogeneous environments. Pages 439–476 in S. A. Levin (ed.), Studies in Mathematical Biology: Part II. Populations and communities. Studies in mathematics 16. Mathematical Association of America, Washington, D.C.

———. 1992. The problem of pattern and scale in ecology. Ecology 73:1943–1967.

Lightfoot, D. C., and W. G. Whitford. 1991. Productivity of creosote bush foliage and associated canopy arthropods along a desert roadside. American Midland Naturalist 125:310–322.

Likens, G. E. 1983. A priority for ecological research. Bulletin of the Ecological Society of America 64:234–243.

Losos, J. B., J. C. Marks, and T. W. Schoener. 1993. Habitat use and ecological interactions of an introduced and a native species of *Anolis* lizard on Grand Cayman, with a review of the outcomes of anole introductions. Oecologia 95:525–532.

Losos, J. B., K. L. Warheit, and T. W. Schoener. 1997. Adaptive differentiation following experimental island colonization in *Anolis* lizards. Nature 387:70–73.

Louda, S. M. 1982. Distribution ecology: variation in plant recruitment over a gradient in relation to insect seed predation. Ecological Monographs 52:25–41.

MacMahon, J. A. 1980. Ecosystems over time: succession and other types of change. Pages 26–58 in R. H. Waring (ed.), Forests: Fresh Perspectives from Ecosystem Analysis. Oregon State University Press, Corvalis.

MacMahon, J. A., and F. H. Wagner. 1985. The Mojave, Sonoran and Chihuahuan deserts of North America. Pages 105–202 in M. Evenari (ed.), Hot Deserts and Arid Shrublands. Elsevier, Amsterdam.

MacNally, R. C. 1995. Ecological Versatility and Community Ecology. Cambridge University Press, Cambridge.

Mandelbrot, B. B. 1983. The Fractal Geometry of Nature. W. H. Freeman, New York.

McIntosh, R. P. 1985. The Background of Ecology: Concept and Theory. Cambridge University Press, Cambridge.

McLaughlin, J. F., and J. Roughgarden. 1993. Species interactions in space. Pages 89–98 in R. E. Ricklefs and D. Schluter (eds.), Species Diversity in Ecological Communities: Historical and Geographical Perspectives. University of Chicago Press, Chicago, Illinois.

McQueen, D. J., M. R. Johannes, J. R. Post, and T. J. Stewart. 1989. Bottom up and top down impacts on fresh water pelagic community structure. Ecological Monographs 59: 289–309.

Meehl, P. E. 1967. Theory-testing in psychology and physics: a methodological paradox. Philosophy of Science 34:103–115.

Menge, B. A. 1992. Community regulation: under what conditions are bottom-up factors important on rocky shores? Ecology 73:755–765.

Menge, B. A., and A. M. Olson. 1990. Role of scale and environmental factors in regulation of community structure. Trends in Ecology and Evolution 5:52–57.

Morris, D. W. 1987. Ecological scale and habitat use. Ecology 68:362–369.

Naeem, S. 1990a. Patterns of the distribution and abundance of competing species when resources are heterogeneous. Ecology 71:1422–1429.

———. 1990b. Resource heterogeneity and community structure: a case study in *Heliconia imbricata phytotelmata*. Oecologia 84:29–38.

Oakes, M. 1986. Statistical Inference: A Commentary for the Social and Behavioural Sciences. Wiley, Chichester.

Oksanen, L., and T. Oksanen. 1981. Lemmings (*Lemmus lemmus*) and grey-sided voles (*Clethrionomys rufocanus*) in interaction with their resources and predators on Finnmarksvidda, northern Norway. Reports of the Kevo Subarctic Research Station 17:7–31.

O'Neill, R. V., R. Gardner, B. T. Milne, M. G. Turner, and M. Jackson. 1991. Heterogeneity and spatial hierarchies. Pages 85–96 in J. Kolasa and S. T. A. Pickett (eds.), Ecological Studies: Vol. 86. Ecological Heterogeneity. Springer-Verlag, New York.

Ostfield, R. S., and C. D. Canham. 1995. Density dependent processes in meadow voles: an experimental approach. Ecology 76:521–532.

Paine, R. T. 1992. Food-web analysis through field measurement of per capita interaction strength. Nature 355:73–75.

Palmer, M. W., and P. S. White. 1994. Scale dependence and the species area relationship. American Naturalist 144:717–740.

Parmenter, R. R., and J. A. MacMahon. 1988. Factors limiting populations of arid-land darkling beetles (Coleoptera: Tenebrionidae): predation by rodents. Environmental Entomology 17:280–286.

Peet, R. K., and N. L. Christensen. 1987. Competition and tree death. BioScience 37:586–596.

Peterson, R. O. 1987. Ecological Studies of Wolves on Isle Royale. Annual Report 1986–1987. Michigan Technical University, Houghton.

Peterson, R. O, R. E. Page, and K. M. Dodge. 1984. Wolves, moose, and the allometry of population cycles. Science 224:1350–1352.

Poincaré, J. H. 1946. The Foundations of Science. Science Press, Lancaster, Pennsylvania.

Polis, G. A. 1980. The effect of cannibalism on the demography and activity of a natural population of desert scorpions. Behavioral Ecology and Sociobiology 7:25–35.

———. 1990. Ecology. Pages 247–293 in G. A. Polis (ed.), Biology of Scorpions. Stanford University Press, Stanford, California.

———. 1991a. Complex trophic interactions in deserts: an empirical critique of food web theory. American Naturalist 138:123–155.

————. 1991b. Desert communities: an overview of patterns and processes. Pages 1–26 in G. A. Polis (ed.), The Ecology of Desert Communities. University of Arizona Press, Tucson.

————. 1993. The ecological importance of desert scorpions. Memoirs of the Queensland Museum 33:401–410.

————. 1994. Food webs, trophic cascades and community structure. Australian Journal of Ecology 19:121–136.

Polis, G. A., and R. D. Holt. 1992. Intraguild predation: the dynamics of complex trophic interactions. Trends in Ecology and Evolution 7:151–154.

Polis, G. A., and S. D. Hurd. 1995b. Extraordinarily high spider densities on islands: flow of energy from the marine to terrestrial food webs and the absence of predation. Proceedings of the National Academy of Science of the USA 92:4382–4386.

————. 1996. Linking marine and terrestrial food webs: allochthonous input from the ocean supports high secondary productivity on small islands and coastal land communities. American Naturalist 147:396–423.

Polis, G. A., and S. J. McCormick. 1987. Intraguild predation and competition among desert scorpions. Ecology 68:332–343.

Polis, G. A., and D. Strong. 1996. Food web complexity and community dynamics. American Naturalist 147:813–842.

Polis, G. A., C. A. Myers, and R. Holt. 1989. The ecology and evolution of intraguild predation: potential competitors that eat each other. Annual Review of Ecology and Systematics 20:297–330.

Polis, G. A., W. B. Anderson, and R. D. Holt. 1997a. Towards an integration of landscape and food web ecology: the dynamics of spatially subsidized food webs. Annual Review of Ecology and Systematics 29:289–316.

Polis, G. A., S. D. Hurd, C. T. Jackson, and F. Sanchez-Piñero. 1997b. El Niño effects on the dynamics and control of a terrestrial island ecosystem in the Gulf of California. Ecology 78:1884–1897.

Polis, G. A., S. D. Hurd, C. T. Jackson, and F. Sanchez-Piñero. In press. Multifactor population limitation: variable spatial and temporal control of spiders on Gulf of California islands. Ecology.

Polis, G. A., R. D. Holt, B. A. Menge, and K. O. Winemiller. 1995. Time, space and life history: influences on food webs. Pages 435–460 in G. A. Polis and K. O. Winemiller, (eds.), Food Webs: Integration of Patterns and Dynamics. Chapman and Hall, New York.

Pounds, J. A., and M. L. Crump. 1994. Amphibian declines and climate disturbance: the case of the golden toad and the harlequin frog. Conservation Biology 8:72–85.

Power, M. E. 1992. Top down and bottom up forces in food webs: do plants have primacy? Ecology 73:733–746.

Power, M. E., W. J. Matthews, and A. J. Stewart. 1985. Grazing minnows, piscivorous bass, and stream algae: dynamics of a strong interaction. Ecology 66:1448–1456.

Quinn, W. H., and V. T. Neal. 1983. Long-term variations in the Southern Oscillation, El Niño, and Chilean subtropical rainfall. Fishery Bulletin 81:363–374.

Ray, C., and A. Hastings. 1996. Density dependence: are we searching at the wrong spatial scale? Journal of Animal Ecology. 65:556–566.

Ricklefs, R. E. 1990. Ecology. W. H. Freeman, New York.

Ricklefs, R. E., and D. Schluter. 1993. Species diversity: an introduction to the problem. Pages 1–10 in R. E. Ricklefs and D. Schluter (eds.), Species Diversity in Ecological Communities: Historical and Geographical Perspectives. University of Chicago Press, Chicago, Illinois.

Risser, P. G., E. C. Birney, H. D. Blocker, S. W. May, W. J. Parton, and J. A. Wiens. 1981. The True Prairie Ecosystem. Hutchinson Ross, Stroudsburg, Pennsylvania.

Robinson, G. R., R. D. Holt, M. S. Gaines, S. P. Hamburg, E. A. Martinko, and S. H. Fitch. 1992. Diverse and contrasting effects of habitat fragmentation. Science 257:524–526.

Root, R. B., and N. Cappuccino. 1992. Patterns in population change and the organization of the insect community associated with goldenrod. Ecological Monographs 62:393–420.

Rotenberry, J. T., and J. A. Wiens. 1985. Statistical power analysis and community-wide patterns. American Naturalist 125:164–168.

Roughgarden, J., S. D. Gaines, and S. W. Pacala. 1987. Supply side ecology: the role of physical transport processes. Pages 491–518 in J. H. R. Gee and P. S. Giller (eds.), Organization of Communities: Past and Present. Blackwell, Oxford.

Sale, P. F., and R. Dybdahl. 1975. Determinants of community structure for coral reef fishes in an experimental habitat. Ecology 56:1343–1355.

Schaffer, W. M. 1985. Order and chaos in ecological systems. Ecology 66:93–106.

Schindler, D. W. 1977. Evolution of phosphorus limitation in lakes. Science 195:260–262.

———. 1987. Detecting ecosystem response to anthropogenic stress. Canadian Journal of Fisheries and Aquatic Science 44 (Suppl. 1): 6–25.

———. 1990. Experimental perturbations of whole lakes as tests of hypotheses concerning ecosystem structure and function. Oikos 57:24–41.

Schoener, T. W. 1983. Field experiments on interspecific competition. American Naturalist 122:240–285.

———. 1988. Leaf damage in island buttonwood, *Conocarpus erectus*: correlation with pubescence, island area, isolation and the distribution of major carnivores. Oikos 53: 253–266.

Schroder, G. D., and M. L. Rosenzweig. 1975. Perturbation analysis of competition and overlap in habitat utilization between *Dipodomys ordii* and *Dipodomys merriami*. Oecologia 19:9–28.

Seely, M. K. 1991. Sand dune communities. Pages 348–382 in G. A. Polis (ed.), The Ecology of Desert Communities. University of Arizona Press, Tucson.

Senn, S. 1994. Fisher's game with the devil. Statistics in Medicine 13:217–230.

Service, S. K., and R. J. Feller. 1992. Long-term trends of subtidal macrobenthos in North Inlet, South Carolina. Hydrobiologia 231:13–40.

Sih, A., P. Crowley, M. McPeek, J. Petranka, and K. Strohmeier. 1985. Predation, competition, and prey communities: a review of field experiments. Annual Review of Ecology and Systematics 16:269–311.

Simberloff, D. S., and E. O. Wilson. 1969. Experimental zoogeography of islands: defaunation and monitoring techniques. Ecology 50:267–278.

Smith, D. E., 1983. Factors controlling tadpole populations of the chorus frog (*Pseudacris triseriata*) on Isle Royale, Michigan. Ecology 64:501–510.

Smith, R. L. 1981. The trouble with "bobos," *Paraleucopsis mexicana* Steyskal, at Kino Bay, Sonora, Mexico (Diptera: Chamaemyiidae). Proceedings of the Entomological Society of Washington 83:406–412.

———. 1990. Ecology and Field Biology. HarperCollins, New York.

Sousa, W. P. 1984. The role of disturbance in natural communities. Annual Review of Ecology and Systematics 15:353–391.

Southwood, T. R. E., and P. M. Reader. 1976. Population census data and key factor analysis for the viburnum whitefly, *Aleurotrachelus jelinekii* (Frauenf.) on three bushes. Journal of Animal Ecology 45:313–325.

Southwood, T. R. E., M. P. Hassell, P. M. Reader, and D. J. Rogers. 1989. Population dynamics of the viburnum whitefly (*Aleurotrachelus jelinekii*). Journal of Animal Ecology 58:921–942.

Spiller, D. A., and T. W. Schoener. 1994. Effects of top and intermediate predators in a terrestrial food web. Ecology 75:182–196.

Strauss, S. Y. 1991. Indirect effects in community ecology: their definition, study and importance. Trends in Ecology and Evolution 6:206–210.

Strong, D. R., Jr., D. Simberloff, L. G. Abele, and A. B. Thistle (eds.). 1984. Ecological Communities: Conceptual Issues and the Evidence. Princeton University Press, Princeton, New Jersey.

Tilman, D. 1988. Ecological experimentation: strengths and conceptual problems. In G. E. Likens (ed.), Long Term Studies in Ecology, Approaches and Alternatives. Springer-Verlag, New York.

———. 1994. Competition and biodiversity in spatially structured habitats. Ecology 75:2–16.

Toft, C. A., and P. J. Shea. 1983. Detecting community-wide patterns: estimating power strengthens statistical inference. American Naturalist 122:618–625.

Tovar, H., V. Guillén, and D. Cabrera. 1987. Reproduction and population levels of Peruvian guano birds, 1980 to 1986. Journal of Geophysical Research 92:14445–14448.

Turchin, P. 1990. Rarity of density dependence or population regulation with lags? Nature 344:660–663.

Tversky, A., and D. Kahneman. 1971. Belief in the law of small numbers. Psychological Bulletin 76:105–110.

Underwood, A. J., and P. S. Petraitis. 1993. Structure of intertidal assemblages in different locations: how can local processes be compared? Pages 39–51 in R. E. Ricklefs and D. Schluter (eds.), Species Diversity in Ecological Communities: Historical and Geographical Perspectives. University of Chicago Press, Chicago, Illinois.

Underwood, T. 1986. The analysis of competition by field experiments. Pages 240–268 in J. Kikkawa and D. J. Anderson (eds.), Community Ecology: Pattern and Process. Blackwell, Melbourne.

Van Buskirk, J. 1992. Competition, cannibalism, and size class dominance in a dragonfly. Oikos 65:455–464.

———. 1993. Population consequences of larval crowding in the dragonfly *Aeschna juncea*. Ecology 74:1950–1958.

Vandermeer, J., B. Hazlett, and B. Rathcke. 1985. Indirect facilitation and mutualism. Pages 326–343 in D. H. Boucher (ed.), The Biology of Mutualism. Oxford University Press, Oxford.

Virkkala, R. 1993. Ranges of northern forest passerines: a fractal analysis. Oikos 67:218–226.

Ward, D., and D. Saltz. 1994. Foraging at different spatial scales: Dorcas gazelles foraging for lilies in the Negev Desert. Ecology 75:48–58.

Wiens, J. A. 1977. On competition and variable environments. American Scientist 65:590–597.

———. 1986. Spatial scale and temporal variation in studies of shrubsteppe birds. Pages 154–172 in J. Diamond and T. J. Case (eds.), Community Ecology. Harper and Row, New York.

———. 1989a. Spatial scaling in ecology. Functional Ecology 3:385–397.

———. 1989b. The Ecology of Bird Communities: Vol. 2 Processes and Variations. Cambridge University Press, Cambridge.

Wiens, J. A., J. F. Addicott, T. J. Case, and J. Diamond. 1986. Overview: the importance of spatial and temporal scale in ecological investigations. Pages 145–153 in J. Diamond and T. J. Case (eds.), Community Ecology. Harper and Row, New York.

Wilbur, H. M., P. J. Morin, and R. N. Harris. 1983. Salamander predation and the structure of experimental communities: anuran responses. Ecology 64:1423–1429.

Wise, D. H. 1981. A removal experiment with darkling beetles: lack of evidence for inter-specific competition. Ecology 62:727–738.

———. 1984. The role of competition in spider communities: insights from field experiments with a model organism. Pages 42–52 in D. R. Strong Jr., D. Simberloff, L. G. Abele, and A. B. Thistle (eds.), Ecological Communities: Conceptual Issues and the Evidence. Princeton University Press, Princeton, New Jersey.

———. 1993. Spiders in Ecological Webs. Cambridge University Press, Cambridge.

Yodzis, P. 1996. Food webs and perturbation experiments: theory and practice. Pages 192–200 in G. A. Polis and K. O. Winemiller (eds.), Food Webs: Integration of Patterns and Dynamics. Chapman and Hall, New York.

14

Using Models to Enhance the Value of Information from Observations and Experiments

ELIZABETH A. MARSCHALL & BERNADETTE M. ROCHE

Ecologists use a number of approaches to understand how complex ecological processes and interactions affect population dynamics. A straightforward approach is to design manipulative experiments that directly measure effects on populations. This requires a replicated manipulation done at an appropriate spatial scale, usually one that encompasses a large portion of a population, and an appropriate temporal scale, usually one that encompasses several generations of a population. This approach limits the populations we can study to those that occur over a small area and those that have short generation times. Some ecologists have recognized these limitations and have chosen to study community and trophic dynamics at small scales (e.g., Morin, Lawler studies of patterns in microbial communities, this volume). Others are successfully taking this direct approach in larger systems (e.g., Brown, studies of desert ecosystems, this volume). But as systems get bigger we lose the ease of replication, and as life cycles get longer we lose the ease of encompassing multiple generations. Our ability to conduct replicated manipulative experiments at large spatial scales and long time scales to assess impacts on whole populations is severely limited by time, systems, and money.

An alternative approach combines models that represent large spatial or temporal scales with smaller scale experiments. Investigators have combined models and experiments in a variety of ways, but in this essay we address specifically the idea of using models to explore cause and effect and experiments to set the relevance of these patterns in a particular system. This may involve either (1) using experiments to interpret model results or (2) using models to assess importance of experimental results. In our examples, we will concentrate on the latter, but these two approaches have in common the idea of combining small-scale (in either time, space, or both) experiments with models addressing processes at larger scales. We begin by measuring some treatment response in experiments and then use models to interpret the importance of these responses at a larger spatial or temporal scale.

Ecologists often assume that strong responses by individuals (e.g., survival, growth rate, or preference responses) in experiments indicate a strong response at the population level to the particular treatment of interest, but this can be misleading. For example, to measure competitive interactions between individuals of two species we often use individuals growth rates and survival probabilities as responses. When we see that the presence of species A causes greatly reduced growth or survival rates of individuals of species B, we naturally conclude that species A has a negative impact on the population of species B. But without understanding what drives the dynamics of species B, we cannot really know whether a reduction in growth or survival, even a strong one, at a particular life stage really has any impact at the population level. Here we present several examples of how coupling population dynamic models with empirical observations has progressed our understanding of what drives population and cohort dynamics in a variety of systems.

Combining Models with Experiments to Understand Population Dynamics

A model is nothing more than a set of assumptions organized to answer specific questions. Models are determined entirely from model assumptions, and model results follow exactly from the model. In this essay, we address the use of empirical information to guide the design of the model, the use of the model to see how patterns in this information lead to patterns in results, and then a return to empirical information to limit the scope of these model results to something applicable to a particular system (Fig. 14-1). As an illustration of how we are using models, we will step through an example addressed in more detail later.

Empirical studies of the interaction between a piscivorous fish (southern flounder, *Paralichthys lethostigma*) and its prey (spot, *Leiostomus xanthurus*) have provided much information on size-specific vulnerability of prey as a function of predator size, growth rates of both prey and predator, and size distributions of prey cohorts under different predation regimes (Barker 1991, Rice et al. 1993, Wright et al. 1993). If prey-to-predator size ratios drive prey vulnerability and if both prey and predator change sizes over time, then the predator will have a dynamic effect on prey cohort size distribution ("General Assumptions," Fig. 14-1). Our goal in this system is to understand cause (e.g., vulnerability as a function of relative size and prey and predator growth rates) and effect (prey cohort size distribution at different points throughout the growing season). To what extent do each of the pieces (and their interactions) drive prey cohort size distribution at the end of the growing season (steps 3–5, Fig. 14-1)? In theory it is possible to answer this question through a large factorial set of experiments, but in reality it is not feasible. Instead, we put all this empirically derived knowledge into a model that allows us to vary each piece individually (in a factorial design of model runs) to assess to what extent each piece drives prey size distribution. The model merely organized our assumptions about how the system works and followed the implications of these assumptions over time. The model can include any values of growth rate and any functions of prey vulnerability (i.e., any set of assumptions). Thus, we can find general cause-and-effect patterns for systems of this type that it would not be possible to derive experimentally for a single system. Once these patterns have been

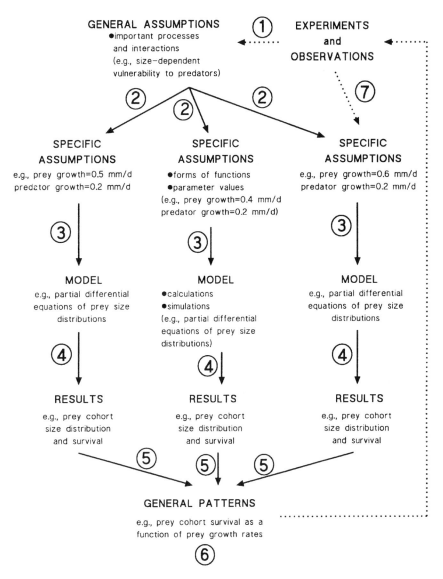

Figure 14-1. Flow diagram of approach that combines models with experiments. Numbers refer to the order in which we refer to each step in our presentation. In our examples, we have focused on steps 1–5.

identified, we need to limit the model assumptions to those values and functions that are representative of the particular system of interest (step 7, Fig. 14-1)—that is, we need to use empirical information to set the relevance of our general model results to our specific system. The model results help us identify which of these assumptions are important to know precisely by showing us which processes and parameters drive the results. In this essay, we concentrate on the first portion of this approach: moving from empirical information to models to understand general patterns (steps 1–5, Fig. 14-1). We demonstrate the utility of this approach by describing examples in which empirical observations at the level of individual organisms or classes of individuals were used to provide a basis for models at the next higher level of organization—the population (see Werner, this volume). These examples show how models can provide insight when experimental studies fall short of testing population-level effects.

Example: Vector Preferences and Pathogen Dynamics

Background

The role of vector behavior in mediating host–parasite interactions has been largely ignored by evolutionary ecologists, with a few notable exceptions (see Kingsolver 1987, Real et al. 1992, Thrall et al. 1993, McElhany et al. 1995). As part of a larger project that investigated the interaction between *Silene alba* (white campion, a weedy perennial plant) and an anther-smut fungus (*Ustilago violacea*), we studied preference behavior of the insect vector of this pathogen and its impact on the epidemiology of host–pathogen interactions. Both male and female *S. alba* infected with *U. violacea* produce anther sacs filled with teliospores of the fungus. Thus, all diseased flowers have the appearance of male flowers with fungal spores replacing pollen grains on the anthers. Flower-visiting insects act as vectors of the disease when they visit diseased flowers and subsequently visit healthy plants (Baker 1947). Diseased flowers differ from healthy flowers in appearance (fungal spores produced in place of pollen are dark purple). Alexander (1990) demonstrated variation in preferences for diseased and healthy plants among individual bumblebees (*Bombus* spp.) foraging in an experimental population of *S. alba*. Our intuition (and that of others before us; e.g., Clay 1987) was that a vector preference for diseased plants would likely enhance the rate at which disease would spread through a *Silene* population and a vector preference for healthy (noninfected) plants would inhibit the rate of disease spread. The strength of the preference, we assumed, would be positively correlated with the strength of the impact of the preference. Thus, we set out to characterize the extent of these vector preferences for healthy or diseased plants through observation and then built an epidemiological model to quantify how preferences of the observed magnitude influence the rate of disease spread in the plant population.

Observations

Bumblebee foraging was observed in an experimental population of healthy and diseased *S. alba* established by Alexander and Antonovics (1995) at Mountain Lake Biological Station (Giles County, Virginia) in 1988. Plants were spaced 0.6 m apart in a hexagonal array with 32 rows and 33 columns for a total of 1,056 possible plant locations. Plants were positioned randomly with respect to disease status and sex. Not all plants were in flower at the time of the study, but there were an average of 654 plants in flower during our observations. The sex and disease status (healthy or diseased) of each flowering plant in the population at the time of each foraging observation were known.

To start each observation, we randomly chose a row and column in the hexagonal array and waited at that location until the first honeybee or bumblebee appeared within approximately a 5-m radius. We followed that individual until it flew out of the plot or disappeared from view, noting disease status and sex of each plant it visited. To determine whether bees were foraging randomly with respect to disease status of plants in the population, we calculated the expected number of visits to healthy and diseased plants based on the total number of flowering plants in each category at the time of the observation.

Bees exhibited nonrandom foraging with respect to three plant categories (healthy male, healthy female, and diseased). Figure 14-2 illustrates data from eight individual foraging sequences observed on one day (July 1988). Most individuals (75%) avoided diseased plants; a few individuals (25%) seemed to be attracted to diseased plants (Fig. 14-2). Although biases were strong, they did not determine all plant visits: individuals showing a bias toward diseased plants also visited healthy plants, and individuals with a bias for healthy male plants also visited healthy female and diseased plants. Thus, either the preference for disease was not absolute or the expression of the preference was imperfect. We built a model to assess the impact of a given individual's variability in plant visits—that is, the impact of this "incomplete" preference.

Epidemiological Model

To assess propensity for a particular behavior to enhance rate of disease spread, we modeled the probability that an insect's next visit would result in an increase in disease occurrence in the plant population under different preference regimes. The probability

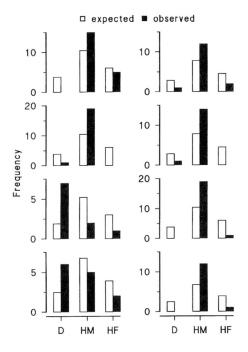

□ expected ■ observed

Figure 14-2. Expected (*open bars*) and observed (*solid bars*) numbers of visits to each of three types of *Silene alba* plants (D = diseased, HM = healthy male, HF = healthy female) by individual bumblebees. Only data from observations of bumblebees visiting a series of at least 10 plants were used. Observed frequencies were different from expected frequencies for all bumblebees (χ^2 test, significance level = .05) except one (*upper right corner of figure*, P = .09). Individuals represented in the two graphs in the lower left corner of the figure visited diseased plants out of proportion to their occurrence in the population. All other individuals apparently avoided diseased plants and visited healthy male plants more frequently than would be expected by chance.

that the next visit would result in a new infection is equal to the probability that the next visit would be to an uninfected plant times the probability that infection occurs in this visit to a healthy plant. The probability of a vector visiting a healthy or diseased plant is dependent on both the proportion of healthy and diseased plants in the population and the vector's preferences. Vectors are assumed to encounter healthy and diseased plants according to their proportional representation in the population, but at each encounter a vector chooses to visit based on its preferences. We let ρ_D = the probability that the next encounter will be with a diseased plant (i.e., ρ_D = the proportion of the population consisting of diseased plants) such that the probability that an encounter is with a healthy plant is $1 - \rho_D$. The probability of visiting a diseased plant given an encounter is η_D, and the probability of visiting a healthy plant given its encounter is η_H. The probability δ_D that a particular visit will be to a diseased plant is the probability of an encounter and visit to a diseased plant as a proportion of the probability of an encounter and visit to either a diseased or healthy plant.

$$\delta_D = \frac{\rho_D \eta_D}{(1 - \rho_D)\eta_H + \rho_D \eta_D} \tag{1}$$

Then the probability δ_H that a particular visit is to a healthy plant is just $1 - \delta_D$.

The last probability we need is the probability that infection occurs when the current visit to a healthy plant is the tth visit to a healthy plant since the last visit to a diseased plant. We assume the probability of transmission declines exponentially and is given by $e^{-\gamma(t-1)}$, where γ is a parameter that controls rate of transmission decay.

We can now calculate the probability that the next visit results in a new infection as

$$\sum_{t=1}^{\infty} \delta_D \delta_H^t \, e^{-\gamma(t-1)} = \frac{\delta_D \delta_H}{1 - \delta_H \, e^{-\gamma}} \tag{2}$$

Using this expression, we explored the rate of disease spread over a continuum of disease concentrations from rare to common, at three rates of transmission decay ($\gamma = 0.001$, $\gamma = 0.1$, and $\gamma = 0.7$) and for three classes of preferences (prefer healthy, prefer disease, and no preference). We mimicked the individual preference patterns observed in our experiments by modeling preferences as "imperfect"—that is, the probability of visiting a plant of nonpreferred disease status given an encounter with that plant was set at .1 rather than zero. We refer to foragers that have a preference for diseased plants as "spore-seeking" and those that have a preference for healthy plants as "spore-avoiding." This terminology was chosen for ease of discussion and was not meant to suggest an understanding of what drives the behavioral preferences in these foragers.

Probability of disease transmission in the next visit (eq. 2) is high for spore-seeking foragers at low concentrations of disease and decreases as diseases concentrations increase (dashed lines, Fig. 14-3). Of course, when disease concentration is zero, the probability of transmission is also zero, but this probability increases very quickly with only a slight increase in disease concentration. For foragers that prefer healthy plants, there is a low probability of transmission during the next visit at low concentrations of the disease, but this probability increases with increasing disease concentration (dotted lines, Fig. 14-3). As a consequence, at high disease concentrations a preference for

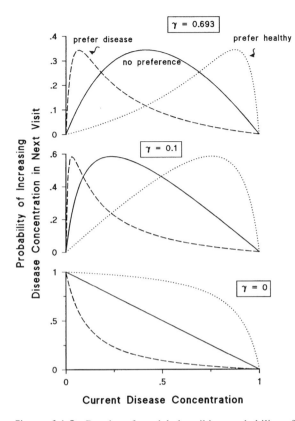

Figure 14-3. Results of model describing probability of
increasing disease concentration in next visit as a function
of current disease concentration (ρ_D) at three levels of γ
(eq. 3). The rate of transmission decay decreases from the
top panel to the bottom panel. (A value of $\gamma = 0$ would
represent the extreme case in which there is no loss of
transmission effectiveness.) Results assume foragers have a
preference for diseased plants ("spore-seeking") (*dashed
lines*), a preference for healthy plants ("spore-avoiding")
(*dotted lines*), or no preference (*solid line*).

healthy plants will more likely result in a transmission than a preference for diseased
plants (compare dotted lines to dashed lines at high disease concentration, Fig. 14-3).
Thus, we are presented with the somewhat surprising result that under certain circumstances attractiveness of diseased plants can actually *reduce* the probability of disease
spread.

The tendency for spore-avoiding foragers to cause a greater increase in pathogen
spread than spore-seeking foragers at even moderate disease concentrations (compare
dotted to dashed lines, Fig. 14-3) or at low levels of transmission decay (bottom panel,
Fig. 14-3) is a result of the differential "carryover" effect of a "mistake" by each of

these two types of foragers. The preference of the vector affects the impact of a single foraging mistake. If a spore-seeking insect makes a mistake, it visits a healthy plant and has a high probability of infecting it. But there is a low probability of the next visit also being a mistake, and so the outcome of a single mistake is generally a single transmission.

In foragers seeking healthy plants, however, a single mistake has a longer carryover effect. If this type of forager makes a mistake, it visits a diseased plant. Again, there is a low probability that the next visit will also be a mistake, and so the next visit will probably be to a healthy plant, resulting in a high probability of transmission. Because the forager prefers healthy plants, it likely will continue visiting them and will continue to have a positive (though declining) probability of infecting them with the pathogen. So a single mistake by a spore-avoiding forager potentially has a greater impact than a single mistake by a spore-seeking forager.

There is also an interactive effect of vector preference and initial disease concentration on the probability of disease spread. This is a consequence of the interaction between disease concentration and probability of making a mistake. If a forager is seeking healthy plants, it still has some low probability of visiting a diseased plant. As the concentration of disease increases, the probability of mistakenly visiting a diseased plant increases, but the probability of subsequently visiting a healthy plant decreases. If we begin at a low concentration of disease, then a given increase in disease concentration will have a greater proportional impact on probability of visiting a diseased plant than on probability of visiting a healthy plant. Consequently, at low disease concentrations we should expect the probability of transmission, which requires a visit to a diseased plant followed by a visit to a healthy plant, to increase with increasing disease concentration (see dotted line, Fig. 14-3).

Although the value of the parameter γ of the negative exponential probability of transmission, given encounter with disease t time units ago, does not affect the qualitative relationship of the curves, it does affect where they cross and how quickly they drop to zero at extremely low and high disease concentrations (compare panels in Fig. 14-3). If this parameter is small (i.e., little decay in transmission rate; bottom panel, Fig. 14-3), then foragers with preferences for healthy plants will cause a higher rate of disease spread than those with preferences for diseased plants at almost all disease concentrations. At large values, however (i.e., rapid decay in transmission rate; top panel, Fig. 14-3), we find that spore-seeking foragers spread the pathogen more rapidly than spore-avoiding foragers at low to moderate concentrations of disease.

Although we began behavioral observations believing that vector preference for diseased plants would enhance disease spread, a model at a longer time scale based on observations that included behavioral variability produced initially counterintuitive results. Ultimately, we could explain in words this effect of vector preference on disease spread (i.e., the initially counterintuitive became intuitive), but the model enabled us to integrate the original set of observations to elucidate the "cause-and-effect" patterns (steps 1 through 6, Fig. 14-1). The actual ecological impact of vector preferences can only be understood by measuring disease concentration and rate of transmission decay (step 7, Fig. 14-1), a result obtained from combining simple observational results with a simple model to "step up" to the next level of organization.

Example: Population Dynamics and Stage-Specific Vital Rates

Observations

The case of the decline of sea turtle populations and the ability of population biologists to have an impact on how these animals are managed (e.g., Crouse et al. 1987, Crowder et al. 1994) is becoming a classic and often cited example of how basic ecology (and, specifically, matrix population models) can serve conservation biology (Heppell et al. in press). It is also an excellent example of how empirical observations alone may mislead us as to what is driving a population's dynamics.

Observations of decreasing numbers of nests and nesting females in the 1970s indicated that certain loggerhead sea turtle (*Caretta caretta*) populations were in steady decline (National Research Council 1990). Data also indicated that hatching mortality was extremely high, due to both natural and anthropogenic causes (National Research Council 1990). Mortality at the egg/hatchling stage is naturally high due to predators on the beach and in the water (Dodd 1988), but human activity may also enhance predator densities through species introductions and food supplementation. In addition, commercial and residential development of nesting beaches can affect egg and hatchling survival through a number of other pathways, including erosion (loss of nesting habitat), compaction of sand by pedestrians and motor vehicles (Mann 1977), and hatchling disorientation due to artificial lighting (McFarlane 1963). Given the magnitude of egg/hatchling mortality, it seemed logical to concentrate conservation efforts on this life history stage. Biologists and conservationists initiated extensive programs of nesting beach preservation, nest and nesting female protection, hatchling protection, and hatchling ''head-starting'' (raising hatchlings in captivity until they are past high mortality stages; National Research Council 1990). However, biologists had also observed sea turtle mortality related to accidental drowning in shrimp trawls (Murphy and Hopkins-Murphy 1989). Survival rates of large juveniles and adults were being affected, though apparently not as severely as the egg and hatchling stages on nesting beaches.

Model

Crouse et al. (1987) built a Lefkovitch model, a stage-based population projection matrix model, from a life table for loggerhead sea turtles nesting in Georgia (Frazer 1986). Crowder et al. (1994) revised this model, partitioning the turtle population into five life stages (eggs/hatchings, small juveniles, large juveniles, subadults, and adults) and using transition rates between these stages from empirical estimates of loggerhead vital rates (Frazer 1983, 1987; Crowder et al. 1994). The resulting population projection matrix (Fig. 14-4, top panel) consisted of constant annual transition rates between life stage classes. This model includes no density dependence in transition rates, but is believed to be an adequate representation of a sea turtle population known to be at low population levels (an example of general assumptions in Fig. 14-1). Using the matrix of transition rates, the sea turtle population was projected over time until a stable stage distribution was reached (i.e., until distribution of the population among life stage classes did not change with time). We refer to population growth rate, stage class

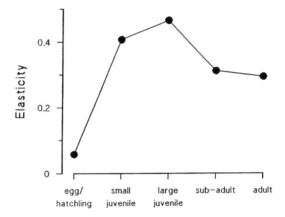

Projection Matrix

eggs/hatchlings	0	0	0	4.665	61.896
small juveniles	.675	.703	0	0	0
large juveniles	0	.047	.657	0	0
sub–adults	0	0	.019	.682	0
adults	0	0	0	.061	.8091

Figure 14-4. Population projection matrix for loggerhead sea turtles (Crowder et al. 1994) and resulting patterns of sensitivity of asymptotic population growth rates to changes in annual survival within each life stage. The value in row j and column i (element a_{ij} represents the transition rate from an individual in class i to class j in one year. Sensitivity is calculated as "elasticity," which is the proportional change in population growth rate given a proportional change in specific survival parameters (a_{ji}/λ) (dλ/da_{ji}, Caswell 1989). This measure of sensitivity allows direct comparisons between parameters of very different magnitudes—for example, comparisons between changes in survival rates (range of 0 to 1) and changes in fecundities (range of 0 to 62). Data from Crowder et al. (1994).

distribution, and stage-specific reproductive values at this time as "asymptotic" values of these responses.

From the resulting population projection matrix (Fig. 14-4), Crowder et al. (1994) measured the sensitivity (by calculating "elasticity"; deKroon et al. 1986, Caswell 1989) of asymptotic population growth rate to changes in vital rates (stage-specific growth, survival, and fecundity) (an example of varying specific assumptions, Fig. 14-1). Population growth rates were most sensitive to changes in annual survival rates within each stage rather than to fecundity values or changes in transition rates between stages (Crowder et al. 1994). Of these stage-specific annual survival rates, population growth rates were most sensitive to changes in survival rates associated with large juvenile turtles and only mildly sensitive to changes in survival in the egg/hatchling stage (Fig. 14-4). In fact, setting survival in this youngest stage at 100% still resulted in a population decline. Sensitivity of asymptotic population growth rate to changes in rate of transition between two stage classes i and j is determined by a combination of the asymptotic proportional representation of stage class i and the asymptotic reproductive values of stage class j. In other words, the impact of a specific transition rate is a result of how many individuals this rate applies to (number in stage i) and the future value of each of these individuals in their new stage class (reproductive value in stage j). It is the high reproductive value of these intermediate stages that gives them such a large impact on the population growth rate.

These results caused biologists to question whether targeting conservation efforts at increasing survival of eggs and hatchlings was the most effective way of improving the outlook for the population (National Research Council 1990). Although mortality rates at these young stages are very high, increasing survival at later stages appears to be more important to the population. This illustrates the point made earlier that without taking into account what drives population dynamics, the magnitude of an effect at the individual level may not necessarily correspond to its importance at the population level.

Example: Dynamic Size-Dependent Predator–Prey Interactions

Background

Many species of marine fish spend the first season or year of their lives in estuaries (Chao and Musick 1977). Consequently, processes and species interactions in these estuaries determine the first-year success of cohorts of these species. Food abundance, physical conditions, and predation pressure within estuaries can drive ultimate numbers and size structure of a cohort of fish.

Wright et al. (1993) studied the impact of a major estuarine predator, southern flounder, on the survival and ultimate size distribution of a common prey, spot. In pond experiments, they found that presence of flounders not only determined the number of young-of-the-year spot that survived through the summer but also determined spot size distribution (Wright et al. 1993). Because piscivorous fish are generally size-biased in their prey selection (Juanes 1994), the number and size of prey that survive a season will depend on the relative sizes of predator and prey (Wilbur 1988).

To assess the differential impact of predators of different sizes, Rice et al. (1993) did pond experiments that used two size classes of young-of-the-year spot and two size classes of flounder. The treatments were designed such that the small size class of spot would be vulnerable to predation by both size classes of flounder, but the large spot would be vulnerable only to predation by the large flounder (Rice et al. 1993). Their results showed that presence of flounders had a significant effect on spot size distributions and that size of flounder had a significant effect on relative survival of small and large spot (Rice et al. 1993). But because of initial conditions in the experimental community (i.e., low food levels for spot) and limitations on length of the experiment, relative growth rates of spot and flounder were probably not representative of their growth in nature. Although the experiment demonstrated the importance of relative sizes of predator and prey in determining ultimate prey survival and size distribution, it also pointed out the need to understand the relationship between static relative sizes and dynamic ones. Both predator and prey grow during a given season in the estuaries; how their relative rates of growth affect the outcome of the size-dependent predation process is important to ultimate success of the prey cohort.

Model

Because growth rates are a difficult parameter to manipulate in an experiment, we developed a model in which we could control growth rates of both predators and prey (Rice et al. 1997). We used a simple partial differential equation model to assess the effects of a growing predator on a size-structured growing prey cohort. In piscivore systems, prey vulnerability is often low at low prey-to-predator size ratios, low at high prey-to-predator size ratios, and high at some intermediate size ratio (Juanes 1994). Using size-selective foraging parameters estimated independently in laboratory experiments for spot and flounder (Barker 1991), we found that we could mimic the results of the original experiment in Rice et al. (1993). But more important, by systematically altering growth rates of predator and prey in the model we were able to assess the impact of the dynamics of relative predator/prey sizes on the survival and size structure of the prey cohort (Rice et al. 1997). For example, if prey begin the season at a size at which they are vulnerable to a size-specific predator and then both predator and prey cohort grow at rates that maintain this vulnerability, the prey cohort will exhibit poor survival. However, it is more likely that young-of-the-year prey will have higher growth rates than their larger predators and thus may be able to grow out of the sizes at which they are vulnerable to predation by these specific predators (highest survival at high prey growth rates and low predator growth rates; top panel, Fig. 14-5). The high survival at very low prey and predator growth rates (left corner, top panel, Fig. 14-5) is a result of prey growing so slowly that in the duration of these model runs they never get large enough to be highly vulnerable to these predators. When we extended the time over which we modeled these dynamics, we saw that slow growth rate ultimately resulted in poor survival. Even with a given set of predator and prey growth rates, the ultimate effect of the dynamic process of size-selective predation coupled with growing predators and prey will be determined by the initial relative sizes of predator and prey (compare solid and dotted lines, bottom panel, 14-5). If prey begin smaller than the relative size at which they are most vulnerable to the predator, then prey growth can actually move prey into a size at which they increase their vulnerability (solid line, bottom panel, fig 14-5).

We were able to use experiments here both to point us toward important questions and to independently estimate model parameters, but we required a model to determine how relative size and growth of predator and prey might dynamically interact to affect cohort size and size structure. Questions of how different sets of assumptions about size dynamics ultimately translate into cohort characteristics could only be answered with the model. The experiments necessary to derive these relationships are not feasible.

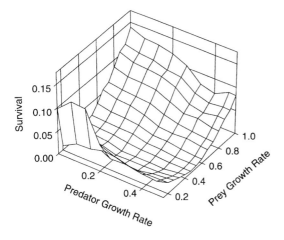

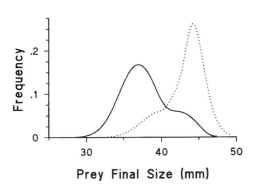

Figure 14-5. Results of a partial differential equation model of the effect of growing predators on size-structured cohorts of growing prey. **Top**: Proportion of cohort surviving to 60 days as a function of prey and predator growth rates. Initial predator and prey size distributions were constant over all model runs. **Bottom**: Size-frequency distribution of 15-d survivors when initial mean prey size is smaller than the most vulnerable size (*solid line*) and when initial mean prey size is larger than the most vulnerable size (*dotted line*). Initial predator size and growth rates of both prey and predator are identical between these two examples.

Using Experiments to Set Relevance of Model Results

In none of the previous examples is the model the final step. Initial empirical observations in complex systems called for models to organize the existing knowledge (steps 1 thorough 3, Fig. 14-1). From these models arose patterns in population dynamics (counterintuitive patterns in some cases), but in each case the actual population-level pattern was determined by the value of particular parameters and forms of particular functions (steps 3 through 6, Fig. 14-1). The next step is to design experiments to estimate these parameters and functions; the model can help identify those processes and parameters that are most important in driving population dynamics in each case (step 7, Fig. 14-1). In the pathogen dynamics example, we must estimate transmission decay rate to predict the relative impact of different vector preferences at any particular disease concentration. In the sea turtle example, the model says that negative impacts on late juvenile/subadult survival will have greater population consequences than similar impacts on eggs and hatchlings. Now we need to estimate the magnitude of negative anthropogenic impacts on these stages. Although this is fairly well quantified for eggs and hatchlings, it has only recently been assessed for later life stages. Survival at this late juvenile/early adult stage is being manipulated in some populations through the use of turtle-excluder devices on shrimp boats. Preliminary analyses (Royle and Crowder 1994) have provided an estimate of the impact humans are having on survival at these life stages. In the predator–prey system, the model clarified the importance of knowing initial relative prey and predator sizes in being able to predict the impact of the predator.

Models in Resource Management Versus Models as Research Tools

Some ecological models are developed to be ends in themselves. This is often the case for models used in resource management. The goal in building these models is to mimic patterns in resource dynamics to help guide resource management decisions. Thus, when we combine experiments with this type of model, it is generally part of an effort to "test" that model. The question to be answered is: Does the model output match patterns observed in nature? This is a valid and important use of models, but it is not the only use.

Ecological models as we have used them in the examples presented here are simply one of many types of tools in research. Here the goal is to see how combinations of assumptions about processes and parameters work together to produce a pattern—that is, to evaluate cause and effect. Once we are confident that the logic is correct (e.g., mathematics or programming are correct) in this type of model, we have a tool that can tell us how a particular set of assumptions produces certain patterns. Our goal in building these models is not simply to test them; our goal is to ask whether our model assumptions are appropriate for a given system in which we would like to apply understanding arising from our model and to design experiments to determine when the particular model is relevant to our particular system.

Summary

Models and experiments as we refer to them here are both manipulations of nature in an attempt to understand cause and effect. They both are limited by the assumptions we make in designing them. Experimental results are driven by how we set up experiments. A common criticism of experiments is that their design does not mimic nature. Model results are also driven by how we set up the models (typically referred to as assumptions). A common criticism of models is that their results do not mimic nature. But neither model nor experiment would have value in explaining mechanisms if it was designed to mimic every aspect of nature. We can only understand cause and effect if we can manipulate the cause, controlling for other sources of variation, and see if the presumed effect changes. That is the utility of both manipulative experiments and models as we have described them in this chapter.

We have chosen to focus this chapter on just a portion of an approach that combines models and experiments. We do not prescribe always alternating model with experiment as if one always requires the other as a next step. Sometimes experimental results can best be extended by doing another experiment, sometimes by building a model. In general, we use experiments to set some bounds on the infinite possibilities of models and we use models to more precisely and completely explore patterns of cause and effect that are not feasible to manipulate empirically.

We have described a tool for combining information from small-scale experiments (e.g., observations at small spatial scales or short time scales, often at the level of an individual organism) with population-level models to assess the importance to the population of impacts on individual vital rates and behaviors. Using models to organize empirical observations, we can ask how different processes interact to produce population-level results. This often leads us to possibilities that we would not have seen considering only relative strength of empirically derived individual responses. And finally, having elucidated the possible outcomes through the model, we need to pinpoint where in the set of possible outcomes our system lies. Again, we can return to experiments to estimate the parameters of importance to set the relevance of our model outcomes to a particular system. This method of combining observations and experiments with models has proved useful in answering questions in systems too large, too rare, too slow, or too complex to be studied efficiently through whole-system experiments.

ACKNOWLEDGMENTS Support for this work was provided by National Science Foundation grants DEB-9410327 to EAM and BSR-891044 to BMR. We thank S. Heppell, D. Crouse, and L. Crowder for sharing an unpublished manuscript. We thank Bill Resetarits, Joe Bernardo, and an anonymous reviewer for very helpful comments on an earlier draft of this manuscript. This article could not have been completed without the generous help of the Loucks family.

Literature Cited

Alexander, H. M. 1990. Dynamics of plant-pathogen interactions in natural populations. Pages 31–45 in J. J. Burdon and S. R. Leather (eds.), Pests, Pathogens, and Natural Plant Communities. Blackwell Scientific, Oxford.

Alexander, H. M., and J. Antonovics. 1995. Spread of anther-smut disease (*Ustilago violacea*) and character correlations in a genetically variable population of *Silene alba*. Journal of Ecology 83:783–794.

Baker, H. G. 1947. Infection of species of *Melandrium* by *Ustilago violacea* (Pers.) Fuckel and the transmission of the resultant disease. Annals of Botany 11:333–348.

Barker, D. L. 1991. Size-dependent responses of southern flounder to its prey. M. S. thesis, North Carolina State University, Raleigh.

Caswell, H. 1989. Matrix Population Models. Sinauer, Sunderland, Massachusetts.

Chao, L. N., and J. A. Musick. 1977. Life history, feeding habits, and functional morphology of juvenile sciaenid fishes in the York River estuary, Virginia. Fishery Bulletin 75:657–702.

Clay, K. 1987. The effect of fungi on the interaction between host plants and their herbivores. Canadian Journal of Plant Pathology 9:380–388.

Crouse, D. T., L. B. Crowder, and H. Caswell. 1987. A stage-based population model for loggerhead sea turtles and implications for conservation. Ecology 68:1412–1423.

Crowder, L. B., D. T. Crouse, S. S. Heppell, and T. H. Martin. 1994. Predicting the impact of turtle excluder devices on loggerhead sea turtle populations. Ecological Applications 4:437–445.

deKroon, H., A. Plaisier, J. V. Groenendael, and H. Caswell. 1986. Elasticity: the relative contribution of demographic parameters to population growth rate. Ecology 67:1427–1431.

Dodd, C. K., Jr. 1988. Synopsis of the Biological Data on the Loggerhead Sea Turtle *Caretta caretta* (Linnaeus 1758). United States Fish and Wildlife Service Biological Report 88(14), Gainesville, Florida.

Frazer, N. B. 1983. Demography and life history evolution of the Atlantic loggerhead sea turtle, *Caretta caretta*, nesting on Little Cumberland Island, Georgia. Ph. D. dissertation, University of Georgia. Athens.

Frazer, N. B. 1986. Survival from egg to adulthood in a declining population of loggerhead turtles (*Caretta caretta*). Herpetologica 42:47–55.

———. 1987. Preliminary estimates of survivorship for juvenile loggerhead sea turtles (*Caretta caretta*) in the wild. Journal of Herpetology 21:232–235.

Heppell, S. S., D. T. Crouse, and L. B. Crowder. In press. Using matrix models to focus research and conservation management efforts. In S. Ferson (ed.), Quantitative Methods in Conservation Biology. Springer-Verlag, Berlin.

Juanes, F. 1994. What determines prey size selectivity in piscivorous fishes? Pages 79–100 in D. J. Stouder, K. L. Fresh, and R. J. Feller (eds.), Theory and Application in Fish Feeding Ecology. University of South Carolina Press, Columbia.

Kingsolver, J. G. 1987. Mosquito host choice and the epidemiology of malaria. American Naturalist 130:811–827.

Mann, T. M. 1977. Impact of developed coastline on nesting and hatching sea turtles in southeastern Florida. M. S. thesis, Florida Atlantic University, Boca Raton.

McElhany, P., L. A. Real, and A. G. Power. 1995. Vector preference and disease dynamics: a study of barley yellow dwarf virus. Ecology 76:444–457.

McFarlane, R. W. 1963. Disorientation of loggerhead hatchlings by artificial road lighting. Copeia 1963:153.

Murphy, T. M., and S. R. Hopkins-Murphy. 1989. Sea Turtle and Shrimp Fishing Interactions: A Summary and Critique of Relevant Information. Center for Marine Conservation, Washington, D.C.

National Research Council. 1990. Decline of the Sea Turtles: Causes and Prevention. National Academy Press, Washington, D.C.

Real, L. A., E. A. Marschall, and B. M. Roche. 1992. Individual behavior and pollination ecology: implications for the spread of sexually transmitted plant diseases. Pages 492–508 in D. L. DeAngelis and L. J. Gross (eds.), Individual Based Models and Approaches in Ecology: Populations, Communities, and Ecosystems. Chapman and Hall, New York.

Rice, J. A., L. B. Crowder, and K. A. Rose. 1993. Interactions between size-structured predator and prey populations: experimental test and model comparison. Transactions of the American Fisheries Society 122:481–491.

Rice, J. A., L. B. Crowder, and E. A. Marschall. In press. Predation on juvenile fishes: dynamic interactions between size-structured predators and prey. Pages 333–356 in R. C. Chambers and E. A. Trippel (eds.), Early Life History and Recruitment in Fish Populations. Chapman and Hall, New York.

Royle, J. A., and L. B. Crowder. 1994. Analysis of Loggerhead Turtle Strandings from South Carolina and Estimation of the Effect of Turtle-Excluder Device Use in Shrimp Nets. Technical Report 113. National Institute of Statistical Sciences, Research Triangle Park, North Carolina.

Thrall, P. H., J. Antonovics, and D. W. Hall. 1993. Host and pathogen coexistence in sexually transmitted and vector-borne diseases characterized by frequency-dependent disease transmission. American Naturalist 142:543–552.

Wilbur, H. M. 1988. Interaction between growing predators and growing prey. Pages 157–172 in B. Ebenman and L. Persson (eds.), Size-structured Populations: Ecology and Evolution. Springer-Verlag, Berlin.

Wright, R. A., L. B. Crowder, and T. H. Martin. 1993. The effects of predation on the survival and size-distribution of estuarine fishes: an experimental approach. Environmental Biology of Fishes 36:291–300.

15

Experimental Approaches to Studying the Population Dynamics and Evolution of Microorganisms

J. A. MONGOLD

Theoretical ecology has provided clear, unambiguous formulations of many of the questions with which ecologists are concerned (reviewed in Roughgarden et al. 1989). Experimental tests of theory in ecology and evolutionary biology have increased in recent decades (for reviews see Stearns 1982, Hairston 1991), yet much of this body of theory remains untested, and there is still a great need for experimental verification (Kareiva 1989). The call for more experimentation in ecology, however, has focused mainly on field experiments (Hairston 1991). Laboratory experiments are sometimes criticized on the grounds that they involve oversimplified environments and utilize organisms whose field relationships are poorly understood. Unfortunately, the richness of field studies comes at the cost of experimental ease and control. Expense, labor, and physical constraints can make it difficult or impossible to obtain sufficient replication for strong statistical inference (Hurlbert 1984). This problem is intensified by the very complexity that makes field experiments so attractive. Uncontrollable variation in weather, habitat quality, abundance of predators, or other aspects of the environment frequently confounds the interpretation of results (see Endler 1986 for a discussion of some of the difficulties to be faced). Laboratory microcosms are, of course, restricted environments, and one must be careful to avoid overgeneralization of conclusions from such studies. The same must be said, however, of a study of a single pond or stretch of streambed. Thus, one must always weigh the costs and benefits of experimental control versus generality (see Morin, this volume, for a discussion of these issues). A multilevel approach achieved by an increased use of laboratory models to complement field observations could be very profitable for the field of ecology.

At almost the opposite end of the spectrum, microbial ecology has traditionally been an extremely reductionist and predominantly lab-oriented science. Because of obvious constraints of size, there are limited opportunities for direct observational studies of microorganisms in their natural habitats. Much of the focus has been on efforts to develop the means to culture organisms and categorize them by their metabolic capabilities or pathogenicity. Many microbial ecologists are now moving beyond the natural history aspects of their science and are beginning to address issues related to population-level processes such as community organization, competitive interactions, and adaptive evolution. In the process, there is new interest in applying existing ecological and evolutionary theory that was developed for plant and animal populations to microbial populations.

The separation of microbial ecology as a distinct discipline from the rest of ecology is merely a historical artifact and an arbitrary division based on the size of the organisms being studied (see Andrews 1991 for an alternative organizational scheme). This segregation, however, is rapidly changing. The use of 16S rDNA sequences to construct phylogenies of microbes (Olsen and Woese 1993), the development of taxa-specific rRNA-targeted probes to study microbial communities in situ (Braun-Howlland et al. 1992), and the ease of experimentation with microbes are drawing more and more ecologists, population geneticists, and evolutionary biologists to exploit microbial systems.

Bacteria, in particular, meet the requirements of a good model system for testing theoretical predictions. They are small and easy to grow in the lab and have short generation times. It is relatively easy to manipulate the genetic composition of a bacterial population and initiate genetically identical replicate populations. While bacteria reproduce asexually, they do sometimes undergo recombination between individuals, and, therefore, the extent of recombination is a potentially manipulable factor that could be incorporated into an experimental design. In addition, for many bacterial species the wealth of physiological and genetic information available makes it possible to infer mechanistic explanations at many different levels. Of course, bacteria clearly would not be useful for studies of processes that are unique to eukaryotes, but they have proved to be useful tools for demonstrating such basic evolutionary principles as the randomness of mutations (Luria and Delbrück 1943, Lederberg and Lederberg 1952; see Sniegowski and Lenski 1995 for a review of recent controversy) and adaptation by natural selection (Lenski et al. 1991).

Microbial ecology today draws together investigators from diverse backgrounds that include traditional microbiology, genetics, ecology, and evolutionary biology. The experimental approaches are at least as diverse. For example, some researchers have taken advantage of the availability of techniques for genetic manipulation to study the physiology and adaptive consequences of naturally occurring enzyme variants (Dykhuizen and Hartl 1980, Dean 1995). Others have followed microbial populations over many generations to observe the origin, nature, and competitive outcome of genetic variation arising in laboratory experiments (for some examples, see Helling et al. 1987, Bull et al. 1991, Lenski 1995).

In this chapter, I will not attempt to review this rapidly growing field but rather will provide a limited and somewhat biased set of examples to demonstrate the utility of microbial systems for addressing issues of broad relevance to contemporary problems in ecology and evolutionary biology. First, I will describe two experimental evolution studies. One is an extremely simple system intended to yield data for testing analytical methods. The other is a study of natural selection in the laboratory designed to estimate the relative importance of adaptation versus chance events and historical constraints during evolution in a novel thermal environment. Second, I will describe two examples of studies that look at interactions between individuals and the effects of the environment on population structure and dynamics.

Long-Term Experimental Evolution Studies

One of the main goals of evolutionary biology is to understand the forces that influence the nature and extent of genetic and phenotypic diversity within and among populations. Because the time scale of evolutionary processes generally precludes direct observation, comparative studies use phylogenetic inference, as well as historical and contemporary information about environmental conditions, to infer the processes that have been operating from the patterns observed among contemporary populations (Huey 1987). The comparative approach, however, has several drawbacks. First, there are problems inherent in reconstructing phylogenetic and environmental histories. Second, it is often difficult to distinguish adaptive from chance differences in traits when the relevant environmental factors are uncertain. Third, without multigenerational data and the ability to manipulate the environment, adaptive differences can be obscured or incorrectly identified by the presence of acclimatory responses to different environments.

One way around these problems is to choose a system to study in which the evolutionary time scale is shorter. Microorganisms, as mentioned earlier, have short generation times and large population sizes and thus are amenable to studies of evolutionary processes. In addition, genetically identical clones can be obtained so that the evolutionary process can be replicated with populations that are initially genetically identical. Clones of individual genotypes can be frozen and resuscitated, unchanged, at later times for analysis. This means that ancestral and derived genotypes can be directly compared in a common environment. In short, microbial systems allow one to use a direct experimental approach to studying very general patterns and processes of evolutionary change. In the following sections, I will present two approaches which have been used to address the problems mentioned previously with respect to comparative studies: verification of phylogenetic reconstruction techniques and estimating the relative importance of adaptation versus chance or historical contingencies.

Experimental Phylogenetics

Phylogenetic information is critical to the evolutionary interpretation of comparative biological studies (Huey 1987, Harvey and Pagel 1991). Since our knowledge of the evolutionary history of taxonomic groups is typically incomplete, most studies rely on various statistical methods of inferring the most likely phylogenetic history from contemporary data (Swofford and Olsen 1990). Unfortunately, every method has inherent biases and incorporates assumptions about the evolutionary process. For this reason, phylogenies must always be viewed as hypotheses with varying levels of support. Since we rarely have full knowledge of the evolutionary history for a set of taxonomic groups, we cannot directly test the strengths of the various methods or validate their assumptions. Computer-simulated data have been useful in testing the sensitivity of algorithms to different assumptions and topologies. However, computer simulations suffer from the same lack of biological knowledge. Hillis and colleagues (Hillis et al. 1992, Bull et al. 1993) put an interesting twist in this field of investigation by substituting a rapidly growing bacteriophage for the computer used to generate a set of phylogenetic data with complete knowledge of the history of the lineages.

A set of nine taxa was constructed by serially propagating wild-type T7 phage growing on *Escherichia coli* W3110. The original line was divided at predefined intervals to yield a symmetric phylogeny with equal distances between nodes. At the end, restriction site maps were constructed for each of the nine taxa using 34 restriction endonucleases. Five methods of phylogenetic reconstruction were evaluated for their ability to predict the correct branching topology, branch lengths, and ancestral states. All methods predicted the correct topology, but they differed in their ability to predict the correct branch lengths.

This is perhaps the most extreme example of a microorganism being used purely to generate data with which to test analytical algorithms. Yet even in this simplest case, the biological model system is more than an extension of a computer simulation. In this example, all assumptions of the model are derived from biological knowledge. The incorporation of the ability to directly test the assumptions makes this type of data highly superior to computer-simulated data.

Effects of History and Chance versus Adaptation in Populations of E. coli

Evolutionary biologists have struggled for decades with the problem of determining the relative importance of the forces of adaptation, historical contingency, and chance events in generating and maintaining diversity. It is difficult to distinguish these forces in nature because evolutionary change occurs as a series of unique events. If, however, one could replicate the evolutionary process for a single genotype, the effect of chance would be reflected in an increase in variation among the replicate populations. This is because differences in the nature and effect of randomly occurring mutations could result in divergence of lines even if they are evolving in a common environment. Similarly, if populations that are evolutionarily adapted to different environments could be experimentally moved to a common environment, the ability of historical constraints to cause the populations to maintain their differences, or even diverge, in phenotype in spite of identical current selective pressures could be observed. The strength of adaptation, on the other hand, would be reflected in a systematic change in all populations that evolve in the common environment.

The contributions of adaptation, chance, and history have been examined in an experiment with *E. coli* in a novel thermal environment (Travisano et al. 1995). The first stage of this experiment was obtaining replicate populations with known differences in their environmental and phylogenetic history. For this, a single genotype that was preadapted to growth at 37°C in glucose media with daily serial transfer (Lenski et al. 1991) became the common ancestor from which all of the experimental lines were derived. This genotype was cloned to initiate 24 replicate populations. These 24 populations were divided into four groups of six, and each group was maintained in one of four thermal environments (32°, 37°, 42°C, and daily alternation between 32° and 42°C) for 2,000 generations (Bennett et al. 1992). At the end of 2,000 generations, all

24 populations were shown to have specifically adapted to their respective thermal environment (Bennett et al. 1992, Bennett and Lenski 1993).

The second stage of this experiment was placing the 24 lines with different histories into a common, and novel, environment. A single clone from each of the populations was used to found 24 new populations, which were propagated in the same environment but now at 20°C, for an additional 1,000 generations (Travisano et al. 1995, Mongold et al. 1996). A diagram of the phylogenetic and thermal history of these lines is presented in Figure 15-1. At the end of 1,000 generations, a clone from each of the 24 populations was chosen at random for analysis. For each population, two traits were measured with replication. Fitness was measured by competition against the ancestor in the 20°C thermal environment, and it was calculated as the ratio of Malthusian parameters (Lenski et al. 1991). The second trait was a morphological character— average cell size. Cell size was not significantly correlated with relative fitness and therefore is expected to be more prone to chance divergence and retention of historical differences.

The effect of adaptation versus the combined effects of chance plus history on relative fitness during 1,000 generations at 20°C is shown in Figure 15-2a. The adaptive response, defined as the mean change in relative fitness from the beginning to the end of this period, was highly significant. The effect of chance plus history was quantified by performing an analysis of variance (ANOVA) to estimate the component of variance that is due to differences among the populations. The effect of chance plus history increased and was statistically significant at the end of the experiment, but the effect of adaptation was significantly greater than that of chance and history (Travisano et al. 1995). Thus, historical differences between populations and random effects on relative fitness were small compared with the overall increase in adaptation of all of the populations to the novel environment. The effect of chance plus history on cell size (Fig.

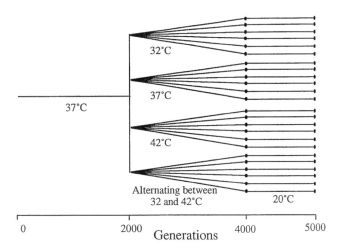

Figure 15-1. Phylogeny, thermal history, and taxonomy of experimental lines used in this study. Culture temperatures are given below each set of lines. Modified from Mongold et al. (1996).

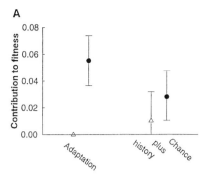
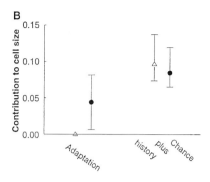

Figure 15-2. Evolution of fitness (**A**) and cell size (**B**) during 1,000 generations at 20°C and relative contributions of adaptation and chance plus history before (Δ) and after (•) 1,000 generations at 20°C. Adaptation is calculated as the mean change in either relative fitness or size from the value at the beginning of period of incubation at 20°C. The contributions of chance and history are quantified by performing an (ANOVA) to estimate the variance component corresponding to chance and historical differences between populations. Error bars represent 95% confidence intervals. Modified from Travisano et al. (1995).

15-2b), however, was highly significant at both the beginning and the end of the 20°C evolutionary period and was a significantly greater influence on the final cell size of the derived populations than was adaptation (Travisano et al. 1995). Thus, for a trait that was not significantly correlated with fitness in this environment, historical effects persisted for at least 1,000 generations.

Generalization from experimental results such as these must, of course, be made cautiously because of the limited range of environmental variation studied, the relatively short time span of the experiment, and the simple set of interacting variables. This experiment, however, is a demonstration of the strength of a simple microbial model system to attack questions that have in the past appeared intractable, including even the role of historical constraints on evolutionary change.

Microbial Population Dynamics and Environmental Interactions

Genetic Exchange within Bacterial Populations

For several decades, scientists have sought to describe the structure of bacterial populations and the extent to which recombination influences the patterns of genetic variation which are observed. The first surveys of genetic variation in natural populations of a bacterial species were performed in the 1970s (Milkman 1975, Selander and Levin 1980). These surveys of electrophoretic variation in enzymes among isolates of the bacterial species *E. coli* showed that there were a large number of variants at most loci. The level of diversity, however, was significantly lower than had been predicted assuming that *E. coli* had an extremely large effective population size (Milkman 1972). They also showed evidence of nonrandom association of alleles at multiple loci—that is, linkage disequilibrium (Selander and Levin 1980). These two observations led Levin

(1981) to hypothesize that bacterial species like *E. coli* consist of a limited number of clonal lineages which persist for long periods of time. Bacteria reproduce asexually and so recombination was assumed to be very rare.

Recently, molecular studies of bacterial population genetics have been extended to the level of DNA sequences (Milkman and Bridges 1993, Guttman and Dykhuizen 1994), and these new results suggest that recombination is more frequent in bacteria than was previously believed. Nonconcordant phylogenies obtained from different parts of the genome indicate that in spite of the asexual nature of bacterial reproduction, linkage relationships between genes are broken down frequently enough that the impact is observed in their pattern of inheritance.

Field studies have therefore led to several quantitative questions. Is recombination between different clonal lineages rare enough to account for the linkage disequilibrium observed between electrophoretic variants at multiple loci? On the other hand, is recombination common enough to prevent the unlimited divergence of lineages and maintain species integrity? And, as a necessary corollary, is recombination among species sufficiently depressed to function in the delineation of bacterial species? It is very difficult, however, to estimate a rate of recombination in the field. The rate is generally so low that a large number of samples need to be collected and analyzed and an impractically large number of genetic markers must be available. For this reason, laboratory estimates are indispensable.

Many bacteria harbor plasmids which are capable of moving genes from one individual to another or even one species to another. Laboratory estimates of the rate at which these elements move (Levin et al. 1979, Levin 1988, Gordon 1992), however, are so low that it is not believed that they could have a significant impact on population structure with respect to the bacterial genome. Natural transformation, although not as taxonomically widespread, is an intrinsic property of many bacterial species (Stewart 1989). There is some evidence that populations of naturally transformable bacteria may be closer to linkage equilibrium (Maynard-Smith et al. 1993). A survey of the naturally transformable species *Haemophilus influenzae* showed less clonality than is typically observed in species such as *E. coli*, suggesting that transformation is important in natural populations (Porras et al. 1986).

Although the timing of genetic exchange by transformable bacteria in the natural environment is not known, it is typically a highly regulated process and in many species appears to be correlated with changes in resource availability (Stewart 1989). Chemostat cultures are particularly useful for studying this relationship because population density can be kept nearly constant while growth rate varies in response to changes in the rate of resource input.

The proportion of competent cells in a chemostat culture of *H. influenzae* was determined by mixing a sample of the chemostat population with genetically marked DNA, incubating for 30 minutes to allow the DNA to be taken up by competent cells, and plating on selective and nonselective media to estimate the fraction of the population which had acquired the genetic marker carried on the transforming DNA (Mongold 1992). The competent fraction of a culture of *H. influenzae* increases when the flow rate of the chemostat is changed, increased or decreased, but then remains at a constant level once the steady-state growth rate and density are established (Fig. 15-3). The level of competence during steady-state growth is higher at the faster growth rate.

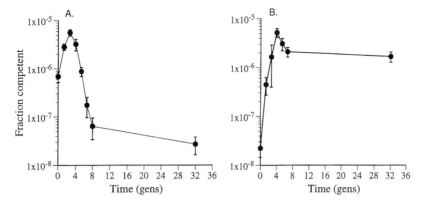

Figure 15-3. Competence in chemostat culture. The level of competence in chemostat cultures was estimated by mixing samples from a chemostat culture with a saturating quantity of DNA extracted from a novobiocin-resistant (NovR) strain of *Haemophilus influenzae*. After 30 minutes' incubation, the samples were plated using an agar overlay protocol (Mongold 1992) to determine the number of NovR genetic transformants/ml. Each data point represents the average ratio of genetic transformants to total cells obtained for replicate chemostat populations sampled over time. Error bars represent the standard error. The rate of flow of resources into the chemostats was adjusted at time zero from (**A**) 0.9 to 0.1 hr^{-1} and (**B**) 0.1 to 0.9 hr^{-1} and then maintained at a constant rate for the entire sampling period.

Similar results that indicated a strong dependence of competence on growth rate and multiple regulatory cues were obtained with Bacillus chemostat (Portolés et al. 1973) and batch (Dorocicz et al. 1993) cultures. The impact of transformation on population structure in this species will, therefore, depend on the level of resources in the environment and their stability. This suggests that future field experiments should concentrate not on attempts to measure recombination but on determining whether the prevailing ecological conditions will result in high levels of competence.

One can also use laboratory estimates to determine the maximum potential rate of gene exchange between pairs of strains if the conditions were optimal. Assuming that recombinants and parental strains are growing at a constant and equal rate, the accumulation of recombinants would be proportional to the equilibrium densities of the donor (P cells/ml) and competent recipient (C cells/ml) populations, the relative transformation efficiencies of the gene used to mark transformants versus that used to mark total competent cells (α), and the overall rate constant of transformation (X) for that donor-recipient pair:

$$\frac{dT}{dt} = \alpha X \hat{C} \hat{P} \tag{1}$$

where $\hat{C}$ is equal to the recipient cell density times the fraction of competence.

In order to detect the production of recombinants between a donor and recipient strain, two strains were chosen that could be distinguished by resistance to a combination of antibiotics. The accumulation of recombinants in the chemostats was monitored by sampling from the chemostats at different time intervals and plating on selective media that only allowed their growth. Recombinants accumulated at a constant rate (Fig. 15-4).

In the experiment shown in Figure 15-4, the rate constant of transformation, X, was estimated to be roughly 1.54×10^{-11} ml cell^{-1} gen^{-1}. In words, X is the probability that a competent cell will acquire a gene by transformation, and it is dependent on both time and the density of genetically distinct potential donor cells. Therefore, with a moderate population density of 1×10^4 donor cells/ml, the probability that a competent cell will be transformed at a particular locus is 1.54×10^{-7} per generation. Since bacteria often live at very high local densities, this process could approach or be higher than the per gene neutral mutation rate estimated for *E. coli* to be approximately 4×10^{-7} per generation (Drake 1970).

Effects of Habitat Structure on Competitive Interactions

The effects of habitat structure on population structure and dynamics are often complex and difficult to handle, both theoretically and experimentally. For this reason, most experimentalists choose the simplest possible habitat that is practical. Even then, the organisms themselves often modify the environment, introducing complexities that yield unexpected competitive interactions (Chao and Ramsdell 1985, Rosenzweig et al. 1994).

Often, however, habitat structure is exactly the parameter of interest. In these cases, microbial ecologists have varied the viscosity of the habitat from that of a well-mixed liquid to that of a solid surface. In two independent studies, one with yeast (Wilke et al. 1992) and one with bacteria (Korona et al. 1994), investigators found that popula-

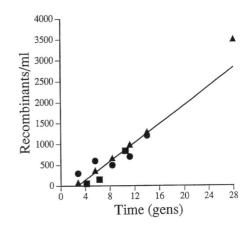

Figure 15-4. Chemostat matings. For each mating experiment, equal volumes of two steady-state chemostat populations were mixed at time zero and allowed to mate. The donor population was resistant to two antibiotics, streptomycin and rifampicin (StrR RifR). The recipient population was resistant to naladixic acid and spectinomycin (NalR SpcR). Data points are the cell densities of streptomycin resistant transformants (NalR SpcR StrR). The different symbols represent the density of recombinants sampled over time from three different experiments. The solid line represents the average linear regression for the three replicates.

tions that evolved on solid surfaces maintained higher levels of polymorphism. These results support the hypothesis that environments with greater diversity of ecological interactions result in greater diversification.

Similarly, varying the habitat from a liquid to a solid surface was used to test the effect of population structure on the level at which selection may act for certain types of genes. Many bacteria produce and release toxins that kill susceptible bacteria, bacteriocins. The bacteria which produce these toxins are resistant to their own toxin. A simple hypothesis for the function of these toxins is that they are used to eliminate competition. However, individuals bear a great cost associated with their production. Only a fraction of the population ever releases toxins at any one time, but those individuals die in the process. Therefore, individual selection should act strongly in the direction of not producing toxins. Chao and Levin (1981) showed that in a well-mixed environment this was indeed the case. A toxin-producing individual was unable to invade the population. Once established, however, a toxin-producing population could maintain itself. When bacteria were grown on surfaces, this frequency dependence was reduced. In that case, the benefits of reduced competition were localized to individuals in the vicinity of the one that died releasing the toxin. These individuals were all likely to be related, clones in fact, of the one that was sacrificed. Thus, the gene for production of the toxin spread, even when it was initially rare, due to selection on the group of related individuals—that is, by kin selection (Hamilton 1964).

Summary

The ability to isolate and observe the dynamics of such central evolutionary processes as natural selection, mutation, and genetic recombination makes laboratory model systems an invaluable tool with which to bridge the gap between theoretical biology and field studies. The use of laboratory experiments, in conjunction with mathematical modeling, can direct future refinements of theoretical models as well as provide estimates of the biologically realistic range of values for model parameters. These joint results should then provide a more complete conceptual framework for the interpretation of observations in specific field situations. In addition, one of the most valuable features of a model system is the power which is gained from the combined information about the genetics, biochemistry, and physiology of the model organism. With a broad base of knowledge, it becomes possible to address questions of mechanism at many different levels. Bacteria, such as *H. influenzae* and *E. coli*, are extremely easy to grow, a great deal is known about their physiology and genetics, their lab environment is easy to manipulate, and they have short generation times which allow multigenerational studies of ecological and evolutionary processes to be completed on a reasonable time scale for a research project.

ACKNOWLEDGMENTS I am grateful to Rich Lenski, Al Bennett, Bruce Levin, Mike Travisano, Jim Bull, Joe Bernardo, and several anonymous reviewers for helpful discussions and editorial assistance during the preparation of this manuscript. I was supported during the preparation of this manuscript by the National Science Foundation Center for Microbial Ecology (Grant BIR-91-20006).

Literature Cited

Andrews, J. H. 1991. Comparative Ecology of Microorganisms and Macroorganisms. Springer-Verlag, New York.

Bennett, A. F., and R. E. Lenski. 1993. Evolutionary adaptation to temperature: II. Thermal niches of experimental lines of *Escherichia coli*. Evolution 47:1–12.

Bennett, A. F., R. E. Lenski, and J. E. Mittler. 1992. Evolutionary adaptation to temperature: I. Fitness responses of *Escherichia coli* to changes in its thermal environment. Evolution 46:16–30.

Braun-Howlland, E. B., S. A. Danielsen, and S. A. Nierzwicki-Bauer. 1992. Development of a rapid method for detecting bacterial cells in situ using 16S rRNA-targeted probes. Biotechniques 13:928–932.

Bull, J. J., I. J. Molineux, and W. R. Rice. 1991. Selection of benevolence in a host-parasite system. Evolution 45:875–882.

Bull, J. J., C. W. Cunningham, I. J. Molineux, M. R. Badgett, and D. M. Hillis. 1993. Experimental molecular evolution of bacteriophage T7. Evolution 47:993–1007.

Chao, L., and B. R. Levin. 1981. Structured habitats and the evolution of anticompetitor toxins in bacteria. Proceedings of the National Academy of Science of the USA 78: 6324–6328.

Chao, L., and G. Ramsdell. 1985. The effects of wall populations on coexistence of bacteria in the liquid phase of chemostat cultures. Journal of General Microbiology 131:1229–1236.

Dean, A. M. 1995. A molecular investigation of genotype by environment interactions. Genetics 139:19–33.

Dorocicz, I. R., P. M. Williams, and R. J. Redfield. 1993. The *Haemophilus influenzae* adenylate cyclase gene: cloning, sequence, and essential role in competence. Journal of Bacteriology 175:7142–7149.

Drake, J. W. 1970. The Molecular Basis of Mutation. Holden-Day, San Francisco, California.

Dykhuizen, D., and D. L. Hartl. 1980. Selective neutrality of 6PGD allozymes in *E. coli* and the effects of genetic background. Genetics 96:801–817.

Endler, J. A. 1986. Natural Selection in the Wild. Princeton University Press, Princeton, New Jersey.

Gordon, D. M. 1992. The rate of plasmid transfer among *Escherichia coli* strains isolated from natural populations. Journal of General Microbiology 138:17–21.

Guttman, D. S., and D. E. Dykhuizen. 1994. Clonal divergence in *Escherichia coli* as a result of recombination, not mutation. Science 266:1380–1383.

Hairston, N. G., Sr. 1991. Ecological Experiments: Purpose, Design, and Execution. Cambridge University Press, Cambridge.

Hamilton, W. D. 1964. The genetical evolution of social behavior, I and II. Journal of Theoretical Biology 7:1–52.

Harvey, P. H., and M. D. Pagel. 1991. The Comparative Method in Evolutionary Biology. Oxford University Press, Oxford.

Helling, R. B., C. N. Vargas, and J. Adams. 1987. Evolution of *Escherichia coli* during growth in a constant environment. Genetics 116:349–358.

Hillis, D. M., J. J. Bull, M. E. White, M. R. Badgett, and I. J. Molineux. 1992. Experimental phylogenetics: generation of a known phylogeny. Science 255:589–592.

Huey, R. B. 1987. Phylogeny, history, and the comparative method. Pages 76–98 in M. E.

Feder, A. F. Bennett, W. W. Burggren, and R. B. Huey (eds.), New Directions in Ecological Physiology. Cambridge University Press, Cambridge.

Hurlbert, S. H. 1984. Pseudoreplication and the design of ecological field experiments. Ecological Monographs 54:187–211.

Kareiva, P. 1989. Renewing the dialogue between theory and experiments in population ecology. Pages 68–88 in J. Roughgarden, R. M. May, and S. A. Levin (eds.), Perspectives in Ecological Theory. Princeton University Press, Princeton, New Jersey.

Korona, R., C. H. Nakatsu, L. J. Forney, and R. E. Lenski. 1994. Evidence for multiple adaptive peaks from populations of bacteria evolving in a structured habitat. Proceedings of the National Academy of Science of the USA 91:9037–9041.

Lederberg, J., and E. Lederberg. 1952. Replica plating and indirect selection of bacterial mutants. Journal of Bacteriology 63:399–408.

Lenski, R. E. 1995. Evolution in experimental populations of bacteria. Pages 193–215 in S. Baumberg, J. P. W. Young, S. R. Saunders, and E. M. H. Wellington (eds.), Population Genetics of Bacteria. Cambridge University Press, Cambridge.

Lenski, R. E., M. R. Rose, S. C. Simpson, and S. C. Tadler. 1991. Long-term experimental evolution in *Escherichia coli*: I. Adaptation and divergence during 2,000 generations. American Naturalist 138:1315–1341.

Levin, B. R. 1981. Periodic selection, infectious gene exchange and the genetic structure of *E. coli* populations. Genetics 99:1–23.

———. 1988. The evolution of sex in bacteria. Pages 194–211 in R. E. Michod and B. R. Levin (eds.), The Evolution of Sex: An Examination of Current Ideas. Sinauer, Sunderland, Massachusetts.

Levin, B. R., F. M. Stewart, and V. A. Rice. 1979. The kinetics of conjugative plasmid transmission: fit of a simple mass action model. Plasmid 2:247–260.

Luria, S. E., and M. Delbrück. 1943. Mutations of bacteria from virus sensitive to virus resistant. Genetics 28:491–511.

Maynard-Smith, J., N. H. Smith, M. O'Rourke, and B. G. Spratt. 1993. How clonal are bacteria? Proceedings of the National Academy of Science of the USA 90:4384–4388.

Milkman, R. 1972. How much room is there left for non-Darwinian evolution? Pages 217–229 in H. H. Smith (ed.), Evolution of Genetic Systems. Gordon and Breach, New York.

———. 1975. Electrophoretic variation in *E. coli* from diverse natural origins. Pages 273–285 in C. L. Markent (ed.), Isozymes, Vol. 4. Academic Press, New York.

Milkman, R., and M. McKane Bridges. 1993. Molecular evolution of the *Escherichia coli* chromosome: IV. Sequence comparisons. Genetics 133:455–468.

Mongold, J. A. 1992. Evolution and population dynamics of natural transformation in *Haemophilus influenzae*. Dissertation, University of Massachusetts, Amherst.

Mongold, J. A., A. F. Bennett, and R. E. Lenski. 1996. Evolutionary adaptation to temperature: IV. Adaptation of *Escherichia coli* at a niche boundary. Evolution 50:35–43.

Olsen, G. J., and C. R. Woese. 1993. Ribosomal RNA: a key to phylogeny. FASEB Journal 7:113–123.

Porras, O., D. A. Caugant, B. Gray, T. Lagergard, B. R. Levin, and C. Svanborg-Eden. 1986. Difference in structure between type b and nontypable *Haemophilus influenzae* populations. Infection and Immunity 53:79–89.

Portolés, A., R. López, and A. Tapia. 1973. An approach to the study of competence regulation in *B. subtilis* growing in a chemostat. Pages 65–80 in L. J. Archer (ed.), Bacterial Transformation. Academic Press, London.

Rosenzweig, F. R., R. R. Sharp, D. S. Treves, and J. Adams. 1994. Microbial evolution in a simple unstructured environment: genetic differentiation in *Escherichia coli*. Genetics 137:903–917.

Roughgarden, J., R. M. May, and S. A. Levin (eds.). 1989. Perspectives in Ecological Theory. Princeton University Press, Princeton, New Jersey.

Selander, R. K., and B. R. Levin. 1980. Genetic diversity and structure in *Escherichia coli* populations. Science 210:545–547.

Sniegowski, P. D., and R. E. Lenski. 1995. Mutation and adaptation: the directed mutation controversy in evolutionary perspective. Annual Review of Ecology and Systematics 26:553–578.

Stearns, S. C. 1982. The emergence of evolutionary and community ecology as experimental sciences. Perspectives in Biology and Medicine 25:621–648.

Stewart, G. J. 1989. The mechanism of natural transformation. Pages 139–164 in S. B. Levy and R. V. Miller (eds.), Gene Transfer in the Environment. McGraw-Hill, New York.

Swofford, D. L., and G. J. Olsen. 1990. Phylogenetic inference. Pages 411–501 in D. M. Hillis and C. Moritz (eds.), Molecular Systematics. Sinauer, Sunderland, Massachusetts.

Travisano, M., J. A. Mongold, A. F. Bennett, and R. E. Lenski. 1995. Experimental tests of the roles of adaptation, chance, and history in evolution. Science 267:87–90.

Wilke, C. M., E. Maimer, and J. Adams. 1992. The population biology and evolutionary significance of Ty elements in *Saccharomyces cerevisiae*. Genetica 86:155–173.

16

The Dual Role of Experiments in Complex and Dynamic Natural Systems

BARBARA L. PECKARSKY

The Role of Observations and Experiments in Ecology

Ecology is the study of patterns and processes that explain the distribution, abundance, and interactions among organisms and the influence of organisms on the flux of energy and materials through ecosystems. A healthy debate has been raging over the past two decades regarding the role and utility of manipulative experiments in ecology (e.g., Simberloff 1978, Diamond and Gilpin 1982, Lehman 1986, Peters 1986, Hairston 1989). Underwood (1990, 1991) illustrates a simple scheme (Fig. 16-1) in which experiments are part of a sequence of approaches that begins with observation (description of patterns, puzzles, or problems in nature) and proceeds to development of explanatory models (theory) and predictions from those models (hypotheses) of potential processes that could explain observed patterns. The role of experiments in this scheme is to evaluate model predictions by methodically testing them with the goal of falsifying their logical antitheses (null hypotheses). Failure to falsify the null hypothesis results in rejection of the model and generation of new models, new hypotheses, and perhaps new observations.

The structure of Underwood's scheme assumes that observations are the necessary first step to generating hypotheses and that the only role of experiments is to test hypotheses. In this approach to ecology, observations reveal patterns in nature and experiments explain those patterns by determining the underlying causes, processes, or mechanisms. I will argue, however, that observations may not always illuminate patterns in nature and that experiments can also be used effectively to generate hypotheses. In complex and dynamic ecological systems, patterns obtained through observation may not be sufficiently clear to generate hypotheses regarding the importance of underlying ecological processes that occur in those systems. In those cases, it may be necessary to use phenomenological experiments (= mechanism-free experiments of Dunham and

311

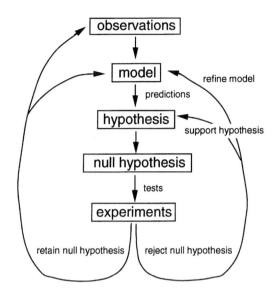

Figure 16-1. The logical components of a falsificationist experimental procedure and their relationships. Redrawn from Underwood (1990).

Beaupre, this volume) to identify cryptic processes that do not leave a distinctive signature in natural, unmanipulated systems. Thus, phenomenological experiments may be a more effective first step toward identifying patterns in nature and generating explanatory hypotheses.

Figure 16-2 illustrates an alternative approach to studying ecology in complex and dynamic natural systems where experiments can take on a dual role. Here I define phenomenological experiments as controlled manipulations of the environment that enable observation of a pattern but do not reveal the mechanism or why it happened. Mechanistic experiments retain their hypothesis-testing role as in Underwood's scheme, providing explanations of phenomena observed in mechanism-free experiments. We are left with the problem of defining the role of observation in this scheme. What do we conclude if patterns observed in natural systems are not consistent with hypotheses generated from experiments conducted under controlled conditions in manipulated systems? If field observations are not consistent with processes demonstrated to occur in experiments, then we fail to verify experimental findings. What, then, is the role of observational data in the study of complex and dynamic systems?

Other ecologists who have grappled with this dilemma strongly criticize the very idea of ecological experimentation, suggesting that inconsistencies between experimental and observational data should always be resolved in favor of observations (Rigler 1982, Peters 1991). Although they also advocate formulation of testable hypotheses, these ecologists argue that experiments are not necessary to test hypotheses and, instead, empirically derived regularities are acceptable explanations (Rigler 1982). In their view, the role of ecology is to make predictions from observational data. Thus, they use the same field data to generate and validate predictions (Peters 1991), stopping after step 3 in Underwood's scheme (Fig. 16-3). This approach has been harshly criticized as having "spawned a horde of sleuths armed with log-log equations in search of para-

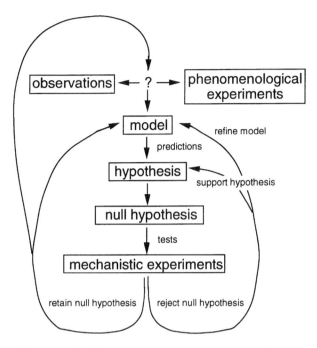

Figure 16-2. The logical components of an alternative approach to ecology in complex and dynamic systems that include phenomenological (hypothesis-generating) and mechanistic (hypothesis-testing) experiments.

meters that describe all aspects of physiology and demography'' (Lehman 1986:1163). In my opinion, this approach is tautological, can never go beyond description, and incorrectly infers process from pattern in natural systems (Miller 1986, Cooper and Dudley 1988).

Most critics of ecological experiments consider any deviation from reality unacceptable (Peters 1986). For example, Harris suggests that ''controlled biomanipulation experiments will always be difficult, if not impossible to perform'' (1994:154), and Pianka states that ''manipulative experiments, even if feasible, are of very limited utility due to indirect effects of complex networks'' (1994:365). An anonymous reviewer

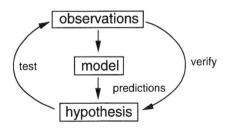

Figure 16-3. The logical components of an experiment-free approach to hypothesis testing in ecology. After Rigler (1982) and Peters (1986).

(1982) of my first National Science Foundation (NSF) proposal wrote: "At the outset, I confess that I'm not an enthusiastic believer in community organization studies involving manipulations (simplification) of community structure." I will argue that realism in experiments is often, but not always, a worthy goal. If the objective of an experiment is to understand how a particular system behaves (generate a pattern), then experimentalists should strive to make experimental systems as natural as possible. If, however, experiments are performed to elucidate general principles or interactions among organisms or between organisms and abiotic parameters, realism may be less important (see case studies for examples).

Ecological systems vary with respect to the difficulties associated with detecting patterns, inferring processes from those patterns, and the ease with which they can be manipulated, replicated, or controlled. The difficulties and challenges intensify as system spatial and temporal heterogeneity and complexity increase. I will use research on headwater streams as examples of systems in which patterns and processes are often cryptic and attributes are not particularly conducive to manipulation. For experiments where reality is an issue, it will be especially difficult to design reliable experiments in such systems. I will describe three case studies to illustrate the challenges and limitations of interpreting observations and conducting manipulative studies in stream ecosystems and a set of guidelines for improving the rigor and validity of manipulations in complex natural systems.

Why I Do Experiments: Case Studies

Experimentation has challenged stream ecologists because the attributes of stream ecosystems are not particularly conducive to manipulation (Hairston 1989). For example, flowing water, spatial habitat heterogeneity, and temporal thermal fluctuations characteristic of streams are difficult to replicate. Further, many stream organisms are highly mobile, live under stones, and are crepuscular or nocturnal. These attributes also make it inherently difficult to observe patterns at temporal and spatial scales relevant to processes occurring in streams. In such complex and variable environments, patterns of species distributions and abundances may not be consistent or readily observable (Peckarsky 1991a). Natural variability (both spatial and temporal) may complicate Underwood's (1990, 1991) proposed scheme of proceeding logically from observations to experimental determination of mechanisms that explain patterns. In fact, processes may occur in complex habitats that cannot be predicted by observation alone due to potentially confounding effects of multiple factors acting in synergistic or nonadditive ways. In such cases, the analysis of pattern may be confusing and inadequate, necessitating the careful design of interactive experiments to determine the relative influence of multiple factors in explaining natural variability. Three examples illustrate the dilemma of interpreting patterns from field observations in headwater streams and highlight how I have used different kinds of ecological experiments to both generate and test hypotheses.

Distribution and Abundance of Predatory Stoneflies

The purpose of the first study was to describe and explain causal factors of the distribution and abundance of predatory stoneflies in headwater streams. Observational data obtained in a Rocky Mountain stream over three summers revealed that predatory stoneflies were randomly distributed (Peckarsky 1988) and more abundant on rocks with a top surface area greater than 200 cm^2 (Peckarsky 1991a). We randomly sampled 270 rocks, recording densities of predatory stoneflies, surface and interstitial current velocity, water depth, water temperature, prey density, density of competitors, algal biomass, and biomass of detritus. Despite the considerable time and expense associated with this study, our horde of ecologists armed with log-log equations found no parameters consistently associated with the preferred microhabitats of stoneflies. Lack of empirically derived regularities prevented the formulation of any acceptable explanations or hypotheses and the drawing of any rigorous conclusions. Thus, this observational approach failed to explain the distribution of stoneflies in streams.

In contrast, by conducting both phenomenological and mechanistic experiments we have developed and evaluated alternative hypotheses to explain the distribution and abundance of predatory stoneflies in streams. By marking rocks of a range of sizes after a stream riffle dried (thus leaving them in their "natural lie" and not altering their natural stability), we determined the probability that different rock sizes would be disturbed by high-water events. Those experiments in natural systems showed that rocks preferred by stoneflies provided more stable habitats than smaller substrates (Peckarsky 1991a), suggesting habitat stability as one hypothesis that potentially explains stonefly distributions and abundance.

We also manipulated densities of predatory stoneflies among replicate enclosures in natural streams and observed the subsequent distribution and abundance of stonefly colonists (Peckarsky and Dodson 1980). Those experiments generated patterns consistent with the hypothesis that predatory stoneflies avoid microhabitats with resident predatory stoneflies. Mechanistic behavioral experiments designed to test this hypothesis enabled us to conclude that predators in aggregations experience interference competition (Peckarsky and Penton 1985, Peckarsky 1991b). Those tests were done under more simplified conditions in flow-through channels placed directly in the stream, providing minimal refuges so that observations of stonefly behavior could be made. Critics of experiments might argue that the behaviors exhibited by stoneflies under those conditions were not indicative of those occurring in nature. However, other mechanistic experiments conducted in more natural conditions added credibility to those conducted under simple conditions. In stream-side replicated circular flow-through channels with natural substrates and stream water, we demonstrated that density of competitors had significant negative effects on stonefly growth and fecundity (Peckarsky and Cowan 1991). Thus, we used different kinds of experiments to generate and test the hypothesis that interactions with competitors influence stonefly distribution and abundance in streams.

Other phenomenological enclosure experiments revealed the pattern that stoneflies did not aggregate on substrates with higher prey densities, suggesting no benefit to predator aggregative behavior (Peckarsky 1988). Mechanistic behavioral experiments that used natural stream water and substrates carried out in arenas allowing observation from above and below rocks provided an explanation for this pattern. Initially dense patches of highly mobile prey were temporally and spatially unstable because prey dispersed from rocks occupied by stoneflies (Peckarsky 1980, 1996). These experimental tests explain why we should not expect to observe natural associations between stonefly and prey densities in the field (Peckarsky 1991a), which was also consistent with hypotheses generated from phenomenological experiments (Peckarsky and Dodson 1980, Peckarsky 1988).

Several conclusions can be drawn from this example. First, observational data did not provide patterns of associations between predatory stonefly distribution and abundance and potential causal factors, such as abiotic variables, prey density, and competitor density. Second, caging manipulations in natural stream systems, using enclosures designed to minimize the difference between natural and enclosed habitats, generated patterns that suggested hypotheses to explain the distribution and abundance of stoneflies. These phenomenological experiments acted as an alternative to direct observations in unmanipulated systems, providing a tool by which patterns of association among variables could be revealed. Realism is an important aspect of these experiments because their goal is to understand how a particular system behaves. Third, mechanistic experiments under more simplified conditions tested hypotheses generated from the cage experiments. While reality was not as critical in these later trials, experiments carried out at different scales and with different levels of reality determined whether the results were robust to increasing complexity of the habitat. Finally, these single-factor experiments must be supplemented with interactive experiments that examine two or more independent variables in a factorial design to rigorously test the relative importance of multiple explanations of patterns of stonefly abundance and distribution in streams.

Selective Predation by Stoneflies

A second study illustrates how conducting mechanistic experiments at different scales and levels of habitat complexity can reveal processes or interactions that are sensitive to the specific conditions of the experiment. Predatory stoneflies were observed in simple arenas placed in streams to determine their encounter rates, attack probabilities, and capture success with alternative mayfly prey species (Peckarsky and Penton 1989). A later study made the same observations in more realistic arenas with natural substrates that could be viewed from above and below (Peckarsky et al. 1994). We also tested directly the effect of increasing refuge availability on selective predation by stoneflies by conducting feeding trials in enclosures in streams (Peckarsky and Penton 1989).

These experiments demonstrated that encounter rates between predators and prey could not be determined accurately without including natural substrates, because refuges were a key attribute of the habitat affecting this parameter. Further, absolute rates of predation on alternative prey species were affected by the presence of refuges. However, the relative rank order of predator preferences for different prey species, the relative attack rates, and capture success were robust to simplification of the experimental habitat. This set of experiments provided information critical for evaluating the validity of results of small-scale mechanistic experiments in the context of patterns observed in field observations and phenomenological experiments. I recommend that, whenever possible, this type of reality check be implemented in mechanistic experiments carried out in simplified environments.

Patterns and Explanations for Variability in Mayfly Size at Emergence

In the third study (ongoing), we seek to describe and explain variability in mayfly size at emergence both temporally and spatially in Rocky Mountain streams. Up to twofold differences in male and female size at emergence of the mayfly, *Baetis bicaudatus*, occur among streams in the same catchment and between years in the same stream (Fig. 16-4). Further, size of individuals at emergence varies temporally within the same population of mayflies emerging from the same stream, with mayflies of the second summer generation emerging at smaller sizes than those of the overwintering generation and overwintering generation mayflies declining in size throughout their emergence period (Fig. 16-5). This variability is important to mayfly population growth because larger females are more fecund (Peckarsky et al. 1993) and larger males may have higher reproductive success (Flecker et al. 1988).

A survey of the sizes of *Baetis* emerging from streams in the East River Valley was conducted during the summers of 1994 and 1995 to establish the patterns of adult size variation across streams throughout time and to seek factors that could account for

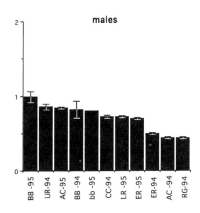

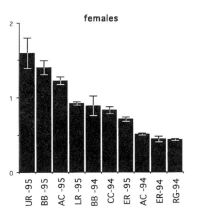

Figure 16-4. Variation in size at maturation of 1994 and 1995 summer generation *Baetis bicaudatus* emerging from different streams in the East River Valley, Gunnison County, Colorado. Mean mass (mg + S.E.) calculated from regressions of males and females from head capsule widths (mm). BB = Benthette Brooke, UR = Upper Rock Creek, AC = Avery Creek, bb = Billy's Brook, CC = Copper Creek, LR = Lower Rock Creek, ER = East River, RG = Rustler's Gulch.

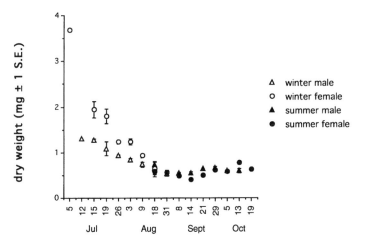

Figure 16-5. Variation in size at maturity over time of emergence of overwintering and summer generation *Baetis bicaudatus* emerging from the East River at the Rocky Mountain Biological Laboratory, summer 1993.

some of that variation. In each stream, throughout the period of growth and development of summer-generation *Baetis*, we measured a number of physical and biological factors that could influence growth rates of mayflies, including water temperature, algal food resources, density of other algal grazers (of the same and other species), density of predators (stoneflies and trout), and prevalence of mermithid nematode parasites (Vance and Peckarsky 1996). None of the parameters measured except presence or absence of trout provided data consistent with any hypotheses that explained the variation in body size within and among these streams. Again, this empirical approach generated few empirical regularities that could provide acceptable explanations or insights into the factors that affected body size in mayflies.

In contrast, single-factor and interactive experiments have demonstrated that many factors—including algal resources, competitor densities, and presence of predators—may cause size variation in *Baetis*. In two very different types of streamside channels using natural stream water and manipulated algal-covered tiles for subtrates, treatments with higher levels of algae and lower mayfly densities produced larger mayflies (Allan and Flecker 1992, Ode and Peckarsky 1995). Single-factor (stoneflies or trout only) and interactive experiments (stoneflies and trout) in streamside channels at two different spatial scales also demonstrated that presence of stoneflies and trout affects mayfly size at emergence (Peckarsky et al. 1993, Peckarsky and McIntosh in press). Mechanistic experiments in observation chambers have provided explanations of these patterns demonstrating that predator avoidance behaviors are disruptive, reducing mayfly access to food, thereby affecting their growth rates and fecundities (Cowan and Peckarsky 1994, Peckarsky 1996). Again, more interactive experiments are necessary to discern the relative importance of each of these factors as a determinant of mayfly size at emergence.

This example leaves us with the perplexing problem of explaining why most associations with causal variables obtained in experiments were not consistent with patterns of variation in size at emergence among natural streams. I would argue that each stream has a different combination of factors that potentially affect mayfly size at emergence, some promoting and some inhibiting growth. When all these factors act in combination, their individual effects can rarely be detected. Furthermore observation of significant associations in complex systems, such as smaller mayflies in trout streams than in fishless streams, suggests that the trout–mayfly interaction is relatively strong in natural streams. Alternatively, we may not have achieved enough replication to detect more subtle patterns in nature. We plan to continue making observations to seek patterns until we have attained acceptable replication to rule out this alternative explanation. But at the moment, we must resolve and interpret the role of observational approach in the ecology of complex systems.

Interpreting Observational and Experimental Data

In stream ecosystems, as in many others with comparable complexity, because of methodological constraints and inherent complexity it may not be possible to observe directly processes that are potentially important in natural streams. Since observers cannot attain resolution at the temporal and spatial scales at which processes may be occurring, we often draw the erroneous conclusion that such processes do not occur or are not important in streams (Peckarsky 1991a). In other words, important processes, such as predation and competition, may occur but may simply not be tractable to observers in unmanipulated systems. Further, even if null observations are correct (e.g., lack of associations between predators and prey), processes (such as predator–prey interactions) may still play an important causal role in the distribution and abundance patterns of organisms. The lack of observable positive or negative associations between predators and prey may simply be due to the confounding influence of a myriad of biological and physical factors on predator and prey distributions. For example, the effect of predators on prey distributions may obscure the effects of prey on predator distributions! Phenomenological experiments such as those described there enable us to identify and then explore cryptic processes that do not leave an obvious mark (observable pattern) in natural systems.

Clearly, my recommendations for what to do if observational and experimental data do not agree run directly counter to those of Peters (1986). I would not rely on observational data as sufficient for generating hypotheses regarding potential causal pathways of species interactions or relationships among variables in complex systems where multiple factors might be operating in unknown ways. I suggest that experiments are necessary for generating hypotheses as well as testing them. These two different kinds of experiments each play an important role in elucidating and explaining patterns in nature.

I am not arguing that field observations be discontinued. On the contrary, the degree to which observations fail to generate regular patterns and deviate from results of phenomenological experiments is interesting and indicative of the complexity of the system. Controlled experiments will enable us to determine whether certain processes (e.g., predation and competition) may occur under manipulated conditions and their relative importance. The most difficult question is whether those processes occur or are important in unmanipulated systems. Just because we are unable to observe natural patterns of distribution and abundance that are consistent with a process does not mean that these patterns do not occur or that they are not important. The consequences may simply not be observable within the context of the other biological and physical factors that influence distributions and abundances of organisms in complex systems.

How To Design Good Experiments in Complex Systems

The attacks on experimental ecology are not without basis (Peters 1986) in that the main criticism of experiments (for being unrealistic) is worthy of serious consideration by experimentalists. The challenge of designing reliable hypothesis-generating and hypothesis-testing experiments in natural systems is one not to be taken lightly. If the ultimate goal is to understand how a particular system behaves, we should simulate as closely as possible the natural conditions and spatial and temporal scale under which species interact and within those conditions manipulate singly and interactively variables hypothesized to influence patterns of species distributions and abundances. The closer we can get to mitigating experimental artifacts, the better the interpretive power of experiments is. Ideally, results of experiments should be compared over various temporal and spatial scales and using a gradient of habitat complexity (Peckarsky et al. 1997). That way, we can determine the scales at which processes may be detected in nature and which processes are sensitive to environmental complexity. Those innovations will facilitate making the leap from experimental results to conclusions relevant to unmanipulated systems.

After having grappled with the agonies of defeated experiments over the past 20 years of attempts to manipulate variables in streams, the thrill of a few victories has given us cause for some optimism about the feasibility of using experiments in complex and dynamic systems. In some cases, progress has been slow, but we have continued to steadily improve and refine our ability to be confident about the validity and power of our experimental tests. Other innovations are being implemented increasingly by a large number of stream ecologists to meet the challenges of experimentation in stream systems. I refer readers to the following studies for creative ideas: Hart 1981, 1986; Lamberti and Resh 1983; Cooper 1984; McAuliffe 1984; Power 1984; Kohler 1985; Sih et al. 1988; Glozier and Culp 1989; Kohler and McPeek 1989; Cooper et al. 1990; Crowl and Covich 1990; Wilzbach 1990; Culp et al. 1991; Resetarits 1991, 1995a,b; Flecker 1992; Pringle and Blake 1994; Feminella and Hawkins 1994; McIntosh and Townsend 1994; Scrimgeour and Culp 1994; Power et al., this volume). Contributions made by these and other stream ecologists are important, because ''anything that improves conditions of experimental designs in complex, very variable, very interactive ecological studies would be a good thing'' (underwood 1991:862).

I conclude with a set of recommendations for aspiring experimental ecologists as counterarguments to Peters's (1986) top five reasons why not to do experiments: (1) only a few experiments will succeed; (2) experiments are too expensive; (3) experiments take too long; (4) experiments exceed the capacity of most labs; (5) the goals of experimental ecologists are elusive and ethereal.

In my opinion, the best studies of ecological systems combine a rigorous knowledge of natural history obtained through careful field observations with carefully crafted experimental tests of falsifiable hypotheses generated from model predictions derived from observations, where possible, and from phenomenological experiments where not possible. Experimentalists must be sensitive to the strengths and limitations of their methods while accepting the challenges of manipulating variables in temporally and spatially complex systems. One should always start with simple designs before moving to complex interactive experiments. Simple experiments may be the most effective in providing clear answers to complex questions (S. I. Dodson, important lesson number 1 to his graduate students). A simple approach also increases the chances that experiments will succeed and reduces the necessity for spending too much money or exceeding the capacity of one's laboratory.

In temporally variable systems, long-term studies should be implemented whenever possible, encompassing multiple seasons, annual cycles, and all stages of organisms with complex life cycles, even though progress may be slow. The benefits of coming closer to the truth outweigh the costs of spending too much time to get there. I think the goal of experimental ecologists should be to understand patterns and processes that occur in complex natural systems and that this goal is neither elusive nor ethereal. Finally, experimental ecologists should keep an open mind at all times (S. I. Dodson, important lesson number 2 to his graduate students), trusting their observations even if they fly in the face of existing dogma. Beginning graduate students and very famous scientists are all operating under the same rules with regard to criteria for drawing valid conclusions and are subject to the same constraints regarding the importance of implementing responsible experimental designs. By keeping open minds we can all contribute new ideas, which is how the science of ecology grows and matures as a discipline.

ACKNOWLEDGMENTS   I'd like especially to thank Stan Dodson, my former adviser, who provided an atmosphere that enabled me to grow as an ecologist and taught me more by example than by formal instruction. Stan exemplified everything I tried to say in this essay about the way ecology should be done and was an invaluable role model who will influence my approach to research and teaching for the rest of my career. Tony Underwood's dogmatism inspired me to stick my neck out more than I probably should have in this essay. But if he doesn't get into too much trouble, why should I? A special thanks to my talented and creative lab group (Angus MacIntosh, Peter Ode, Sarah Vance, Chester Anderson, David Lytle, Jessamy Rango, and Gail Blake), who critically read earlier drafts of this article and contributed many excellent ideas to help make it better. Also, thanks to Bill Resetarits and four anonymous reviewers for helping me recognize the most important contribution I had to make to this symposium volume. The illustrations were done by Gail Blake, and the data in this essay were obtained with support from the National Science Foundation.

Literature Cited

Allan, J. D., and A. S. Flecker. 1992. Resource depression and intraspecific competition in the mayfly *Baetis tricaudatus*. Bulletin of the North American Benthological Society 7:94.

Cooper, S. D. 1984. The effects of trout on water striders in stream pools. Oecologia 63: 376–379.

Cooper, S. D., and T. L. Dudley. 1988. Interpretation of "controlled" vs. "natural" experiments in streams. Oikos 52:357–361.

Cooper, S. D., S. J. Walde, and B. L. Peckarsky. 1990. Prey exchange rates and the impact of predators on prey populations in streams. Ecology 71:1503–1514.

Cowan, C. A., and B. L. Peckarsky. 1994. Feeding and positioning periodicity of a grazing mayfly in a trout stream and a fishless stream. Canadian Journal of Fisheries and Aquatic Sciences 51:450–459.

Crowl, T. A., and A. P. Covich. 1990. Predator-induced life-history shifts in a freshwater snail. Science 247:949–951.

Culp, J. M., N. E. Glozier, and G. J. Scrimgeour. 1991. Reduction of predation risk under the cover of darkness: avoidance response of mayfly larvae to a benthic fish. Oecologia 86:163–169.

Diamond, J. M., and M. E. Gilpin. 1982. Examination of the "null" model of Conner and Simberloff for species co-occurrences on islands. Oecologia 52:64–74.

Feminella, J. W., and C. P. Hawkins. 1994. Tailed frog tadpoles differentially alter their feeding behavior in response to non-visual cues from four predators. Journal of the North American Benthological Society 13:310–320.

Flecker, A. S. 1992. Fish predation and the evolution of invertebrate drift periodicity: evidence from neotropical streams. Ecology 73:438–448.

Flecker, A. S., J. D. Allan, and N. L. McClintock. 1988. Male body size and mating success in swarms of the mayfly *Epeorus longimanus*. Holarctic Ecology 11:280–285.

Glozier, N. E., and J. M. Culp. 1989. Experimental investigations of diel vertical movements by lotic mayflies over substrate surfaces. Freshwater Biology 21:253–260.

Hairston, N. G., Sr. 1989. Ecological Experiments: Purpose, Design, and Execution. Cambridge University Press, Cambridge.

Harris, G. P. 1994. Pattern, process and prediction in aquatic ecology: a limnological review of some general ecological problems. Freshwater Biology 32:143–160.

Hart, D. D. 1981. Foraging and resource patchiness: field experiments with a grazing stream insect. Oikos 37:46–52.

———. 1986. Do experimental studies of patch use provide evidence of competition in stream insects? Oikos 47:123–125.

Kohler, S. L. 1985. Identification of stream drift mechanisms: an experimental and observational approach. Ecology 66:1749–1761.

Kohler, S. L., and M. A. McPeek. 1989. Predation risk and the foraging behavior of competing stream insects. Ecology 70:1811–1825.

Lamberti, G. A., and V. H. Resh. 1983. Stream periphyton and insect herbivores: an experimental study of grazing by a caddisfly population. Ecology 64:1124–1135.

Lehman, J. T. 1986. The goal of understanding in limnology. Limnology and Oceanography 31:1160–1166.

McAuliffe, J. R. 1984. Competition for space, disturbance, and the structure of a benthic stream community. Ecology 65:894–908.

McIntosh, A. R., and C. R. Townsend. 1994. Interpopulation variation in mayfly anti-predator tactics: differential effects of contrasting predatory fish. Ecology 75:2078–2090.

Miller, J. C. 1986. Manipulations and interpretations in tests for competition in streams: "controlled" vs. "natural" experiments. Oikos 47:120–123.

Ode, P. R., and B. L. Peckarsky. 1995. The role of food limitation and competitor density in larval mayfly ecology. Bulletin of the North American Benthological Society 12:162.

Peckarsky, B. L. 1980. Predator-prey interactions between stoneflies and mayflies: behavioral observations. Ecology 61:932–943.

———. 1988. Why predaceous stoneflies do not aggregate with their prey. Verhandlungen der Internationale Vereinigung für theoretische und angewandte Limnologie 23:2135–2140.

———. 1991a. Habitat selection by stream-dwelling predatory stoneflies. Canadian Journal of Fisheries and Aquatic Science 48:1069–1076.

———. 1991b. Mechanisms of intraspecific interference between stream-dwelling stonefly larvae. Oecologia 85:521–529.

———. 1996. Alternative predator avoidance syndromes in stream-dwelling mayflies. Ecology. 77:1888–1905.

Peckarsky, B. L., S. D. Cooper, and A. R. McIntosh. 1997. Extrapolating from individual behavior to populations and communities in streams. Journal of the North American Benthological Society 16:375–390.

Peckarsky, B. L., and C. A. Cowan. 1991. Consequences of larval intraspecific interference to stonefly growth and fecundity. Oecologia 88:277–288.

Peckarsky, B. L., and S. I. Dodson. 1980. An experimental analysis of biological factors contributing to stream community structure. Ecology 61:1283–1290.

Peckarsky, B. L., and A. R. McIntosh. In press. Fitness and community consequences of avoiding multiple predators. Oecologia

Peckarsky, B. L., and M. A. Penton. 1985. Is predaceous stonefly behavior influenced by competition? Ecology 66:1718–1728.

———. 1989. Mechanisms of prey selection by stream-dwelling stoneflies. Ecology 70:1203–1218.

Peckarsky, B. L., C. A. Cowan, M. A. Penton, and C. R. Anderson. 1993. Sublethal consequences of stream-dwelling predatory stoneflies on mayfly growth and fecundity. Ecology 74:1836–1846.

Peckarsky, B. L., C. A. Cowan, and C. R. Anderson. 1994. Consequences and plasticity of the specialized predatory behavior of stream-dwelling stonefly larvae. Ecology 75:166–181.

Peters, R. H. 1986. The role of prediction in limnology. Limnology and Oceanography 31:1143–1159.

———. 1991. A Critique for Ecology. Cambridge University Press, Cambridge.

Pianka, E. R. 1994. Evolutionary Ecology. HarperCollins, New York.

Power, M. E. 1984. Habitat quality and the distribution of algae-grazing catfish in a Panamanian stream. Journal of Animal Ecology 53:357–374.

Pringle, C. M., and G. A. Blake. 1994. Quantitative effects of atyid shrimp (Decapoda: Atyidae) on the depositional environment in a tropical stream: use of electricity for experimental exclusion. Canadian Journal of Fisheries and Aquatic Sciences 51:1443–1450.

Resetarits, W. J., Jr. 1991. Ecological interactions among predators in experimental stream communities. Ecology 72:1782–1793.

———. 1995a. Competitive asymmetry and coexistence in size-structured populations of brook trout and spring salamanders. Oikos 73:188–198.

———. 1995b. Limiting similarity and the intensity of competitive effects on the mottled sculpin, *Cottus bairdi*, in experimental stream communities. Oecologia 104:31–38.

Rigler, F. H. 1982. Recognition of the possible: an advantage of empiricism in ecology. Canadian Journal of Fisheries and Aquatic Sciences 39:1323–1331.

Scrimgeour, G. J., and J. M. Culp. 1994. Foraging and evading predators: the effect of predator species on a behavioral trade-off by a lotic mayfly. Oikos 69:71–79.

Sih, A., J. W. Petranka, and L. B. Katz. 1988. The dynamics of prey refuge use: a model and tests with sunfish and salamander larvae. American Naturalist 132:463–483.

Simberloff, D. S. 1978. Using island biogeographic distributions to determine if colonization is stochastic. American Naturalist 112:713–726.

Underwood, A. J. 1990. Experiments in ecology and management: their logic, functions and interpretations. Australian Journal of Ecology 15:365–389.

———. 1991. The logic of ecological experiments: a case history from studies of the distribution of macro-algae on rocky intertidal shores. Journal of the Marine Biological Association of the United Kingdom 71:841–866.

Vance, S. A., and B. L. Peckarsky. 1996. The infection of nymphal *Baetis bicaudatus* (Ephemeroptera) by the mermithid nematode *Gasteromermis* spp. Ecological Entomology 221:377–381.

Walde, S. J., and R. W. Davies. 1984. The effect of intraspecific interference on *Kogotus nonus* (Plecoptera) foraging behavior. Canadian Journal of Zoology 62:2221–2226.

Wilzbach, M. A. 1990. Nonconcordance of drift and benthic activity in *Baetis*. Limnology and Oceanography 35:945–952.

17

Design, Implementation, and Analysis of Ecological and Environmental Experiments

Pitfalls in the Maintenance of Logical Structures

A. J. UNDERWOOD

There has been an explosion of numbers of ecological experiments during the last 20 to 30 years. There has, however, also been ample evidence of problems with experimental and analytical procedures, so that several assessments of ecological experiments have been critical, rather than appreciative (e.g., Dawkins 1981; Underwood 1981, 1986, 1988; Preece 1982; Hurlbert 1984; Seaman and Jaeger 1990; Snaydon 1991; Bennington and Thayne 1994; Hamilton 1994).

One of the worrisome aspects of the numerous problems in design, analysis, or interpretation is the focus on statistical details—particularly where such details might help solve problems with the data—rather than ecological (or simply logical) issues. This is a broad assertion, but this essay is intended to demonstrate that our collective increased skills with manipulations of data, complex and imaginative analyses, and increased adoption of computer-intensive randomization and bootstrapping are admirable, but not satisfactory. Increased technology and numerical expertise are not sufficient to solve problems of designing and interpreting ecological experiments. My central theme is that in many experimental studies too little logical structure is identifiable before the experiment. Therefore, saving the analytical structure afterward, by whatever means, including sophisticated ones, will not cause coherent and steady advance in ecological understanding.

This discussion will focus on the need for an explicit logical structure or framework in any experimental study. The framework chosen can vary but must clearly identify the purpose of the experimental manipulation and how it would be interpreted. Obviously, this draws attention to the need for very explicit hypotheses before the experiment can be planned. Further, however, this discussion will dwell on the need for the hypotheses being tested to be properly constructed based on the theory or model being examined. The theory or model, in turn, must be proposed to describe or explain some pattern or observation. Because these three steps (observations or pattern, theory or explanation, prediction or hypothesis) become confused in some studies (at least to the readers), there are problems with designing the appropriate experiment and subsequently analyzing and interpreting the data. It is particularly difficult when the observations and model proposed to explain them are really quantitative, but the hypotheses tested in the experimental part of the study are purely qualitative or verbal. Not only does this lead to problems for interpretation, but it also makes calculation of the power of the relevant statistical component of analyses impossible. This will be illustrated for two different types of study.

Then, aspects of the spatial and temporal scales of ecological studies will be considered to examine problems with the connection between the scale of observation, those of models proposed to account for them, and the scale of experimental or sampling studies done to test hypotheses. Finally, because interpretations of ecological phenomena are made difficult by the lack of repeated experimentation, some aspects of the need for multiple experiments and multiple sites will also be discussed. These considerations lead to the conclusion that we must be much more prepared to own up to the shortcomings of experimental procedures and much less happy to attempt definitive interpretations from short-term or small-scale studies. This is not likely to be new— but the fact that problems keep appearing indicates that the issues need to be raised repeatedly.

Making Hypotheses Explicit

A Philosophical Framework

There are probably as many attempts to define an appropriate philosophical framework for ecological experiments as there are ecologists. Nevertheless, it is worth expounding one here with a plea that if this is considered inappropriate, then each person so considering it must be able to articulate the framework in which the experimental work is done. The structure of any experimental test of logically derived hypotheses must fit into the broader philosophical framework that links theory, experiment, and interpretation. The details of the following philosophical position may be found in Underwood (1990), with ecological case histories also in Underwood (1991b). Alternatives abound, but particularly relevant discussion is available in Caswell (1976), Dayton (1979), Simberloff (1980), Quinn and Dunham (1983), Connor and Simberloff (1986), Fagerstrom (1987), Loehle (1987, 1988), Mentis (1988), and Bourget and Fortin (1995).

In Underwood (1990, 1991b, 1997), I argued that we proceed in ecology by making observations about nature and then attempting to explain them by proposing theories or models. Usually, several possible models will be equally valid as explanations of some set of observations (e.g., Chamberlin 1965). We therefore need some discriminatory procedure to distinguish among alternative models. Therefore, we deduce from each model a specific hypothesis (or set of hypotheses) which predicts events in some as yet undocumented scenario, if the model is correct. If a model is incorrect, its predictions will not come true and it will therefore fail—provided that it is, in fact, tested. Experiments are, therefore, tests of hypotheses. An experiment is the test that occurs when the circumstances specified in a hypothesis are created so that the validity of the predictions can be examined. Because of the logical nature of ''proof'' (see Lemmon 1971, Hocutt 1979), we must usually attempt to turn a hypothesis into its opposite (the ''null'' hypothesis) and to do the experiment in an attempt to disprove the null hypothesis and thereby provide empirical and logical support for the hypothesis and model. The details of this procedure, as used in ecology and environmental science, are expounded in full elsewhere (Underwood 1990, 1991b, 1995, 1997).

Statistical procedures are usually necessary to help decide whether to reject or to retain the stated null hypothesis. This causes two very different problems for the experimental ecologist. First, statistical null hypotheses are often quite different from logical null hypotheses, causing immense potential for confusion and quite invalid (i.e., illogical) inferences in certain types of experiment. The details of these problems are discussed in Underwood (1990, 1994c) and will not be considered further here.

Second, statistical analyses used to help make decisions about rejection or retention of a null hypothesis almost invariably require assumptions—often quite strict assumptions—about the data gathered during an experiment.

Both strands of experimental ecology—the logical framework in which interpretations and conclusions are to be made and the statistical procedures used to help reach a decision—require great care in the design, establishment, and maintenance of ecological experiments. Some of these issues are illustrated in the following section in an attempt to increase the discussion of experiments during their planning.

An Example from Predator–Prey Studies

Consider observations like those in Figure 17-1a. Where predators are numerous, there are fewer prey; where there are fewer predators, prey are abundant (Spearman's rank correlation coefficient = .69, 10 df, $P < .05$).

One model that can explain this pattern is that the predators cause the trend by eating more prey where they (the predators) are numerous. There are several other possible explanatory models. For example, the predators and prey may have different requirements for some aspect of habitat, so that they tend to be relatively more abundant in different parts of the studied area.

To determine whether the first of these models is realistic, several hypotheses might be proposed. The one considered here is that removal of predators from the areas where they are numerous should lead, after some specified time, to greater numbers of prey compared with controls where predators are not removed. Under some circumstances, this is not a particularly compelling hypothesis. The observed pattern may only have been noted at the end of the process, so that no further change due to predation will now occur. The hypothesis about predation would be tested at an inappropriate time.

To keep the discussion simple, assume that prey recruit in spring and that removal of predators is an easy chore, requiring no controls for fences or whatever form of installation is used to keep them out. An experiment is done with four independent replicate plots where predators are removed and four control plots where they are present. The replicates are, of course, independently and randomly scattered around the study area. Suppose the experiment runs for a few months from before to after spring, so that it ends at the time of year when the observations were originally made. At the end, there were more prey where predators had been removed (Fig. 17-1). This was

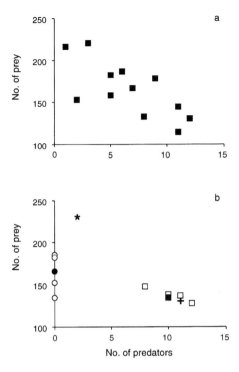

Figure 17-1. An experiment on effects of predators. Observations (**a**) show a negative correlation between numbers of prey and numbers of predators. The outcome of an experimental removal of predators is shown in (**b**). Open circles are four experimental replicates where predators were removed; mean number of prey (164, S.E. 12) is shown as the filled circle. Open squares are four controls (with predators present), with mean number of prey (138, S.E. 4) shown as the filled square. The asterisk and the cross represent the mean numbers of prey at large (i.e., 10 to 12 per sampling unit) and small (1 or 2) numbers of predators, respectively, from the original observations in (a).

significant by a *t* test, $P < .05$ (ignoring the heterogeneity of variances in the two treatments for the sake of keeping the discussion simple). As a result, we should reject the null hypothesis and deduce that predators indeed reduce the numbers of prey by eating them.

This is, however, an ambiguous interpretation because it is possible that recruitment of prey differed in plots with and those without predators. In other words, differences in numbers of prey may be caused by recruits or immigrants being deterred from establishing themselves in areas where there are numerous predators. Predators would therefore be implicated as the cause of the observed difference in numbers of prey, but not through killing/consumption of prey as specified in the model. To sort this out would have required the eight plots to have the predators removed until recruitment (or immigration) was finished. Four of them would then be altered to allow access to the predators when the experiment started. In such an experiment, there is no reason to presume any differences in recruitment or migration between "control" and "experimental" plots before the experiment starts (and, indeed, this could be checked before the predators are allowed in).

We could proceed with the interpretation that the significant difference supports the model—that is, the difference is due to mortality caused by predators. At one level, this would be satisfactory. We have results that could be used as part of a paper in a prestigious journal (these data support a currently popular theory that predators are important; see Dayton 1979). In the spirit of trying to arrive at a realistic understanding of how the system being investigated actually functions (which was, presumably, the real point of the study), the result and conclusion are, however, *not* satisfactory.

Note the relationship between the experimental results and the proposed explanation for the original observations. The mean values from the experiment are plotted in Figure 17-1b, with the means of the original data, so that the difference in mean number of prey between experimental and control plots is indicated in relation to the original pattern.

Note that the mean number of prey in the controls (where predators were not removed) matches the numbers originally observed. The mean number of prey where there are no predators is, however, smaller than the numbers observed in nature. Predation on its own does *not* explain the original observations. The model is wrong or inadequate, even though predation is a demonstrably significant process that influences the numbers of prey in the area studied.

Thus, the experimental result is significant, but the hypothesis and model cannot be considered corroborated. The results do not support the model that the original correlation is due to predation. More complex models are needed, which must incorporate other processes and relationships. These will lead to different hypotheses about the outcome of removing predators in different parts of the habitat, where there are different densities of prey. Alternatively, more complex hypotheses could be proposed about the numbers of prey which should survive when numbers of predators are manipulated to mimic the numbers found in different parts of the habitat. This would require experimental manipulations to remove different numbers of predators in various places to create a gradient of numbers of predators in each different part of the habitat (see also Underwood and Denley 1984 on the need for such experiments).

A logical approach to such an experiment would be to use all the available information to make the hypothesis much more quantitatively predictive. If predators are responsible for the observed correlation, their removal should result in numbers of prey equivalent to those originally observed where there were no predators. If, as before, predators are removed from the areas where they are most dense, the mean difference in numbers of prey should be an increase from about 130 to about 230 (as shown in Fig. 17-1b). Thus, the predicted increase as an alternative to the null hypothesis is about 100. This difference (100) is known as the "effect size" and was estimable from the original observations.

Recognizing the nature and details of the pattern being explained, rather than blindly focusing on a simple model, would have three advantages. First, it would prevent illogical conclusions about the outcome. Second, it would cause increased focus on the magnitude and therefore the ecological importance of the effect of removal of prey, rather than on the simple fact of statistical significance. Third, it would also provide the most crucial component of anticipatory calculation of the power of the test of this ecological hypothesis (e.g., Cohen 1977, Toft and Shea 1983, Underwood 1990). Calculation of power of any statistical test requires assessment of the effect size—that is, the difference among treatments anticipated under some particular hypothesis if the null hypothesis is, in fact, incorrect.

An Example from Environmental Rehabilitation

It is crucial for any experiment that the hypothesis (or hypotheses) be identified wisely and well. This will be illustrated here by an example of environmental sampling to determine the success of a program of rehabilitation of a polluted site (e.g., USEPA 1985, McDonald and Erickson 1994). The object of the study is to ascertain when an area used for mining can be considered "recovered" or back to normal after the disturbance to vegetation caused by mining. At the start of rehabilitation, there are few plants in the area mined because the mine removed them. Work is done to replant the appropriate terrestrial vegetation so that the original biomass or cover or diversity of species will be re-created. If sufficient biomass, cover, or species are later found in the area, the site will be declared rehabilitated.

The usual approach is based on the original biomass of plants in the mined site being different from that in a control site (or, better, a series of control sites; see Underwood 1992). The typical hypothesis can therefore be stated as "the ratio of the average biomass of plants in the disturbed site to that in the control(s) will be less than unity." The null hypothesis is that the ratio of mean biomasses will be equal to (or greater than) 1.

When plants in the previously mined and the control sites are sampled, the hypothesis of no difference can be tested. If the null hypothesis continues to be rejected in statistical tests, the mined site cannot be declared rehabilitated and the program of remediation must continue. This classical view leads to a very serious, if not environmentally disastrous, problem.

If poor sampling is done, leading to very imprecise estimates of the mean values of the variable sampled, it is easy to demonstrate that there is no difference between the control and disturbed sites, even if they are very different. This is illustrated in Figure 17-2a (see McDonald and Erickson 1994).

In contrast, if the mean value of the chosen variable really is very similar between the disturbed and control sites, but sampling is very precise, the statistical test will keep rejecting the null hypothesis (Fig. 17-2b). Under these circumstances, expensive remediation will have to be continued, even though there is no persistent reduction of plant biomass compared with controls.

Obviously, there will be considerable financial pressure on those responsible for rehabilitation to use sloppy and imprecise methods of sampling. Such a situation is not only irrational but also indefensible.

The problem is the defined hypothesis and null hypothesis. What is needed is a reversal of the hypothesis. This could only be accomplished by defining what minimal mean biomass of plants would be acceptable as an indication that the site is minimally recovered. This is illustrated in Figure 17-2c, where it was decided that if the biomass of plants was about three-quarters of that in an undisturbed control area, this would represent "recovery." Once this has been defined, it is then only possible to use precise sampling to demonstrate equivalence of abundance of biomass in the two areas, as shown in the lower parts of Figure 17-2. If imprecise estimates are now obtained from poorly designed sampling, the data will not cause rejection of the null hypothesis. The disturbed site will continue to be declared unrehabilitated (McDonald and Erickson

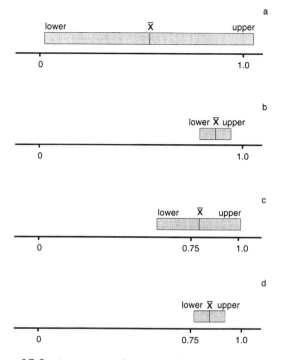

Figure 17-2. Assessment of recovery of a disturbed site. In all four diagrams, the ratio of biomass of plants in a previously mined area to that in a control area is shown as the theoretical value of 1, which represents no difference, that is, complete recovery of the population in the previously mined area. In each case, a sample is taken, with mean abundance of plants, $\bar{X}$, and upper and lower boundaries of confidence limit, as shown. In (**a**) and (**c**), sampling is sloppy, precision is not great, and confidence limits are quite large. In (**b**) and (**d**), sampling is precise and confidence limits are small. Under a traditional null hypothesis (the disturbed and control sites do not differ), situation (a) would erroneously retain the null hypothesis—solely because of sloppy sampling. The two sites are clearly different. In contrast, under the traditional null hypothesis, situation (b) would cause rejection of the null hypothesis, even though the two sites are really similar. This is because the sampling is so precise that small differences seem to be important. In (c) and (d), the null hypothesis is that the ratio of mean numbers between the two sites is less than 0.75, the minimal ratio being considered to represent recovery of the population (this was arbitrarily chosen for this example). Now, sloppy sampling in (c) leads to retaining the null hypothesis (as it should). Precise sampling allows rejection of the null hypotheses and therefore a correct declaration that the two sites are similar and recovery is satisfactory. After McDonald and Erickson (1994).

1994). Only by having appropriate precise estimates will it be possible to demonstrate that the minimal conditions required to demonstrate rehabilitation have been met.

This example serves notice that considerable care must be put into the appropriate methods and logics of environmental sampling before ecologists can have a useful role in decision making about environmental issues (see also Peters 1991 and Shrader-Frechette and McCoy 1994). Unless clear thinking prevails, there is no future for sensible ecological contributions to such issues.

Designing Experiments to Be Appropriate

Spatial Scales of Processes

One area of concern in designing any ecological experiment is that there are severe logistic constraints on experiments at large spatial scales. It is quite difficult to do any experiment over a large scale, even if the cost and effort were not great. It is therefore often tempting (or necessary) to do the study over a small scale which is hoped to represent events over the larger scale.

To illustrate one aspect of this, consider an environmental study which provides a large-scale experiment. It has been proposed that a jetty be built near a sea grass bed in a bay, somewhere along a relatively unspoiled coastline. The developers have suggested that any impacts on biological systems in the sea grass bed will be negligible, if they occur at all. In contrast, local environmentalists claim that operation of the new jetty will cause contamination and an impact on fauna in the sea grass. In formal terms, they have observed that in other areas with commercial jetties there are fewer species and smaller abundances of fish and invertebrates. Their model to explain such observations is that the operation of a jetty causes an environmental impact, and they predict that a new jetty will cause declines in numbers and richness of species. The relevant null hypothesis is that there will be no decline (or, for logical completeness, there may be increases in populations). The prospect of increased numbers of animals is also a potential environmental impact (Underwood 1991a, 1992), so, technically, the appropriate null hypothesis is that there will be no change and any alternative will be evidence of an impact caused by the jetty.

Of course, the proponents assert that they will manage the operations carefully and predict that there will be no environmental impact. Their hypothesis is, of course, the null hypothesis derived from the environmentalists' model.

After due debate and argument, some authority approves the jetty, subject to appropriate "monitoring" to detect any consequent impact. In ecological experimental terms, someone authorizes the experiment to test the null hypothesis. To collect data for this test, therefore, requires sampling in the immediate vicinity of the new jetty and in several replicate control or reference areas of similar sea grass elsewhere in the bay (as in Fig. 17-3a). The protocol for such an experiment and the justification and rationale for selecting several controls are explained in full elsewhere (Underwood 1992, 1993, 1994a).

Suppose, however, that there is no real knowledge of the relevant spatial scale of any disturbance due to the jetty. Perhaps, despite previous experience, a mishap during unloading freight would cause widespread chemical contamination, as illustrated in

a

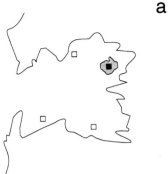

b

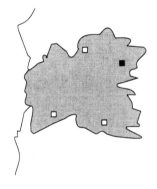

Figure 17-3. Sampling designed to detect environmental impacts due to waterborne contaminants in an estuary. In (**a**), the potential impact is believed to be of small scale (the shaded area around a jetty (*black symbol*). Therefore, control or reference areas can be elsewhere in the estuary (*as shown by the empty symbols*). In (**b**), in contrast, the impact turns out to be of large scale, affecting the entire estuary. The reference sites are also affected, yet the comparison between reference sites and the jetty would reveal no differences and therefore no impacts. Sampling at the appropriate spatial scale is crucial; where the scale is not known, sampling must be planned very carefully to avoid misinterpretations (see also Underwood 1992, 1994c).

Figure 17-3b, leading to decreases in populations. All the control sites would then be affected. Quite erroneously, the null hypothesis of no change (no environmental impact) due to the jetty would be retained. There would be no difference in abundances of species near the jetty from those found elsewhere. The general decline in populations would be attributed to some larger scale ecological process and not a result of impact due to the jetty.

As explained elsewhere (Underwood 1992, 1994a, 1995), uncertainty about the scale of relevant processes requires experimental designs that include controls at several spatial scales. The analysis and interpretation of such experiments are then necessarily more complicated, but such complication is much more easy to deal with than a completely illogical outcome (as in Fig. 17-3a).

Relevant Temporal Scales of Sampling

As with spatial scales, temporal patterns in ecological measures also require careful consideration of the appropriate scales or intervals of sampling. Among others, Stewart-

Oaten et al. (1986) have discussed the necessity to attempt randomization of sampling through time in order to avoid coincidence with cyclic phenomena.

There is, however, the associated problem that temporal sampling to detect long-term (e.g., seasonal) patterns of change in numbers of animals or plants also requires appropriate temporal replication. This is illustrated in detail in Underwood (1994a, b). Usually, seasonal or other intervals are examined by sampling to estimate abundances (or whatever variable) once in each season. The replication (random quadrats, cores, sets, traps, etc.) is almost always spatial—several replicated units are sampled at the one time. Thus, some estimate is made of spatial variation, but nothing unconfounds temporal differences. If there are shorter term (i.e., smaller scale) fluctuations in numbers of the relevant organism, the samples can only reveal that there is a difference from one time to another—not that there is a difference from one season to another (see Hurlbert 1984 for related discussions). Proper tests of hypotheses about annual, seasonal, or other relatively long periods require appropriate independent replicated sampling at shorter intervals. Then, the average conditions at one relevant time (season, year, etc.) can be compared and any differences validly interpreted with respect to a hypothesis about the longer time scale. Shorter term fluctuations can be separated from longer term changes (e.g., Morrisey et al. 1992) if the latter are real. Alternatively, if there is as much short-term as long-term change, hypotheses about the latter will not be interpreted incorrectly unless the former are also investigated.

Preserving Logical Structures

Lack of Replication

It is still the case that many tests in ecological experiments are confounded by the lack of appropriate replication. Comparisons of abundances, sizes, feeding, or any ecological processes between two locations cannot be interpreted as being due to the particular cause specified in the model that led to the comparison. Such models lead to a specific prediction (i.e., hypothesis) of difference between the two places in some variable because of some specified cause. To make a valid conclusion about the reasons for an observed difference between two locations, it is compulsory to demonstrate that all other possible explanations (including any ecological process that causes different outcomes in two geographically distinct areas) are wrong. In most ecological studies, such a demonstration would necessitate averaging all other conditions except that specified in the hypothesis (e.g., the removal of predators in the example discussed previously) over a series of representative locations scattered and interspersed at the same spatial scales as those that separate different experimental treatments. This has been discussed in great detail with numerous relevant ecological examples by Hurlbert (1984).

A related phenomenon is the erroneously termed "natural" experiment used by some ecologists (e.g., the discussion by Diamond 1986). "Natural" experiments consist of comparisons of unmanipulated areas. As an example, models about the role of predation can be tested by proposing the hypothesis that there will be more prey animals in areas where predators are sparse then where they are abundant. A natural experiment would be to find areas with naturally small or large numbers of predators and to compare the abundances of prey in the two sets of areas. As documented in full elsewhere

(Connell 1974, Underwood 1986), these comparisons are useless for making logical inferences about hypothesized causes of difference between sites or circumstances. Obviously, any difference in numbers of prey may be due to predators or to any other processes operating differently in the two types of area. Many other environmental and ecological processes may cause differences in numbers of prey species, which will therefore inevitably be correlated with numbers of predators. Note that the two types of area were chosen *because* they differ (one type has few predators). The potential difference may be correlated with numerous processes that also cause differences in numbers of prey.

As discussed elsewhere (Underwood 1990), the emphasis on these being "natural" draws attention away from the fact that they are not experimental tests of hypotheses. By changing the emphasis in the phrase used, we can deprive them of their status. Thus, by removing the "success-grammar" (Stove 1980) involved in focusing on "natural" rather than "experimental," the lack of logical structure of these comparisons might be made more obvious (Underwood 1990).

Other Confounding due to Lack of Controls

Another area of concern in the design and interpretation of ecological experiments is the issue of controls for an experimental manipulation. As an example, consider an experiment on aspects of density dependence. Previous observations have demonstrated that animals tend to move away from crowded parts of the habitat, so that their dispersion is fairly uniform. One model to explain this is that rates of migration from any place are density-dependent; where there are more individuals, the rate of departure is faster. There are other relevant models; for example, emigration may be at the same rate, but immigration may be slower to areas with relatively large density. From this specified model, one sensible hypothesis is that if the local density of animals is increased, the animals will emigrate at a faster rate than in similar areas where there are naturally small densities. The null hypothesis is that the rates of emigration in the two sets of conditions will be equal or the rate of emigration will be greater from the areas with unchanged (relatively small) densities.

It is traditional to call the set of replicate areas with increased density the experimental treatment and to call the other set the control. This is not a convincing terminology for this type of experiment. Both sets of conditions are necessary under the hypothesis—because it predicts the contrast in dispersive behavior under the two sets of conditions.

If the proportions of animals leaving the plots were recorded throughout the experiment and differ, on average, so that a greater proportion left the experimental plots, the null hypothesis would be correctly rejected and the hypothesis of greater dispersal would be supported. The logical conclusion would be to support the model of density-dependent migration being the cause of the observed dispersion.

The conclusion about the model is, however, irrational, even though the hypothesis is correctly supported. The introduction of new individuals to increase densities confounds any interpretation in terms of the model. A list of some other factors that might equally well explain the observed larger proportional rate of migration is given in Table 17-1. The design of the experiment is unsatisfactory because all of these (and perhaps

Table 17-1. Confounding influences requiring controls in an experimental test of the hypothesis that increased density of limpets will lead to increased (density-dependent) emigration from plots

	Control initial density	Experimental increased density	
Potential influence on behavior	Original animals	Original animals	Introduced animals
Handling	Untouched	Untouched	Handled
Microhabitat	Normal	Normal	Random
Orientation	Normal	Normal	Random
Site on shore	Familiar	Familiar	Unfamiliar
Surrounding individuals	Familiar	Unfamiliar	Unfamiliar
Density	Natural	Enhanced	Enhanced

Control plots have undisturbed animals at natural densities. Experimental plots have undisturbed animals at natural densities and introduced animals (transplanted from elsewhere) to increase densities. The introduced animals have been handled, put in strange surroundings, in the wrong orientation with respect to such features as gradients across the habitat, and all animals in the experimental plots now have unfamiliar surrounding individuals. If more animals leave the experimental plots, this may be due to any of the confounding factors rather than the effect of increased density as specified in the hypothesis. The relevant controls for this experimental manipulation were discussed in full in Underwood (1988), where the influence of confounded effects of unfamiliar surroundings was more important for limpets than was increased density.

others) are uncontrolled. Controls are needed to demonstrate that (a) no confounding influence (i.e., outcome of a different model which also predicts faster emigration from the experimental plots) can also explain the observed results and (b) the results from the experimental plots can be properly related to those for undisturbed (natural) animals which were the ones for which the hypothesis was proposed.

In some ideal (and theoretical) world, increased densities can be achieved without disturbing individuals. For some experiments to test hypotheses about density-dependent processes in animals and plants, it is possible to grow animals in situ from eggs or to sow seeds at different densities so that no uprooting of individuals is necessary (e.g., Trenbath 1974, in contrast to Kroh and Stephenson 1980). Otherwise, the real world intrudes.

Thus, the experiment must have the experimental conditions of enhanced density, natural conditions, and controls for the confounding influences (as indicated for the example in Table 17-1). Further discussion of these issues may be found in Dayton and Oliver (1980), Underwood (1986, 1992) and, particularly with respect to experiments that involve transplants of animals or plants, Chapman (1986), Underwood (1988), and Chapman and Underwood (1992).

Nonindependence of Ecological Data

Because data from ecological experiments are always variable and usually complex, decisions about rejection or retention of null hypotheses must be based on statistical, probabilistic procedures. These procedures inevitably involve assumptions about the nature of the data. Among important assumptions, most statistical procedures assume that data will be independent (uncorrelated) from one replicate to another within experimental treatments and from one experimental treatment to another (e.g., Gurevitch and Chester 1986, Crowder and Hand 1990).

There are many ways in which ecological data may become nonindependent. The most common forms of nonindependence discussed have been temporal or serial autocorrelation (e.g., Swihart and Slade 1985, 1986) and spatial autocorrelation (Cliff and Ord 1973, Legendre and Fortin 1989). These are however, largely statistical properties of measurements, although they relate to aspects of life history, aggregation, and so on.

More relevant for biologists, many aspects of behavior, ecological interactions, and so on of animals also cause positive or negative correlations among measurements. There is no room here to discuss the details of how these occur and what their consequences are for analyses. Ecological examples of various types of nonindependence were described in detail in Underwood (1994c). Broadly, there are four types of nonindependence in planned experiments. There can be correlation among replicates within experimental treatments or among different treatments. Each of these correlations can be positive or negative.

As an example, consider an experiment to test hypotheses about differences in feeding by animals kept on one of three different diets. If the animals are kept in groups (in pens or aquaria) and have feeding behavior that influences other, nearby individuals, the amount of time spent feeding or the frequency of bouts of feeding or other aspects of feeding will be influenced for some or all individuals. If some animals are dominant and spend a lot of time at the food and fight with or threaten other animals so that they keep away, there will be a negative correlation in rates of feeding among the individuals. Some will feed often and have unlimited access to food; others will feed rarely. Such negative correlation increases the variance among individuals in each set of replicate animals compared with the variance if the animals were kept separate. Separate animals could feed *ad lib*—independently of the threats of dominance of other individuals. So, if replicate individuals are sampled in each group kept together, their rates of feeding will not be independent. Whether this matters to statistical tests and their interpretation depends entirely on the hypotheses being tested (see Underwood 1994c for details). The point is that it is the behavior of the animals and not abstract statistical properties of the numbers themselves that causes the nonindependence.

Because many biological processes lead to nonindependence, ecological experiments need to be planned very carefully in order to prevent correlations or nonindependence. Consider the experiment on grazing in Figure 17-4a. There are three treatments: natural, unmanipulated areas, N; experimental areas, E, where grazers have been removed and fences placed around to prevent them returning; and controls (C), where grazers have not been removed, but fences have been placed round the plots to control for effects on plants due to the presence of fences themselves.

One replicate of each treatment is placed fairly close to each other, and sets of replicates are placed across the area (Fig. 17-4a). The number of plants in each plot would be recorded during the experiment to test the hypothesis that removal of grazers will cause an increase in densities of plants.

Now, as would be usual in most field experiments, suppose that natural differences in survival or recruitment of plants vary very little at a small spatial scale like that from treatment to treatment in a block. Variation in survival or recruitment at the larger scale

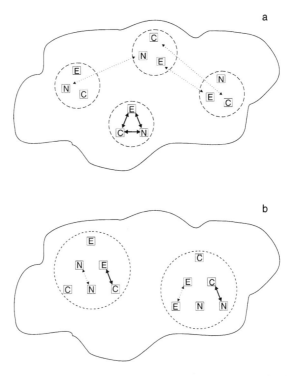

Figure 17-4. Diagram illustrating two types of randomized block design. In (**a**), three treatments, N (natural), E (experimental removal of predators), and C (controls with fences), are applied four times, once in each "block" (*dashed circles*). Spatial variation among replicates (*dotted arrows*) is at a larger scale than among treatments (*solid arrows*) so that intrinsic variation between and among replicates and treatments are not likely to be equal; see text for details. In (**b**), two replicates are placed in each "block." Spatial variation between replicates and treatments is now at the same scale and therefore likely to be equal. Design (a) is unreplicated within "blocks."

from block to block is greater. Increasing variance with increasing distance apart of sample units is a common phenomenon in ecological experiments (Goodall 1974, Bell et al. 1993).

If the mean numbers of plants in the three treatments are simply compared by analysis of variance (or any of its alternatives based on rank orders), the replicates in each treatment must be assumed to be independent. The analysis includes estimation of variance among replicates within treatments and of the same variance, but calculated from differences among the means of each treatments. If there is no effect on numbers of plants due to fences or to removal of grazers (the null hypothesis is true), these two estimates of spatial variance are assumed to be equal. In this experiment, they will not necessarily be equal. If, for example, spatial variance increases with increasing distance and there are no differences among the three treatments, the variance estimated among replicates will be larger than that estimated among means of treatments. So, if the null hypothesis is true, the latter will be smaller than the former. Alternatively, if the null hypothesis is false and there are differences among treatments, these will not appear as significant unless the differences are very large. Thus, the null hypothesis will be less likely to be rejected than is assumed in the structure of the analysis (see particularly Underwood 1994c).

The solution usually proposed is to treat the data as a randomized block experiment. In the design often recommended, "blocks" are arbitrarily or randomly chosen pieces of the area. One replicate of each treatment is placed in each block, as in Figure 17-4a. The variation attributable to differences among the blocks is then estimated separately from that among treatments and from what remains due to differences among replicates of each treatment. In theory, therefore, the variation among blocks is removed from that among replicates, making the comparison among treatments more readily interpretable.

The use of this analysis is, however, now dependent on the assumption that differences among experimental treatments are themselves independent of differences from one block to another. This is usually stated as the assumption that there is no *interaction* between treatments and blocks. In many (probably most) ecological experiments, this is unlikely to be realistic. For this experiment, suppose that the densities of grazers naturally and consistently vary from one block to another. Removing them where they are sparse will make little difference to the numbers of plants. In contrast, removing grazers from areas where they are abundant will probably have very large effects on the survival of plants. Therefore, the difference between natural (N) and experimental (E) plots will be very different from one block to another. This is a statistical interaction and would demonstrate that differences among treatments cannot be independent of differences from one block to another. Such interactions violate the assumption of the unreplicated randomized block design (only one unit of each treatment in each block).

To avoid making this assumption and to allow tests for the presence of interactions between blocks and treatments, the solution is to arrange the blocks and replicates so that there are replicates of each treatment in each block, as in Figure 174b. Now the estimate of spatial variance among replicates is at the same scale and therefore presumably similar to the spatial differences among treatments. If there are no differences among treatments, the two variances (among replicates and among treatments) will both be calculated at the smaller scale within blocks. Differences among blocks (i.e., the larger spatial scale) are still being satisfactorily separated from the variation among replicates, but there is a true measure of variation among replicates in each treatment in each block to allow a test for the presence of interactions between treatments and blocks. In this example, the workloads for the two experimental designs are identical.

The effects of these different types of correlations on analyses of variance of experimental data are summarized in Table 17-2. To avoid such correlations in ecological experiments requires careful planning. It also requires careful thought about the natural history, behavior, and ecology of the animals and plants being studied, so that potential behavioral or distributional correlations can be anticipated. Note that these aspects of biology and ecology are not the normal background knowledge of statistical advisers,

Table 17-2. Different patterns of nonindependence and their effects on analyses of variance of results of experiments; for full details of the types of nonindependence, see Underwood (1994c)

	Nonindependence is:	
	Among treatments	Within treatments (i.e. among replicates)
Positive correlation	Increased Type II error	Increased Type I error
Negative correlation	Increased Type I error	Increased Type II error

Note that most statistical procedures have similar assumptions about independence of data. Increased Type I error is increased likelihood of rejecting a null hypothesis when it is true. The converse is increased Type II error—retaining a null hypothesis when it is false.

who will remain unaware of the potential violations of assumptions unless the ecologist doing the experiment points them out (Underwood 1994c).

Where the nature and magnitude of correlation are known or can be estimated from the experimental data, it is sometimes possible to solve the problem for statistical analyses. Nevertheless, to discover correlations in the data requires that the experimenter realizes it is necessary to search for them. It is better to plan the experiment (e.g., to plan the spacing of replicates and the various treatments) to avoid the problems rather than to attempt to resolve them subsequently.

Constraining Interpretations

Extrapolation

Many ecological experiments—however well they are done—are done at a few sites (usually only one) or at few times (usually only one). As a result, ecology continues to be dominated by very singular case histories. Theories must inevitably be constructed by extrapolation from relatively small experiments with few replicates in one place and at one time.

The dangers of extrapolation from the laboratory to the field have been fairly well understood. It appears that we still have a long way to go before extrapolation from one to many sites and then to general rules of ecology is equally cautiously done. This was well put by Järvinen: "A typical good ecological study, giving a generous estimate, covers 3 years (out of the three and a half billion possible) and 1 km² (out of the more than 10^8 possible on earth). If the available area-time space is represented as a cube with sides of 1 m, the volume of a good ecological study constitutes a cube with sides about 1 micrometer. . . . [This lead to] generalizations based on a few species out of the few million extant ones" (1986:331).

So we need to be more cautious about generalizations (see some examples in Underwood and Denley 1984, Foster 1990). We need to be much more willing to do studies in more than one site and to do studies again (Connell and Sousa 1983). Otherwise, attempts to construct a more thoughtful ecology from solid case studies, as strongly advocated by Shrader-Frechette and McCoy (1994), will inevitably founder.

Their "logic of case studies" will only be meaningful if the case studies are strongly logical and based on robust and reliable data.

Multiple Experimental Sites

There are considerable problems in attempting to generalize from one geographical region to another. These have been summarized in the context of ecological experiments by Underwood and Petraitis (1993). One of the first things needed for comparisons of experimental studies is adequate replication and repetition from place to place within any area to ensure that comparisons among areas are meaningful. This is also true for any attempt to make a generalization about some process or its relative importance to other processes from one area to another or one type of habitat to another.

Thus, thought most be given to the nature of repeating experiments in more than one place. This seems to have caused some difficulty, judging by recent literature (McKone and Lively 1993, Lively and McKone 1994). Whenever an experimental study needs to be done at more than one place to provide some generality or simply to provide replication of the treatments, the nature of the experimental design must be considered very carefully. In all instances, the requirements of the design are determined by the nature of the hypothesis to be examined. For example, consider the experiment on removal of herbivores discussed by McKone and Lively (1993). They proposed an experiment in which herbivores were removed from 10 plants and 10 plants were kept undisturbed as a control, in each of three replicate sites. The sites were randomly chosen to represent different parts of a habitat, and presumably the experiment was done in several places to provide some spatial repeatability of the experiment, in order to increase its generality.

In their discussion, McKone and Lively (1993) identified (correctly) that the appropriate analysis was a "mixed-model" analysis of variance, with the experimental treatment being fixed, with one degree of freedom for comparing the two treatments (removal and control) and the sites representing a random factor (being chosen essentially randomly from those available). The problem for them arose because there was a significant interaction between the treatments and sites. In fact, the interaction identified that there is no simple effect of removing herbivores because the effect of such removal differed among sites (i.e., was site-dependent) as pointed out by Greenwood (1994) and in the previous discussion of randomized block experiments. McKone and Lively (1993) recommended that the two treatments should be compared in each site by a further partitioning of the analysis of variance. This is, in essence, Scheffé's (1953) test for comparing multiple means. Greenwood (1994) recommended the same procedure, but in the form of the corresponding t test. Neither author addressed the problems of increased probability of Type I error (increased chance of rejecting a null hypothesis when it is true) implicit in the multiple use of t tests or F ratios that they recommended. A preferable procedure would provide protection against excessive Type I error. In this case, the Student-Newman-Keuls procedure would be perfectly appropriate (see Underwood 1981, 1997). In other cases, where there are more treatments to be compared, there may be superior alternatives, as reviewed extensively by Day and Quinn (1989). There has been substantial discussion of the problems of the use of mixed models in experimental ecology—for example, Underwood (1997) and Bennington and Thayne (1994).

The discussion summarized here may have caused confusion because the various authors referred to their mixed-model designs and analyses as a "nested design," but nested designs were not being advocated in the McKone-Lively-Greenwood debate. A nested design for their experiment would have different sites for the removal of herbivores and for the controls—so that the sites are nested in each treatment. The experiment described by McKone and Lively (1993) was orthogonal or factorial—all experimental treatments were present in every site.

There are fundamental differences between the nature of the hypotheses for which it is appropriate to use a nested as opposed to a factorial experimental design. In the context of generalization from experimental results, there is great value in using factorial, mixed-model designs when the issue to be examined experimentally is whether some process is consistent across several sites and, if not, what differences occur. Thus, the null hypothesis that there is no difference in the biomass of plants when herbivores are removed can be tested in an experiment with removals and controls in the same set of sites. The difference between the two treatments is then analyzed. If there is an interaction, there is evidence that there is a different result for the test in some of the sites—removing herbivores does not cause a constant difference in mean biomass. The test of the null hypothesis should then proceed by an appropriate multiple comparison in each site. This will determine whether the null hypothesis should be retained or rejected in each site and, where there are differences between the treatments, provide the difference and its confidence limits.

In contrast, where there is a hypothesis that there is a "global" effect of herbivory, the null hypothesis is that there will be no difference between the mean biomass of plants when herbivores are removed and the mean biomass controls, despite any (i.e., over and above any) variation from one site to another. Thus, variation among sites must not be so large or so contrary that there is no measurable large difference between the two treatments. In this case, the appropriate design would be a nested one—three sites are to be chosen for the removal of herbivores and three different ones for the controls. Averaging for each of the treatments then involves averaging across a new, independent set of sites. There will be no correlation between the two treatments in the two sites. The difference between the two treatments in each site is irrelevant and therefore has not been examined in the experiment. There are no possible interactions between the two treatments and any differences among sites.

The latter experiment is useful for a restricted range of hypotheses. Usually, we should probably be more interested in testing an hypothesis in several sites to determine the generality of the outcome. Thus, the former design is usually more appropriate. Note, however, that the decision is not arbitrary but totally dependent on the hypothesis being addressed. Where hypotheses are not made clear (e.g., McKone and Lively 1993, Greenwood 1994), it is not possible to consider how best to do the experiment. Where there is no clear hypothesis, it is impossible to determine how appropriate any of the alternative designs might be. Nor, of course, is it possible to determine how the results should be interpreted—except in the general sense that they reveal something about herbivory.

Multiple Experiments and Their Interpretation

Whatever the hypothesis being tested, there is increasing need for multiple tests, repeated in many sites and, wherever possible, independently at many times. Among other reasons for this need is the notion of repeatability. Despite cynicism that ecology is always singular (based on the fact that most studies are singular, so there is little possibility of contradiction), a criterion of repeatability is a widespread safeguard against incautious belief in erroneous experimental results in many sciences. The best examples of repeatability are those done by critics of the original experiments and, ideally, those done by different people under different circumstances. When very different conditions reveal the same processes operating at similar magnitudes, there is more substance to our claim to understand what is happening in the world at large. This is particularly informatively discussed by Collins (1985).

Almost always, small experiments done independently in lots of places and at many different times will be more useful for providing generality than will single large experiments with an obsession for statistical power in their design. Discovering a significant difference several times in different experiments will almost always provide more faith in its realism and validity than discovering it once with smaller probability of Type I error! Note that tests at different times and places have all the weight of discovering similarity of processes using different cohorts of organisms (perhaps differing genetically in addition to other historical issues), different geography, and different mixes of all the participants. Any process that operates at all convincingly under these different circumstances presumably means something important for ecological understanding. Of course, it is easier to repeat experimental studies in areas of science like physics and chemistry, where the material to be used is likely to be very similar, if not identical, from one continent to another, provided the conditions of the experiment have been described coherently. In ecology, in contrast, the organisms, habitats, weather, and so forth are all so different from one country to another that such repeatability is not going to happen. Therefore, caution about generalizations should be more a catchword of ecologists and one that identifies us as the ones doing the hard science!

It is worth briefly resurrecting two methods for comparing the results of several experiments done to test a single null hypothesis. First, suppose two small independent experiments have been done to test a hypothesis about some process—for example, predation and its influence on a rare species. Each experiment was done with limited numbers of prey (to avoid adding ecological experiments to the ever-expanding list of proximate causes of extinction of rare species). Neither experiment had large power. Suppose that, in each experiment, the probability of getting the observed result if the relevant null hypothesis were true was not significant but quite small ($P < .10$ and $P < .06$, respectively). To make a single interpretation, we need to combine these two independent probabilities in a meta-analysis. The results can be combined using Fisher's (1935) formula:

$$C = -2\sum_{i=1}^{k} log_e P_i$$

where P_i is the probability associated with test i and there are k independent tests or experiments. C is distributed as X^2 with $2k$ degrees of freedom. In this example, $C = 10.23$ with 4 degrees of freedom, which is significant ($P < .05$). Thus, over the two small experiments, there is evidence that the null hypothesis should be rejected.

A second method for combining experimental results in more complex situations was used by Underwood and Chapman (1992) for a series of experimental manipulations of microhabitat for intertidal snails. In each experiment, there were the same six treatments. Each time the experiment was done, the analysis of variance revealed significant differences, but, in multiple comparisons, we were unable to identify which means differed. So, over a series of 12 independent repeats of the experiment—that is, the experiment was done 12 times over a period of several years—we ranked the order of means in each experiment and then examined the concordance of order of results. The number of times each treatment appeared first, second, and so on in order of magnitude of mean was then analyzed by the procedure described by Anderson (1959). In Chapman and Underwood (1992), the results were strikingly clear-cut. Thus, although no single experiment could ever be unambiguous, the set of repeated experiments was remarkably straightforward to interpret.

Although there are numerous problems involved in comparing experiments from one time or place to another, the available procedures should be much more widely used. This would allow separate studies to serve as independent reinforcement of accepted ideas and would reveal when previously exposed notions are incorrect.

Conclusions

Own up to Problems

One of the first steps in improving experimental approaches to ecology is to describe carefully the potential problems for any interpretation. Circumstances conspire to prevent the best-laid plans from being realized. Thus, there may be a loss of replication because of some disaster. Nevertheless, the results may be informative and should be used to indicate to others why a more definitive study is necessary or how to design a better attempt. This would be a much more useful end point if the problems for any interpretation were identified clearly rather than being buried. "There were no replicates, causing potential pseudoreplication (see Hurlbert 1984), but nevertheless the results indicate . . ." is a common method for dealing with illogicality. Authors who cite Hurlbert would do better if they had read his paper! If there is a good reason for describing results of unreplicated, uncontrolled, or unreliable experiments, the onus is on the experimenter to explain the reason rather than leaving the reader to guess. If a study done in Europe or the United States or Australia is to be interpreted by someone elsewhere who cannot know the biology or geography or history of the region, the onus is on the experimenter to explain the possible problems. Otherwise, those elsewhere must guess or refuse to accept the findings if there is any doubt at all about their validity.

The Myth of Definitive Results

From the previous considerations, we must recognize that virtually all our work is preliminary. Even though the results are published in good journals, that does not make the conclusions or the theories based on them correct. It does mean that the concepts are current, the thinking clear enough to satisfy the lottery of reviewers, and the result sufficiently convincing to those who accepted them.

Nevertheless, the whole thrust of the framework of philosophy summarized at the start of this essay is that models that have stood some testing should then be probed by more serious tests. Experimental results must be used to feed back to the models and theories. Attempts must be made to refine the models or theories to make them more precise or to expand and increase them to make them more general. Further hypotheses must be proposed and further experimental tests done. This is the way of all science. There is no point in sitting back and assuming that the work is over because some small-scale or short-term experiment is apparently convincing.

One advantage of constantly re-probing experimental analyses of hypotheses derived from current models is that the people doing the work usually find the problems before readers of the results do. As a result, there is some satisfaction in improving our understanding of our own branch of ecology rather than having to defend ourselves from the attacks of critics who have found the flaws in the experimental evidence that underpins the laurels on which we rest. It is better (and often thought to be admirable) to be self-critical rather than be criticized. So, accept that nearly all ecological work is, in some way, preliminary and should be used as a guide to something more integrated or larger scale. That way, there will be less need to attempt to defend shoddy experiments—because our careers and reputations will not be based on them. Instead, a research program of sustained critical evaluation of ideas, using increasing amounts of experimentation will be the stuff of progress. The hallmark of progressive ideas is that they progress. Given that there is a good chance we are wrong quite often, we should be prepared to discover how wrong as fast as possible.

The starting basis for achieving more rapid and directional progress in ecology is thinking more clearly about the nature of the experiments done to test our ideas. The first step is ensuring that there is a logical coherent structure in the way the study is designed and planned. Then, this logical structure must be maintained by practical considerations of the problems of experimental procedures so that the logic is not lost in the experiment. Both arms of this commonsense program will help ensure that the results of experiments can be interpreted in a sensible and practical manner and that they are not going to disappear as soon as they are held up to the light of critical appraisal. By defining carefully the logical framework in which ecological ideas are evaluated and by ensuring that the logical structure is maintained throughout each phase of a study we may provide every chance for optimism in the future capacity of ecology to solve real problems and to progress (contra the criticisms leveled by such notables as Sagoff 1985 and Peters 1991). Experimental design and the logical structure of hypotheses are intimately connected. Demonstrating and nurturing their connections can only help and may be the only method for ensuring that we put *logical* back into *ecological*.

ACKNOWLEDGMENT   I thank Dr. M. G. Chapman for excellent advice and help in the preparation of this essay and for teaching experimental design with me. I am grateful to the Australian Research Council for a Special Investigator's Award to support this research, to Dr. M. Beck and the referees for useful comments, and to V. Mathews and Dr. M. G. Chapman for help with the preparation of the figures.

Literature Cited

Anderson, R. L. 1959. Use of contingency tables in the analysis of consumer preference studies. Biometrics 15:582–590.

Bell, G., M. J. Lechowicz, A. Appenzeller, M. Chandler, E. DeBlois, L. Jackson, B. Mackenzie, R. Preziosi, M. Schallenberg, and N. Tinker. 1993. The spatial structure of the physical environment. Oecologia 96:114–121.

Bennington, C. C., and W. V. Thayne. 1994. Use and misuse of mixed model analysis of variance in ecological studies. Ecology 75:717–722.

Bourget, E., and M. Fortin. 1995. A commentary on current approaches in the aquatic sciences. Hydrobiologia 300/301:1–16.

Caswell, H. 1976. The validation problem. Pages 313–325 in B. C. Patten (eds.), Systems Analysis and Simulation in Ecology, Vol. 4. Academic Press, New York.

Chamberlin, T. C. 1965. The method of multiple working hypotheses. Science 148:754–759.

Chapman, M. G. 1986. Assessment of some controls in experimental transplants of intertidal gastropods. Journal of Experimental Marine Biology and Ecology 103:181–201.

Chapman, M. G., and A. J. Underwood. 1992. Experimental designs for analyses of movements by molluscs. Pages 169–180 in J. Grahame, P. J. Mill, and D. G. Reid (eds.), Proceedings of the Third International Symposium on Littorinid Biology. Malacological Society of London.

Cliff, A. D., and J. K. Ord. 1973. Spatial Autocorrelation. Pion, London.

Cohen, J. 1977. Statistical Power Analysis for the Behavioral Sciences, rev. ed. Academic Press, New York.

Collins, H. M. 1985. Changing Order: Replication and Induction in Scientific Practice. Sage, London.

Connell, J. H. 1974. Ecology: field experiments in marine ecology. Pages 21–54 in R. Mariscal (ed.), Experimental Marine Biology. Academic Press, New York.

Connell, J. H., and W. P. Sousa. 1983. On the evidence needed to judge ecological stability or persistence. American Naturalist 121:789–824.

Connor, E. F., and D. Simberloff. 1986. Competition, scientific method and null models in ecology. American Scientist 75:155–162.

Crowder, M. J., and D. J. Hand. 1990. Analysis of Repeated Measures. Chapman and Hall, London.

Dawkins, H. C. 1981. The misuse of t-tests, LSD and multiple-range tests. Bulletin of the British Ecological Society 12:112–115.

Day, R. W., and G. P. Quinn. 1989. Comparisons of treatments after an analysis of variance. Ecological Monographs 59:433–463.

Dayton, P. K. 1979. Ecology: a science or a religion? Pages 3–18 in R. J. Livingstone (ed.), Ecological Processes in Coastal and Marine Systems. Plenum, New York.

Dayton, P. K., and J. S. Oliver. 1980. An evaluation of experimental analyses of population and community patterns in benthic marine environments. Pages 93–120 in K. R. Tenore and B. C. Coull (eds.), Marine Benthic Dynamics. University of South Carolina Press, Columbia.

Diamond, J. M. 1986. Overview: laboratory experiments, field experiments and natural experiments. Pages 3–22 in J. M. Diamond and T. J. Case (eds.), Community Ecology. Harper and Row, New York.

Fagerstrom, T. 1987. On theory, data and mathematics in ecology. Oikos 50:258–261.

Fisher, R. A. 1935. The Design of Experiments. Oliver and Boyd, Edinburgh.

Foster, M. S. 1990. Organization of macroalgal assemblages in the Northeast Pacific: the assumption of homogeneity and the illusion of generality. Hydrobiologia 192:21–34.

Goodall, D. W. 1974. A new method for analysis of spatial pattern by random pairing of quadrats. Vegetatio 29:135–146.

Greenwood, J. J. D. 1994. Statistical analysis of experiments conducted at multiple sites. Oikos 69:334.

Gurevitch, J., and S. T. Chester, 1986. Analysis of repeated measures experiments. Ecology 67:251–255.

Hamilton, N. R. S. 1994. Replacement and additive designs for plant competition studies. Journal of Applied Ecology 31:599–603.

Hocutt, M. 1979. The Elements of Logical Analysis and Inference. Winthrop, Cambridge.

Hurlbert, S. J. 1984. Pseudoreplication and the design of ecological field experiments. Ecological Monographs 54:187–211.

Järvinen, O. 1986. The neontologico-paleontological interface of community evolution: how do the pieces in the kaleidoscopic biosphere move? Pages 331–350 in D. M. Raup and D. Jablonski (eds.), Pattern and Process in the History of Life. Dahlem Konferenzen, Springer-Verlag, Berlin.

Kroh, G. C., and S. N. Stephenson. 1980. Effects of diversity and pattern on relative yields of four Michigan first year fallow field plant species. Oecologia 45:366–371.

Legendre, P., and M-J. Fortin. 1989. Spatial pattern and ecological analysis. Vegetatio 80: 107–138.

Lemmon, E. J. 1971. Beginning Logic. Nelson, Surrey.

Lively, C. M., and M. J. McKone. 1994. Choosing an appropriate ANOVA for experiments conducted at few sites. Oikos 69:335.

Loehle, C. J. 1987. Hypothesis testing in ecology: psychological aspects and the importance of theory maturation. Quarterly Review of Biology 62:397–409.

———. 1988. Philosophical tools: potential contributions to ecology. Oikos 51:97–104.

McDonald, L. L., and W. P. Erickson. 1994. Testing for bioequivalence in field studies: has a disturbed site been adequately reclaimed? Pages 183–197 in D. Fletcher and B. J. Manly (eds.), Statistics in Ecology and Environmental Monitoring. University of Otago Press, Dunedin, New Zealand.

McKone, M. J., and C. M. Lively. 1993. Statistical analysis of experiments conducted at multiple sites. Oikos 67:184–186.

Mentis, M. T. 1988. Hypothetico-deductive and inductive approaches in ecology. Functional Ecology 12:1–5.

Morrisey, D. J., A. J. Underwood, L. Howitt, and J. S. Stark. 1992. Temporal variation in soft-sediment benthos. Journal of Experimental Marine Biology and Ecology 164:233–245.

Peters, R. H. 1991. A Critique for Ecology. Cambridge University Press, Cambridge.

Preece, D. A. 1982. The design and analysis of experiments: what has gone wrong? Utilitas Mathematica 21:201–244.

Quinn, J. F., and A. E. Dunham. 1983. On hypothesis testing in ecology and evolution. American Naturalist 122:602–617.

Sagoff, M. 1985. Fact and value in environmental science. Environmental Ethics 7:99–116.

Scheffé, H. 1953. A method of judging all contrasts in the analysis of variance. Biometrika 40:87–104.

Seaman, J. W., and R. G. Jaeger. 1990. Statisticae dogmaticae: a critical essay on statistical practice in ecology. Herpetologica 46:337–346.

Shrader-Frechette, K. S., and E. D. McCoy. 1994. What ecology can do for environmental management. Journal of Environmental Management 41:29–307.

Simberloff, D. 1980. A succession of paradigms in ecology: essentialism, materialism and probabilism. Pages 63–99 in E. Saarinen (ed.), Conceptual Issues in Ecology. Reidel, Dordrecht.

Snaydon, R. W. 1991. Replacement or additive designs for competition studies. Journal of Applied Ecology 28:930–946.

Stewart-Oaten, A., W. M. Murdoch, and K. R. Parker. 1986. Environmental impact assessment:"pseudoreplication" in time? Ecology 67:929–940.

Stove, D. 1980. Popper and After: Four Modern Irrationalists. Pergamon, Oxford.

Swihart, R. K., and N. A. Slade. 1985. Testing for independence in animal movements. Ecology 66:1176–1184.

———. 1986. The importance of statistical power when testing for independence in animal movements. Ecology 67:255–258.

Toft, C. A., and P. J. Shea. 1983. Detecting community-wide patterns: estimating power strengthens statistical inference. American Naturalist 122:618–625.

Trenbath, B. R. 1974. Biomass productivity of mixtures. Advances in Agronomy 26:177–210.

Underwood, A. J. 1981. Techniques of analysis of variance in experimental marine biology and ecology. Annual Reviews of Oceanography and Marine Biology 19:513–605.

———. 1986. The analysis of competition by field experiments. Pages 240–268 in J. Kikkawa and D. J. Anderson (eds.), Community Ecology: Pattern and Process. Blackwell, Melbourne.

———. 1988. Design and analysis of field experiments on competitive interactions affecting behaviour of intertidal animals. Pages 333–358 in G. Chelazzi and M. Vannini (eds.), Behavioural Adaptation to Intertidal Life. Plenum, New York.

———. 1990. Experiments in ecology and management: their logics, functions and interpretations. Australian Journal of Ecology 15:365–389.

———. 1991a. Beyond BACI: experimental designs for detecting human environmental impacts on temporal variations in natural populations. Australian Journal of Marine and Freshwater Research 42:569–587.

———. 1991b. The logic of ecological experiments: a case history from studies of the distribution of macro-algae on rocky intertidal shores. Journal of the Marine Biological Association of the United Kingdom 71:841–866.

———. 1992. Beyond BACI: the detection of environmental impact on populations in the real, but variable, world. Journal of Experimental Marine Biology and Ecology 161:145–178.

———. 1993. The mechanics of spatially replicated sampling programmes to detect environmental impacts in a variable world. Australian Journal of Ecology 18:99–116.

———. 1994a. On beyond BACI: sampling designs that might reliably detect environmental disturbances. Ecological Applications 4:3–15.

———. 1994b. Spatial and temporal problems with monitoring. Pages 101–123 in P. Calow and G. E. Petts (eds.) Rivers Handbook, Vol. 2. Blackwell Scientific, London.

———. 1994c. Things environmental scientists (and statisticians) need to know to receive (and give) better statistical advice. Pages 33–61 in D. Fletcher, and B. J. Manly (eds.),

Statistics in Ecology and Environmental Monitoring. University of Otago Press, Dunedin, New Zealand.

————. 1995. Ecological research and (and research into) environmental management. Ecological Applications 5:232–247.

————. 1997. Ecological Experiments: Their Logical Design and Interpretation using Analysis of Variance. Cambridge University Press, Cambridge.

Underwood, A. J., and M. G. Chapman. 1992. Experiments on topographic influences on density and dispersion of *Littorina unifasciata* in New South Wales. Pages 181–195 in J. Grahame, P. J. Mill, and D. G. Reid (eds.), Proceedings of the Third International Symposium on Littorinid Biology. Malacological Society of London.

Underwood, A. J., and E. J. Denley. 1984. Paradigms, explanations and generalizations in models for the structure of intertidal communities on rocky shores. Pages 151–180 in D. R. Strong, D. Simberloff, L. G. Abele, and A. Thistle (eds.), Ecological Communities: Conceptual Issues and the Evidence. Princeton University Press, Princeton, New Jersey.

Underwood, A. J., and P. S. Petraitis. 1993. Structure of intertidal assemblages in different locations: how can local processes be compared? Pages 38–51 in R. E. Ricklefs and D. Schluter (eds.), Species Diversity in Ecological Communities: Historical and Geographical Perspectives. University of Chicago Press, Chicago, Illinois.

USEPA. 1985. Short-Term Methods for Estimating the Chronic Toxicity of Effluents and Receiving Waters to Freshwater Organisms. EPA/600/4–85/014, Environmental Monitoring and Support Laboratory, Cincinnati, Ohio.

18

The Motivation for and Context of Experiments in Ecology

TIMOTHY WOOTTON & CATHERINE A. PFISTER

Over the past three decades, experimentation in ecology has become increasingly common (Connell 1972, 1983; Paine 1977; Schoener 1983; Hairston 1989). The experimental method has provided ecologists with a technique for examining the origins of pattern and the role of various processes in natural systems. As experimentation has proliferated in ecology, specific prescriptions for the design and implementation of experiments have emerged (e.g., Bender et al. 1984, Hurlburt 1984, Underwood 1986, Hairston 1989, Underwood and Petraitis 1993, Wootton 1994b). While there is merit in prescribing rigorous requirements for the design of experiments, it should not be carried out so as to limit the range of possible experimental approaches that can be usefully applied. For this reason, it is helpful to keep in mind the varied uses of experiments in ecology. In this chapter, we examine what we consider to be the three primary motivations for the use of experiments in ecology, illustrating them with examples drawn both from our work and from the wider literature. We then discuss how different philosophical approaches and different questions of interest in ecology call for different types of (equally valid) experimental designs and advocate applying experiments in an integrated program of ecological study.

The Motivation for Experiments

For our purposes, we will consider an experiment to involve manipulating one or more factors in a system, while the effects of other factors are either minimized or unmanipulated. This is best done in such a way that other factors do not vary systematically with the manipulation, although occasionally it may be useful to manipulate a number of factors at once simply to introduce variation into the study system (for example, imposing disturbance to initiate a successional sequence). Laboratory experiments often hold other factors as constant as possible, while field experiments usually allow other factors to vary naturally, employing randomization techniques to minimize the effects of confounding variables. Regardless of the type or specific goals of a given manipulation, all properly designed experiments provide unambiguous knowledge of the primary source of variation in a factor and therefore isolate the effect of that manipulated factor on the study system. Our definition of experiment follows closely that of Paine (1994), with the exception that we do consider some mensurative endeavors as experiments. We see three general reasons why experiments are conducted in ecology; we list them and discuss their contributions to ecology in the following sections.

Experiments as Means of "Seeing What Happens"

Perhaps the most common type of experiment currently conducted in ecology is one where a factor is suspected to have some effect on the system and is manipulated to see what happens (curiosity-driven experiments). For example, Paine (1966, 1974) experimentally removed the starfish *Pisaster ochraceous* from stretches of rocky shores of Washington State and compared the resulting set of sessile species to those on unmanipulated stretches of shore to "see what would happen." He found that the diversity of organisms attached to or foraging on the rock surface was reduced in the absence of starfish because a major prey of the starfish, the mussel *Mytilus californianus*, took over most of the available bare space when released from control by its predator. Likewise, Wilbur and his colleagues (e.g., Wilbur 1972, Morin 1983, Alford and Wilbur 1985, Wilbur 1987, Wilbur and Fauth 1990, Leibold and Wilbur 1992) have systematically manipulated a wide array of biological and physical factors to determine the nature of their effects in artificial pond communities. These studies have found, among other things, that size structure, phenology, the composition of predators and competitors, nutrients, and habitat permanence can all play roles in affecting community and ecosystem patterns in small ponds. In our own work, birds in rocky intertidal communities have been manipulated to determine what direct and indirect effects they might have on associated invertebrates and algae (Wootton 1992, 1993a, 1994a, 1995).

There are several benefits of conducting experiments that are motivated simply by curiosity about their outcome. First, they can sometimes generate great insight into the biological and physical processes operating in ecological systems and spawn theoretical and methodological advances in the field. For example, experimental removals of starfish (Paine 1966, 1974) yielded the idea of keystone species and have subsequently inspired a great deal of investigation into food web structure, indirect effects, and the relative strength of species interactions. Experimental explorations in the laboratory of the effects of habitat arrangements on predatory and herbivorous mites by Huffaker (1958) spawned a heightened awareness of the role that spatial structure can play in stabilizing the interactions among species. Paine and Vadas (1969) manipulated sea urchins, and their observations prompted an initial formulation of the intermediate disturbance hypothesis. In our own work, outcomes derived from manipulating intertidal birds have prompted the development of a mechanistic framework for identifying potential indirect effects (Wootton 1992, 1993a) and of statistical approaches to disentangling indirect effects (Wootton 1994a).

Second, such curiosity-driven experiments more commonly provide an increase in our understanding of a specific system that is not possible by simple observation. Often there is insufficient variation in a study system to derive insights from observational data alone, and when there is only natural variation several factors may often covary and therefore be difficult to tease apart. As more and more pieces of a given study system are uncovered through experimental manipulation, we gain a better idea of the rules under which that system operates and, consequently, what types of theory might be applied effectively to understand and predict the dynamics of the system. Toward this end, it is likely that this experimental approach will be most effective when applied to specific model systems, in much the same way that developmental biologists concentrate their experimental effort on the fruit fly *Drosophila melanogaster*, the nematode *Caenorhabditis elegans*, the sea urchins *Strongylocentrotus* spp., the zebra fish *Brachydanio rerio*, and the house mouse, *Mus musculus*. When experiments are conducted haphazardly across different systems, we are left with a partial understanding of numerous systems, no particularly complete understanding of any, and no idea of the relative importance of different species or processes. Thus, exploratory ("see what happens") experimental work will be most effective when concentrated on several model systems of relatively high experimental tractability such as old-fields/grasslands (e.g., Tilman and Wedin 1991, Goldberg and Barton 1992), desert communities (e.g., Davidson et al. 1984, Brown and Heske 1990), benthic marine communities (e.g., Connell 1961, Paine 1966, Dayton 1971, Sutherland 1974, Menge 1976, Lubchenco 1978, Sousa 1979, Underwood et al. 1983, Pfister 1995), and freshwater communities (e.g., Wilbur 1972, 1987; Werner and Hall 1976; Morin 1983; Werner 1984; Schindler et al. 1987; Leibold 1989; Power 1990; Carpenter and Kitchell 1993; Wootton and Power 1993).

Third, "see what happens" experiments can be closely allied with important applied ecology issues. In these cases, the experiment is motivated by a conservation-related concern and is often performed at relatively small spatial and temporal scales in the hopes that the results will be concordant with those at larger scales. For example, Peterson et al. (1987) used a field experiment to examine the effects of commercial clam raking (a process that disturbs marine soft sediment habitats) on subsequent recruitment of the clam *Mercenaria mercenaria*, as well as the effect on other organisms in shallow estuarine areas. Similarly, Brosnan and Crumrine (1994) experimentally manipulated human "trampling" effects in marine rocky intertidal plots either 20 × 20 cm or 20 × 30 cm to glean insight into how tourism in coastal areas might affect intertidal organisms. Experiments like these can be important to both guide management decisions and guide future sampling and monitoring efforts when answers are required more quickly than useful theory can be developed.

Despite the benefits of curiosity-driven experiments just described, there are potential problems with relying solely on this type of experiment in ecology (see also Power et al., this volume). The popularity of such experiments arises in part because no thought about conceptual generality is required, and therefore they are relatively easy or less risky to employ. Consequently, these experiments may not provide any general insights by themselves. We have also noticed a recurring misconception about what simple experiments to "see what happens" can do. Specifically, in interacting with colleagues and speakers at national meetings, we often hear statements such as: "We must conduct an experiment to predict the effect of factor X on the environment." However, experiments do not make predictions; they test predictions (be they as unsophisticated as "factor X has some unspecified effect"). Predictions are only derived from some underlying theory, be it as simple as "if X happens once, I predict that it will happen again." Although this simple prediction might seem reasonable, repeated experiments can often yield surprisingly different outcomes (e.g., Power et al. 1995), suggesting that a more comprehensive and insightful theory would be useful.

Experiments as a Means of Measurement

Often experiments are done to explore the specific functional relationship between the manipulated factor and some response variable of interest. Experimental manipulation often allows the investigator to explore a wider range of variation than is naturally present and therefore obtain a more precise estimate of the functional relationship. In this case, the goal usually is to estimate parameters or identify the shape of their relationship so that reasonable constraints on theory can be applied, rather than to test a specific hypothesis, and consequently this type of experiment does not adhere to the standard formulas of experimental design. Typically, there are a number of different treatment levels used in this approach, and the concept of an unmanipulated control, which is critical for many other types of experiments, is hard to apply or is arbitrary. In ecology, this approach has been underutilized, but as ties between empiricism and theories become more pervasive, we expect these experiments to become increasingly important in ecology.

Perhaps an area where mensurative experiments (defined as experiments where measurements are made on manipulated "treatments" in the absence of a control; sensu Hurlburt 1984) have been most important in determining a functional relationship is in the area of density dependence. There has been a plethora of experiments in ecology designed to detect whether intraspecific competition is an important determinant of organism fitness (e.g., Harper 1977), and the commercial interest in how plant yield is related to initial planting density has been central to this literature. Thus, not only are plant ecologists interested in whether negative density dependence can be demonstrated experimentally in natural populations (Schmitt et al. 1987, Reed 1990), but there is also deep interest in whether the slope of the relationship between plant fitness or yield and planting density is identifiable repeatedly as $-\frac{1}{2}$ on a log scale (Weller 1987, Osawa and Sugita 1989). In addition to the obvious benefits of knowing the strength of negative density dependence in agricultural plots, the strength of negative density dependence critically affects the dynamics of populations (May 1974).

As ecologists have become increasingly interested in the interconnectedness of populations, mensurative experiments that focus on dispersal, gene flow, and migration have increased. For example, Morris (1993) used an experimental array to quantify pollen dispersal by honeybees. The data generated by this array were used to fit diffusion models to pollen dispersal and are ultimately being used to predict how pollen moves as a function of interplant distance, thereby providing insight into how genes from genetically engineered plants might spread. Similarly, Herzig (1995) manipulated densities of beetle larvae and access to mates to understand the relationship between these factors and rates of beetle emigration. Here the goal was to explore experimentally the factors related to local outbreaks and rapid spread in this leaf-feeding beetle.

Mensurative experiments have also become more important in the investigation of the dynamics of food webs in several ways. First, models with different assumptions about the shape of the functional response (Holling 1959) can yield very different predictions about the consequences of manipulating top consumers and limiting nutrients in food webs (Rosenzweig 1973, Oksanen et al. 1981, Arditi and Ginzberg 1989, Schmitz 1992). Therefore, to guide the development of food web models, experiments that vary prey and predator density are required to examine the shape of functional response surfaces (e.g., Chant and Turnbull 1966, Eveleigh and Chant 1982, Katz 1985). Second, a basic knowledge of the magnitude and distribution of the strength of species interactions is required to place reasonable constraints on food web models. Recent manipulations of various invertebrate grazers on algae (Paine 1992), of various arthropod prey on mantid predators (Fagan and Hurd 1994), and of intertidal avian predators on various invertebrate prey (Wootton 1997) have been conducted to estimate interaction strength. One key result of these studies is that the distributions of interaction strengths appear skewed toward weak interactions (Power et al. 1996), rather than normally distributed as is assumed in many analyses of randomly constructed food webs (e.g., May 1973, Gilpin 1994).

Sometimes experiments that simply introduce variation into a system may be of use in estimating important ecological relationships. A prime example of this approach is the application of PULSE experiments (Bender et al. 1984) to estimate species interaction strength. In PULSE experiments, particular species are perturbed temporarily to move the community from any putative equilibrium, and the short-term rate of change in other species is measured subsequently. We believe that approaches that introduce variation into a system can be extended profitably by taking advantage of the longer term dynamics introduced by either PULSE or PRESS (sensu Bender et al. 1994) experimental manipulations. For example, manipulating bird predators produces a ''signal'' of known origin that is transmitted to other members of the community, and by applying path analysis techniques the relative strength of interactions among unmanipulated species can be obtained (Wootton 1994a). Such an approach has the potential to be used in conjunction with time-series data of the rates of change of different members of the community through time following the manipulation to provide estimates of the strength of species interactions (Wootton 1994c, Pfister 1995, Laska and Wootton in press).

Experiments as a Means of Testing Ecological Theory

Perhaps the most powerful, or efficient, use of experiments in any scientific discipline is to test theory. Not only do the results of these experiments yield the insights discussed previously for curiosity-driven experiments, but they also usually yield insights into approaches or theories that might (or might not) be applied generally to novel conditions found in other systems or when the particular study system is altered. Experimental tests of theory can address two basic issues: the accuracy of the predictions made by the theory and the adequacy of the assumptions underlying theory.

Testing Predictions   When applying experiments to theory, ecologists usually test predictions of the theory. Typically, predictions are more interesting to examine because science is at its best when it predicts previously unestablished results. Testing predictions has been carried out in a variety of contexts. For example, the intermediate disturbance hypothesis (Paine and Vadas 1969, Horn 1975, Connell 1978) predicts maximal species diversity at moderate levels of disturbance. Competitive exclusion through the monopolization of resources is hypothesized to be disrupted by mortality that arises from disturbance, but disturbance cannot be too high or some species may be unable to maintain positive growth rates in the face of such high mortality. Sousa (1979) conducted an experiment designed specifically to test a prediction of this theory by examining the diversity of algal assemblages on intertidal boulder fields. Smaller rocks are less resistant to wave forces, and consequently the organisms living on them experience higher disturbance rates as a result of rolling. By experimentally stabilizing rocks, Sousa effectively reduced disturbance rates, and diversity increased as a result, in accordance with the model. This example can be used to illustrate the power of linking experimental tests to theory. Because of the underlying theory used to motivate the test, Sousa's results potentially have implications for our understanding of the effects of disturbance in a variety of forms on a range of other systems. In the absence of this theoretical link, they would only show that rolling rocks have an effect on diversity in an intertidal boulder field in California.

Experimental tests of theoretical predictions can also be usefully applied to theories that examine interactions among species at multiple trophic levels with variations in productivity (e.g., Rosenzweig 1973, Oksanen et al. 1981, Arditi and Ginzberg 1989). Such models are interesting because they provide an avenue to link ecosystem-level features (variations in productivity) with community-level processes (trophic interactions). Depending on the assumptions about the relationship between predator consumption rates and both prey and predator densities, different predictions are obtained regarding how different trophic levels should change as productivity changes. For example, if per capita consumption rates are functions of prey densities, increasing productivity will only increase certain types of consumers and producers (Rosenzweig 1973, Oksanen et al. 1981), whereas if per capita consumption rates are functions of the ratio of prey to predator densities, all trophic groups should change together (Arditi and Ginzberg 1989, Schmitz 1992). Experimental manipulations are required to test these predictions because observational comparisons across systems are confounded by changes in species composition, which distort the predictions of the models (Mittelbach et al. 1988, Persson et al. 1988, Leibold 1989). Experimental manipulations of productivity in a river, designed specifically to test these predictions, show that producers and predators, but not grazers, increase with increasing productivity (Wootton and Power 1993). Experimentally elevating levels of limiting nutrients or light in ponds, streams, and rocky intertidal shores increases grazers but not algae (Leibold 1991, Leibold and Wilbur 1992, Hill et al. 1995, Wootton et al. 1996b). These results are more in accordance with the predictions of prey-dependent models than those of ratio-dependent models.

Experiments have also been applied extensively to test predictions from life history theory. Experimental manipulations of offspring number (e.g., Vander Werf 1992,

Young 1996) have been done in an effort to test Lack's (1947) hypothesis that clutch size has evolved toward the number which produces the most offspring. Similarly, experimental manipulations of the level of commitment to reproduction (e.g., Snow and Whigham 1989, Pfister 1992) have been a commonly used means of detecting whether there is a cost of reproduction, as proposed by Williams (1966). Experiments have been especially powerful in this context, since erroneous correlations can be estimated when treatments are not randomly assigned among individuals (Reznick 1985, Pease and Bull 1988). Indeed, in a summary of published results, Reznick (1985) found a marked difference in the proportion of studies that found reproductive costs with correlative versus experimental methods. Correlative, nonexperimental results (''phenotypic correlations'') demonstrated a cost with a lower frequency (22 of 33) than that of studies that manipulated experimentally the level of reproductive investment (17 of 20).

Experimental tests of theoretical predictions can be especially useful when applied to tractable systems as a proxy for intractable systems. Thus, we advocate increased validation of theory or observational approaches in systems where experiments are easily performed before applying such techniques in experimentally intractable situations. We illustrate this point by considering how observational approaches have been used to gain insight or make predictions about the strengths of species interactions in natural communities. As a first example, one common approach assumes communities are at equilibrium and uses regressions of species abundances in different locations to derive an index of interspecific interactions (Schoener 1974, Hallett and Pimm 1979, Pimm 1985). When applied to an assemblage of tide pool fishes that are amenable to experimentation, this technique predicted mutualistic interactions among species after factoring out habitat variation (Pfister 1995). In contrast, an experimental manipulation designed in part to test this approach demonstrated strong asymmetric competitive effects among species (Pfister 1995), effects predicted using an alternative, nonequilibrial approach that estimates rates of population change over time as a function of initial community composition.

As a second example, another approach toward evaluating community structure imposes an experimental perturbation on the community, allows unmanipulated members of the community to respond, and applies path analysis, a technique designed to disentangle the relative strengths of different direct and indirect pathways among variables (Wootton 1994a). By following the signal of the perturbation through the community, path analysis can predict other potentially strong interactions in the community. Results of experimental manipulations of birds in rocky intertidal communities were analyzed using this technique, which predicted the effects of other types of manipulations involving acorn barnacles, goose barnacles, snails, and birds (Wootton 1994a). Followup experimental manipulations tested the predictions of the approach, and the results indicated that the experimental/path analysis approach has some validity (Wootton 1994a).

A third example involves combining observational data on feeding rates, predator behavior, predator density, and prey density to produce a theoretically appropriate measure of per capita interaction strengths among consumers and resources. Experimental manipulations of intertidal birds (gulls, oystercatchers, and crows) that tested relevant predictions for a variety of different prey quantitatively fit the predicted interaction strengths, indicating that certain types of observational measures might be of use in identifying strongly interacting species in experimentally intractable communities (Wootton 1997).

Testing Assumptions   An often ignored but potentially powerful use of experiments to test theory is to evaluate the assumptions rather than the predictions of theory. Perhaps the best illustration of the power of experimentally testing assumptions is work examining Darwin's theory of evolution by natural selection (Darwin 1859). Authors have argued (Waddington 1957; Birch and Ehrlich 1967; Peters 1976, 1991) that Darwin's theory should be discarded because it makes either untestable or circular predictions. Why, then, does this theory still retain its central place in biology? The weakness in the circularity/testability argument is that it supposes that only the predictions of theories are open to testing when, in reality, evaluating the assumptions of theories provides a far more stringent test (see Dayton 1973, Holt 1977). Darwin's theory is actually very testable. It rests on three assumptions: (1) variation exists between individuals in the traits they possess; (2) the variation is (at least to some extent) heritable; and (3) differences in traits produce differences in survival and/or reproduction (i.e., natural selection exists). All of these assumptions have been overwhelmingly supported. Common garden experiments, classical breeding experiments, and experimental molecular genetics have clearly demonstrated that assumptions (1) and (2) are true (Futuyma 1986). Numerous ecological and morphological experiments have generally demonstrated that assumption (3) is true (e.g., Endler 1986). These experimental tests of assumptions, rather than predictions, leave little doubt as to the applicability of Darwin's theory to biological systems.

Testing assumptions has also played an important role in community ecology. For example, a fundamental assumption of many models is that competition plays an important role in structuring communities (e.g., Lotka 1925, MacArthur and Levins 1967, Schoener 1974, Diamond 1975, Roughgarden 1976, Tilman 1982, Chesson 1983). In response to criticism that such an assumption might be unwarranted (Connor and Simberloff 1979, Strong et al. 1979), a number of experiments have been conducted specifically to test this assumption in a variety of communities (e.g., Pfister 1995, reviewed in Connell 1983, Schoener 1983, Goldberg and Barton 1992, Gurevitch et al. 1992). In reading the recent literature, however, our impression has been that the motivation for experimental tests of competition has increasingly drifted away from testing assumptions of ecological theory. Instead, recent experiments have tended to ask only, What happens when a potential competitor is removed from a particular community? Ideally, the results of such experiments should be pursued for their broader implications—for example, probing whether predictions from different competition-based models are upheld and, if they are not, determining the underlying mechanisms (e.g., Creese and Underwood 1982, Schmitt 1996).

A related consideration is that finding negative results in competition experiments does not necessarily mean that competition is unimportant in determining community structure. Most experiments that test for competition examine the simple prediction that reducing one species will increase the population of its competitor. When interspecific competition operates in conjunction with other processes, however, a variety of compensatory effects may hide the negative effects of competition, even when interspecific competition is strong (Fig. 18-1; see Davidson et al. 1984, Fairweather 1990 for ex-

EXPLOITATIVE COMPETITION WITH PREDATOR-RESISTANT
AND PREDATOR-SUSCEPTIBLE SPECIES

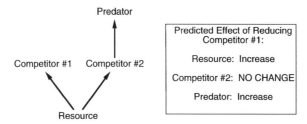

Figure 18-1. An example where competition would play an important role in community structure, but manipulations of one competitor would fail to reveal its negative effects on the other competitor. If competitor #1 was reduced in abundance experimentally, competitor #2, rather than increase, would experience no change in abundance, while the predator would increase. The increase in predator abundance compensates for the increased productivity in competitor #2 that results from competitive release.

amples). Thus, finding a negative result may permit the conclusion that a model based solely on interspecific competition is inadequate, but it may not necessarily refute models with the assumption that competition is one of several important interactions that are occurring (e.g., Leibold 1996).

Another area where tests of assumptions might be profitably applied in community ecology revolves around the issue of how multispecies models can be constructed. The simplest approach is to examine linked dynamical equations describing direct interactions among species (sensu Wootton 1994c), such as the community matrix approach used by Levins (1969), May (1973), Levine (1976), Pimm (1982), Yodzis (1988), and others. A fundamental assumption of these models is that the per capita effects of one species on another are unaffected by the abundance of other members of the community (i.e., instantaneous effects of each species can be added together in differential equations without including "higher order" terms that incorporate simultaneously the abundances of more than one interacting species). Early examinations tested predictions of models with this assumption and, when predictions deviated from the results, concluded that the assumption was incorrect (e.g., Wilbur 1972, Neill 1974, Morin et al. 1988, Pfister and Hay 1988, Wilbur and Fauth 1990, Wootton 1992, 1993a). However, a number of alternative factors (nonlinear pairwise interactions, asymmetries in experimental and theoretical time scales, theoretically inappropriate data transformations, and other types of indirect effects) make interpretation of deviations between observed and predicted results very difficult (Case and Bender 1981, Pomerantz 1981, Adler and Morris 1994, Billick and Case 1994, Wootton 1994b). An alternative approach is to focus on the mechanistic interpretation of the assumption (that the effects of one species on another are not modified by the abundance of other members of the community; see Wootton 1993a, 1994b, 1994c) and conduct experiments that directly address this assumption.

This involves preventing other members of the community from changing the abundance of the focal pair of interacting species and observing whether the net effects of the interaction change as a result. For example, Werner (1992) and Wissinger and McGrady (1993) were able to change the interactions between frog tadpoles and odonate nymphs, respectively, by introducing a predatory odonate species that was prevented from feeding. An additional issue raised in work that examines this assumption is whether it is necessarily a good idea to discard a model even if the assumptions are demonstrated to be incorrect. In many cases, the critical question is whether failure of the assumptions severely affects model predictions, since virtually all models have some simplifying, incorrect assumptions. For example, Morin et al. (1988) found patterns in experimental pond communities that indicated that effects of two species on a third target species were not independent but pointed out that the additional explanatory value of accounting for the interactive effects of the two species was slight.

In addition, testing the assumptions of ecological theory may often be possible when testing predictions is intractable, just as in the case of evolution by natural selection. For example, theories that address how species coexist may fall into this category. In the case of the ''storage effect'' model for coexistence (Warner and Chesson 1985, Chesson 1994), the assumptions are relatively straightforward and testable: (1) the species compete; (2) species experience effective bouts of recruitment at different times; (3) species have relatively high adult survival, and competition and environment variation affect survival relatively little; and (4) species can increase from low density. However, the prediction of the storage effect model is that species coexist locally due to the preceding four criteria being met. To test the prediction of the storage effect model ignores other potential mechanisms of coexistence.

Testing the assumptions of ecological theory may also be more feasible ethically. For example, a number of authors have recently used metapopulation models to explore coexistence among competing species (Tilman 1994, Tilman et al. 1994, Kareiva and Wennergren 1995). In these models, it is assumed that there is a trade-off between competitive ability and colonization success. Superior competitors always replace inferior competitors in ''patches'' of solitary individuals, and inferior competitors are capable of colonizing empty patches at a greater rate. Another assumption involves the initial distribution of species abundance (e.g., a geometric series of species abundance was used by Tilman et al. 1994). One of the predictions made by these models is that increasing habitat destruction or degradation results in an increased probability of loss of the superior competitor (Tilman et al. 1994). Clearly, in many systems testing the prediction of how habitat loss affects species abundance patterns may be prohibitive if manipulating experimentally the loss of habitat is unethical or impossible. Thus, for systems in which the predictions of the theory have the most conservation importance, testing assumptions of the models may be the only way to evaluate their relevance.

Testing the assumptions of theory may be as or more rigorous than testing the predictions. Darwin had enough natural history insight to derive a theory based on several simple (and ultimately upheld) assumptions. However, many ecological theories do not share these qualities, and often these theories consist of a number of simplifying assumptions. In addition, one problem with testing the assumptions of a theory is that the status of the theory is unclear if we cannot uphold experimentally one or several of the assumptions. Consider the storage effect model again. If we test the assumption that species compete in a particular system and find out they do not, then this finding would indicate strongly that we should not use the storage effect model. If, however, we find that adult mortality can be relatively high at times, in contradiction to a storage effect model assumption, then do we dispense with the model? The latter scenario does not yield a clear answer. Minimally, experimental tests of assumptions should reveal to us where theory needs modification.

We believe that experimental tests of the underlying assumptions of ecological theory are underutilized and offer one instance where they could be utilized more. The theory that subpopulations might be distributed as metapopulations has yielded a variety of metapopulation models that describe the occupancy of discrete habitat through time with differential equations (Levins 1969, Hanski 1982, Gotelli 1991). Although a variety of assumptions underlie all models, the assumptions that distinguish among models are how they relate extinction and colonization rates to habitat occupancy. Although several investigators have used census data to explore how extinction and colonization rates relate to occupancy (Gotelli and Kelley 1993, Hanski et al. 1995, Pfister in press), we are aware of no studies that have manipulated experimentally either extinction, colonization, or occupancy to determine the relationship. As the use of previously formulated metapopulation models grows (Gotelli and Kelley 1993, Tilman 1994, Hanski et al. 1995, Pfister in press) and the derivation of new metapopulation models increases (Hanski and Gyllenberg 1993), empirical tests of metapopulation model assumptions are sorely needed.

The Context of Experiments

The Interplay of Questions and Experimental Approaches

Aside from differences in experimental approaches that arise from the three general approaches outlined previously, differences may also arise because of different specific questions of interest, even when studying the same general phenomenon. One example is in the study of interspecific competition. Ideally, experimental treatments that test for the effects of competition should be designed such that each target species of interest is examined at a control density (preferably that found in the natural environment), at a control density in combination with a putative competitor species (again, preferably at its natural density), and as a single species at a density equal to the total density in the mixed species treatment. This design permits detection of interspecific competitive intensity, intraspecific competitive intensity, and the relative strength of inter-to intraspecific competition (Connell 1983, Underwood 1986; see, e.g., Creese and Underwood 1982, Fauth et al. 1990). In some circumstances, however, the experimental logistics are such that not all treatments can be employed while attaining replication for adequate statistical power. Under these circumstances, the experimental design depends critically on the investigator's interests. If the investigator is initially interested in whether interspecific competition occurs, then treatments should use the control and mixed species treatments (an additive design). If the investigator is more concerned with the relative intensity of interspecific competition to intraspecific competition, which has implications for species coexistence, then treatments should include the mixed species and high-density treatments (a substitutive design). Either design can be appropriate, depending on the underlying motivation of the investigator, but care should be taken not to overextend the results (for example, concluding that competitive exclusion might result on the basis of the additive design alone). With logistic constraints, the sequential implementation of the two designs might be appropriate, since it is usually not particularly interesting to examine the relative strength of intraspecific and interspecific competition unless interspecific competition is demonstrated. For example, Werner and Hall (1976) conducted an additive experiment which indicated that interspecific competition among sunfish (*Lepomis* spp.) occurs, whereas Werner and Hall (1977) report a substitutive experiment where the relative impact of inter- to intraspecific competition was similar on green sunfish (*Lepomis gibbosus*) but was much higher on bluegill sunfish (*Lepomis macrochirus*).

Similarly, different types of experimental approaches are best suited to alternative strategies that attempt to gain general insights about ecological communities. One possible strategy to address this problem seeks generality through repetitive experimentation across communities and asks what percentage of community variation can be explained by a particular factor (Weldon and Slauson 1986, Underwood and Petraitis 1993). This approach dictates the use of multifactorial experimental manipulations of as many variables as possible, conducted across a number of communities. Complex analyses of variance (ANOVAs) are then applied to estimate the fraction of variation explained by any manipulated factor, which provides an index of the ''average'' importance of that factor, and to estimate the fraction of variation accounted for by the site by variable interaction terms, which provide indices of how variable the importance of the manipulated factor may be.

An alternative strategy might be to investigate intensively a particular community, develop a detailed mechanistic understanding of its workings, and then explore theoretical and methodological approaches that might generally be applied to other communities. Therefore, in this strategy generality is sought through the development of a flexible theoretical framework that synthesizes a variety of general ecological processes. Obviously, this strategy requires intensive investigation of a target community rather than cross-community experimental comparisons of portions of the community.

Either strategy is valid, given an investigator's interests and philosophy. We personally have opted for the second strategy, for a variety of reasons but particularly because it offers opportunities for generalization to a wider range of communities. First, potential generalization is not limited by the type of organisms present. Thus, the importance of limpets might be determined for a range of rocky intertidal communities, but the result is irrelevant to grasslands, whereas a general theoretical framework for grazing developed by studying limpets might be applicable to bison in grasslands, too. Because generalization in the second approach requires development of process-based theory, it offers the possibility of prediction in novel situations that the first approach does not. On the other hand, it might be countered that the first approach is desirable because its generalizations are unconstrained by particular theoretical constructs. Such constructs are present implicitly, however, as the basis of statistical comparisons (e.g., ANOVA) used to make the generalizations, and the conclusions drawn may be affected by these underlying theoretical assumptions (e.g., Adler and Morris 1994, Billick and Case 1994, Wootton 1994b).

The Place of Experiments in Ecology

Finally, it is worth considering where experiments fit into the general realm of ecological endeavor. In some ways, one's perspective is shaped by the differing approaches to generality outlined previously. The value of current ecological work (as assessed by the content of leading ecological journals) seems to be determined primarily on the basis of whether or not it is experimental (see Power et al., this volume). In some ways, this reflects the philosophy of determining generality through the partitioning of natural variance as outlined previously, although the heterogeneity of experimental design makes it difficult to compare results across systems (Underwood and Petraitis 1993). Meta-analysis of independent experiments may be one potentially usefully strategy in addressing this problem (Gurevitch et al. 1992, Wooster 1994, Arnqvist and Wooster 1995). We find, however, that some of the most compelling work in ecology takes a more pluralistic approach that combines experiments, observational information, and theory (see Leibold and Tessier, this volume, for more detailed discussion of the approach). This orientation is more in keeping with the second strategy to obtain generality outlined previously and is often necessary because of logistical constraints on experimentation. For example, Werner and colleagues have developed theory that ties behavioral ecology to species interactions, evaluated its assumptions in small-scale laboratory experiments and medium-scale experimental pond experiments, and tied these results to patterns of variation in the community structure of small lakes and ponds (e.g., Werner and Hall 1974, 1976, 1977; Werner 1984, 1992; Werner and Gilliam 1984; Mittelbach et al. 1988; McPeek 1990; Mittelbach and Osenberg 1993; and references therein). Similarly, the work of Carpenter, Kitchell, and their colleagues unites bioenergetic food web theory, small-scale experiments that calibrate the models and verify their assumptions, whole-lake experiments to test model predictions, and cross-lake comparisons to glean general patterns (see review, in Carpenter and Kitchell 1993). In each of these examples, combining methodologies that individually have different shortcomings (unverified underlying processes, somewhat unnatural small-scale experimental conditions, and lack of replication in experiments under more natural conditions) produces a much deeper understanding of the system.

We have found such an approach profitable in our own work. For example, experiments can rigorously evaluate important local interactions (for example, competition among tidepool fishes or effects of consumers on intertidal succession), but their larger scale implications cannot be assessed without the integration of theory (storage effect [Chesson 1983, 1984, 1994; Warner and Chesson 1985] or local regional models of species interactions [Horn 1975]). Critical, theoretically relevant observational data are also needed, such as patterns of recruitment variation in relation to local or regional abundances, because it is impossible to manipulate large-scale dispersal processes or regional sources of reproduction that affect the local recruitment of marine larvae (Pfister 1993, 1996; Wootton 1993b). Similarly, linking dynamic models of food webs with small-scale laboratory experiments to verify assumptions, mesocosm experiments to evaluate critical processes and predictions, and cross-river comparisons to evaluate the generality of the predicted patterns has proven powerful in investigating the interplay of disturbance, productivity, and species interactions in rivers in the western United States (Wootton and Power 1993, Power et al. 1995, Wootton, et al. 1996a). In summary, then, we believe that experiments are critical to furthering our ecological knowledge. However, focusing only on experimentation is far less effective than integrating experiment, theory, and observation in a synthetic framework.

ACKNOWLEDGMENTS   We are grateful for the perspectives on experimentation we have encountered from numerous colleagues over the years. Among those who have been particularly influential to our thinking are C. Adler, M. Hay, P. Kareiva, L. Johnson, M. Leibold, M. Power, W. Sousa, R. Steneck, and particularly R. Paine. We thank B. Resetarits and J. Bernardo for the invitation to add our thoughts on experimentation to this volume. J. Bernardo, D. Branstretor, J. Chase, L. Cochran-Stafira, M. Leibold, R. Paine, B. Resetarits, and J. Tsao provided helpful comments on the manuscript. This work was supported in part by grants from the Mellon Foundation and the National Science Foundation (DEB93-17980).

Literature Cited

Alder, F. R., and W. F. Morris. 1994. A general test for interaction modifications. Ecology 75:1552–1559.

Alford, R. A., and H. M. Wilbur. 1985. Priority effects in experimental communities: competition between *Bufo* and *Rana*. Ecology 66:1097–1105.

Arditi, R., and L. R. Ginzburg. 1989. Coupling in predator–prey dynamics: ratio-dependence. Journal of Theoretical Biology 139:311–326.

Arnqvist, G., and D. Wooster. 1995. Meta-analysis: synthesizing research findings in ecology and evolution. Trends in Ecology and Evolution 10:236–240.

Bender, E. A., T. J. Case, and M. E. Gilpin. 1984. Perturbation experiments in community ecology: theory and practice. Ecology 65:1–13.

Billick, I., and T. J. Case. 1994. Higher order interactions: what are they and how can they be detected? Ecology 65:1–13.

Birch, L. C., and P. R. Ehrlich. 1967. Evolutionary history and population biology. Nature 214:349–352.

Brosnan, D. M., and L. L. Crumrine. 1994. Effects of human trampling on marine rocky shore communities. Journal of Experimental Marine Biology and Ecology 177:79–97.

Brown, J. H., and E. J. Heske. 1990. Mediation of a desert-grassland transition by a keystone rodent guild. Science 250:1705–1707.

Carpenter, S. R., and J. F. Kitchell (eds.). 1993. The Trophic Cascade in Lakes. Cambridge University Press, Cambridge.

Case, T. J., and E. A. Bender. 1981. Testing for higher-order interactions. American Naturalist 118:920–929.

Chant, D. A., and A. L. Turnbull. 1966. Effects of predator and prey densities on interactions between goldfish and *Daphnia pulex* (de Geer). Canadian Journal of Zoology 44:285–289.

Chesson, P. L. 1983. Coexistence of competitors in a stochastic environment: the storage effect. Lecture Notes in Biomathematics 52:188–198.

———. 1984. The storage effect in stochastic population models. Lecture Notes in Biomathematics 54:76–89.

———. 1994. Multispecies competition invariable environments. Theoretical Population Biology 45:227–276.

Connell, J. H. 1961. The influence of interspecific competition and other factors on the distribution of the barnacle *Chthamalus stellatus*. Ecology 42:710–723.

———. 1972. Community interactions on marine rocky intertidal shores. Annual Review of Ecology and Systematics 3:169–192.

———. 1978. Diversity in tropical rain forests and coral reefs. Science 199:1302–1310.

———. 1983. On the prevalence and relative importance of interspecific competition: evidence from field experiments. American Naturalist 122:661–696.

Connor, E. F., and D. Simberloff. 1979. The assembly of species communities: chance or competition? Ecology 60:1132–1140.

Creese, R. G., and A. J. Underwood. 1982. Analysis of inter- and intra-specific competition amongst intertidal limpets with different methods of feeding. Oecologia 53:337–346.

Darwin, C. 1859. The Origin of Species. John Murray, London.

Davidson, D. W., R. S. Inouye, and J. H. Brown. 1984. Granivory in a desert rodent ecosystem: experimental evidence for indirect facilitation of ants by rodents. Ecology 65:1780–1786.

Dayton, P. K. 1971. Competition, disturbance and community organization: the provision and subsequent utilization of space in a rocky intertidal community. Ecological Monographs 41:351–389.

———. 1973. Two cases of resource partitioning: making the right prediction for the wrong reasons. American Naturalist 107:662–670.

Diamond, J. 1975. Assembly of species communities. Pages 342–444 in M. Cody and J. Diamond (eds.), Ecology and Evolution of Communities. Harvard University Press, Cambridge, Massachusetts.

Endler, J. A. 1986. Natural Selection in the Wild. Princeton University Press, Princeton, New Jersey.

Eveleigh, E. S., and D. A. Chant. 1982. Experimental studies on acarine predator-prey interactions: the effects of predator density on prey consumption, predator searching efficiency, and the functional response to prey density (Acarina: Phytoseiidae). Canadian Journal of Zoology 60:611–629.

Fagan, W. F., and L. E. Hurd. 1994. Hatch density variation of a generalist arthropod predator: population consequences and community impact. Ecology 75:2022–2032.

Fairweather, P. G. 1990. Is predation capable of interacting with other community processes on rocky reefs? Australian Journal of Ecology 15:453–464.

Fauth, J. E., W. J. Resetarits Jr, and H. M. Wilbur. 1990. Interactions between larval salamanders: a case of competitive equality. Oikos 58:91–99.

Futuyma, D. J. 1986. Evolutionary Biology, 2nd ed. Sinauer, Sunderland, Massachusetts.

Gilpin, M. 1994. Community-level competition: asymmetrical dominance. Proceedings of the National Academy of Sciences of the USA 91:3252–3254.

Goldberg, D. E., and A. M. Barton. 1992. Patterns and consequences of interspecific competition in natural communities: a review of field experiments with plants. American Naturalist 139:771–801.

Gotelli, N. J. 1991. Metapopulation models: the rescue effect, the propagule rain, and the core-satellite hypothesis. American Naturalist 138:768–776.

Gotelli, N. J., and W. G. Kelley. 1993. A general model of metapopulation dynamics. Oikos 68:36–44.

Gurevitch, J., L. L. Morrow, A. Wallace, and J. S. Walsh. 1992. A meta-analysis of competition in field experiments. American Naturalist 140:539–572.

Hairston, N. G., Sr. 1989. Ecological Experiments: Purpose, Design, and Execution. Cambridge University Press, Cambridge.

Hallett, J. G., and S. L. Pimm. 1979. Direct estimation of competition. American Naturalist 113:593–599.

Hanski, I. 1982. Dynamics of regional distribution: the core and satellite species hypothesis. Oikos 38:210–221.

Hanski, I., and M. Gyllenberg. 1993. Two general metapopulation models and the core-satellite species hypothesis. American Naturalist 142:17–41.

Hanski, I., J. Poyry, T. Pakkala, and M. Kuussaari. 1995. Multiple equilibria in meta-population dynamics. Nature 377:618–621.

Harper, J. L. 1977. Population Biology of Plants. Academic Press, San Diego, California.

Herzig, A. L. 1995. Effects of population density on long-distance dispersal in the goldenrod beetle *Trirhabda virgata*. Ecology 76:2044–2054.

Hill, W. R., M. G. Ryon, and E. M. Schilling. 1995. Light limitation in a stream ecosystem: responses by primary producers and consumers. Ecology 76:1297–1309.

Holling, C. S. 1959. The components of predation as revealed by a study of small mammal predation of the European pine sawfly. Canadian Entomologist 91:293–320.

Holt, R. 1977. Predation, apparent competition, and the structure of prey communities. The-oretical Population Biology 12:197–229.

Horn, H. 1975. Markovian properties of forest succession. Pages 196–211 in M. L. Cody and J. M. Diamond (eds.), Ecology and Evolution of Communities. Harvard University Press, Cambridge, Massachusetts.

Huffaker, C. B. 1958. Experimental studies on predation: dispersion factors and predator-prey oscillations. Hilgardia 27:343–383.

Hurlburt, S. H. 1984. Pseudoreplication and the design of ecological field experiments. Eco-logical Monographs 54:187–211.

Kareiva, P., and U. Wennergren. 1995. Connecting landscape pattern to ecosystem and population processes. Nature 373:299–302.

Katz, C. H. 1985. A nonequilibrium marine predator–prey interaction. Ecology 66:1426–1438.

Lack, D. 1947. The significance of clutch size, parts I and II. Ibis 89:302–352.

Laska, M. S., and J. T. Wootton. In press. Theoretical concepts and empirical approaches to measuring interaction strength. Ecology.

Leibold, M. A. 1989. Resource edibility and the effects of predators and productivity on the outcome of trophic interactions. American Naturalist 134:922–949.

———. 1991. Trophic interactions and habitat segregation between competing *Daphnia* species. Oecologia 86:510–520.

———. 1996. A graphical model of keystone predators in food webs: trophic regulation of abundance, incidence and diversity patterns in communities. American Naturalist 147:784–812.

Leibold, M. A., and H. M. Wilbur. 1992. Interactions between food-web structure and nu-trients on pond organisms. Nature 360:341–343.

Levine, S. H. 1976. Competitive interactions in ecosystems. American Naturalist 110:903–910.

Levins, R. 1968. Evolution in Changing Environments: Some Theoretical Explorations. Princeton University Press, Princeton, New Jersey.

———. 1969. Some demographic and genetic consequences of environmental heterogeneity for biological control. Bulletin of the Entomological Society 15:237–240.

Lotka, A. J. [1925] 1965. Elements of mathematical biology. Dover, New York.

Lubchenco, J. 1978. Plant species diversity in a marine intertidal community: importance of herbivore food preference and algal competitive abilities. American Naturalist 112:23–39.

MacArthur, R., and R. Levins. 1967. The limiting similarity, convergence, and divergence of coexisting species. American Naturalist 101:377–385.

May, R. M. 1973. Stability and Complexity in Model Ecosystems. Princeton University Press, Princeton, New Jersey.

———. 1974. Biological populations with non-overlapping generations: stable points, stable cycles and chaos. Science 186:645–647.

McPeek, M. A. 1990. Behavioral differences between *Enallagma* species (Odonata) influencing differential vulnerability to predators. Ecology 71:1714–1726.

Menge, B. A. 1976. Organization of the New England rocky intertidal community: role of predation, competition and environmental heterogeneity. Ecological Monographs 46: 335–393.

Mittelbach, G. G., and C. W. Osenberg. 1993. Stage-structured interactions in bluegill: consequences of adult resource variation. Ecology 74:2381–2394.

Mittelbach, G. G., C. W. Osenberg, and M. A. Leibold. 1988. Trophic relations and ontogenetic niche shifts in aquatic ecosystems. Pages 219–235 in B. Ebenman and L. Persson (eds.), Size-structured Populations. Springer-Verlag, Berlin.

Morin, P. J. 1983. Predation, competition, and the composition of larval anuran guilds. Ecological Monographs 53:119–138.

Morin, P. J., S. P. Lawler, and E. A. Johnson. 1988. Competition between aquatic insects and vertebrates: experimental measures of interaction strength and higher order interactions. Ecology 69:1401–1409.

Morris, W. F. 1993. Predicting the consequences of plant spacing and biased movement for pollen dispersal by honey bees. Ecology 74:493–500.

Neill, W. E. 1974. The community matrix and interdependence of the competition coefficients. American Naturalist 108:399–408.

Oksanen, L., S. D. Fretwell, J. Arruda, and P. Niemela. 1981. Exploitation ecosystems in gradients of primary productivity. American Naturalist 118:240–261.

Osawa, A., and S. Sugita. 1989. The self-thinning rule: another interpretation of Weller's results. Ecology 70:279–283.

Paine, R. T. 1966. Food web complexity and species diversity. American Naturalist 100:65–75.

———. 1974. Intertidal community structure: experimental studies on the relationship between a dominant competitor and its principal predator. Oecologia 15:93–120.

———. 1977. Controlled manipulations in the marine intertidal zone, and their contributions to ecological theory. Pages 245–270 in Changing Scenes in Natural Sciences, 1776–1976. Special Publication no. 12, Academy of Natural Sciences of Philadelphia.

———. 1992. Food-web analysis through field measurement of per capita interaction strength. Nature 355:73–75.

———. 1994. Marine Rocky Shores and Community Ecology: An Experimentalist's Perspective. Ecology Institute, Oldendorf/Luhe, Germany.

Paine, R. T., and R. L. Vadas. 1969. The effects of grazing by sea urchins, *Strongylocentrotus* spp., on benthic algal populations. Limnology and Oceanography 14:710–719.

Pease, C. M., and J. J. Bull. 1988. A critique for measuring life history trade-offs. Journal of Evolutionary Biology. 1:293–303.

Persson, L., G. Andersson, S. F. Hamrin, and L. Johansson. 1988. Predator regulation and primary production along the productivity gradient of temperate lake ecosystems. Pages 45–65 in S. R. Carpenter (ed.), Complex Interactions in Lake Communities. Springer-Verlag, New York.

Peters, R. H. 1976. Tautology and evolution in ecology. American Naturalist 110:1–12.

———. 1991. A Critique for Ecology. Cambridge University Press, Cambridge.

Peterson, C. H., H. C. Summerson, and S. R. Fegley. 1987. Ecological consequences of mechanical harvesting. Fishery Bulletin 85:281–298.

Pfister, C. A. 1992. Costs of reproduction in an intertidal kelp: patterns of allocation and life history consequences. Ecology 73:1586–1596.

———. 1993. The dynamics of fishes in intertidal pools. Ph.D. dissertation, University of Washington, Seattle.

———. 1995. Estimating competition coefficients from census data: a test with field manipulations of tidepool fishes. American Naturalist 146:271–291.

———. 1996. The role and importance of recruitment variability to a guild of tidepool fishes. Ecology 77:1928–1941.

———. In press. Extinction, Colonization and species occupancy in tidepool fishes. Oecologia.

Pfister, C. A., and M. E. Hay. 1988. Associational plant refuges: convergent patterns in marine and terrestrial communities result from differing mechanisms. Oecologia 77:118–129.

Pimm, S. L. 1982. Food Webs. Chapman and Hall, London.

———. 1985. Estimating competition coefficients from census data. Oecologia 67:588–590.

Pomerantz, M. J. 1981. Do "higher-order interactions" in competition systems really exist? American Naturalist 117:583–591.

Power, M. E. 1990. Effects of fish on river food webs. Science 250:811–814.

Power, M. E., M. S. Parker, and J. T. Wootton. 1995. Disturbance and food chain length in rivers. Pages 286–297 in G. A.. Polis and K. O. Winemiller (eds.), Food Webs: Integration of Patterns and Dynamics. Chapman and Hall, New York.

Power, M. E., D. Tilman, J. A. Estes, B. A. Menge, W. J. Bond, L. S. Mills, G. Daily, J. C. Castilla, J. Lubchenco, and R. T. Paine. 1996. Challenges in the quest for keystones. BioScience 46:609–620.

Reed, D. C. 1990. An experimental evaluation of density dependence in a subtidal algal population. Ecology 71:2286–2296.

Reznick, D. 1985. Costs of reproduction: an evaluation of the empirical evidence. Oikos 44:257–267.

Rosenzweig, M. L. 1973. Exploitation in three trophic levels. American Naturalist 107:275–294.

Roughgarden, J. 1976. Resource partitioning among competing species—a coevolutionary approach. Theoretical Population Biology 9:388–424.

Schindler, D. W., K. H. Mills, D. F. Malley, D. L. Findlay, J. A. Shearer, I. J. Davies, M. A. Turner, G. A. Linsey, and D. R. Cruikshank. 1987. Long-term ecosystem stress: the effects of years of experimental acidification on a small lake. Science 228:1395–1401.

Schmitt, J., J. Eccleston, and D. W. Ehrhardt. 1987. Dominance and suppression, size-dependent growth and self thinning in a natural *Impatiens capensis* population. Journal of Ecology 75:651–655.

Schmitt, R. J. 1996. Exploitation competition in a mobile grazers: trade-offs in use of a limited resource. Ecology 77:408–425.

Schmitz, O. J. 1992. Exploitation in model food chains with mechanistic consumer-resource dynamics. Theoretical Population Biology 41:161–183.

———. 1993. Trophic exploitation in grassland food chains: simple models and a field experiment. Oecologia 93:327–335.

Schoener, T. W. 1974. Competition and the form of habitat shift. Theoretical Population Biology 6:265–307.

————. 1983. Field experiments on interspecific competition. American Naturalist 122:240–285.

Snow, A. A., and D. F. Whigham. 1989. Costs of flower and fruit production in *Tipularia discolor* (Orchidaceae). Ecology 70:1286–1293.

Sousa, W. P. 1979. Experimental investigations of disturbance and ecological succession in a rocky intertidal algal community. Ecological Monographs 49:227–254.

Strong, D. R., L. A. Szyska, and D. Simberloff. 1979. Tests of community-wide character displacement against null hypotheses. Evolution 33:897–913.

Sutherland, J. P. 1974. Multiple stable points in natural communities. American Naturalist 108:859–873.

Tilman, D. 1982. Resource Competition and Community Structure. Princeton University Press, Princeton, New Jersey.

————. 1994. Competition and biodiversity in spatially structured habitats. Ecology 75:2–16.

Tilman, D., and D. Wedin. 1991. Dynamics of nitrogen competition between successional grasses. Ecology 72:1038–1049.

Tilman, D., R. M. May, C. L. Lehman, and M. A. Nowak. 1994. Habitat destruction and the extinction debt. Nature 371:65–66.

Underwood, A. J. 1986. The analysis of competition by field experiments. Pages 240–258 in J. Kikkawa and D. J. Anderson (eds.), Community Ecology: Pattern and Process. Blackwell Scientific, Oxford.

Underwood, A. J., and P. S. Petraitis. 1993. Structure of intertidal assemblages in different locations: How can local processes be compared? Pages 39–51 in R. E. Ricklefs and D. Schluter (eds.), Species Diversity in Ecological Communities. University of Chicago Press, Chicago, Illinois.

Underwood, A. J., E. J. Denley, and M. J. Moran. 1983. Experimental analyses of the structure and dynamics of mid-shore rocky intertidal communities in New South Wales. Oecologia 56:202–219.

VanderWerf, E. 1992. Lack's clutch size hypothesis: an examination of the evidence using meta-analysis. Ecology 73:1699–1705.

Waddington, C. H. 1957. The Strategy of the Genes. Allen and Unwin, London.

Warner, R. P., and P. L. Chesson. 1985. Coexistence mediated by recruitment limitation fluctuations: a field guide to the storage effect. American Naturalist 125:769–787.

Welden, C. W., and W. L. Slauson. 1986. The intensity of competition versus its importance: an overlooked distinction and some implications. Quarterly Review of Biology 61:23–43.

Weller, D. E. 1987. A reevaluation of the $-3/2$ power rule of plant self thinning. Ecological Monographs 57:23–43.

Werner, E. E. 1984. The mechanisms of species interactions and community organization in fish. Pages 360–382 in D. R. Strong Jr., D. Simberloff, L. G. Abele, and A. B. Thistle (eds.), Ecological Communities: Conceptual Issues and the Evidence. Princeton University Press, Princeton, New Jersey.

————. 1992. Individual behavior and higher-order species interactions. American Naturalist 140:S5–S32.

Werner, E. E., and J. F. Gilliam. 1984. The ontogenetic niche and species interactions in size-structured populations. Annual Review of Ecology and Systematics 15:393–425.

Werner, E. E., and D. J. Hall. 1974. Optimal foraging and the size selection of prey by the bluegill sunfish *Lepomis macrochirus*. Ecology 55:1042–1052.

————. 1976. Niche shifts in sunfishes: experimental evidence and significance. Science 191:404–406.

————. 1977. Competition and habitat shift in two sunfishes (Centrarchidae). Ecology 58: 869–876.

Wilbur, H. M. 1972. Competition, predation, and the structure of the *Ambystoma-Rana sylvatica* community. Ecology 53:3–21.

————. 1987. Regulation of structure in complex systems: experimental temporary pond communities. Ecology 68:1437–1452.

Wilbur, H. M., and J. E. Fauth. 1990. Experimental aquatic food webs: interactions between two predators and two prey. American Naturalist 135:176–204.

Williams, G. C. 1966. Natural selection, the costs of reproduction, and a refinement of Lack's principle. American Naturalist 10:687–690.

Wissinger, S., and J. McGrady. 1993. Intraguild predation and competition between larval dragonflies: direct and indirect effects on shared prey. Ecology 74:207–218.

Wooster, D. 1994. Predator impacts on stream benthic prey. Oecologia 99:7–15.

Wootton, J. T. 1992. Indirect effects, prey susceptibility, and habitat selection: impacts of birds on limpets and algae. Ecology 73:981–991.

————. 1993a. Indirect effects and habitat use in an intertidal community: interaction chains and interaction modifications. American Naturalist 141:71–89.

————. 1993b. Size-dependent competition: effects on the dynamics versus the endpoint of mussel bed succession. Ecology 74:195–206.

————. 1994a. Predicting direct and indirect effects: an integrated approach using experiments and path analysis. Ecology 75:151–165.

————. 1994b. Putting the pieces together: testing the independence of interactions among organisms. Ecology 75:1544–1551.

————. 1994c. The nature and consequences of indirect effects in ecological communities. Annual Review of Ecology and Systematics 25:443–466.

————. 1995. Effects of birds on sea urchins and algae: a lower-intertidal trophic cascade. Écoscience 2:321–328.

————. 1997. Estimates and tests of per-capita interaction strength: diet, abundance, and impact of intertidally foraging birds. Ecological Monographs 67:45–64.

Wootton J. T., and M. E. Power. 1993. Productivity, consumers and the structure of a river food chain. Proceedings of the National Academy of Science of the USA 90:1384–1387.

Wotton, J. T., M. S. Parker, and M. E. Power. 1996a. Effects of disturbance on river food webs. Science 273:1558–1561.

Wootton, J. T., M. E. Power, R. T. Pame, and C. A. Pfister. 1996b. Effects of productivity, consumers, competitors, and El Niño events on food chain patterns in a rocky intertidal community. Proceedings of the National Academy of Science of the USA 93:13855–13858.

Yodzis, P. 1988. The indeterminacy of ecological interactions as perceived through perturbation experiments. Ecology 69:508–515.

Young, B. E. 1996. An experimental analysis of small clutch size in tropical house wrens. Ecology 77:472–288.

19

The Logic, Value, and Necessity of Grounding Experiments in Reality

An Essential Link in the Inferential Chain Back to Nature

JOSEPH BERNARDO

Experiments are an essential part of, and serve multiple roles in, ecological research (Lawton, this volume; Morin, this volume; Wooten and Pfister, this volume). In addition, experiments vary in their levels of realism, in their intimacy with natural variation, and in the sophistication of their designs along these continua. However, all appropriately replicated experiments share the key property of involving directed (intentional) manipulations of one or more supposed causal variables with the expressed purpose of estimating the impact of those manipulations upon one or more focal variables. It is from this purposeful manipulation of hypothesized causal factors that experiments derive their property of allowing unambiguous partitioning of trait variance among putative causes. This property is the unique contribution that experiments make to ecological and evolutionary inference, and it is the reason that many ecologists consider experiments indispensable in any inferential chain that hypothesizes causation.

Experiments have played a long and influential role in ecological research. Nonetheless, while some of the truly classic works of ecology were experimental studies (e.g., Tansley 1917; Turresson 1922a,b, 1925, 1930; Gause 1934; Clausen et al. 1940, 1947, 1948; Park 1948, 1954), it is only in the last two decades that experimentation has become a dominant paradigm in ecological research. This is evidenced by trends in mainstream ecological journals (quantitative reviews, Hairston 1989a, Tilman 1989b, Stiling 1994). Analogous detail from the National Science Foundation (NSF) and other agencies is not as readily available, but a large leap is not required to hypothesize that trends in published ecological research reflect, to some degree, patterns in funding. There can be little debate that experimentation has been and continues to be fundamental to the development of ecological knowledge, and I will take it as a guiding principle that experiments comprise an essential part of any ecological research program that also requires conceptual grounding. Other authors have made similar observations (Tilman 1989b; Werner, this volume).

The Continuing Refinement of Experimental Method and Practice in Ecology

Despite their long history and extensive use in contemporary ecological research, the practice of experimentation is still being refined. The widespread use of experiments in ecological research has spawned several evaluations of how they are done and what they tell us (e.g., Hurlbert 1984, Diamond 1986, Hairston 1989a,b, Tilman 1989a,b, Lubchenco and Real 1992). Some issues have been explored in detail, but others have not received much attention. For example, many authors have discussed the role of experiments in ecological inference as a confirmatory tool (see Loehle 1987), possibly because experiments have a strong traditional link to theoretical or conceptual approaches (Fig. 19-1). This view traces at least to Gause's (1934) classic experiments which were directed at testing explicitly the Lotka-Voltera models of population growth and species interactions. Many ecologists continue to see experimentation as a servant of theory (Karieva 1989, Scheiner 1993), and testing theory remains a valuable role for experimentation (see Lawler, this volume; Lawton, this volume; Morin, this volume). But experiments serve other roles in making ecological inferences that are not directly connected to theory testing (see following discussion; Marschall and Roche, this volume; Wooten and Pfister, this volume), and some of these will be explored here. One issue that has been explored in some detail is methodological difficulties with experiments, or what may be called *mechanical aspects* of experimentation—these aspects pertain to all experiments regardless of system or question (Connell 1983, Hurlbert 1984, Underwood 1986, Hairston 1989a, and many works on experimental design: Fisher 1935, Cochran and Cox 1950, Cox 1958, Winer 1971, Lindman 1992, Manly 1992, and many others). These include the use of experimental designs appropriate to the hypotheses posed, treatment dispersion (the need to assign treatments randomly to experimental units), the need for replication of treatments, the need to ensure independence among replicate experimental units, and so on.

In this essay, I draw attention to some less-explored aspects of making ecological inferences from experiments to which I refer as *cogitative* aspects of experimentation—less obvious issues attendant to experimentation which must be thought about carefully and deliberately. Specifically, I explore difficulties attendant to both the design and interpretation of field experiments that lack grounding in empirical field data, or what is variously called descriptive or natural history data. Hence this essay concerns a specific—but rather large—subset of ecological experiments: those whose goal is to make inferences back to nature from field experiments. Such experiments are a common paradigm in modern ecological research (Hairston 1989a). As Lawton (this volume) and Morin (this volume), among others, have noted, field experiments are often perceived as a superior source of strong inference about ecological processes (compared to laboratory experiments, for example) precisely because they are conducted in the field. I will argue that ecologists should not assume that experiments conducted in the field are superior ipso facto. I contend that many field experiments actually yield little insight into ecological processes in nature and that such experiments may often mislead because of inattention to details which are needed to connect the experimental conditions to those in the natural system about which the experiment is intended to inform. A key point I will argue is that although these cogitative aspects of experimentation

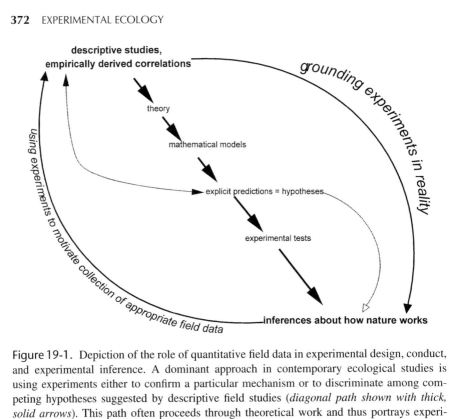

Figure 19-1. Depiction of the role of quantitative field data in experimental design, conduct, and experimental inference. A dominant approach in contemporary ecological studies is using experiments either to confirm a particular mechanism or to discriminate among competing hypotheses suggested by descriptive field studies (*diagonal path shown with thick, solid arrows*). This path often proceeds through theoretical work and thus portrays experiments as a tool for testing theory and field data as a motivation for theory (as described by Scheiner 1993 or Karieva 1989, for example). However, experiments may be used to validate a causal pathway hypothesized on the basis of field sampling without requiring theory (*connection via dashed arrows*), or experiments may implicate specific factors that might explain a particular pattern of variation in nature that then needs to be verified with field sampling (*curved solid arrow on left of figure*). Both of these approaches to experimentation in ecological inference are extremely informative yet are far less commonly employed by ecologists, hence the vague nature of the arrow that connects hypotheses to inferences on the bottom right of the figure. Two key points I hope to illustrate with this figure are that experiments are part of a process of scholarly inference, but also that field data relate to experiments in many ways other than the one illustrated by the main axis of the figure.

may not be as widely considered by experimentalists as are mechanical issues, they are as critical to making *sound* inferences from experiments. Sound inferences derive from a comprehensive research program that requires not just conceptual grounding—a point that is widely accepted—but empirical rooting as well. A corollary of this argument is that no matter their statistical power or creativity of their designs and no matter how intriguing their findings might be, experiments alone provide only weak evidence for the importance of any ecological process in natural systems. I argue that experiments must be accompanied by intimate understanding of trait and environment distributions in nature so as to connect the conditions of the experiment, and thus its scope of inference, to the natural world.

Another, derivative concern I raise is that lack of such connections leads to several kinds of insidious inference that are not often recognized because the fact that an experiment was conducted in the field gives such studies and all ancillary interpretations made thereof an unwarranted aura of realism and accuracy. Such unjustified confidence results in misleading inferences that add little to ecological knowledge and diminish the value of careful experiments and cautious, reasoned inference.

To provide clarity to my arguments, I first explore the relationship between "natural history" and "ecology," arriving at a definition for the kinds of data that I see as relevant to the conduct of realistic and meaningful field experiments. Next, I develop a rationale for grounding experiments in reality at a variety of levels when they are conducted for the purposes of making inferences about natural populations or communities. In conjunction, I describe several ways in which the powerful role of experiments can be misapplied in this kind of inference, and I offer concrete examples of how these pitfalls can be avoided. Thus, I illustrate how appropriate grounding adds to the strength of experimental inference and how lack of such grounding can lead to questionable, if not erroneous, conclusions about the role of a particular factor as an explanation of natural patterns.

Changing Role of "Natural History Data" in Ecology

The definition and practice of ecology have evolved rapidly in the last century, and so, too, have the kinds of data that ecologists gather and the perceptions of how different kinds of data inform about ecological principles. Descriptive data in particular have changed in prominence as a source of information about how ecological systems work. An older view at one extreme can be traced in part to Bates, who saw population and community ecology and systematics as parts of a larger field of natural history, with no distinction between these areas. He wrote: " 'Ecology' is erudite and profound; while 'natural history' is popular and superficial. Though, as far as I can see, both labels apply to just about the same package of goods." (1950:7). Others shared Bates's view: "Ecology is basically just an elegant word meaning natural history" (Poole 1974: 3). These ideas of ecology as natural history simply reflected the long, slow development of ecology as a conceptual and manipulative science from its deep roots in descriptive studies of nature, which can be traced back several hundred years (see review in McIntosh 1985). Even by the middle of this century ecology remained a largely descriptive field, despite the theoretical efforts of Volterra (1926), Pearl (1927), Lotka (1932), Nicholson (1933), and others and the early experiments of Tansley (1917), Gause (1934), Clausen et al. (1940, 1947, 1948), Park (1948, 1954), and others. Lacking any competing paradigm for ecology, Bates obviously should have arrived at a definition of natural history as a purposeful scientific enterprise: "I defined natural history as biological investigation at the level of the individual organism: the study of the relations of organisms among themselves and with the physical environment, and of their organization into populations and communities" (241). However, his definition clearly deviates from the modern perception of natural history as a casual pursuit.

Marginalization of "Natural History Data" in Ecological Research

The collection and interpretation of natural history data are not part of the mainstream culture of contemporary ecological research. This is clear from reviews of current published research which show that natural history or descriptive studies do not find homes in the major ecological journals (Hairston 1989a, Tilman 1989b, Stiling 1994); it is also clear to anyone who has submitted a paper containing descriptive data as one of its major components. Experiments, on the other hand, occur in over 60% of published papers according to the most recent review (Stiling 1994). It was not until the experimental revolution (in ecology) of the 1960s and 1970s that the collection of descriptive natural history data began to be placed in opposition to experimentation as an approach within the larger context of ecological research. Indeed, the term *natural history* now connotes another extreme in contemporary ecological circles, sometimes deemed less rigorous than experimental or theoretical approaches. If, as Tansley (quoted by Varley 1957, Tilman 1989a, b) and others have argued, observation, theory, and experiment should be equal components of ecological research (see also Fig. 19-1), why should there exist such a disparity between the major kinds of empirical papers being published?

Several factors probably contribute to this trend. First, fewer descriptive studies are initiated than just a few decades ago, perhaps reflecting a lack of straight organismal courses (the long-forgotten "ologies") in biology departments. Second, experiments are often the first kind of study graduate students are pushed to conduct, in part because of a "theory-testing bias" in ecology (Loehle 1987). Third, it is unlikely that purely descriptive approaches are supported by granters as much as experimental "tests"-of-theory kinds of proposals. However, each of these explanations has its roots, in my opinion, in a fourth explanation: There is clearly a cultural bias between the muddy boots school (the old-fashioned naturalists) and the tweed jacket school (the theoreticians). The rapid development of experimentation as the dominant empirical paradigm has contributed markedly to this rift and thus to the marginalization of natural history or observational approaches in ecology. Several recent authors (McIntosh 1985, Hairston 1989a) have treated this topic more extensively, and it is not hard to see from their reviews how this segregation developed. Although experiments, including some conducted in the field (Clausen et al. 1940, 1947, 1948), have enjoyed a long history in ecology, perhaps the watershed event came in the field-experimental studies of Connell (1961) and Paine (1966) during the 1960s, in which it was clearly shown for the first time that ecological *theories* could be tested experimentally *in the field*.

It did not take long for ecologists of all flavors to realize that major advances might be made by testing theory with field experiments (Fig. 19-1). In this context, natural history or observational approaches languished, and they were increasingly seen as the poor cousin of experimentation as a means of "testing theory" (extensive examples and discussion are in McIntosh 1985, Hairston 1989a). Thus, a key reason that natural history data are not widely regarded as a critical pursuit in modern ecological research is that experimentation is seen by many workers as the critical and *only* tool to test theory (e.g., Karieva 1989, Scheiner 1993; see also Wise 1993: 266; for a different view see Tilman 1989a, b and Fig. 19-1). In this light it is not surprising that natural history data and the collection of quantitative field data are now of tertiary interest (at best) to many editors and grantors.

Another part of the reason that "natural history data" are not prominent in contemporary ecological papers published in the leading journals is that in current use *natural history* is a nebulous term that encompasses any and all observations one might make in the field; it connotes the descriptive tradition of the nascent biological disciplines (Stein 1966; Greene 1986, 1994; Greene and Losos 1988). Natural history observations are often not quantitative but anecdotal and thus not tremendously informative. Any trained biologist who goes afield cannot help but do "natural history." And there can be little disagreement that such careful observations, in the tradition of the Victorian naturalists, are intriguing and potentially valuable to the ecologist (see example in Greene 1994). However, there is more to using field data in experimentation that goes beyond the ease with which casual "natural history" observations may be made or the truism that such anecdotes may be illuminating about ecological processes. Because this distinction is not generally appreciated, clarification of terms is necessary for the purposes of this essay.

The kind of information I require is better described as "quantitative field data," which I define as parameter estimates and their associated confidence intervals, derived from observational, mensurative, or other descriptive methods. Thus, noting that salamanders at higher elevations in the southern Appalachians appear to be larger than their counterparts at lower elevations is a natural history observation, but estimating the regression of salamander body size on elevation (e.g., Hairston 1949, Martof and Rose 1963) is a distinct kind of quantitative field data which immediately have great value to both the demographer (Tilley 1980) and the experimentalist (Bernardo 1994).

The Decisive Contribution of Quantitative Field Data to Experimental Method

Purposeful manipulation is a defining feature of experimentation (Fisher 1935). Thus, by their nature experiments are artificial—there is no such thing as a "natural experiment" (Diamond 1986). But this artificiality does not obviate the need to connect experiments to the natural systems about which inferences are to be made. How can this connection be made? In what ways are the connections informative? What are the limits to inference from experiments that lack such connections? Here I explore some of the ways in which quantitative field data enhance the conceptualization, design, and interpretation of experiments, in the hopes of contributing to the refinement of experimentation as a key inferential tool in ecological research.

Choice of Experimental Factors

Ecological experiments are not really expensive within the realm of research that biologists conduct, but they certainly are time-consuming and labor-intensive to execute. A complete factorial design sees the number of treatments grow exponentially as the number of factors incorporated into the design increases. A fully crossed two-factor design with four replicates requires 16 experimental units, and these numbers quickly mushroom: a complete four-factor design contains 16 treatments requiring 64 units.

This size may still be tractable but approaches the logistic limit for many systems and budgets (time or money). Nevertheless, it is not hard to imagine more than four factors which might contribute to the kinds of variables biologists often assay, such as growth rate.

This multicausal nature of most ecological phenomena presents an unavoidable philosophical challenge to the experimentalist—namely, choosing which variables to manipulate in an experiment. The process of variable choosing is a critical phase of experimental design because *a factor that is not allowed into the design cannot produce statistically attributable variance in the experiment.* Interestingly, this issue is one of the reasons that experiments have achieved preeminence over purely correlational approaches. Correlations are typically bivariate and nondirectional and, hence, indeterminate. Thus, correlational evidence is always subject to the criticism that another, unconsidered, variable is driving the association. But the same is true of unconsidered variables or of variables that simply were not represented in an experiment due to the reasonable logistic constraints of an experiment.

Consider the situation faced by the biologist studying local determinants of plant growth. It is reasonable to think about manipulating various micronutrient levels, singly or in concert, or levels of herbivory, plant density, light environment and moisture regime, or to consider genetic variation as possible sources of variation in growth among individuals in the field. We know that plants need nutrients, light, water, and space to grow and that some plants are intrinsically better than others given the same conditions for growth. Many published experimental studies of plant growth show that manipulations of any of these variables are likely to produce statistically significant variance in growth in an experiment. Parallel treatments would occur to the animal ecologist—food levels, conspecific density, predation regime, temperature, and perhaps social context. Again, abundant published studies of animals (e.g., of frog larvae) show the effects that each of these variables can have on growth and development. However, the key issue is: *In the system under study,* which of these things are likely candidate causes of variation in growth? Which variables are less important in this particular case, and *why is that interesting*?

These vignettes intimate one of the most important reasons to ground experiments in quantitative field data: correlations developed on the basis of field sampling *provide a logical objective way of choosing which factors ought to be manipulated* in an experiment. Theory, too, suggests which variables might draw the attention of the experimentalist, although many ideas formalized by theory also arose from correlational results. How does field sampling aid in the choice of experimental factors?

Suppose, for example, that the individual growth rates of weeds in a population in a plowed, fertilized, irrigated field in full sun are to be studied. What limits plant growth under these conditions? The efficient experimentalist might greatly benefit from a few weeks of descriptive sampling before initiating an experiment. Soil samples for nutrient analysis, estimates of light incidence and soil moisture level, and quadrat sampling estimates of plant density or even percentage of cover could be made quickly, even coarsely, to assay how much spatial variation there is among the microsites in which different individual plants occur. Clearly, such sampling would not rule out any of these factors as a determinant of variation in among-individual plant growth, but it would provide a meaningful and objective way of choosing from the laundry list of possible factors which could be manipulated in the experiment. This is because one could easily conduct a descriptive analysis that asks, Which variable or variables explain(s) the bulk of the total variance in plant size in the field? These variables might then be targeted for further study in an experiment.

The amphibian ecologist likewise might conduct several days of box sampling (Morin 1983, Harris et al. 1988), minnow trapping (Resetarits and Fauth, this volume), quadrat sampling (Bernardo 1994), or Zippen removals (Hairston 1987, Resetarits 1991). Any of these simple exercises could yield quantitative and therefore *objective* estimates of population sizes for target species or functional groups—for example, invertebrate and vertebrate predator densities, conspecific densities, competitor densities, prey densities, and so on. Water samples could be rapidly tested with aquarium water test kits to ascertain in a coarse survey the availability of fundamental nutrients for algal growth. Even crudely analyzed diatometer studies would quickly connect the resource base of an experiment with that found in natural ponds or streams. Certainly such estimates are preferable to no information at all, and the motivated ecologist could easily conduct more careful analyses.

Barring some objective choosing method, the biologist is forced to attribute variation in the response variable—in our example, growth rate of a plant or a tadpole—only to those factors included in the experiment. Only a minority of the papers I read and review provide a clear explanation or justification for the choice of variables or treatment levels, and fewer still use field data as the objective guide in this process. While it is true that journals such as *Ecology* do not encourage such detail (in their instructions for authors), this decision process is a key element of the logical structure of an experiment and should therefore be clearly articulated in any experimental paper. Hence, the lack of such information in so many published papers intimates that this process of variable choice is not given serious attention by many ecologists, although it is perhaps one of the most critical phases of experimental design.

Determination of Treatment Levels and Experimental Densities

If the choice of experimental factors is sometimes haphazard or undirected, so is the choice of treatment levels often arbitrary—but this need not be. A common choice is a presence-absence (+/0) experiment in which a potential competitor, predator, or nutrient/food treatment is used or not in different treatments. Such experiments are valuable to test important hypotheses: Do these two species compete or not (e.g., Fauth et al. 1990, Bernardo et al. 1995)? Are these organisms nutrient- or food-limited or not (e.g., Bernardo 1994)? A challenge for such designs, however, is that a single treatment level is potentially unrealistic, especially when presented in all-or-none fashion.

One solution that requires no information about natural conditions is simply to represent a range of conditions in the experiment. For instance, in studies of density-dependence (experimental factor) of tadpole growth (response variable), Wilbur and colleagues have conducted many studies in which initial tadpole densities range from very low ("unnaturally low") to very high ("unnaturally high"). Such an experiment requires much more work and many more experimental units than a simple high/low experiment because each density constitutes a treatment; thus, an experiment involving five density treatments each replicated three times comprises 15 experimental units. There are several key advantages of such an experiment in which multiple treatment levels are used. First, the mere inclusion of multiple treatment levels, especially if dispersed nonlinearly on density (e.g., $1\times$, $2\times$, $4\times$, $8\times$, $16\times$; see Wilbur 1977), are likely to include relevant natural densities. Further, the use of multiple levels that span a large part of the range of possible real densities enables the biologist to estimate with a high degree of statistical confidence the functional relationship between density and growth.

Despite these valuable features of such a design, however, there remains a nontrivial gap in the interpretative extension of the experimental results for describing density-dependence of tadpole growth in the field: Without quantitative field data about the densities actually experienced by wild tadpoles, the insights into the density-dependence of larval growth gained from the experiment are academic, because there is no objective way of identifying which part of the estimated function actually applies to the natural populations about which inferences are inevitably made (e.g., Wilbur 1976, 1977). A common retort that I have received in response to this criticism is that the experimental conditions are sufficiently broad so as to encompass natural conditions. Indeed, many of these sorts of papers contain statements such as "the experimental conditions span the range of conditions I have observed in the field." Yet this justification begs the question of why any experiment is conducted whose goal—stated or not—is to make inferences about ecological processes in nature. What is germane when making eco-logical interpretations of a field experiment is not that a particular experimental con-dition can be found in nature—even extreme conditions are sometimes found in the field—but we would like to know something about the probability distribution of field conditions: Is an experiment addressing a rare or common event? It is also possible that not all of the experimental treatment levels evinced an effect—under such circum-stances, this issue is more acute because it becomes essential to know if that part of the experimental parameter space in which an effect was seen actually overlaps with conditions in the field. A replicated program of field sampling through time is the only method for obtaining this kind of information. For instance, do predators commonly release tadpole populations from competition by cropping total tadpole densities (Morin 1981, 1983; Fauth and Resetarits 1991; Resetarits and Fauth, this volume), or are tad-pole populations seldom so crowded in the field? It seems trivial to argue that such ecological phenomena are worthy of study if we do not know anything about their spatiotemporal prevalence or the magnitude of long-term impacts of rare events.

Even the choice of densities for background organisms that are not the focus of the study is relevant in the design of meaningful ecological experiments. This is because prey densities or species composition may alter the nature of a species interaction between two predators (discussion in Bernardo et al. 1995). An example is Schluter's (1994) attempt to muster experimental evidence for character displacement. In his study, Schluter tested the idea that the most morphologically similar individuals from each of two synthesized morphotypes of fish (a "limnetic" and a "benthic" morph) were the most likely to compete for invertebrate prey resources and, therefore, the most likely to exhibit differential survival based on morphology. In other words, he expected that natural selection on trophic morphology, mediated through prey resource competition, was a mechanism that he could demonstrate in a field experiment. Fundamental to this premise is the notion that the ecological resource base against which the experimental test was performed was actually relevant to that experienced by fish in nature. To meet this goal, Schluter used large naturalistic ponds "seeded" with invertebrates from natural ponds. However, the ponds in which the experiment was conducted, while large and apparently realistic, had been allowed to develop for two years after inoculation without the presence of any fish. A rich literature had shown clearly many years earlier that the size and structure of a zooplankton assemblage are affected dramatically by the presence of fish predators (e.g., Brooks and Dodson 1965). In other words, the kinds of resources for which the fish might actually compete in nature were not adequately or realistically represented in the experimental system (Bernardo et al. 1995) and hence the species composition and density of the prey resource base used in the experiment were likely not ecologically relevant to the interaction between the fish species in nature.

Establishing Experimental Factors and Treatment Levels Using Field Data: An Example from Desmognathus Salamander Reproductive Allocation

In this section, I describe an experiment I conducted to examine determinants of reproductive allocation in a small Appalachian salamander, *Desmognathus ochrophaeus*. This study illustrates several ways in which quantitative field data aid the design and conduct of an experiment whose purpose is to understand a particular pattern in nature.

The *D. ochrophaeus* species complex has attracted a great deal of interest from evolutionary ecologists for a half-century because different populations are remarkably variable in body size, developmental timing, and reproductive traits. There is also striking clinal variation in adult body size along small spatial scales of just a few kilometers, with larger animals occurring at higher elevations. The larger body size is achieved via delayed maturation at high (relative to low) elevation sites (Tilley 1980), and this difference in maturation timing appears to be largely genetically based (Bernardo 1994).

A rich life history theory provides the conceptual basis for wanting to study this naturally variable system; fitness is understood to be extremely sensitive to variation in age at maturity, clutch size, and egg size (Bernardo 1993, 1996). Because my interest is in understanding what this variation means for fitness in real organisms in the natural populations, I have taken a field-based approach that uses both in situ experiments and quantitative sampling. In my research program, field experiments are the culmination of (often) several years of field sampling. Field sampling plays several key roles in my research program. One purpose is to establish phenotypic patterns that exist in nature using replicated sampling within and among years. This sampling program is labor-intensive in that one year's fieldwork may only result in a single regression line with confidence intervals. I see this information as essential for designing an experiment appropriate to evaluate the hypotheses of interest.

One example centers on patterns of reproductive investment and how this investment is partitioned between the size and number of eggs a female produces. Coarse scale quantitative field sampling over 2 years showed that large females not only produce larger clutches but also make significantly larger eggs than do smaller females (Bernardo unpublished data). Reproducing females vary by sixfold in wet mass, and they produce eggs that vary sixfold in dry mass. Fecundity also varies among females by a factor of 5. Interestingly, however, clutch size and egg size positively covary, and, further, these clutch phenotypes are also positively correlated with female size: Larger females produce not only larger clutches (a common pattern among ectotherms) but substantially larger eggs as well. Life history theory tells us we should expect a trade-off between the clutch phenotypes of fecundity and propagule size (reviewed in Bernardo 1996), so this pattern of allocation is unusual. As an example, consider that in the extreme the largest females could increase their fecundity by nearly sixfold if egg size did not vary. It seems that in this system variation in female body size overwhelms the intuitive, theoretical trade-off between propagule size and number. Another complication is that larger females live in resource-poor habitats compared to the sites in which the early-maturing, smaller females occur (Bernardo 1994). Thus, understanding why females produce the clutch phenotypes that they do involves examining the synergistic impact of food resource level and female body size on the bivariate relationship between fecundity and egg size.

In order for any experimental analysis of this problem to be relevant to interpreting patterns of variation in nature, several key connections via field data were required. First, it was necessary that the experiment reflect the distributions of female phenotypes encountered in the field. This required extensive fieldwork to estimate body size distributions of actively reproducing females in multiple populations. Females were collected from 10 populations, then they were measured and weighed. At least 4 but usually 10–45 females were obtained from each population, for a total of 200 females. Figure 19-2 summarizes the ranges of body sizes of reproductive females within and among the 10 populations that were actually used as source material for the experimental analysis of reproductive allocation. Females were then sorted by size, and four size classes were designated such that equal numbers of females from all parts of the entire size distribution were represented in the experiment (gray bands in Fig. 19-2). These groups were then assembled so that within each size class females were drawn so as to include as many source populations as possible, to avoid confounding of a particular body size phenotype with a particular, possibly genetically distinctive, population.

Another important connection to the field involved linking resource levels used in the experiment to some objective standard that could be applied to the resource environments experienced by females in many natural populations. One way to do this would be estimating food levels in many natural populations (as in Bernardo 1994) and somehow mimicing those levels in the experiment. The weakness of this approach, however, is that it provides little information about the relative energy budgets of females from different populations. Females from different populations certainly experience different resource environments, but because they are ectotherms, they also incur different metabolic costs depending on relative temperatures of their native sites (Dunham et al. 1989). Also, the metabolic demand of these females is tied to their body size. For all of these reasons, an absolute level of prey abundance as an experimental treatment level would not be strictly comparable among females of different body sizes or among different populations. That is, the interpretation of different fixed resource

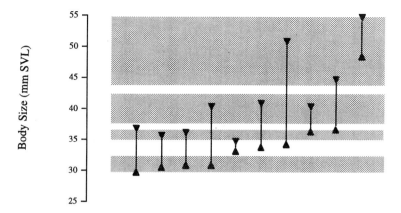

Figure 19-2. Ranges of sizes of actively reproductive females in 10 populations from which experimental females were drawn. The gray boxes illustrate the size range for each of the body size classes defined for the experiment. Note that multiple populations were represented within each size class. See text for detailed description of experiment.

level treatments in an experiment would not translate back to understanding how resource availability in the field affects reproductive allocation. Although this was not an issue intrinsic to the experimental females, in that they were all raised in a common environment during the experiment, I wanted to be able to make inferences to the natural populations among which productivities, metabolic demand, body sizes, and, hence, maintenance costs vary. How could a food treatment be structured so that it was truly relevant and hence appropriate for making comparisons among females and among populations?

My approach to this problem was to use food treatment levels that were scaled to the metabolic demand of nongravid females at the mean daily temperature they were experiencing in the field common garden study (approx. 15°C). I estimated total demand based on measurements in Fitzpatrick (1973). He estimated that such females require 3.6 calories/g/per day for maintenance at 15°C. This was taken as a maintenance ration and was thus one treatment level. Other treatment levels were factors of 0.8, 2, and 4 times the maintenance ration. I raised females individually, so it was possible to both keep track of and administer rations to individual females. The experimental ration was delivered as a count of *Drosophila virilis*, a large fruit fly whose energetic content had been determined as 3 calories/fly (Jaeger and Barnard 1981). Using these four ration levels and the four body size classes described here, I achieved a meaningful and well-grounded design with which to evaluate the synergistic roles of female size and resource level on reproductive phenotypes (Fig. 19-3).

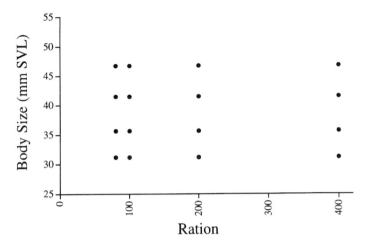

Figure 19-3. Final design of reproductive allocation experiment. Each dot represents a treatment combination. The mean body size of females in each size class was determined by stratified subsampling within each pool of females from each size class as described by the field distributions of body size illustrated in Figure 19-2. The ration levels were in terms of percentage of maintenance ration, as described in the text.

Phenology of an Experiment

The experiment just described also illustrates another critical issue that concerns the importance of the timing of an experiment: experiments should be conducted during natural growing seasons, and dynamic treatments (in which a treatment purposely changes during the course of the experiment) should mimic natural variation in that variable in nature. My study of reproductive allocation in female salamanders occurred in the late summer, as brooding females were naturally leaving their just hatched clutches and beginning to forage again. Why should the timing of the experiment be so important?

A previous study (Fraser 1980) of the effects of food levels on patterns of reproductive allocation in a plethodontid salamander did not detect an effect of increased food level on clutch size. However, Fraser's study was conducted outside the natural phenology of vitellogenesis. As far as is known, all plethodontid salamanders exhibit an annual reproductive cycle producing a single clutch of eggs (at most) per year. What Fraser did not know, but what his experiment suggested, was that female plethodontids physiologically commit to a particular clutch size via follicle recruitment very soon after laying a clutch of eggs. In other words, an experiment in which the effects of nutritional levels on clutch size could be assayed would need to be conducted at a time in a female's vitellogenic cycle at which she could physiologically respond to the food treatment by adjusting clutch size, a point Fraser recognized. I drew on this result in the design of my study. I reasoned that because females do not forage during brooding (Forester 1981), the nutritional environment females experience after brooding was most likely to "set" their clutch size. I knew from Fraser's study that even a few months after the beginning of postovipositional foraging clutch size appears to have been determined. Thus, I began my experiment at about the time that females emerge from brooding.

Another good example of the importance of grounding experiments in an appropriate natural phenology comes from studies of amphibian metamorphosis in response to pond drying. Amphibians that breed in temporary ponds have had to contend evolutionarily with the disappearance of their habitat during development of the larvae—temporary ponds, by definition, dry out. Newman (1987) showed, for instance, that in a desert pond system larval spadefoot toads have, on average, just 8 days to grow and metamorphose; failure to do so results in intense selection as the slow growers dry into a tad-mat in the bottom of a dry pool. Wilbur and Collins (1973) recognized that despite this kind of selection for rapid metamorphosis, larvae should spend as much time in the pond habitat as possible, because every day spent during this rapid growth phase may persistently impinge on adult fitness components such as survival to and timing of first breeding. They argued that phenotypic plasticity in the timing of metamorphosis was the most likely strategy to maximize fitness. Wilbur (1987) and Semlitsch and Wilbur (1988) conducted experiments in which the metamorphic responses to pond drying were explored. Although the studies were conducted in field mesocosms (artificial ponds) and not in natural ponds, realism was given to the experimental drying regime by using a schedule of drying based on that of a natural pond from which one of the populations used in the experiment was drawn. It is not unreasonable to argue that this is a realistic treatment profile and, therefore, that the responses of the focal organisms in the experiment are also relevant to understanding how pond drying affects real organisms in the field.

Duration of an Experiment

Another way in which quantitative field data aid in the conduct of experiments concerns the duration of an experiment and how it relates to the magnitude of change in the response variables. It seems a trivial point, but the duration of experimental treatments must be sufficient either to evince sufficient variance in the response variables among treatments so that treatment effects can be detected given the power of the design or to give the experimentalist confidence that there are not, in fact, real effects of the treatments. That is, the experimentalist would like to avoid commission of Type II error, accepting a null hypothesis that is actually false, because the experiment did not proceed long enough for treatment means to diverge. Importantly, this problem is related to effect strength, because weak effects of treatments will require more time to detect than strong effects. How might quantitative data enable the experimentalist to judge objectively whether an experiment has run long enough to allow a sufficient change in the response variables so that it may be detected by the experiment?

Suppose a linear estimate of body size is chosen as a response variable to an experimental treatment of a putative competitor or to gauge nonlethal effects of a predator. Further, suppose that the focal organism grows 3 to 5 mm in a year in nature and, lastly, that one's measurement error is 0.5 mm. (Note that measurement error arises not just from the accuracy of the measuring device [one can measure a linear distance within 0.01 mm with a caliper under a dissecting microscope] but also from the repeatability of the measurement by the observer, which is seldom as small in magnitude as the error of the measuring device.) How long must experimental treatments last in order to detect an effect body size?

Quantitative field data on growth of natural organisms—even based on cohort average growth rates—would provide a simple way of deciding how long to conduct the experiment. Consider an experiment examining growth in larval salamanders (Beachy 1994). This study purported to examine the impact of two predatory species of larval salamanders on the growth of larvae of a third species (*Eurycea wilderae*). It also examined density-dependent aspects of growth in the *Eurycea*. Beachy concluded that "the lack of a density effect on growth and survival of the prey [Eurycea] coupled with the non-significant interaction among the predators, suggests that competition and mutualism among larval salamanders may not be viable mechanisms of population regulation in this community" (1994:133). This conclusion may be correct, but his experiment did not last long enough for Beachy to detect a growth response in the prey, and this fact can be established by examination of previously published data on growth rates of larval *Eurycea* in nature (Bruce 1982, 1985, 1988).

Beachy allowed the experiment to run for 30 days. In several studies of *Eurycea* larvae in the same region, Bruce found that, based on inspection of body size histograms during the same part of the year in which Beachy did his experiment, mean growth rates based on cohort averages are in the range of 1–2 mm/month. Because Beachy used a size-structured cohort of individuals within each experimental unit but did not follow growth of individuals in his design, it is not clear that he could detect such small growth increments in his short experiment. Further, because there was no growth in his low-density treatments that contained *Eurycea* alone, either, one must ask why the growth expected on the basis of Bruce's field data did not occur in Beachy's study (see following discussion). Beachy's conclusions may be correct, but his experiment does not provide compelling evidence that the result is real and thus must be interpreted cautiously.

Experimental studies of *Eurycea* growth yield several additional lessons relevant to this essay. There are now five separate experiments by three investigators over several years in which the growth of larval two-lined salamanders (*Eurycea bislineata* complex) was scored as a response variable to the presence of competitors or predators (Resetarits 1991; Beachy 1993, 1994; Gustafson 1993, 1994). Populations used in these different experiments all occur in the southern Appalachians but are currently considered distinct species (Jacobs 1987). This distinction is not accepted universally because species boundaries have not been adequately established (Petranka 1997) and the genetic structure of the groups is currently under study (Bernardo and Ryan unpublished data). This geographic variation is both interesting and relevant to the interpretation of the results of these five experiments. In two of the five studies, conducted on the Blue Ridge form (*E. bislineata wilderae*), no evidence of predator effects on growth rate of *Eurycea* larvae was found. However, three studies conducted on the more northerly form (*E. bislineata* ssp., *E.b. cirrigera*) do reveal such nonlethal effects of predators on the growth of the larval *Eurycea*!

At least three explanations for these contradictory results exist. First, if the experiments were all taken at face value (i.e., if it were assumed that the experiments were all suitably designed, appropriately analyzed, and so on), then we might conclude that there is geographic variation in the nature of species interactions between the predatory species and its prey or, perhaps, that there are simply genetic differences among *Eurycea* populations in their growth responses under stressful conditions. That is, we might conclude that there are real, adaptive differences in the growth responses of *Eurycea* from different regions that differ in the composition of the salamander fauna. Perhaps southern populations of *Eurycea*, because they have evolved in the presence of a dense and diverse assemblage of other salamanders, are resistant to the effects of competition, but more northerly populations with a different evolutionary experience have not evolved this tolerance (see, e.g., Fauth, this volume). This interpretation is bolstered by the presumed genetic differences between the two subspecies used and by the fact that the southern form occurs in the most diverse and most dense salamander ensemble in the world.

Second, perhaps the designs of the studies, some methodological distinctions among them, or year-to-year differences in the times at which they were conducted caused the different responses. For example, Resetarits (1991) was studying the impact of trout and spring salamanders, whereas none of the other studies included fish predators. Such differences might be termed *benign differences*, because each worker had slightly different goals and they worked independently.

Third is the least interesting (at least biologically) explanation for these different results, which, however, seems the most likely: the extreme experimental conditions of some studies were simply unlikely to produce realistic growth responses of the *Eurycea* larvae. Figure 19-4 summarizes experimental conditions in the five studies. Both the sizes of the experimental arenas and the numbers of larval salamanders used per experimental unit positively covaried by two orders of magnitude among these five studies. The result is that the effective density of larvae spanned three orders of magnitude. Not surprisingly, the results and ecological interpretations made by the authors of these studies differed markedly. Those experiments conducted near the range of natural larval densities (rightmost points nearest the reference lines in Fig. 19-4) did in fact find effects of the treatments on *Eurycea* growth. Those experiments conducted rather far out of the range of natural densities did not detect treatment effects. The obvious explanation is that in those studies larvae were so crowded that none of them grew, even in the controls, and so there was no way that a growth effect of the treatment could have been detected! The importance of this artifact is that it affects the conclusions drawn by the investigators and the interpretations that one might easily make in comparing the results of these various studies, particularly if the worker comparing across the studies is not familiar with the organism in question. In other words, this artifact affects the state of ecological knowledge. Mere acknowledgment of already published quantitative field data would have obviated this problem.

Validation of Responses of Controls

This discussion leads to the last major point about the value of using quantitative field data throughout the process of experimentation: quantitative field data give the experimentalist an objective way of assaying how experimental conditions actually compare to natural conditions. The *Eurycea* example illustrates a very dangerous in-

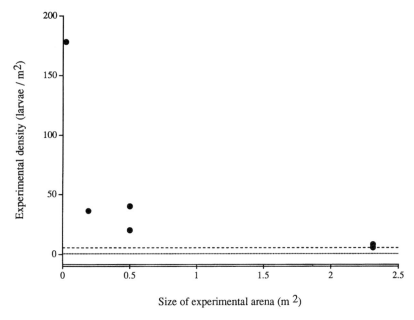

Figure 19-4. Comparison of experimental densities and the size of experimental units in five studies of larval growth in salamanders of the *Eurycea bislineata* complex (Resetarits 1991; Beachy 1993, 1994; Gustafson 1993, 1994). The two dashed lines show field estimates of *Eurycea* larval density based on studies of Bruce (1982: *fine dash*; 1988: *coarse dash*). Note that the x-axis is displaced below 0 so that the baseline low densities estimated by Bruce can be illustrated. The leftmost symbols show conditions in two experiments by Beachy (1993, 1994) which used extraordinarily high densities of *Eurycea* in small to very small enclosures. Under these conditions, no growth effects of any of the treatments were detectable. Beachy estimated natural densities of *Eurycea* from pools at the base of a weir in an experimental watershed, which had unnaturally high densities caused by the accumulation and trapping of larvae in the weir pool. The two nearly overlapping points at the right side of the graph are experiments by Resetarits (1991) and Gustafson (1994, experiment 1) in an array of large experimental stream mesocosms. Note that their experimental densities are comparable to those estimated by Bruce from field data. Both of these authors did detect treatment effects on *Eurycea* growth relative to control conditions. The intermediate values come from Gustafson (1993, 1994 experiment 2). Growth responses of *Eurycea* in these studies were mixed.

terpretive pitfall of ecological experiments that can be avoided by using quantitative field data in the planning of an experiment. But in addition to using this information in planning the experiment, the ecologist would still like to know how the conditions of an experiment actually compared to those typically found in nature. One bridge that can be built is between the performance of the control organisms and that of free-ranging organisms in the field. Suppose an experiment asks whether growth or survival of a focal species is affected by the presence of a second species. It is against the performance of controls that the efficacy of experimental treatments is judged. As we have seen in the *Eurycea* example, the control conditions of an experiment are implicitly assumed to represent a benign natural state. The *Eurycea* experiments all asked whether the presence of predators affected the growth of larvae compared to that of larvae grown in the absence of predators. A seemingly ignored issue with regard to the statistical analysis of experiments is that this comparison is sensitive to the performance of the controls—the statistical test is a relativistic comparison that depends on the intrinsic conditions of an experiment. Hence, the performance of the control must be gauged to determine whether it is indeed representative of natural conditions.

The experimentalist can conduct several kinds of analyses to tackle this question. For instance, Resetarits (1991) was interested in whether large predatory trout grew normally in his experimental streams. Were the experimental streams an adequate mesocosm in which to study ecological interactions that involved such a large, active predator? Resetarits computed and illustrated a mass–standard length regression for his source trout population and then illustrated the values of his experimental trout on that relationship (his Fig. 4). As he put it, this analysis showed that "the final relative condition of all the experimental trout was within the range exhibited by wild-caught trout from the same population, indicating that *Salvelinus* in the experimental streams experienced conditions for growth comparable to those in natural streams." A more detailed use of field data of this sort is found in another paper that explores competition among species of benthic stream fish (Resetarits 1997). Here Resetarits used similar mass–standard length regressions to explore in detail the responses of juvenile fish in his experiments in an effort to understand the nature of their responses to competition.

I performed a similar analysis of condition to estimate the performance of controls in a study of salamander growth (Bernardo 1994, fig. 2). However, I was able to take the analysis a step further. Because I was studying populations that had been studied earlier by Tilley (1980), I could explicitly test for a difference in growth between my control conditions and natural growth by computing an analysis of variance using growth rates of free-ranging individuals estimated by Tilley. This analysis did not reveal a statistical difference between these growth rates (Bernardo 1994, fig. 1), providing confidence in the realism of the experimental system and, therefore, the relevance of the results of the experiments to the natural populations about which I wanted to make inferences.

Discussion

Experiments have become the dominant paradigm of contemporary mainstream ecological research (see data in Hairston 1989a, Tilman 1989b, Stiling 1994), a key reason for which is that the multiple causality inherent in biological systems (Dunham et al. 1978, Quinn and Dunham 1983, Tilman 1989b) stifled progress based on correlational analyses (Varley 1957). Yet the very reasons that experiments have had such great impact in ecological research also contribute to the philosophical challenges associated with their design, conduct, and interpretation.

Experiments *are capable* of providing a unique type of evidence in any inferential enterprise, provided that well-understood and described standards of experimental design and analysis are followed (Fisher 1935 and many others). However, a variety of other issues attend the design, conduct, and interpretation of field experiments besides mechanical concerns, and these have been given far less attention by ecologists. These cogitative issues, as I have called them, may be tedious to consider, but ecologists need to begin to pay them greater heed if experimentation is going to continue to develop as a key paradigm of ecology.

Previously I suggested that quantitative field data are the evidentiary component required to connect inferences from experiments with natural processes. These connections occur at a variety of levels—from conceiving of an appropriate experiment, to picking the treatment levels and other experimental conditions, to the interpretation of the realism and, hence, the relevance and generality of the results. Quantitative field data also serve to root experimental conditions directly in natural conditions, ensuring that the parameter space of the experiment is within that found in nature. But in addition to these many useful contributions of quantitative field data to the enterprise of making inferences about ecological processes from field experiments, quantitative field data provide the experimentalist an extremely powerful tool for making connections between experimental *results* and natural processes.

I have argued that many experiments whose purpose is to explain ecological processes in nature actually yield little insight into natural processes due to a lack of appropriate grounding of experimental treatments and conditions in natural variation. Hence, I suggest that one of the greatest abuses of experimentation in ecological research is the acceptance of experimental results as relevant to natural processes when the design or conduct of the experiment allows no such extension.

The mere use of an experimental approach alone should no longer be taken as an adequate criterion for judging the validity of inferences that authors make based on their results. A field experiment that lacks connections to the field via quantitative field data is subject to the very same (if not a more stringent) criticism as that which is so often leveled against simple mathematical models—so what? This criticism may be more difficult for experimentalists to swallow because at least many modelers do acknowledge the limitations of their models. Most experimental ecologists never constrain their explanatory zeal.

Yet simply taking an experimental approach remains sufficiently compelling that most experimental studies, and inferences made therefrom, continue to be accepted at face value, despite poignant comments over a decade ago by Connell (1983) and Underwood (1986) along these lines. Connell stated:

> The idea of using field experiments . . . is not new . . . , but unless they are designed and executed so as to test relevant hypotheses (which many are not), they are of limited use. In fact, since field experiments are often regarded as the *ne plus ultra* ["no more beyond"; see Medawar 1984:61–63] of ecological research, poor ones can be worse than useless since the conclusions based on them are often accepted with little question. When field experiments were few the tendency was to use this small amount of information rather uncritically. Now there are plenty of field experiments to choose from and I feel that it is time to reexamine the evidence. (1983:662).

While such critical evaluation of the process of doing science is often controversial (Power et al. 1995), it is nonetheless an essential part of the process (Leefer 1948, Connell 1983).

It is remarkable that, more than a decade after Connell's fair critique, the major ecological journals, while stressing the need for appropriate designs and analysis of ecological experiments (e.g., Fowler 1990), provide no standards nor do they routinely search for evidence in the review process that an experiment was conducted under natural conditions. Again, this test applies to papers whose goal is to make statements about natural systems, a point that may be easily judged by reading a paper's discussion. Because this is part of the logical structure of an experiment, it should be an explicit part of any paper that reports an experiment. This is because the validity of an experiment—that is, the quality of its data for adding to ecological knowledge—depends critically on cogitative aspects of the experiment.

Taking this argument a step further, I suggest that desirable attributes of careful experiments—such as objective identification of cause and effect and statistical rigor— and, hence, their great explanatory power are misapplied when those properties are extended to embrace an inferential chain that is longer than that encompassed by the experiment. Further, I contend that field experiments have a greater burden to be realistic than laboratory experiments because the expressed purpose of conducting experiments in the field is to make realistic inferences back to specific natural systems (e.g., Blaustein et al. 1994, Schluter 1994); laboratory experiments often have a very different goal (Lawler, this volume; Morin, this volume). When making statements about natural processes is the goal of the research (as judged by the discussion of a manuscript or by the rationale or significance sections of a grant proposal), the researcher has a nontrivial burden to ensure that the experimental conditions are grounded in nature. Experiments that lack such grounding in reality contribute little to the goal of understanding how nature works, but worse, they may actively mislead us in our pursuit of this goal. Quantitative field data provide the unbiased and valuable insights necessary to the process of making reasonable inferences about ecological processes from field experiments.

ACKNOWLEDGMENTS   I wish to thank my mentors, whose views about experimental approaches in ecology have affected mine. Several of my committee members (Henry Wilbur, Nelson Hairston Sr, and Janis Antonovics) have greatly influenced my thinking in different ways. I especially thank Art Dunham and Bill Resetarits, with whom I have had many challenging discussions about the role of experiments in ecological inference for many years. I doubt any of these people will agree with all of the arguments made here. The three years of fieldwork preceding my experimental studies were funded by a variety of small awards from the American Museum Theodore Roosevelt Memorial Fund, the North Carolina Wildlife Resources Commission, the Explorers Club, and Sigma Xi. The generous support of the Highlands Biological Foundation, an arm of the Highlands Biological Station, enabled me to conduct extensive fieldwork for lengthy periods. Because of this fundamental role of the biological field station in the pursuit of basic descriptive studies, I dedicate this essay to the founders of the Highlands Biological Station (C. Foreman, C. Pope, E. E. Reinke, and W. C. Coker), who had the foresight to establish and endow this unique facility dedicated to the study of the ecology of the southern Appalachian ecosystem. Mark Rausher, as director of graduate studies at Duke University, accommodated requests for lengthy stays in the field which were also critical for extensive fieldwork. Experimental aspects of my research have been supported by NSF grants BSR-90-01587, BIR-94-11048, and DEB-94-07844. The essay was written with the support of the BIR and DEB awards.

Literature Cited

Bates, M. 1950. The Nature of Natural History. Scribner's, New York.

Beachy, C. K. 1993. Guild structure in streamside salamander communities: a test for interactions among larval plethodontid salamanders. Journal of Herpetology 27:465–468.

———. 1994. Community ecology in streams: effects of two species of predatory salamanders on a prey species of salamander. Herpetologica 50:129–136.

Bernardo, J. 1993. Determinants of maturation in animals. Trends in Ecology and Evolution 8:166–173. (Note correction in 8:227.)

———. 1994. Experimental analysis of allocation in two divergent, natural salamander populations. American Naturalist 143:14–38.

———. 1996. The particular maternal effect of propagule size, especially egg size: patterns, models, quality of evidence and interpretations. American Zoologist 36:216–236.

Bernardo J., W. J. Resetarits Jr. and A. E. Dunham. 1995. Criteria for invoking character displacement. Science 268:1065–1066.

Blaustein, A. R., P. D. Hoffman, D. G. Hokit, J. M. Kiesecker, S. C. Walls, and J. B. Hays. 1994. UV repair and resistance to solar UV-B in amphibian eggs: A link to population declines? Proceedings of the National Academy of Science of the USA 91:1791–1795.

Brooks, J. L., and S. I. Dodson. 1965. Predation, body size and composition of plankton. Science 150:28–35.

Bruce, R. C. 1982. Larval periods and metamorphosis in two species of salamanders of the genus *Eurycea*. Copeia 1982:117–127.

———. 1985. Larval period and metamorphosis in the salamander *Eurycea bislineata*. Herpetologica 41:19–28.

———. 1988. An ecological life table for the salamander *Eurycea wilderae*. Copeia 1988:15–26.

Clausen, J., D. D. Keck, and W. M. Heisey. 1940. Experimental studies on the nature of species: I. Effects of varied environments on western North American plants. Carnegie Institution of Washington Publication 520.

———. 1947. Heredity of geographically and ecologically isolated races. American Naturalist 81:114–133.

———. 1948. Experimental studies on the nature of species: III. Environmental responses of climatic races of *Achillea*. Carnegie Institution of Washington Publication 581:1–129.

Cochran, W. G., and G. M. Cox. 1950. Experimental Designs. Wiley, New York.

Connell, J. H. 1961. Effects of competition, predation by *Thais lapillus* and other factors on natural populations of the barnacle *Balanus balanoides*. Ecological Monographs 31:61–104.

———. 1983. On the prevalence and relative importance of interspecific competition: evidence from field experiments. American Naturalist 122:661–696.

Cox, D. R. 1958. Planning of Experiments. Reissued 1992. Wiley, New York.

Diamond, J. M. 1986. Overview: laboratory experiments, field experiments, and natural experiments. Pages 3–22 in J. Diamond and T. J. Case (eds.), Community Ecology. Harper and Row, New York.

Dunham, A. E., D. W. Tinkle, and J. W. Gibbons. 1978. Body size in island lizards: a cautionary tale. Ecology 59:1230–1238.

Dunham, A. E., B. W. Grant, and K. L. Overall. 1989. Interfaces between biophysical and physiological ecology and the population ecology of terrestrial vertebrate ectotherms. Physiological Zoology 62:335–355.

Fauth, J. E., and W. J. Resetarits Jr. 1991. Interactions between the salamander *Siren intermedia* and the keystone predator *Notophthalmus viridescens*. Ecology 72:827–838.

Fauth, J. E., W. J. Resetarits Jr. and H. M. Wilbur. 1990. Interactions between larval salamanders: a case of competitive equality. Oikos 58:91–99.

Fisher, R. A. 1935–1960. The Design of Experiments. Oliver and Boyd, London.

Fitzpatrick, L. C. 1973. Energy allocation in the Allegheny mountain salamander, *Desmognathus ochrophaeus*. Ecological Monographs 43:43–58.

Forester, D. C. 1981. Parental care in the salamander *Desmognathus ochrophaeus*: female activity pattern and trophic behavior. Journal of Herpetology 15:29–34.

Fowler, N. 1990. The 10 most common statistical errors. Bulletin of the Ecological Society of America 71:161–164.

Fraser, D. F. 1980. On the environmental control of oocyte maturation in a plethodontid salamander. Oecologia 46:302–307.

Gause, G. F. 1934. The Struggle for Existence. Williams and Wilkins, Baltimore, Maryland. Reprinted, 1971. New York.

Greene, H. W. 1986. Natural history and evolutionary biology. Pages 99–108 in M. E. Feder and G. V. Lauder (eds.), Predator-Prey Relationships: Perspectives and Approaches from the Study of Lower Vertebrates. University of Chicago Press, Chicago, Illinois.

———. 1994. Systematics and natural history, foundations for understanding and conserving biodiversity. American Zoologist 34:48–56.

Greene, H. W., and J. B. Losos. 1988. Systematics, natural history, and conservation. BioScience 38:458–462.

Gustafson, M. P. 1993. Intraguild predation among larval plethodontid salamanders: a field experiment in artifical stream pools. Oecologia 96:271–275.

———. 1994. Size-specific interactions among larvae of the plethodontid salamanders *Gyrinophilus porphyriticus* and *Eurycea cirrigera*. Journal of Herpetology 28:470–476.

Hairston, N. G. 1949. The local distribution and ecology of the plethodontid salamanders of the southern Appalachians. Ecological Monographs 19:47–73.

———. 1987. Community Ecology and Salamander Guilds. Cambridge University Press, Cambridge.

———. 1989a. Ecological Experiments: Purpose, Design, and Execution. Cambridge University Press, Cambridge.

———. 1989b. Hard choices in ecological experimentation. Herpetologica 45:119–122.

Harris, R. N., R. A. Alford, and H. M. Wilbur. 1988. Density and phenology of *Notophthalmus viridescens dorsalis* in a natural pond. Herpetologica 44:234–242.

Hurlbert, S. H. 1984. Pseudoreplication and the design of ecological field experiments. Ecological Monographs 54:187–211.

Jacobs, J. F. 1987. A preliminary investigation of geographic genetic variation and systematics of the two-lined salamander *Eurycea bislineata* Green. Herpetologica 43:423–446.

Jaeger, R. G., and D. E. Barnard. 1981. Foraging tactics of a terrestrial salamander: choice of diet in structurally simple environments. American Naturalist 117:639–664.

Karieva, P. 1989. Renewing the dialogue between theory and experiments in population ecology. Pages 68–88 in J. Roughgarden, R. M. May, and S. A. Levin (eds.), Perspectives in Ecological Theory. Princeton University Press, Princeton, New Jersey.

Leefer, G. W. 1948. The need for destructive criticism. Journal of the Australian Institute of Agricultural Science 14:33–35.

Lindman, H. R. 1992. Analysis of Variance in Experimental Design. Springer-Verlag, Berlin.

Loehle, C. 1987. Hypothesis testing in ecology: psychological aspects and the importance of theory maturation. Quarterly Review of Biology 62:397–409.

Lotka, A. J. 1932. The growth of mixed populations: two species competing for a common food supply. Journal of the Washington Academy of Sciences 22:461–469.

Lubchenco, J., and L. A. Real. 1992. Manipulative experiments as tests of ecological theory. Pages 715–733 in L. A. Real and J. H. Brown (eds.), Foundations of Ecology: Classic Papers with Commentary. University of Chicago Press, Chicago, Illinois.

Manly, B. F. J. 1992. The Design and Analysis of Research Studies. Cambridge University Press, Cambridge.

Martof, B. S., and F. L. Rose. 1963. Geographic variation in southern populations of *Desmognathus ochrophaeus*. American Midland Naturalist 69:376–425.

McIntosh, R. P. 1985. The Background of Ecology: Concept and Theory. Cambridge University Press, New York.

Medawar, P. 1984. The Limits of Science. Oxford University Press, New York.

Morin, P. J. 1981. Predatory salamanders reverse the outcome of competition among three especies of anuran tadpoles. Science 212:1284–1286.

———. 1983. Predation, competition, and the composition of larval anuran guilds. Ecological Monographs 53:119–138.

Newman, R. A. 1987. Effects of density and predation on *Scaphiopus couchi* tadpoles in desert ponds. Oecologia 71:301–307.

Nicholson, A. J. 1933. The balance of animal populations. Journal of Animal Ecology 2:131–178.

Paine, R. T. 1966. Food web complexity and species diversity. American Naturalist 100:65–75.

Park, T. 1948. Experimental studies of interspecies competition: I. Competition between populations of the flour beetles *Tribolium confusum* Duvall and *Tribolium castaneum* Herbst. Ecological Monographs 18:267–307.

———. 1954. Experimental studies of interspecies competition: II. Temperature, humidity, and competition in two species of *Tribolium*. Physiological Zoology 27:177–238.

Pearl, R. 1927. The growth of populations. Quarterly Review of Biology 2:532–548.

Petranka, J. 1997. The Salamanders of the United States and Canada. Smithsonian Institution Press, Washington, D.C.

Poole, R. W. 1974. An Introduction to Quantitative Ecology. McGraw-Hill, New York.

Power, M. E., D. Tilman, S. R. Carpenter, N. Huntly, M. Leibold, P. Morin, B. A. Menge, J. A. Estes, P. R. Ehrlich, M. Hixon, D. M. Lodge, M. A. McPeek, J. E. Fauth, D. Reznick, L. B. Crowder, S. J. Holbrook, B. L. Peckarsky, D. E. Gill, J. Antonovics, G. A. Polis, D. B. Wake, G. Orians, E. D. Ketterson, E. Marschall, and S. P. Lawler. 1995. The role of experiments in ecology [Letter]. Science 270:561.

Quinn, J. F., and A. E. Dunham. 1983. On hypothesis testing in ecology and evolution, American Naturalist 122:602–617.

Resetarits, W. J., Jr. 1991. Ecological interactions among predators in experimental stream communities. Ecology 72:1782–1793.

———. 1997. Interspecific competition and qualitative competitive asymmetry between two benthic stream fish. Oikos 78:429–439.

Scheiner, S. M. 1993. Introduction: theories, hypotheses, and statistics. Pages 1–13 in S. M. Scheiner and J. Gurevitch (eds.), Design and Analysis of Ecological Experiments. Chapman and Hall, New York.

Schluter, D. 1994. Experimental evidence that competition promotes divergence in adaptive radiation. Science 266:798–801.

Semlitsch, R. D., and H. M. Wilbur. 1988. Effects of pond drying time on metamorphosis and survival in the salamander *Ambystoma talpoideum*. Copeia 1988:978–983.

Stein, J. (ed.). 1966. The Random House Dictionary of the English Language. Unabridged ed. Random House, New York.

Stiling, P. 1994. What do ecologists do? Bulletin of the Ecological Society of America 75: 116–121.

Tansley, A. G. 1917. On competition between *Galium saxatile* L. (*G. hercynicum* Weig.) and *Galium sylvestre* poll. (*G. asperum* Schreb.) on different types of soil. Journal of Ecology 5:173–179.

Tilley, S. G. 1980. Life histories and comparative demography of two salamander populations. Copeia 1980:806–821.

Tilman, D. 1989a. Discussion: population dynamics and species interactions. Pages 89–100 in J. Roughgarden, R. M. May, and S. A. Levin (eds.), Perspectives in Ecological Theory. Princeton University Press, Princeton, New Jersey.

Tilman, D. 1989b. Ecological experimentation: strengths and conceptual problems. Pages 136–157 in G. E. Likens (ed.), Long-Term Studies in Ecology. Springer-Verlag, New York.

Turesson, G. 1922a. The species and the variety as ecological units. Hereditas 3:100–113.

———. 1922b. The genotypical response of the plant species to the habitat. Hereditas 3: 211–350.

———. 1925. The plant species in relation to habitat and climate. Hereditas 6:147–236.

———. 1930. The selective effect of climate upon the plant species. Hereditas 14:99–152.

Underwood, T. 1986. The analysis of competition by field experiments. Pages 240–268 in J. Kikkawa and D. J. Anderson (eds.), Community Ecology: Pattern and Process. Blackwell Scientific, London.

Varley, G. C. 1957. Ecology as an experimental science. Journal of Animal Ecology 26: 251–261.

Volterra, V. 1926. Fluctuations in the abundance of a species considered mathematically. Nature 118:558–560.

Wilbur, H. M. 1976. Density-dependent aspects of metamorphosis in *Ambystoma* and *Rana sylvatica*. Ecology 57:1289–1296.

———. 1977. Density-dependent aspects of growth and metamorphosis in *Bufo americanus*. Ecology 58:196–200.

———. 1987. Regulation of structure in complex systems: experimental temporary pond communities. Ecology 68:1437–1452.

Wilbur, H. M., and J. P. Collins. 1973. Ecological aspects of amphibian metamorphosis. Science 182:1305–1314.

Winer, B. J. 1971. Statistical Principles in Experimental Design, 2d ed. McGraw-Hill, New York.

Wise, D. H. 1993. Spiders in Ecological Webs. Cambridge University Press, New York.

20

Investigating Geographic Variation in Interspecific Interactions Using Common Garden Experiments

JOHN E. FAUTH

The operational definitions that ecologists use influence research in subtle and often unintended ways (Gould 1977, Janzen 1980, Mills et al. 1993, Billick and Case 1994, Fauth et al. 1996). For example, many ecologists have a working definition of *species* that is distinctly typological. An individual belongs to a discrete set, and that set has particular attributes (its niche) that distinguish it from other such sets. Included in this notion of the niche is the organism's place in the biotic environment, including its relationship to competitors, predators, and prey (Elton 1927, Whittaker 1967). This nonhistorical (or "nondimensional"; Mayr 1942, 1963; see also Frost et al. 1992) species concept subtly predisposes ecologists to assume that biotic interactions are constant over time and space (Aarssen 1983:719). However, even simple models predict that selection should alter the dynamics of biotic interactions. Past selection may influence contemporary interactions by altering fundamental parameters such as the intrinsic rate of natural increase, saturation densities, and competitive ability (MacArthur and Wilson 1967, Gill 1972), and historical factors may leave unseen marks on ecological communities (Connell 1980; Avise 1989, 1994; Ricklefs and Schluter 1993). Whether interactions between species are constant throughout most of their range or vary from location to location remains an open question of fundamental importance in community ecology (Travis 1996).

For the past 6 years, my research has focused on investigating the extent and implications of geographic variation in competitive and predator–prey interactions. During that time, common garden experiments have proven effective tools for exposing variation in the intensity of interactions within amphibian assemblages. In a common garden experiment, populations from several different sources are raised at a single site. Whether natural or artificial, the "common garden" is simply a single environment; plant ecologists have long considered natural communities "gardens" (Turesson 1922, Wells 1932), and this broad definition remains useful. If the organism is manipulated appropriately in a common garden experiment, phenotypic responses can be partitioned into those underlain by genotype, environment, and genotype by environment interactions (Falconer 1981). While common garden designs are widely used by plant ecologists (Jinks 1954, Norrington-Davies 1966, Pickett and Bazzaz 1976, Mack and Harper 1977, Solbrig and Simpson 1977, Turkington and Harper 1979, Lenhart 1988, Bergelson 1994, Kärkkäinen et al. 1996), such experiments are rarely performed by vertebrate ecologists, who favor reciprocal transplants (Ballinger 1979, Berven et al. 1979, Hairston 1980a,b, Berven 1982a,b, Bernardo 1994), which are just coordinated common garden experiments (another way of thinking about this is that common garden experiments are degenerate forms of a transplant experiment, having just one site). Reciprocal transplants are difficult to do with mobile animals, so few researchers have attempted the large transplant experiments required to test for geographic variation in competitive ability or predator–prey coadaptation (Roughgarden 1983). Yet common garden designs are powerful experimental tools which can provide the statistical power required for rigorous hypothesis tests. They often are more tractable than reciprocal transplant experiments because a single common garden is easier to manage than multiple sites.

For example, Hairston's classic (1980a,b) transplant experiment demonstrating α selection in a Great Smoky Mountains population of the salamander *Plethodon jordani* required 12 treatments in two mountain ranges 50 km apart. Logistic constraints limited Hairston to just two or three replicates per treatment at each site, the bare minimum required for hypothesis testing. In contrast, a common garden experiment that tests the same null hypotheses requires just eight treatments (Fig. 20-1), allowing more replicates to be deployed for the same effort, thus maximizing statistical power. One drawback is that a single common garden cannot mimic complex clinical gradients such as those of interest in many life history studies (Ballinger 1979, Berven et al. 1979, Hairston 1980a,b, Berven 1982a,b, Bernardo 1994). However, this is not a major liability in most

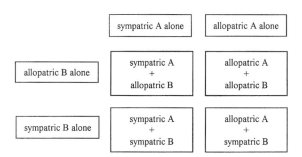

Figure 20-1. The minimum design for testing evolutionary outcomes of species interactions using a common garden experiment. Interacting species (A and B) with the appropriate geographic distributions are raised together and alone in experimental treatments. A priori, allopatric populations are considered ''evolutionarily naive''; sympatric populations are ''evolutionarily experienced.'' Independent assessment of the direction of evolution often can be derived from biogeography, systematics, or molecular genetics.

studies of competition, predation, and other biotic interactions where the focus is on genotypic variation rather than environmental effects and genotype–environment interactions (e.g., Nishikawa 1985, 1987; Travis et al. 1985; Alford 1986).

My research on geographic variation in interspecific interactions between amphibians relies on common garden experiments conducted in artificial mesocosms of the sort pioneered by Morin (1981, 1983) and Wilbur (1987). Artificial mesocosms are well suited for common garden experiments because they retain much of the complexity of the natural environment, allow precise control of experimental conditions, and permit use of the well-replicated experimental designs required for the evaluation of evolutionary hypotheses (e.g., Hairston 1973, Connell 1980, Harris et al. 1990, Semlitsch et al. 1990, Bernardo 1994). Artificial mesocosms are usually designed to mimic general features of natural systems (Fauth et al. 1990, Fauth and Resetarits 1991) and thus reduce environmental differences associated with specific environments: the ''home turf advantage'' that reciprocal transplant experiments are designed to estimate. Within-treatment variation also tends to be low in artificial mesocosm experiments, permitting detection of even small treatment effects with the three to eight replicates most ecologist use (Gurevitch et al. 1992).

The examples in the following discussion used common garden experiments of similar design (Fig. 20-1). In each case, I capitalized on overlapping species' distributions to test the null hypothesis that there were no differences in the responses of allopatric and sympatric populations to ecological interactions. The distributions themselves offered no hint that populations had diverged; it was just a convenient way to secure populations with and without historical experience interacting with their competitive or predatory counterpart. This approach (e.g., Brodie and Brodie 1990, Fauth 1990) differs from that of others (e.g., Hairston 1973, 1980a,b; Ballinger 1979; Berven et al. 1979; Berven 1982a,b; Bernardo 1994) who used experiments to test ecological hypotheses suggested by biogeographic patterns. In a sense, my experiments were shots in the dark, but they were made knowing that failure to detect responses to strong interspecific interactions would also require an explanation (Futuyma and Slatkin 1983).

Microgeographic Variation in Competitive Ability in the American Toad, *Bufo americanus*

To determine whether interactions were constant or varied on a microgeographic scale, I investigated competition between tadpoles of the American toad (*Bufo americanus*) and the pickerel frog (*Rana palustris*) in an array of experimental mesocosms. These anurans co-occur throughout the eastern United States (Conant and Collins 1991), often breeding synchronously in farm ponds, millponds, and beaver ponds. *B. americanus* also breeds in temporary ponds, ditches, and stream overflow pools that are not used by *R. palustris*. Tadpoles of the two species compete (Alford 1989, Wilbur and Fauth 1990), and because toads of the genus *Bufo* are well noted for returning to their breeding sites (Bogert 1960, Heusser 1960, Tracey and Dole 1969, Breden 1987), gene flow among populations may be low (Waldman et al. 1992). Consequently, evolutionary responses to competition might occur quickly and on a microgeographic scale. Indeed, Alford (1989) suggested that the low level of interspecific competition he observed between North Carolina populations of *B. americanus* and *R. palustris* might reflect coadaptation to avoid competition.

I tested this hypothesis in an array of artificial ponds (the methodology has been presented in detail elsewhere—e.g., Morin 1983, Fauth 1990, Wilbur and Fauth 1990). Artificial pond communities were constructed in plastic pools 1.4 m in diameter that held approximately 400 L of water. Pools were positioned in a rectangular array next to a woodlot and filled with tap water. Dry leaf litter collected from the margin of a natural pond in Durham County, North Carolina, was randomly assigned and added to each pool (experiment 1: 2 kg; experiment 2: 1 kg). A suspension of pond water was collected with a 400-μm mesh net from several anuran breeding ponds, and aliquots of these inocula were randomly assigned and added to each experimental unit to establish an aquatic food web. In experiment 2, 20-g portions of commercial rabbit food were randomly assigned and added to each pool to provide an initial burst of nutrients to the food web. The experimental array was exposed to natural photoperiods and temperatures. As constructed, these experimental pools mimicked a general anuran breeding pond rather than any specific pond, making them analogous to the garden plots used by plant ecologists: homogeneous, interchangeable units designed to minimize environmental differences, allowing detection of genetic differences among populations.

Pools were allowed 2–4 weeks for a food web to form; then experimental treatments (see following discussion) were established by introducing randomized groups of tadpoles. The experimental pools were natural enough to attract large numbers of breeding dragonflies, adult aquatic beetles and hemipterans, and ovipositing treefrogs. To prevent these unwanted species from colonizing the experimental ponds, tightly fitting lids of fiberglass window screen covered them throughout the experiment. The lids also prevented experimental animals from escaping after metamorphosis. Pools were checked daily for metamorphosing amphibians, which were held in the laboratory until they resorbed their tails, then weighed to the nearest 0.1 mg. *R. palustris* were weighed individually and released at their pond of origin; *B. americanus* were anesthetized in chloretone and weighed in groups of up to 20 individuals. After all *B. americanus* had transformed or died, I drained the pools and collected the remaining tadpoles of *R. palustris*. Ending the experiments immediately after the last toad metamorphosed prevented differences among *R. palustris* populations from diminishing due to lack of competitive pressure.

Identification of Syntopic and Allotopic Breeding Populations

Eight breeding populations of *B. americanus*, all within a 3.1-km radius in Durham and Orange Counties, North Carolina, were included in this study (Table 20-1). Ponds were defined as syntopic breeding sites if both *B. americanus* and *R. palustris* oviposited in them for 3 consecutive years. Ponds were defined as allotopic breeding sites if *B. americanus* oviposited and *R. palustris* were never heard calling or observed breeding. The exception was BP, a beaver pond in which a few *R. palustris* were heard calling but female frogs, egg masses, and tadpoles were never found despite extensive searches. I was unable to identify breeding ponds used only by *R. palustris*.

Table 20-1. Distances (km) among breeding aggregations of *Bufo americanus*. *B. americanus* was syntopic with *Rana palustris* in four of the eight study ponds: Couch's Pond (CP), Intercom Pond (IP), Moriah Pond (MP), and Whitfield Pond (WP).

	Sympatric populations				Allopatric populations			
	CP	IP	MP	WP	BP	GC	NHC	PP
CP	—	1.0	5.3	4.4	2.3	4.5	3.8	6.1
IP		—	5.1	4.0	3.0	5.3	3.4	6.0
MP			—	1.3	4.5	5.5	1.8	1.0
WP				—	4.1	5.6	0.7		2.3
BP					—	2.3	3.5	2.7
GC						—	5.1	5.6
NHC							—	2.6

All four are man-made ponds that contain fish, including largemouth bass (*Micropterus salmoides*), sunfish (*Lepomis* sp.), and mosquito fish (*Gambusia affinis*). The remaining ponds were allotopic sites: Beaver Pond (BP), Golfcourse Ponds (GC), New Hope Creek Overflow Pools (NHC), and Powerline Pond (PP). BP is permanent and contains fish, NHC sometimes contains fish (*Esox* sp. and *Lepomis* sp.) but generally dries in midsummer, and PP is temporary. The GC population consisted of two adjacent ponds: a man-made pond containing *G. affinis* and *Lepomis* sp. and a natural temporary pond that had large breeding populations of *Pseudacris triseriata*, *P. crucifer*, *Ambystoma opacum*, and *A. maculatum*.

Experimental Designs

Experiment 1   Experiment 1 had a replicated, blocked, incomplete factorial design with within-block replication of selected treatments (Tables 20-2, 20-3). There were four treatments: 500 tadpoles of syntopic *B. americanus* raised alone, 500 tadpoles of syntopic *B. americanus* raised with 100 tadpoles of *R. palustris*, 500 tadpoles of allotopic *B. americanus* raised with 100 tadpoles of *R. palustris*, and 100 tadpoles of *R. palustris* raised alone. The initial tadpole densities (325 *B. americanus*/m² and 65 *R. palustris*/m²) were within the range of densities found in nature (Alford 1989; J. E. Fauth, personal observation).

Insufficient numbers of hatchlings prevented me from replicating the "allotopic tadpoles of *B. americanus* raised alone" treatment in experiment 1. Treatments that contained both *B. americanus* and *R. palustris* were each replicated twice within a block; treatments with either species alone were replicated once per block (Tables 20-2, 20-3). The entire design was replicated six times in a total of 36 experimental pools grouped into six blocks that allowed me to reduce and to statistically account for spatial variation. Syntopic *B. americanus* and *R. palustris* were from eggs laid in CP and IP; allotopic *B. americanus* were from GC and NHC. Samples from all viable egg masses in each source pond were collected and mixed to ensure that representative genetic variation was maintained in the experimental populations. Based on estimates of clutch size in these two anurans, the minimum number of clutches collected was 2–30 for *B. americanus* and 3–12 for *R. palustris*. These estimates are probably accurate for *R. palustris*, which lays its eggs in long strings, with multiple females often intertwining their clutches.

Table 20-2. Responses of *Bufo americanus* in experiment 1

Block	*Rana*	*Bufo* population	No. of metamorphs	Mass	Larval period
I	Absent	Sympatric (IP)	295	0.0657	53.5
	Present	Sympatric (IP)	333	0.0600	52.7
	Present	Sympatric (IP)	334	0.0611	48.3
	Present	Allopatric (GC)	290	0.0657	47.8
	Present	Allopatric (NHC)	245	0.0706	50.9
II	Absent	Sympatric (IP)	399	0.0634	50.3
	Present	Sympatric (IP)	384	0.0537	59.2
	Present	Sympatric (IP)	371	0.0562	59.2
	Present	Allopatric (GC)	254	0.0527	62.2
	Present	Allopatric (NHC)	260	0.0727	48.1
III	Absent	Sympatric (CP)	325	0.0675	40.9
	Present	Sympatric (CP)	320	0.0682	43.2
	Present	Sympatric (IP)	385	0.0584	45.0
	Present	Allopatric (GC)	244	0.0637	45.6
	Present	Allopatric (NHC)	238	0.0701	42.4
IV	Absent	Sympatric (CP)	284	0.0686	44.0
	Present	Sympatric (CP)	296	0.0671	42.8
	Present	Sympatric (IP)	323	0.0601	46.8
	Present	Allopatric (GC)	203	0.0535	46.7
	Present	Allopatric (NHC)	237	0.0583	49.5
V	Absent	Sympatric (IP)	373	0.0652	42.0
	Present	Sympatric (IP)	357	0.0613	42.6
	Present	Sympatric (IP)	314	0.0615	42.9
	Present	Allopatric (GC)	300	0.0595	43.1
	Present	Allopatric (NHC)	252	0.0663	43.0
VI	Absent	Sympatric (IP)	374	0.0577	44.6
	Present	Sympatric (IP)	308	0.0524	43.7
	Present	Sympatric (IP)	246	0.0641	43.5
	Present	Allopatric (GC)	391	0.0682	41.7
	Present	Allopatric (NHC)	259	0.0720	43.0

Source populations of *B. americanus* were nested within blocks (Table 20-2); this procedure allowed conservative testing of the null hypothesis of no difference between syntopic and allotopic populations. Tadpoles of *R. palustris* from the two syntopic ponds were mixed to create a uniform experimental population; this allowed the expression of general competitive differences among *B. americanus* populations but prevented estimating population-specific responses. To estimate specific competitive ability (synonymous with the specific combining ability of plant ecologists), a larger experimental design is required (see experiment 2), resulting in fewer replicates of each treatment. In experiment 1, which was a first attempt at detecting intraspecific variation in competitive ability, I decided to use more replicates of fewer experimental treatments

Table 20-3. Responses of *Rana palustris* in experiment 1

Block	*Bufo* population	Total no.	No. of metamorphs	Tadpole mass
I	Absent	76	6	1.5373
	Sympatric (IP)	95	1	0.4895
	Sympatric (IP)	76	0	0.6133
	Allopatric (GC)	73	2	0.7172
	Allopatric (NHC)	82	1	0.8230
II	Absent	73	4	1.3803
	Sympatric (IP)	69	4	0.5521
	Sympatric (IP)	80	1	0.6778
	Allopatric (GC)	82	0	0.6697
	Allopatric (NHC)	81	2	0.9627
III	Absent	67	6	1.4303
	Sympatric (CP)	78	1	0.9964
	Sympatric (IP)	79	2	0.4129
	Allopatric (GC)	78	6	0.9703
	Allopatric (NHC)	73	3	0.9943
IV	Absent	93	2	1.3345
	Sympatric (CP)	75	1	0.9220
	Sympatric (IP)	85	2	0.6791
	Allopatric (GC)	90	0	0.7522
	Allopatric (NHC)	89	2	0.8029
V	Absent	65	9	1.7684
	Sympatric (IP)	79	4	0.7929
	Sympatric (IP)	81	5	0.6470
	Allopatric (GC)	62	4	1.0891
	Allopatric (NHC)	86	0	0.9898
VI	Absent	84	11	1.5916
	Sympatric (IP)	73	1	0.8386
	Sympatric (IP)	85	2	0.7155
	Allopatric (GC)	86	7	1.1003
	Allopatric (NHC)	89	0	0.7013

Abbreviations are as in Table 20-1. Total no. = number of metamorphs plus number of tadpoles. Mean tadpole mass is in grams.

to ensure detecting even small differences among populations in metamorphic responses (Fauth 1990).

Experiment 2   Experiment 2 had a 5 × 3 × 4 factorial design, where treatments were all possible combinations of five breeding populations of *B. americanus* (MP, WP, BP, PP, or none) raised with three breeding populations of *R. palustris* (MP, WP, or none), except the null treatment (pools lacking both species) was not included (Tables 20-4, 20-5). Each of the 14 treatments was replicated four times in a total of 56 pools. Initial densities of tadpoles in experiment 2 differed from those in experiment 1; the density of *B. americanus* was reduced to 250 tadpoles/pool (162/m²) because intraspecific com-

Table 20-4. Responses of American toad (*Bufo americanus*) populations to treatments in experiment 2

Bufo population	*Rana* population	No. of metamorphs	Mass	Larval period
WP	WP	123.7 ± 15.2	146 ± 10	48.1 ± 3.6
WP	MP	127.5 ± 13.6	143 ± 8	47.0 ± 4.2
WP	Absent	144.7 ± 14.6	171 ± 12	46.0 ± 2.3
MP	WP	103.3 ± 23.5	138 ± 10	47.9 ± 4.7
MP	MP	98.3 ± 23.2	135 ± 12	47.3 ± 2.9
MP	Absent	106.0 ± 21.5	164 ± 22	47.5 ± 3.0
PP	WP	91.7 ± 29.4	117 ± 10	49.0 ± 4.8
PP	MP	99.5 ± 12.3	130 ± 7	49.5 ± 3.0
PP	Absent	133.7 ± 11.5	157 ± 8	47.1 ± 2.6
BP	WP	156.3 ± 15.5	153 ± 5	46.8 ± 3.8
BP	MP	141.2 ± 22.9	143 ± 15	48.1 ± 4.3
BP	Absent	184.5 ± 10.4	178 ± 10	46.8 ± 3.5

Entries are the means (± 1 S.E.) of four replicate populations. Mean mass at metamorphosis is in milligrams; larval period is in days. Abbreviations are as in Table 20-1.

petition had been severe in experiment 1, while the density of *R. palustris* was increased to 140 tadpoles/pool ($91/m^2$) to place greater interspecific competitive stress on *B. americanus* (Fauth 1990).

This experimental design allowed me to test for differences in general and specific competitive ability among American toad populations. If syntopic *B. americanus* had adapted to competition with *R. palustris*, then syntopic populations of *B. americanus* (MP, WP) should do better than the allotopic populations (BB, PP) when raised with *R. palustris*. If *B. americanus* had adapted in response to specific competitive attributes

Table 20-5. Responses of pickerel frog (*Rana palustris*) populations to treatments in experiment 2

Rana population	*Bufo* population	No. of froglets	Froglet mass	No. of tadpoles	Tadpole mass
WP	WP	3.3 ± 2.9	0.458 ± 0.023	117.5 ± 9.4	0.890 ± 0.077
WP	MP	6.7 ± 1.6	0.388 ± 0.007	109.0 ± 2.7	0.931 ± 0.051
WP	PP	6.7 ± 1.5	0.458 ± 0.048	113.5 ± 4.2	0.946 ± 0.045
WP	BP	4.7 ± 2.8	0.404 ± 0.003	117.2 ± 4.5	0.918 ± 0.087
WP	Absent	11.5 ± 2.4	0.472 ± 0.030	106.2 ± 4.3	1.024 ± 0.040
MP	WP	9.7 ± 1.2	0.557 ± 0.019	95.5 ± 5.0	1.085 ± 0.097
MP	MP	13.7 ± 5.3	0.615 ± 0.060	93.0 ± 9.6	1.132 ± 0.031
MP	PP	8.5 ± 2.4	0.613 ± 0.037	101.7 ± 4.9	1.031 ± 0.029
MP	BP	10.3 ± 2.7	0.530 ± 0.020	97.7 ± 6.6	0.886 ± 0.120
MP	Absent	12.7 ± 4.2	0.630 ± 0.027	90.0 ± 10.7	1.257 ± 0.077

Entries are the means (± 1 S.E.) of four replicate populations. Froglet and tadpole masses are in grams. Abbreviations are as in Table 20-1.

of their syntopic *R. palustris* population, they should do better with tadpoles from their natal pond than with those collected from another pond.

Response Variables and Analyses

Survival (number of metamorphs produced), mean mass at metamorphosis, and duration of the larval period were selected as response variables because they are good estimators of adult fitness components in amphibians (Smith 1987, Semlitsch et al. 1988, Berven 1990, Goater 1994, Pechmann 1994, Scott 1994) and are sensitive to competition (Wilbur 1977, 1980). In anurans, competition is generally expressed as reduced survival and mass at metamorphosis and/or increased larval periods (Wilbur 1980). Separate analyses were conducted on each response variable because they are known to independently affect different components of fitness (age at first reproduction and fecundity: Smith 1987, Semlitsch et al. 1988, Berven 1990, Pechmann 1994, Scott 1994). This fact argues against using a single multivariate test, which is designed to hold the experimentwise Type I error rate constant when multiple response variables are correlated. The low number of univariate tests I used, coupled with the independent carryover effects of larval responses on adult fitness, limits the increase in Type I errors above $\alpha = 0.05$, while simultaneously providing more powerful hypothesis tests (decreasing the probability of a Type II error; see Sokal and Rohlf 1995).

Significant population effects in the analyses would indicate that there were genetic differences among populations, arising from either heritable genetic variation or a nongenetic maternal component. The latter arise from prenatal influences of the mother on her offspring, such as egg size (Falconer 1981, Bernardo 1996). Nongenetic maternal effects might indicate that females breeding in different ponds had experienced different terrestrial environments, evidence of breeding philopatry.

Expansion of the artificial pond array in experiment 2 forced me to place many pools into a shaded portion of the Duke Zoology Field Station. Spatial blocks were inadequate to minimize and account for the complex spatial variation in water temperature, which has a profound effect on growth rates and competition in larval amphibians (Duellman and Trueb 1986, Leimberger 1989), so temperature was added as a covariate in the analyses. ANCOVA removed variation in the response variables due to water temperature (measured on day 19 of the experiment) that otherwise would have contributed to the residual error, allowing me to better detect population-level differences in survival, growth, and duration of the larval period.

Results of Experiment 1

Tadpoles from syntopic populations of the two anuran species competed in the experimental ponds. *R. palustris* caused a significant decrease in the mean mass of syntopic *B. americanus* (Table 20-2). Syntopic *B. americanus* attained a mean mass of 64.7 mg when raised alone but only 60.3 mg when *R. palustris* was present ($F_{1,11} = 4.80$, $P < .05$). Syntopic *B. americanus* also affected *R. palustris*: syntopic toad tadpoles caused highly significant decreases in the mean mass of pickerel frog tadpoles ($F_{1,11} = 83.21$, $P < .0001$) and the number of metamorphs that emerged before the ponds were drained ($F_{1,11} = 20.21$, $P < .001$). Ponds in which *R. palustris* were raised alone produced an average of six froglets and the remaining tadpoles averaged 1.507 g, but treatments in which *R. palustris* were raised with syntopic *B. americanus* yielded an average of just two froglets and the tadpoles averaged just 0.695 g, a 54% decrease in mean mass (Table 20-3).

There were significant differences in survival between syntopic and allotopic *B. americanus* populations raised with *R. palustris* ($F_{1,21}$ = 13.89, P < .01, Table 20–2). An average of 335 toadlets emerged from syntopic populations, but only 264 emerged from allotopic populations. There were no differences in survival between source ponds nested within populations, but there were significant differences in their mean mass at metamorphosis ($F_{2,21}$ = 10.14, P < .05). Toadlets from GC and IP were smaller (605.5 and 600.6 mg, respectively) than toadlets from NHC (683.3 mg) and CP (678.5 mg) (Table 20-2). Toadlets from CP and IP differed in mass despite the proximity of the ponds (1 km; Table 20-1) and their similarity in age and construction. Conversely, toadlets from NHC and CP transformed at nearly identical sizes despite the 3.8 km between them and their different pond origins and faunas.

Plant ecologists often use the responses of the plants themselves to assess environmental effects. The advantage of this "phytometer" approach is that the plants often transduce subtle or complex environmental influences into simple, detectable responses. I used the responses of *R. palustris* in a similar fashion, as bioassays to determine the mechanism of competitive evolution in *B. americanus* populations. If syntopic *B. americanus* had adapted via niche diversification, competition with *R. palustris* should be reduced and pickerel frog tadpoles should grow larger and exhibit higher survival when raised with syntopic rather than allotopic *B. americanus*. However, if syntopic populations of *B. americanus* were superior competitors, then pickerel frog tadpoles should grow larger and exhibit higher survival with allotopic *B. americanus* because the latter are weaker competitors.

There was a significant effect of *B. americanus* population (syntopic vs. allotopic) on the mean mass of *R. palustris* tadpoles but no effect on survival or the number of metamorphs ($F_{1,17}$ = 7.99, P < .05; Table 20-3). *R. palustris* tadpoles raised with allotopic *B. americanus* attained a mean mass of 0.882 g, but those raised with syntopic tadpoles attained a mass of only 0.695 g, a 21% reduction. This suggests that syntopic *B. americanus* are superior interspecific competitors than are allotopic *B. americanus*.

Competitive ability itself may be partitioned into two components: pressure and resistance (or effect and response; Goldberg and Fleetwood 1987, Goldberg and Landa 1991; see also Joshi and Thompson 1995). Competitive pressure is the ability of a species to decrease the growth or survival of a competitor, while competitive resistance is the ability of a species to negate the effects of competitors on itself. Competitive pressure is measured as the per capita effect of a species on the responses of competitors, while competitive resistance is measured by the per capita effect of competitors on the responses of the focal species. I examined these two components of competitive ability by regressing the mass of *R. palustris* tadpoles on the number of *B. americanus* metamorphs produced by each experimental population. The per capita effect of syntopic and allotopic *B. americanus* was identical (Fig. 20-2). On average, an additional surviving tadpole from either toad population caused the same 2.2-mg reduction in the mean mass of pickerel frog tadpoles. The greater net competitive effect of syntopic *B. americanus* arises simply from their higher survival, which is consistent with increased

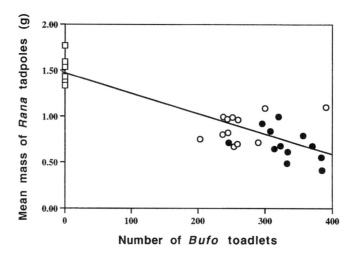

Figure 20-2. Mean mass of pickerel frog (*Rana palustris*) tadpoles as a function of the number of toadlets produced by American toads (*Bufo americanus*) in experiment 1. *Open squares*: syntopic *R. palustris* raised alone; *open circles*: syntopic *R. palustris* raised with allotopic *B. americanus*; *solid circles*: syntopic *R. palustris* raised with syntopic *B. americanus*. The per capita effect of syntopic and allotopic *B. americanus* toadlets on the mean mass of *R. palustris* tadpoles was identical, as indicated by the common regression line.

competitive resistance. However, the same result would arise if the differences were not caused by competition with *R. palustris* but by another mortality factor in the artificial ponds. Lower survival of allotopic *Bufo* when raised alone in the artificial ponds would be evidence for such an effect. I was unable to replicate the allotopic *Bufo* alone treatment in the main array because not enough tadpoles were available. Instead, this possibility was evaluated in experiment 2.

Results of Experiment 2

There were significant differences among *B. americanus* populations in survival (ANCOVA with water temperature as the covariate, $F_{3,36} = 35.64$, $P < .0001$; Table 20-4). Bonferonni-adjusted *t* tests revealed that the two closest populations, MP and PP (Table 20-4), did not differ in the mean number of metamorphs produced (103 and 108, respectively), but they differed significantly from WP populations (132). BP (161) differed significantly from both WP and the MP/PP group. There also were significant differences among toad populations in larval period (ANCOVA $F_{3,36} = 3.13$, $P < .05$; Table 20-4), but not in mass at metamorphosis.

R. palustris caused significant decreases in the survival (ANCOVA $F_{1,36} = 25.26$, $P < .0001$) and mass (ANCOVA $F_{1,36} = 20.53$, $P < .0001$) of *B. americanus* in experiment 2 (Table 20-4). The four toad populations produced an average of 142 metamorphs with a mean mass of 167.7 mg when raised alone. In the presence of *R.*

palustris, however, only 118 toadlets emerged and their mean mass was just 138.2 mg. The effects of *R. palustris* density and *B. americanus* population were additive (all *P* >> .05).

There were significant differences between populations of *R. palustris* in mass at metamorphosis (ANCOVA with water temperature as the covariate, $F_{1,34}$ = 25.86, *P* < .0001), the mean number of tadpoles surviving (MP: 95.6, WP: 112.7; ANCOVA $F_{1,34}$ = 25.86, *P* < .0001) and their mean mass (MP: 1.0785 g, WP: 0.9420 g; AN-COVA $F_{1,34}$ = 7.82, *P* < .0085) (Table 20-5).

B. americanus caused a decline in the number of *R. palustris* that completed metamorphosis but did not affect their mass (Table 20-5). The two *R. palustris* popu-lations produced an average of 12 froglets when raised alone but only 8 when raised with *B americanus*. The effects of *R. palustris* population and *B americanus* density were additive.

The preceding analyses revealed general patterns in the abundance and growth of *B. americanus* and *R. palustris*. To gain a more explicit response of *B. americanus* to the addition of *R. palustris*, I also measured the viability of each toad population. Within each block, this was calculated as the number of toadlets produced in pools with *R. palustris* divided by the number of toadlets in pools with only *B. americanus* ANCOVA revealed significant effects of water temperature, spatial blocks, and toad populations on viability (all *P* ≥ .05); the two syntopic toad populations had the highest values (Fig. 20-3). The average mass of *R. palustris* tadpoles in competition with *Bufo* also

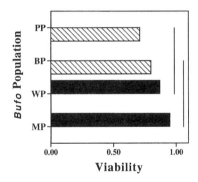

Viability

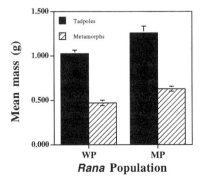

Rana **Population**

Figure 20-3. (**Top**) Mean viability of experi-mental populations of *Bufo americanus*. Viabil-ity was defined as the number of metamorphs produced in pools with *Rana palustris* divided by the number produced in pools with *B. ameri-canus* alone within the same spatial block. Dif-ferences among populations were statistically significant (ANCOVA F_3,20 = 3.05, *P* < .05). Lines with the same letter connect populations that were not significantly different according to Bonferroni-adjusted *t* tests. (**Bottom**) Mean mass of surviving tadpoles and metamorphic *R. palustris* from breeding ponds 1.3 km apart. Differences among breeding ponds were statisti-cally significant for both traits (*P* << .01).

varied with *R. palustris* population (ANCOVA $F_{1,20}$ = 4.97, $P < .04$). On average, tadpoles from MP weighed 0.2443 g, but those from WP weighed just 0.0923 g.

Identifying the Mechanism Underlying Local Genetic Differentiation

I found significant differences among local breeding aggregations of both *B. americanus* and *R. palustris* in a variety of metamorphic traits. In experiment 25 there were differences between *R. palustris* populations in every response measured, although the breeding ponds were only 1.3 km apart. There also were differences among *B. americanus* populations in survival, viability, mass at metamorphosis, and duration of the larval period in both experiments. In some cases, the breeding ponds were only 1 km apart (e.g., CP and IP in experiment 1). In this region of North Carolina, there were many intervening ponds that could have served as stepping-stones for dispersal. Yet both species appear to have differentiated on a local scale.

Is the observed variation adaptive? Alford (1989) suggested that *B. americanus* and *R. palustris* from the same pond had coadapted to reduce competition. He based this idea on the low level of interspecific competition he observed in experiments conducted in artificial ponds. Previous experiments showed that tadpoles of *B. americanus* compete strongly with tadpoles of a close relative of the pickerel frog: the southern leopard frog (*Rana utricularia* = *Rana sphenocephala*; see Alford and Wilbur 1985, Wilbur and Alford 1985). Alford expected the ecologically similar tadpoles of *R. palustris* to compete just as strongly with *B. americanus*, but they did not. He suggested that *R. utricularia*, which overlaps less with *B. americanus* in selection of larval habitats than does *R. palustris*, was under less pressure to evolve differences in resource or microhabitat use or to reduce the effects of chemical inhibition. Consequently, competition between *B. americanus* and *R. utricularia* was more intense (Alford 1989).

My results suggest that Alford was correct: local populations of *B. americanus* have coadapted to *R. palustris*. Syntopic populations of *B. americanus* consistently did better than their "evolutionarily naive" allopatric counterparts in the presence of *R. palustris*. The exact probability that the two syntopic populations of *B. americanus* would have the highest mean performance (measured by survival in experiment 2 and viability in experiment 2) in both experiments is $[(\frac{2}{4})(\frac{1}{3})]^2 = .027$, or greater than that expected by chance alone. However, the mechanism of coadaptation is not avoidance of competition. Instead, syntopic *B. americanus* are competitively superior to their allopatric counterparts because they are more resistant to competition. Using the responses of *R. palustris* as a bioassay, I found that their tadpoles performed worse with syntopic than with allotopic *B. americanus* in experiment 1, the result expected if syntopic toad tadpoles were better competitors than allotopic tadpoles. However, regression analysis revealed that the per capita effect of syntopic and allotopic *B. americanus* was identical; the greater net effect of syntopic toad tadpoles was simply a consequence of their high survival. Greater resistance to chemical growth inhibitors produced by *R. palustris* tadpoles (Richards 1958, Steinwascher 1978, Beebee 1991) likely explains this result.

To show convincingly that these two anurans have coadapted, either to maintain competitive equality or to reduce competition, one must (1) compare the responses of syntopic and allotopic *R. palustris* to *B. americanus* and (2) demonstrate that the responses are under genetic control. Unfortunately, I was unable to find allotopic *R. palustris* breeding aggregations within my study area. However, another experiment in artificial ponds (Wilbur and Fauth 1990) confirmed that tadpoles of *B. americanus* and *R. palustris* were near equal competitors. The tadpoles used in this experiment came from syntopic breeding ponds, mainly IP and WP. Both species showed reduced survival when reared with the other, indicating they did compete. However, in tanks without predators their combined abundance could be accurately estimated by summing their abundance in competitor-free control tanks, indicating neither species held a competitive advantage.

At least some of the metamorphic traits I measured are likely to be under genetic control. Additive genetic variation for metamorphic traits is common among anuran taxa, including the families Hylidae (Travis 1980, 1981; Travis et al. 1987; Woodward et al. 1988), Pelobatidae (Woodward 1986, 1987; Newman 1988a,b), Bufonidae (Mitchell 1990), and Ranidae (Plytycz et al. 1984). In addition, Alford (1986) has demonstrated heritable variation in competitive ability between half-sib families of *Hyla chrysoscelis*. It seems likely that mass at metamorphosis, duration of the larval period, and the other larval and metamorphic traits I measured are heritable in *B. americanus* and *R. palustris*.

Variation in Interspecific Interactions Across a Geographic Boundary

Several experiments have shown that salamanders and anurans are adapted to local selection regimes (e.g., Berven et al. 1979; Hairston 1980a,b; Berven 1982; Harris et al. 1990; Semlitsch et al. 1990; Bernardo 1994, this volume). If local adaption is common in amphibians, it should be relatively easy to detect variation in interspecific interactions at larger geographic scales.

In a series of recent experiments (Fauth, unpublished data), I used an array of artificial stream banks to investigate geographic variation in the responses of northern dusky salamanders (*Desmognathus fuscus*) to their larger predatory congeners: the blackbelly salamander (*D. quadramaculatus*) and the seal salamander (*Desmognathus monticola*). The stream banks (modified from the artificial streams of Resetarits [1991]) were designed to mimic the headwater streams these salamanders inhabit in the southern Appalachians. Each experiment had four replicates of eight treatments; six *D. fuscus* from each of four populations were raised in the presence or absence of a single predatory congener. The populations were from two stream drainages at Mountain Lake Biological Station (Giles County, Virginia), one in the New River (Mississippi) drainage and one in the James River (Atlantic) drainage. *D. quadramaculatus*, the largest and most aggressive member of the genus (Hairston 1987), occurs in the New River drainage but not across the Eastern Continental Divide in the James River drainage, while *D. monticola* occurs in both drainages (Hutchison 1956; Fauth, personal observation). Two additional populations of *D. fuscus* came from regions distant from Mountain Lake: one from central Ohio (Licking County), well outside the range of any other desmognathine salamanders, and one from Whitetop Mountain (Smyth County, Virginia), well within the range of both larger congeners (Conant and Collins 1991).

The responses of *D. fuscus* to its larger congeners were measured by microhabitat distribution, survival, and growth. The null hypothesis was that all populations of *D. fuscus* would respond in a similar fashion to their larger congeners. The null hypothesis has been falsified repeatedly in several different experiments; populations of *D. fuscus* differed in their behavioral and life historical responses in a predictable manner. *D. fuscus* sympatric with the specific predator population used in each experiment consistently displayed the greatest avoidance or suffered the least mortality and nonlethal injuries (Fig. 20-4). These common garden experiments used adults of all three species, so differences among populations of *D. fuscus* may be due to genetic differences, learning, or a combination of the two. In any case, the differences suggest that populations of *D. fuscus* have adapted (behaviorally or genetically) to the aggressiveness of their own particular assemblage of predators.

Conclusions

The species is the fundamental unit of taxonomy and systematics, and ecologists are properly concerned about correctly identifying the organisms they study and understanding the contributions of genotype and environment to phenotypic variation. However, when populations within a species appear morphologically similar and no other evidence suggests that populations have diverged, ecologists often assume that species will interact in the same way in different locations. This is an important presumption that often goes untested.

My experiments demonstrate that there is substantial interpopulational variation in competitive and predator–prey interactions among amphibian species. Local populations of the American toad (*B. americanus*) appear to have evolved increased competitive resistance to tadpoles of the pickerel frog (*R. palustris*). Selection in ponds where both species breed apparently has favored the survival of genotypes that have superior ecological combining ability: "they are good competitors and at the same time, good neighbors" (Allard and Adams 1969:630). *B. americanus* tadpoles from syntopic populations consistently outperformed their allotopic counterparts under competition with *R. palustris*, but their per capita effect on pickerel frog tadpoles was identical. This result agrees well with Alford's (1989) suggestion that *B. americanus* and *R. palustris* have coevolved to reduce competition.

Similarly, populations of the northern dusky salamander (*D. fuscus*) appear to have coevolved with their predatory congeners, the blackbelly salamander (*D. quadramaculatus*) and the seal salamander (*D. monticola*). Reciprocal selection apparently has favored the matching of predator aggressiveness with prey elusiveness, the coevolutionary arms race envisioned by Van Valen (1973). When raised with a nonaggressive population of *D. quadramaculatus* from the Mountain Lake region, only the allopatric population of *D. fuscus* remained in the aquatic microhabitats preferred by their larger congener. But because this particular population of *D. quadramaculatus* does not attack and consume *D. fuscus* with any regularity (at least in laboratory experiments), this niche overlap had no fitness consequence. However, when the same populations of *D. fuscus* were raised with more aggressive *D. quadramaculatus* from the Whitetop Mountain region, the result was entirely different. The two sympatric populations of *D. fuscus* had higher survival than did their allopatric counterparts, suggesting that selection had

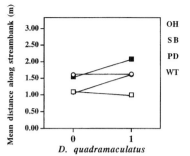

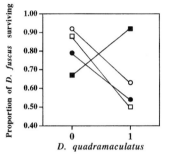

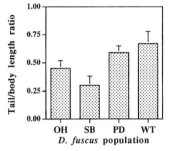

Figure 20-4. Responses of four populations of northern dusky salamanders (*Desmognathus fuscus*) to their larger predatory congeners (*D. quadramaculatus* or *D. monticola*) in three different common garden experiments. Populations of *D. fuscus* are from either near the Eastern Continental Divide or far away and sympatric or allopatric with *D. quadramaculatus*. Pond Drain (PD): near, sympatric; Sartain Branch (SB): near, allopatric; Whitetop Mountain (WT): far, sympatric; Central Ohio (OH): far, allopatric. (**Top**) Mean distance (m) along the stream bank gradient occupied by *D. fuscus* populations when raised alone or with a single nonaggressive *D. quadramaculatus* from Mountain Lake. The main effects of *D. fuscus* population, *D. quadramaculatus* density, and their interaction were all significant ($P < .05$). Only the evolutionarily "naive" population from central Ohio did not exhibit avoidance behavior, but there was no detectable fitness consequence. (**Middle**) Proportion of *D. fuscus* surviving without damage in the presence and absence of a single aggressive *D. quadramaculatus* from Whitetop Mountain. The population by *D. quadramaculatus* interaction was statistically significant ($P < .05$). The interaction between the Whitetop Mountain populations of *D. fuscus* and *D. quadramaculatus* was a mutualism; in the other three cases, it was predation. The two allopatric populations of *D. fuscus* had significantly lower survival when raised with *D. quadramaculatus* than did the two sympatric populations. The sympatric *D. fuscus* population from Pond Drain (PD) suffered significantly more nonlethal damage (mainly autotomized tails) than did its counterpart from Whitetop Mountain (WT). (**Bottom**) Extent of damage inflicted on the four *D. fuscus* populations by *D. monticola* from Pond Drain. The length of tail remaining, relative to body length, in damaged individuals differed significantly among populations ($P = .05$), with salamanders from allopatric populations (OH, SB) autotomizing more of their tail than did individuals from sympatric populations (WT, PD).

favored more elusive or better defended genotypes. However, individuals from the sympatric population of *D. fuscus* from Mountain Lake, which co-occur with a less aggressive population of *D. quadramaculatus*, still suffered considerable tail damage, suggesting their defense mechanism was not as effective as that of their Whitetop Mountain counterparts. A similar result was obtained when the predator was *D. monticola* from Pond Drain instead of *D. quadramaculatus*. Only the prey populations sympatric with that particular predator population, or an equally aggressive one from Whitetop Mountain, avoided damage. The results of these three common garden experiments suggest that prey populations have evolved defenses to counter their specific predator populations. Only by raising populations with and without a history of coexistence in a common garden experiment was Van Valen's (1973) Red Queen exposed.

My experiments show that local adaptation can take place on a fine spatial scale, indicating that selection can overcome the homogenizing influence of gene flow. In *Desmognathus* the Eastern Continental Divide forms a natural boundary between populations of *D. fuscus* with and without evolutionary experience with *D. quadramaculatus*, but in my anuran system the syntopic and allotopic ponds were interspersed. The differences I found in the metamorphic responses of *B. americanus* and *R. palustris* populations imply that these anurans are highly philopatric: independent evidence from recent studies supports this view. Berven and Grudzien (1990) showed that not even one of 11, 195 marked adult wood frogs (*Rana sylvatica*) migrated from one breeding pond to another, and just 18% of the juveniles bred in a nonnatal pond. Using demographic methods, they showed that genetic neighborhoods had a radius of roughly 1.2 km, suggesting breeding populations separated by greater distances would exchange few genes and could show detectable genetic differentiation. I was able to detect such differences between populations of *R. palustris* separated by about this distance (1.3 km). Working with *Bufo woodhousei fowleri* (which interbreeds with *B. americanus* in some regions), Breden (1987) found that most metamorphs returned to their natal pond to breed; dispersers were generally found in adjacent ponds, and few individuals migrated long distances. Working with populations of *B. americanus* in the northeastern United States, Waldman et al. (1992) used mitochondrial DNA to demonstrate that the toads were highly philopatric but avoided incestuous matings by discriminating between the vocalizations of kin and nonkin. They suggested that philopatry, in concert with inbreeding avoidance, could result in "optimal outbreeding" that facilitates local adaption. Their five breeding populations were within a 1-km radius, an even smaller area than in my study (3.1 km) and well within the traversal capabilities of individual toads (Waldman et al. 1992 and references therein). Other aquatic animals show local adaptation at similar spatial scales (e.g., Lively 1989).

If genetic differentiation can take place at such fine levels, then variation in competitive and predator–prey interactions should be detectable at larger geographic scales. Experimental evidence, while still scant, suggests this is true. Hairston (1980a,b, 1987) demonstrated the evolution of α selection in populations of *P. jordani* 50 km apart; I have shown coevolution in another predator–prey system using larval salamanders (*Ambystoma opacum* and *A. maculatum*) from populations 350 km apart (Fauth 1990). Most recently Kurzava (1994; see also Morin, this volume) has shown differences in predatory capabilities of newt populations separated by 750 km. The results are consistent with a coevolutionary scenario; a subspecies of newt (*Notophthalmus viridescens viri-*

descens) that co occurs with *B. americanus* inflicted heavier mortality on their tadpoles than did a subspecies (*N. v. dorsalis*) from outside the range of this toad, but sympatric with three other *Bufo*. These studies suggest that communities may be coevolving units. After a half-century of heated debate about the nature of ecological communities (Clements 1916, Gleason 1926, Whittaker 1967), experimentalists have yet to test this fundamental principle of the Clementsian model.

Interactions between species are not immutable; they vary spatially and temporally as species adapt to one another and to their environment. The spatial scale of variation may range from hundreds of kilometers, as in *Notophthalmus* (Kurzava 1994; Morin, this volume) and *Ambystoma* (Fauth 1990), to just a kilometer or two, as in *Desmognathus* and *Bufo* (this study). The temporal scale of variation may range from thousands of years, the time required for *D. fuscus* to colonize glaciated terrain in Ohio, to just decades, the time required for farm pond populations of *B. americanus* and *R. palustris* to diverge in important life history traits. Adaptation may take the form of genetic change, phenotypic plasticity, or learning, but the important point is that adaptation can occur on a variety of spatial and temporal scales.

For decades plant ecologists have used experiments in common gardens as a tool for deciphering the role of genetics and environment in molding local adaptation. Animal ecologists have been slow to adopt this approach, perhaps because of the daunting logistics required for field experiments. Yet when used in concert with artificial mesocosms, common garden experiments can reveal previously hidden interpopulational variation in species interactions. By incorporating molecular genetics into studies of geographic variation, researchers can identify the genetic mechanisms that contribute to local adaptation, thereby linking evolutionary processes with their resulting patterns. Phylogenetic inference may then be used to identify where ecological interactions are likely to be constant and where they are likely to differ.

ACKNOWLEDGMENTS   I thank H. M. Wilbur, P. Marino, two anonymous reviewers, and especially J. Bernardo for their insightful comments. Financial support for the common garden experiments on anurans was provided by a grant from Sigma Xi, the Scientific Research Society. The experiments on *Desmognathus* were supported by grants to the author from the Carolinas–Ohio Science Education Network and the Denison University Research Foundation, a Pratt Postdoctoral Fellowship from Mountain Lake Biological Station, and National Science Foundation grants BSR-88-17732 (to W. J. Resetarits Jr. and H. M. Wilbur) and DEB-92-07192 (to H. M. Wilbur; R. G. Jaeger and J. E. Fauth, senior investigators). A special thanks goes to C. J. Gill, C. M. Caruso, J. Freed, A. Ehlers, S. M. Welter, and J. A. Herbert for their indomitable efforts on the artificial stream bed experiments.

Literature Cited

Aarssen, L. 1983. Ecological combining ability and competitive combining ability in plants: toward a general evolutionary theory of coexistence in systems of competition. American Naturalist 122:703–731.

Alford, R. A. 1986. Effects of parentage on competitive ability and vulnerability to predation in *Hyla chrysoscelis* tadpoles. Oecologia 68:199–204.

————. 1989. Competition between larval *Rana palustris* and *Bufo americanus* is not affected by variation in reproductive phenology. Copeia 1989:993–1000.

Alford, R. A., and H. M. Wilbur. 1985. Priority effects in experimental pond communities: competition between *Bufo* and *Rana*. Ecology 66:1097–1105.

Allard, R. W., and J. Adams. 1969. Population studies in predominately self-pollinating species: I. Intergenotypic competition and population structure in barley and wheat. American Naturalist 103:621–645.

Avise, J. C. 1989. Gene trees and organismal histories: a phylogenetic approach to population biology. Evolution 43:1192–1208.

————. 1994. Molecular markers, natural history and evolution. Chapman and Hall, New York.

Ballinger, R. E. 1979. Intraspecific variation in demography and life history of the lizard, *Sceloporus jarrovi*, along an altitudinal gradient in southeastern Arizona. Ecology 60: 901–909.

Beebee, T. J. C. 1991. Purification of an agent causing growth inhibition in anuran larvae and its identification as a unicellular unpigmented alga. Canadian Journal of Zoology 69:2146–2153.

Bergelson, J. 1994. The effects of genotype and the environment on costs of resistance in lettuce. American Naturalist 143:349–359.

Bernardo, J. 1994. Experimental analysis of allocation in two divergent, natural salamander populations. American Naturalist 143:14–38.

————. 1996. Maternal effects in animal ecology. American Zoologist 36:83–105.

Berven, K. A. 1982a. The genetic basis of altitudinal variation in the wood frog *Rana sylvatica*: I. An experimental analysis of life history traits. Evolution 36:962–983.

————. 1982b. The genetic basis of altitudinal variation in the wood frog *Rana sylvatica*: II. An experimental analysis of larval traits. Oecologia 52:360–369.

————. 1990. Factors affecting population fluctuations in larval and adult stages of the wood frog (*Rana sylvatica*). Ecology 71:1599–1608.

Berven, K. A., and T. A. Grudzien. 1990. Dispersal in the wood frog (*Rana sylvatica*): implications for genetic population structure. Evolution 44:2047–2056.

Berven, K. A., D. E. Gill, and S. J. Smith-Gill. 1979. Countergradient selection in the green frog, *Rana clamitans*. Evolution 33:609–623.

Billick, I., and T. J. Case. 1994. Higher order interactions in ecological communities: what are they and how can they be detected? Ecology 75:1529–1543.

Bogert, C. M. 1960. The influence of sound on the behavior of amphibians and reptiles. Pages 137–320 in W. E. Lanyon and W. N. Taylor (eds.), Animal Sounds and Communication. American Institute of Biological Sciences Publication no. 7.

Breden, F. 1987. The effect of post-metamorphic dispersal on the population genetic structure of Fowler's toad, *Bufo woodhousei fowleri*. Copeia 1987:386–395.

Brodie, E. D., III, and E. D. Brodie Jr. 1990. Tetrodotoxin resistance in garter snakes: an evolutionary response of predators to dangerous prey. Evolution 44:651–659.

Clements, F. E. 1916. Plant succession: analysis of the development of vegetation. Carnegie Institute of Washington Publication 242.

Conant, R., and J. T. Collins. 1991. A Field Guide to the Reptiles and Amphibians of Eastern and Central North America. Houghton Mifflin, Boston, Massachusetts.

Connell, J. H. 1980. Diversity and the coevolution of competitors, or the ghost of competition past. Oikos 35:131–138.

Duellman, W. E., and L. B. Trueb. 1986. Biology of Amphibians. McGraw-Hill, New York.

Elton, C. 1927. Animal Ecology. Sidgwick and Jackson, London.

Falconer, D. S. 1981. Introduction to Quantitative Genetics, 2nd ed. Longman, New York.

Fauth, J. E. 1990. Ecological and evolutionary interactions between larval amphibians. Ph.D. dissertation, Duke University, Durham, North Carolina.

Fauth, J. E., and W. J. Resetarits Jr. 1991. Interactions between the salamander *Siren intermedia* and the keystone predator *Notophthalmus viridescens dorsalis*. Ecology 72:827–838.

Fauth, J. E., W. J. Resetarits Jr., and H. M. Wilbur. 1990. Interactions between larval salamanders: a case of competitive equality. Oikos 58:91–99.

Fauth, J. E., J. Bernardo, M. Camara, W. J. Resetarits Jr., J. Van Buskirk, and S. A. McCollum. 1996. Simplifying the jargon of community ecology: a conceptual approach. American Naturalist 147:282–286.

Frost, D. R., A. R. Kluge, and D. M. Hillis. 1992. Species in contemporary herpetology: comments on phylogenetic inference and taxonomy. Herpetological Review 23:46–54.

Futuyma, D. J., and M. Slatkin. 1983. Introduction. Pages 1–13 in D. J. Futuyma and M. Slatkin (eds.), Coevolution. Sinauer, Sunderland, Massachusetts.

Gill, D. E. 1972. Intrinsic rates of increase, saturation densities, and competitive ability: I. An experiment with *Paramecium*. American Naturalist 106:461–471.

Gleason, H. A. 1926. The individualistic concept of the plant association. Torrey Botanical Club Bulletin 53:7–26.

Goater, C. P. 1994. Growth and survival of postmetamorphic toads: interactions among larval history, density, and parasitism. Ecology 75:2264–2274.

Goldberg, D. E., and L. Fleetwood. 1987. Competitive effect and response in four annual plants. Journal of Ecology 75:1131–1143.

Goldberg, D. E., and K. Landa. 1991. Competitive effect and response: hierarchies and correlated traits in the early stages of competition. Journal of Ecology 79:1013–1030.

Gould, S. J. 1977. Ontogeny and Phylogeny. Belknap, Cambridge, Massachusetts.

Gurevitch, J., L. L. Morrow, A. Wallace, and J. S. Walsh. 1992. A meta-analysis of competition in field experiments. American Naturalist 140:539–572.

Hairston, N. G. 1973. Ecology, selection and systematics. Breviora 414:1–21.

———. 1980a. Evolution under interspecific competition: field experiments on terrestrial salamanders. Evolution 34:409–420.

———. 1980b. The experimental test of a guild: salamander competition. Ecology 62:65–72.

———. 1987. Community Ecology and Salamander Guilds. Cambridge University Press, Cambridge.

Harris, R. N., R. D. Semlitsch, H. M. Wilbur, and J. E. Fauth. 1990. Local variation in the genetic basis of paedomorphosis in the salamander *Ambystoma talpoideum*. Evolution 44:1588–1603.

Heusser, H. 1960. Über die Beziehungen der Erdkröte (*Bufo bufo* L.) zu ihren Laichplatz: II. Behaviour 16:93–109.

Hutchison, V. H. 1956. An annotated list of the amphibians and reptiles of Giles County, Virginia. Virginia Journal of Science 7:80–86.

Janzen, D. H. 1980. When is it coevolution? Evolution 34:611–612.

Jinks, J. L. 1954. The analysis of continuous variation in a diallel cross of *Nicotiana rustica* varieties. Genetics 39:767–788.

Joshi, A., and J. N. Thompson. 1995. Alternative routes to the evolution of competitive ability in two competing species of *Drosophila*. Evolution 49:616–625.

Kärkkäinen, K., V. Koski, and O. Savolainen. 1996. Geographical variation in the inbreeding depression of Scots pine. Evolution 50:111–119.

Kurzava, L. M. 1994. The structure of prey communities: effects of predator identity and geographic variation in predators. Ph.D. dissertation, Rutgers University, New Brunswick, New Jersey.

Leimberger, J. D. 1989. The effect of short-term temperature fluctuations on interactions within an amphibian assemblage. Ph.D. dissertation, Duke University, Durham, North Carolina.

Linhart, Y. 1988. Intrapopulational differentiation in annual plants: III. The contrasting effects of intra-and interspecific competition. Evolution 42:1047–1064.

Lively, C. M. 1989. Adaptation by a parasitic trematode to local populations of its snail host. Evolution 43:1663–1671.

MacArthur, R. H., and E. O. Wilson. 1967. The Theory of Island Biogeography. Princeton University Press, Princeton, New Jersey.

Mack, R. N., and J. L. Harper. 1977. Interference in dune annuals: spatial pattern and neighbourhood effects. Journal of Ecology 65:345–363.

Mayr, E. 1942. Systematics and the Origin of Species. Columbia University Press, New York, New York.

———. 1963. Animal species and evolution. Belknap, Boston, Massachusetts.

Mills, L. S., M. E. Soulé, and D. F. Doak. 1993. The keystone species concept in ecology and conservation. BioScience 43:219–224.

Mitchell, S. L. 1990. The mating system genetically affects offspring performance in Woodhouse's toad (*Bufo woodhousei*). Evolution 44:502–519.

Morin, P. J. 1981. Predatory salamanders reverse the outcome of competition among three species of anuran tadpoles. Science 212:1284–1286.

———. 1983. Predation, competition, and the composition of larval anuran guilds. Ecological Monographs 53:119–138.

Newman, R. A. 1988a. Genetic variation for larval anuran (*Scaphiopus couchii*) development time in an uncertain environment. Evolution 42:763–773.

———. 1988b. Adaptive plasticity in development of *Scaphiopus couchii* tadpoles in desert ponds. Evolution 42:774–783.

Nishikawa, K. C. 1985. Competition and the evolution of aggressive behavior in two species of terrestrial salamanders. Evolution 39:1282–1294.

———. 1987. Interspecific aggressive behavior in salamanders: species-specific interference or misidentification? Animal Behaviour 35:263–270.

Norrington-Davies, J. 1967. Application of diallel analysis to experiments in plant competition. Euphytica 16:391–406.

Pechmann, J. H. K. 1994. Interactions between larval history and adult reproductive success in frogs. Ph.D. dissertation, Duke University, Durham, North Carolina.

Pickett, S. T. A., and F. A. Bazzaz. 1976. Divergence of two co-occurring successional annuals on a soil moisture gradient. Ecology 57:169–176.

Plytycz, B., J. Dulak, and E. Pecio. 1984. Genetic control of length of the larval period in *Rana temporaria*. Folia Biologica 32:155–166.

Resetarits, W. J., Jr. 1991. Ecological interactions among predators in experimental stream communities. Ecology 72:1782–1793.

Richards, C. M. 1958. The inhibition of growth in crowded *Rana pipiens* tadpoles. Physiological Zoology 31:138–151.

Ricklefs, R. E., and D. Schluter (eds.). 1993. Species Diversity in Ecological Communities: Historical and Geographic Perspectives. University of Chicago Press, Chicago, Illinois.

Roughgarden, J. 1983. Coevolution between competitors. Pages 383–403 in D. J. Futuyma and M. Slatkin (eds.), Coevolution. Sinauer, Sunderland, Massachusetts.

Scott, D. E. 1994. The effect of larval density on adult demographic traits in *Ambystoma opacum*. Ecology 75:1383–1396.

Semlitsch, R. D., D. E. Scott, and J. H. K. Pechmann. 1988. Time and size at metamorphosis related to adult fitness in *Ambystoma talpoideum*. Ecology 69:184–192.

Semlitsch, R. D., R. N. Harris, and H. M. Wilbur. 1990. Paedomorphosis in *Ambystoma talpoideum*: maintenance of population variation and life history pathways. Evolution 44:1604–1613.

Smith, D. C. 1987. Adult recruitment in chorus frogs: effects of size and date at metamorphosis. Ecology 68:344–350.

Sokal, R. R., and F. J. Rohlf. 1995. Biometry, 3rd ed. W. H. Freeman, New York.

Solbrig, O. T., and B. B. Simpson. 1977. A garden experiment on competition between biotypes of the common dandelion (*Taraxacum officinale*). Journal of Ecology 65:427–430.

Steinwascher, K. 1978. Interference and exploitation competition among tadpoles of *Rana utricularia*. Ecology 59:1039–1046.

Tracey, C. R., and J. W. Dole. 1969. Orientation of displaced California toads, *Bufo boreas*, to their breeding sites. Copeia 1969:693–700.

Travis, J. T. 1980. Phenotypic variation and the outcome of interspecific competition in hylid tadpoles. Evolution 34:40–50.

———. 1981. Control of larval growth variation in a population of *Pseudacris triseriata* (Anura: Hylidae). Evolution 37:496–512.

———. 1996. The significance of geographical variation in species interactions. American Naturalist 148:51–58.

Travis, J. T., W. H. Keen, and J. Julianna. 1985. The effects of multiple factors on viability selection in *Hyla gratiosa* tadpoles. Evolution 39:1087–1099.

Travis, J. T., S. B. Emerson, and M. Blouin. 1987. A quantitative-genetic analysis of larval life-history traits in *Hyla crucifer*. Evolution 41:145–156.

Turesson, G. 1922. The species and the variety as ecological units. Hereditas 3:100–113.

Turkington, R., and J. L. Harper. 1979. The growth, distribution and neighbour relationships of *Trifolium repens* in a permanent pasture. Journal of Ecology 67:245–254.

Van Valen, L. 1973. A new evolutionary law. Evolutionary Theory 1:1–30.

Waldman, B., J. E. Rice, and R. L. Honeycutt. 1992. Kin recognition and incest avoidance in toads. American Zoologist 32:18–30.

Wells, B. W. 1932. The natural gardens of North Carolina. University of North Carolina Press, Chapel Hill.

Whittaker, R. H. 1967. Gradient analysis of vegetation. Biological Review 49:207–264.

Wilbur, H. M. 1980. Complex life cycles. Annual Review of Ecology and Systematics 11:67–93.

———. 1987. Regulation of structure in complex systems: experimental temporary pond communities. Ecology 68:1437–1452.

Wilbur, H. M., and R. A. Alford. 1985. Priority effects in experimental pond communities: responses of *Hyla* to *Bufo* and *Rana*. Ecology 66:1106–1114.

Wilbur, H. M., and J. E. Fauth. 1990. Experimental aquatic food webs: interactions between two predators and two prey. American Naturalist 135:176–204.

Woodward, B. D. 1986. Paternal effects on juvenile growth in *Scaphiopus multiplicatus* (the New Mexican spadefoot toad). American Naturalist 128:58–65.

———. 1987. Paternal effects on offspring traits in *Scaphiopus couchi* (Anura: Pelobatidae). Oecologia 73:626–629.

Woodward, B. D., J. T. Travis, and S. Mitchell. 1988. The effects of the mating system on progeny performance in *Hyla crucifer* (Anura: Hylidae). Evolution 42:784–794.

21

Revelations and Limitations of the Experimental Approach for the Study of Plant–Animal Interactions

ROBERT J. MARQUIS & CHRISTOPHER J. WHELAN

Interactions between plants and animals are pervasive in nature. Perhaps for this reason, a rich natural history lore exists for plant–animal interactions from virtually every natural habitat on earth. Despite the encyclopedic description available for a large number of interactions of plants and animals, experimental investigation of these interactions has been attempted only rather recently. We feel the use of experiments in the field of plant–animal interactions has led to both greater establishment of causal relationships and increased fundamental understanding of these interactions. We emphasize, however, the necessity of a vital observational component to this and indeed to all ecological disciplines.

Plant–animal interactions encompass a variety of ecological relationships. For example, antagonistic interactions are pervasive and often dramatic and include folivory, nectar robbing, and seed predation. Also widespread are mutualistic interactions such as seed dispersal, pollination, and protection by the third trophic level, whose participants include both invertebrates and vertebrates. For example, in some tropical forests over 70% of the woody plant species may enlist vertebrates to disperse seeds (Willson et al. 1989). Plants and animals are also linked in webs of trophic interactions and thus allow us to examine both direct interactions, the focus of much of ecology, and indirect interactions, for which much less information is available.

Here we describe field experimental approaches we have taken in three subdisciplines under the umbrella of plant–animal interactions. Specifically, we describe studies of interactions between a plant species and its herbivores, interactions between fruiting plants and vertebrate seed dispersers, and a tritrophic interaction that involves insectivorous birds, leaf-consuming insects, and a temperate, deciduous tree. For each example, we discuss revelations gleaned from the experiments, as well as limitations inherent in either interpretation of results or applications of the experimental techniques. While ecology can progress greatly from field experimentation, not all systems are amenable to experimentation, nor does experimentation provide answers to all pertinent questions. Understanding the advantages as well as the limitations of experimental approaches is critical to optimize their usefulness.

Selected Investigations: Does Leaf Herbivory Affect Plant Fitness?

Over the last 30 years, the field of plant–herbivore interactions has provided a testing ground for potential coevolutionary relationships between plants and animals. Observations of host-plant use by various herbivore species and distribution patterns of secondary chemicals among plant species suggested that herbivores and plants have had reciprocal evolutionary impacts (Ehrlich and Raven 1964). However, an important piece of evidence for such a coevolutionary scenario was missing until the mid-1980s—that is, the selective impact of herbivores on plant fitness in natural settings.

Because so few studies were conducted and the data available did not correctly address the issue, opinions regarding the effects of herbivory on plant fitness ranged from negative (Levin 1976b) to zero (Jermy 1984) to positive (Owen and Wiegert 1981). Experimental data were limited to agricultural and greenhouse settings (reviewed by Belsky 1986, Krischik and Denno 1983, Marquis 1992b). Data from agricultural field/greenhouse environments left uncertain herbivore impact in nature because of the great difference in the two environments. Compensatory ability (regrowth following damage) of plants may depend on resource availability (Whitham et al. 1991), which often varies greatly between managed and natural settings due to differences in the level of plant competition, soil nutrients, and water. Correlative data linked host plant preference (level of defoliation) to decreased growth in temperate trees (Kulman 1971). However, these correlative data can be misleading, particularly if herbivores choose plants nonrandomly. If herbivores attack more vigorous individuals, then the impact of herbivory is underestimated, while if stressed trees are attacked preferentially, then the impact of herbivory is overestimated. There are now data demonstrating that herbivore attack is often nonrandom, sometimes concentrated on stressed individuals (e.g., Lewis 1984) and, in other systems, on the more vigorous members of a population (e.g., Price 1991). Because we are not able to predict the nature of such preferences for unstudied systems, the value of correlative data remains limited.

The impact of leaf herbivores on plant fitness can be determined experimentally in at least three ways: (1) decreasing herbivore damage by removing herbivores with insecticides or by hand removal, (2) adding damage by either adding herbivores directly or preventing access to plants by predators, and (3) creating various levels of damage by artificially damaging plants. Impact of each method can be compared to the appropriate controls. Each method has its own advantages and drawbacks. Methods (1) and (2) measure the effects of naturally produced herbivory but require either additional controls (1) or may not allow control of the exact level of damage produced by herbivores (2). Method (3) does allow precise control of damage, but simulated damage may differ in impact from that produced by herbivores themselves (Baldwin 1990).

Marquis (1984, 1987, 1988, 1992a) was among the first to use an experimental approach to define the impact of an herbivore guild on plant fitness in its natural habitat. A tropical system was chosen because "herbivore pressure" was hypothesized to be greater in tropical than temperate systems (Doutt 1960, MacArthur 1969, Baker 1970,

Levin 1976a). Conditions in lowland tropical wet forests made the use of insecticide sprays and addition of insects relatively impractical as ways to examine effects of leaf herbivory. Frequent rains reduce effectiveness of insecticides; adding herbivores is an option, but numbers of any one insect herbivore species are usually so low in rain forests that it would be impossible to control the kind and level of damage in such an experiment. Thus, Marquis chose to damage plants artificially. If artificial damage simulates the kinds of damage that occurs naturally, then the results are relevant to the impacts of natural herbivory.

Marquis (1984) investigated the impact of leaf herbivores on plant fitness in *Piper arieianum*, an understory shrub of tropical wet forest. Damage was controlled precisely using artificial herbivory. Natural damage, including rates of leaf area loss, was previously monitored and assigned to particular herbivores. Experimental damage was applied to simulate the levels and kinds of damage produced naturally by the herbivore guild. Specifically, leaf area was removed using a hole punch to simulate damage by a common leaf-feeding weevil. Leaf area was removed over a 2-week period, and unequal amounts of leaf tissue were removed from each leaf to simulate naturally uneven distributions of damage within plant crowns. Damage summed to 10%, 30%, or 50% of a plant's total leaf area, for plants of each of three size classes based on total leaf number. A complete defoliation treatment (100%) was included for medium plants to determine the effects of leaf-cutting ant (*Atta*) attack. Sample size was 30 per treatment per size class. Plant size was included as a factor because an earlier experiment demonstrated that impact of leaf damage varied with plant size in this species (unpublished). The effects of this single bout of simulated leaf herbivory on growth (total stem length produced) and seed production were followed for 2 subsequent years.

During the first year that followed simulated damage, seed production decreased in all three size categories with increasing levels of damage (Fig. 21-1). Differences were significant between control and the 30% treatment but not for the 10% removal treatment and controls. Effects on growth during the first year mirrored those for seed production, except that growth was not significantly affected at any defoliation level in large plants.

The effects of leaf removal persisted through the second year for small and medium plants, again with significant decreases in growth and seed production at 30% and 50% leaf removal compared to controls. Large plants recovered more quickly: there were no differences in seed production or plant growth at any of the defoliation levels compared to controls during the second year.

Revelations and Limitations

This experiment demonstrated that leaf area removal can significantly reduce plant growth and reproduction for 2 years after a single defoliation. Because natural leaf damage varies greatly among individuals and is similar in magnitude to simulated damage, these results suggested that leaf herbivores are a relatively strong selective force for *P. arieianum*. Thus, these results provided some of the first evidence that herbivores have the potential to effect evolutionary changes in plant traits. Because this experiment was conducted in nature, Marquis concluded that leaf herbivory is important relative

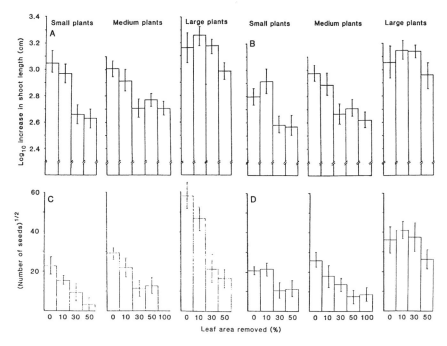

Figure 21-1. The effects of varying levels of leaf area removal on growth and reproduction for three size classes of plants of *Piper arieianum*. (**A**) Total growth during the first year after a single defoliation. (**B**) Growth during the full 2-year period after defoliation. (**C**) Seed production during the first year after defoliation. (**D**) Seed production during the second year after defoliation. Standard errors of the means are shown. From Marquis (1984). Reprinted with permission from Science 226: 538. Copyright 1984 American Association for Advancement of Science.

to other potential selective factors because those factors acted simultaneously on the experimental plants.

The strength of these conclusions lies in the fact that they were firmly grounded in natural history. Marquis (1984) took care to measure the distribution of leaf damage within and between plants, to measure rates of leaf area loss over time, and to observe the kinds of damage caused by different insect species. This allowed him to simulate accurately the kinds of damage produced naturally and to conclude that his results are relevant to those produced by the natural herbivore guild for the plant in its native habitat.

Nevertheless, conducting such an experiment in the wild has limitations. It is important to plant-herbivore ecologists to determine the impact of relatively low damage levels (0–10%) on plant fitness. Does fitness decrease monotonically with increasing damage, is there a threshold of damage needed to reduce fitness, or do plants over-compensate at lower damage levels? Tolerance is probably greatest at lower levels of damage, and lower levels of damage are more common. The problem with experimental defoliations conducted in the wild (and other experiments in which the focal effect may be small but critical) is that other uncontrolled environmental factors may swamp the effects of low damage, reducing the statistical power of the experiment. Thus, in the experiment using *P. arieianum* there were no significant differences between control and 10% defoliated plants even though values for the latter were intermediate between controls and 30% defoliated plants (Fig. 21-1). Larger sample sizes might be a solution but not always feasible (e.g., plant density for any one species in rain forests is often very low). Another alternative is to move the experiment to the greenhouse, where results would reveal how plants respond to low damage under "ideal" conditions. This greenhouse experiment would be informative in that one could be certain that low damage does have an effect at least under some circumstances. The experimenter could then return to the field to determine what other factors might be mitigating the impact of the herbivory on plant fitness. However, the greenhouse conditions (e.g., differences in light level and pot effects) will always leave uncertain the degree to which impact on fitness measured in the greenhouse is relevant to nature.

Selected Investigations: What Plant or Fruit Characteristics Influence Diet Choice in Birds?

Early theoretical consideration of the interactions of fruit-producing plants and their animal dispersers conjectured strong reciprocal selection by both partners in this mutualism (Snow 1971, McKey 1975). If this is true, plant traits should evolve to attract high-quality dispersers while concomitantly discouraging poor-quality dispersers. In turn, high-quality dispersers would evolve morphologies, physiologies, and behaviors that enhance their suitability as dispersers. These seminal papers sparked a remarkably active period of empirical investigation, which led to a modified perspective. In light of empirical evidence, based primarily on detailed observational investigations (e.g., Howe 1977, Wheelwright 1983, Denslow 1987), a subsequent paradigm of "diffuse" coevolution emerged.

While the results of observational studies conflicted with earlier expectations of coevolution of fruiting plants and vertebrate dispersers, more controlled laboratory experiments conducted in aviaries demonstrated highly sophisticated decision-making processes of avian dispersers (e.g., Moermond and Denslow 1983, Levey et al. 1984). These highly controlled experimental investigations revealed behavior by bird dispersers suggestive of selective foraging, a necessary condition for the original theory of reciprocal selection. In contrast, observational studies often found fruit removal to be "more a function of avian biology than of most characteristics of the fruiting plant" (Willson and Whelan 1993). Clearly, resolving these contrasting perspectives presents a challenge to both theoretical and empirical investigations of the evolutionary ecology of plant–vertebrate disperser interactions.

Willson and Whelan developed an approach intermediate to that of field observation versus aviary experimentation (e.g., Willson and Whelan 1989, Whelan and Willson 1994). Specifically, free-ranging birds were offered fruit in artificial "infructescences" (hereafter referred to as fruit displays) in a variety of field situations. These displays consisted of wood doweling with thin wire hooks on which up to 20 fruits were impaled (see Whelan and Willson 1994 for details). The displays were constructed to resemble a raceme. All experiments described in this paper consisted of 10 displays for each experimental treatment, each attached to a tree or shrub at approximately 10-m intervals along the eastern edge of Trelease Woods, a 24-ha woodlot in Champaign County, Illinois. Many aspects of fruit display could be controlled—for example, fruit abundance, proximity to perches, ease of removal by birds, placement in plant, height above ground, and so on. This allowed control and manipulation of a limited variety of fruit/ plant characteristics, while the birds feeding from the displays were in all other respects experiencing a natural environment. Thus, this experimental approach increased the ability to detect the basis of selective foraging over that of purely observational studies in which no fruit or plant traits had been manipulated. Moreover, foraging choices were made by birds in the context of their natural living situation—for example, in the presence of predators and competitors. Thus, many of the unknown influences and confounding effects of the artificial environment of a flight cage were removed.

Both accessibility to fruit from a perch and display attraction or conspicuousness were manipulated. In another set of trials, various fruits of contrasting pericarp chemistry were offered to determine the effect of such chemistry on choice. Results were analyzed to test differences between treatments in (1) mean proportion of displays with at least one fruit removed (HITS) and (2) mean proportion of fruit removed from all displays (FRUIT) (Figs. 21-2, 21-3, 21-4; details: Whelan and Willson 1994).

Fruit Accessibility

Moermond and Denslow (1983) demonstrated that physical accessibility of fruits influences fruit selection and differences in accessibility between fruits can override initial preferences based on size, nutritional reward, and so on. Infructescence position varies greatly relative to perches between plant species, and different bird species demonstrate different abilities to take fruits on the wing. Thus, trials which consider preference based only on reward may be misleading because they ignore the costs of collecting the reward. By manipulating accessibility between fruits of different species, Whelan and Willson could "behaviorally titrate" the relative preferences of birds for the different fruit species, taking into account accessibility. One goal was to determine with a simple experiment if such techniques could be used in the field with free-ranging birds. The experiment consisted of two display treatments, containing fruits of *Phytolacca americana* (pokeweed). An "accessible" treatment consisted of displays attached to the upper surface of a horizontal branch. An "inaccessible" treatment consisted of displays hung vertically from the tips of the same branches as the first treatment. Each display held 10 fruits. Accessible displays had at least one fruit removed more often than inaccessible displays ($F_{1,4}=11.6$, $P < .05$; Fig. 21-2a), and more total fruits were removed from accessible displays than inaccessible displays ($F_{1,4}=19.2$, $P < .05$; Fig. 21-2a). Although the results were as predicted, this was an important demonstration (see following discussion).

Display Attraction

Willson and Thompson (1982) pointed out that for many fruiting plant species, colorful fruit are often found with structures (e.g., infructescence rachis and aril) that

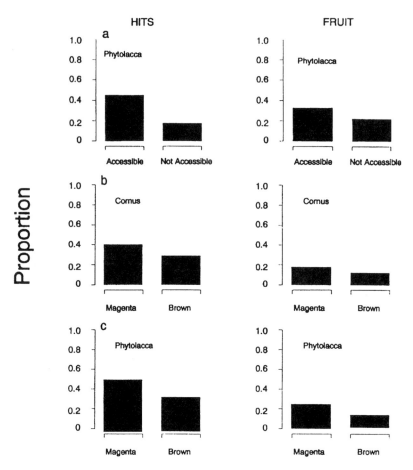

Figure 21-2. Proportion of artificial displays having ≥ 1 fruit removed (HITS) and the proportion of all available fruit taken (FRUIT) in experiments examining effect of (**a**) fruit accessibility, (**b**) infructescence color for *Cornus drummondii*, and (**c**) infructescence color for *Phytolacca americana*. From Whelan and Willson (1994). Reprinted with permission from Oikos 71: 142 (1994).

differ in color from both the fruit and the foliage. They referred to such structures as "bicolored fruit displays." Two hypothesized functions of such displays include increasing the attractiveness or "signal character" of the fruit in order to entice dispersers and indicating something about fruit age and/or quality. Whelan and Willson (1994) tested the first hypothesis by offering fruit of a given species on artificial displays that were either (1) stained a bright magenta color to simulate the natural color or (2) left the original brown color of the wood doweling. The experiments used *Cornus drummondii* (roughleaf dogwood) and *P. americana*. All displays held 10 fruits and were attached to horizontal branches in accessible positions. For both plant species, more bicolored displays had ≥ 1 fruit removed ($F_{1,33}$ = 3.58, P = .067; Fig. 21-2b–c: HITS),

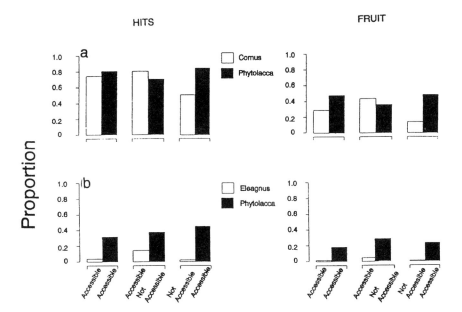

Figure 21-3. (**a**) Proportion of artificial displays having ≥ 1 fruit removed (HITS) and (**b**) proportion of all available fruit taken (FRUIT) in experiments that examine the strength of preference of migratory birds for a variety of plant species pairs. From Whelan and Willson (1994). Reprinted with permission from Oikos 71: 145 (1994).

and more total fruit removed from all the displays ($F_{1,33}$ = 5.94, P < .05; Fig. 21-2b–c: FRUIT). This result is consistent with the hypothesis that bicolored displays increase fruit attractiveness.

Strength of Fruit Preferences

Having demonstrated that free-ranging migrant birds will preferentially take accessible over inaccessible fruit of a given species, Whelan and Willson then manipulated accessibility to measure relative preferences for fruit of different plant species in the following experiment in the autumns of 1987 and 1988. Fruits of two species were offered in three treatments of paired displays: both species 1 and species 2 equally accessible; species 1 accessible and species 2 inaccessible; species 1 inaccessible and species 2 accessible. Ten displays for each fruit species were used for each treatment, and each display held 10 fruits of one of the species. Results for the following pairs of species are reported here: *C. drummondii* versus *P. americana* and *Eleagnus umbellata* (autumn olive) versus *P. americana*.

In the comparison of *Cornus* and *Phytolacca* (Fig. 21-3a), *Phytolacca* was preferred over *Cornus* when both were equally accessible; preference was enhanced when *Phytolacca* was accessible and *Cornus* was not; but preference was reversed when *Cornus* was accessible and *Phytolacca* was not (species by treatment interaction: $F_{2,10}$ = 5.02,

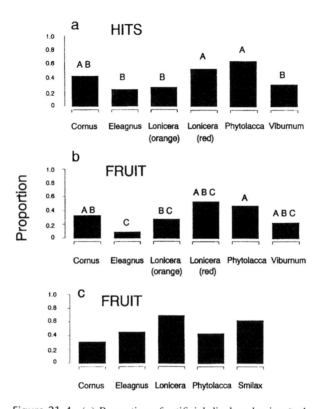

Figure 21-4. (**a**) Proportion of artificial displays having ≥ 1 fruit removed (HITS) and (**b**) proportion of all available fruit taken (FRUIT) in experiments that examine effect of pericarp chemistry on bird fruit preferences conducted from 4 to 6 September 1988. (**c**) Proportion of all available fruit taken (FRUIT) in experiments that examine effect of pericarp chemistry on bird fruit preferences conducted from 13 to 22 October 1988. In a and b, capital letters over the bars indicate statistically significant differences among species at $P < .05$, Tukey's HSD multiple range test. In c, proportions of fruit were found to differ with a chi-squared test ($\chi^2 = 17.92$, df = 4, $P < .001$). From Whelan and Willson (1994). Reprinted with permission from Oikos 71: 146 (1994).

$P < .01$). This result indicates that the initial preference is weak and easily reversed. In contrast, in the comparison of *Eleagnus* and *Phytolacca* (Fig. 21-3b), the initial preference of *Phytolacca* over *Eleagnus* when both were equally accessible was not altered by changes in relative accessibility, indicating that the initial preference for *Phytolacca* is strong and not easily reversed (species by treatment interaction: $F_{2,25} = 1.74$, $P > .05$).

Pericarp Chemistry

It is widely known that fruits vary greatly in the chemical makeup of their pericarp (Stiles 1980, Herrera 1982a,b, Johnson et al. 1985, Snow and Snow 1988), and it was suggested that pericarp chemistry likely influences fruit selection by birds. Whelan and Willson tested the hypothesis that migrant birds prefer fruits whose pericarp is high in lipid (Stiles 1980, 1993; Herrera 1982a) by "smorgasbord" experiments, using fruit from several plant species representing a broad range of pericarp composition (Table 21-1). Ten fruits of each species were presented on dowel displays, and species were alternated at 10-m intervals. The experiment was repeated twice, with mostly the same species offered each time (Table 21-1).

For both experiments, preference for individual species was strong (Fig. 21-4) but did not correspond clearly to gross chemical differences, such as percent lipid and percent sugar. The results do not support the contention that lipid is the major chemical determinant of fruit choice in migrant birds.

Revelations and Limitations

These field experiments provide evidence that fruit presentation and some fruit characteristics can influence diet choice in free-ranging migrant birds. Specifically, accessible fruits are removed faster than inaccessible fruits, fruits of bicolored displays experience greater removal rates than those of monocolored displays, and high-lipid fruits are not necessarily favored. By experimentally manipulating particular aspects of fruit presentation or characteristics while simultaneously controlling other characteristics, the impacts of the specific factors accessibility and color display were isolated. In contrast, in unmanipulated observational studies (e.g., Denslow 1987, Willson and

Table 21-1. Ranges of selected pericarp characteristics of fruit species used in experiment that examines effect of pericarp chemistry on bird foraging preferences

Plant species	% Lipid[1]	% Sugar[1]
Cornus drummondii[2,3]	32–63	13–26
Elaeagnus umbellata[2,3]	3–4	32–48
Lonicera maackii[2,3,4]	3	32
Phytolacca americana[2,3]	0.73	53
Smilax hispida[3]	1.5	30
Viburnum recognitum[2]	32–55	5–7

[1] Percent dry weight of pericarp of arbitrarily chosen fruits from a number of individual plants.
[2] Used in first pericarp experiment, 4–6 September 1988.
[3] Used in second pericarp experiment, 13–22 October 1988.
[4] Two color morphs (red and orange) used in first experiment.

Whelan 1993), two or more factors that potentially affect fruit choice often vary simultaneously, thus masking the individual effects of the factors under investigation. These experiments suggest that aspects of fruit presentation are important determinants of choice by birds and thus support the view that these plant traits may be adaptive characters influenced by selection imposed by bird foraging.

The methodology described in these experiments provides a means for conducting behavioral titrations in the field with free-ranging animals. By determining the ranking of food items, this methodology can be used to test aspects of the plant–animal interaction from both the animal (e.g., optimal foraging models) and plant (e.g., relative competitive abilities of different fruits for dispersers) perspective. By repeating this experiment in various locations, geographic variation in these relationships can be quantified.

Unfortunately, the technique of presenting fruit on artificial displays has some serious limitations. For one, the technique works well only when or where fruit-consuming birds are relatively abundant. Such locations or times cannot be predicted with certainty, so pilot studies may be useful to avoid loss of time and effort. Also, results may be highly dependent on the bird species present and may thus vary somewhat in space and/or time (Whelan and Willson 1994). In addition, the range of manipulable characteristics is limited. For instance, in the experiment that examined effects of pericarp chemistry on fruit choice, we used fruit species that represented a range of pericarp chemistries. But the fruit also varied in color and volume and, presumably, other characteristics, such as number of seeds, pulp to seed ratio, and so on. While artificial displays do offer the opportunity to exert greater control over unmanipulated observational studies, complete control is frequently difficult to achieve. Thus, such experimentation can quite reasonably be viewed as a useful intermediary between the tightly controlled conditions of the aviary and the uncontrolled conditions of purely observational studies.

Selected Investigations: Does Bird Predation of Leaf-Consuming Insects Benefit Host Plant Species?

Following publication of Hairston et al.'s (1960) controversial hypothesis concerning population regulation with respect to trophic level, much effort was expended debating the evidence for "top-down" versus "bottom-up" control in a wide variety of ecosystems. Experimental demonstrations of trophic cascades in aquatic systems supported the view of top-down control (e.g., Power et al. 1985, Power 1990), but whether such effects would be seen in terrestrial ecosystems, or in speciose aquatic systems for that matter, remained controversial (Strong 1992). We conducted an exclusion experiment (Marquis and Whelan 1994) to determine if insectivorous birds participate in a trophic cascade through their predation on leaf-damaging insects in a temperate, deciduous forest. Although many studies demonstrated that birds can, under some circumstances, depress the population size of insects (see review in Marquis and Whelan 1994), whether these direct effects also indirectly benefit the host-plant species was not previously examined. Only Atlegrim (1989) had demonstrated that exclosure of birds affects subsequent damage by leaf-chewing insects.

The experiment took place at the 809-ha Tyson Research Center (Eureka, Missouri). The deciduous forest at Tyson is dominated by white and red oak (*Quercus alba* and *Q. rubra*, respectively), and 31 bird species are common breeding season residents. Ninety saplings (1–5 m in height) were marked in March 1989, in groups of three. Members of a triplet were within 20 m of each other and were similar in stem diameter, height, and crown diameter. Each member of a triplet was randomly assigned to one of three treatments: control (no manipulation), cage (to exclude birds but not insects), and spray (spray of insecticide to reduce damage). Cages consisted of frames to which one layer of monofilament fish netting was tied (holes 3.5 cm diameter). There was no measurable effect of the netting on light levels within the cage. For the spray treatment, resmethrin (a synthetic pyrethroid, both non-phosphorus- and non-nitrogen-based) was sprayed once a week on the foliage top and bottom just enough to wet the foliage. The goal of this last treatment was to determine the effect of insect feeding on plant growth in the presence of birds. Insects on all plants were censused without removal every 2 to 4 weeks, and leaf damage was estimated in October of both 1989 and 1990 just before leaf fall. Growth parameters measured were length of new twigs added, average leaf size, and total number of leaves after leaf and twig expansion were completed in the following year. Regression analysis was used to relate leaf number and twig length to leaf biomass and twig biomass. Additional information about the site and methods can be found in Marquis and Whelan (1994).

In 1989 the number of insects encountered over the season on sprayed plants was one-fifth that found on control plants, while caged plants had twice as many insects as controls (Fig. 21–5A). As a result, sprayed plants lost 6% of their total leaf area by the end of the 1989 season, controls about 13% of their final leaf area, and caged plants approximately 26% of their final leaf area (Fig. 21–5C). Resulting biomass production in 1990 was least in the most heavily damaged plants: caged plants produced significantly less biomass than sprayed plants, with controls intermediate in biomass production between the two treatments (Fig. 21–5E). Differences in total biomass production were due to differences in leaf biomass production and not to effects of the treatments on twig biomass production. Thus, the treatment effect was manifested as a decrease in photosynthetic area, not in a change in the support biomass for leaf area.

Effects of the experiment on insect abundance (Fig. 21–5B) and leaf damage (Fig. 21–5D) in 1990 were similar to those in 1989, and insect abundance and damage were higher across all treatments in the second year. However, insect numbers and resulting damage were greatest in caged plants and lowest in sprayed plants. As a result, leaf biomass production in 1991 was significantly one-third less in caged plants than in control and sprayed plants (Fig. 21–5F).

Revelations and Limitations

This experiment demonstrated that foraging by insectivorous birds on herbivorous insects affects plant growth. Thus, over the long term, forest productivity may be affected by changing population levels of avian predators (Terborgh 1989, Hagan and Johnston 1992). Further, because insectivorous birds often demonstrate foraging preferences for different tree species (e.g., Robinson and Holmes 1984, Holmes and Schultz

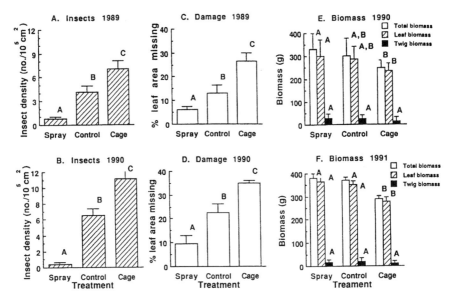

Figure 21-5. Treatment effects on insect numbers, leaf damage, and biomass production. Letters indicate treatment differences at $P < .05$. (**A**) Total insect numbers summed over the 1989 season. (**B**) Total insect numbers summed over the 1990 season. (**C**) Leaf damage in 1989. (**D**) Leaf damage in 1990. (**E**) Biomass production in 1990. (**F**) Biomass production in 1991. Different letters above the bars indicate significant differences among treatments at $P < .05$ and among treatments within a biomass category at $P < .05$ in graphs E and F. Error bars represent 1 S.E. of the mean. From Marquis and Whelan (1994). Reprinted with permission from Ecology 75: 2011 (1994).

1988, Peck 1989), changes in the abundance of birds may affect relative tree species composition of a forest as well. Birds may have similar impacts on plant growth in a wide variety of systems (temperate and tropical forests, deciduous and evergreen forests, and forests and grasslands) because in all but one system (sagebrush community; Wiens et al. 1991) in which cages were used to control access of birds to plants, bird foraging had a significant impact on insect abundance (see Marquis and Whelan 1994).

This experiment further demonstrated that bird foraging affects herbivore guild structure. Birds not only reduce the numbers of insects on white oak but also influence the relative distribution of species by selectively avoiding hairy and spiny caterpillars (Marquis and Whelan 1994). Attempts to understand the forces that structure guilds within communities of herbivores that share a host plant (Root and Cappuccino 1992) may fall short if the role of the third trophic level is omitted. Results of the caging experiment suggest that this would be true for oaks.

The results of this experiment were dramatic and unexpected for a number of reasons. Previous experimental demonstration of trophic cascades had been limited to aquatic systems, and there was some suggestion that third-trophic-level impact on the plant populations might not occur outside an aquatic arena (Strong 1992). Obvious terrestrial systems where trophic cascades might occur involve vertebrate herbivores (e.g., moose, rabbits, voles, wildebeests, and snow geese), which, because of their population size, can have dramatic effects on the vegetation. Thus, changes in population sizes of wolves (*Canis lupus*) were related to tree ring growth in balsam fir (*Abies balsamifera*) through the impact of wolves on their moose prey (*Alces alces*; McLaren and Peterson 1994). Similarly, introduction and subsequent control of rinderpest in Africa caused dramatic changes in vegetation through its impact on wildebeest populations (Sinclair 1979). Changes in population size of vertebrate predators of wildebeest might exert similar effects.

Impact of the third trophic level is not necessarily related to individual body size or total biomass of third-trophic-level participants, nor to the body size of the herbivores on which they prey. Holmes and Sturges (1975) suggested birds might play some as yet unknown but important role in ecosystem structure, but Wiens (1973) rejected this idea because birds in general represent a small portion of the total biomass of most ecosystems. Our study helps resolve this controversy and suggests that impact on ecosystem function and structure is related more to function of the focal species than to its biomass (see also Paine 1980). Data from other systems that involved parasitic wasps (Nafus 1991, Gomez and Zamora 1994), insect predators (Louda 1982, Stamp and Bowers 1996), and lizards (Spiller and Schoener 1990, Dial and Roughgarden 1995) suggest that small members of the third trophic level may have significant impact on plant fitness in terrestrial ecosystems.

Our results suggest that many revelations about ecological systems may be realized by rethinking previous experimental approaches. Such revelations can come from adding additional treatments to a methodology previously developed or taking additional data previously omitted. Note that no new experimental technique was developed for this study, other than measuring the effects of the caging treatment down to the plant level and taking care that the caging material employed had little impact on light levels within the cages. Thus, it was not so much that revelation occurred through development of an experimental procedure but that it occurred through a change in perspective: from the insect or predator point of view to that of the plant.

Our experimental technique is limited as it does not allow us to determine the relevant impact of all third-trophic-level participants, at least at the scale at which we conducted the experiment. Arthropod (spider and insect) predators and parasitoids also likely exert some effect on the leaf-chewing insect fauna of white oak. Although we demonstrated that avian predation is important, that does not preclude the possibility that these other third-trophic-level participants likewise play important roles and could compensate for decreased avian predation due to declining bird populations. Whether such compensation could occur could be tested experimentally but would require exclosures built at the scale of a block of forest and might require several years to detect numerical responses of the alternative caterpillar predators and/or parasites. Marquis is presently conducting a two-factor experiment at the scale of individual saplings that incorporates hand harvesting of arthropod predators in addition to cage exclusion of birds. Parasitoids are much more problematic, as a mesh fine enough to exclude them would also exclude their insect hosts and likely affect light levels. Furthermore, because both the oak herbivore community and the associated parasitoid guild are speciose, there would be few or no times when exclosures built to limit access by parasitoids would not also prevent herbivore colonization.

General Discussion

As we stated in the beginning of this essays, experiments in ecology are extremely useful, but they do not represent the only road to understanding. Experiments may suffer from artificiality, poor grounding in the basic natural history of the system under investigation, and overgeneralization of results. However, we emphasize that with careful planning such issues are surmountable. As demonstrated by Marquis's experiments on herbivory in *Piper*, a solid, basic understanding of natural history can guide experimental design. In this way, natural history observations and experimentation should be viewed not as alternative research programs but as complementary research tools.

Natural history observation identifies interesting or pressing questions or problems and should be used to guide the design of experiments, which are essential to identify causal mechanistic relationships or processes. Such observation can play both a pre- and postexperimentation role. Prior to experimentation, observation of nature can provide inspiration to conjecture how the world works (e.g., Gilbert 1980). In turn, conjecture provides the basis for experimentation to determine causal relationships and to eliminate incorrect hypotheses. Correlations between environmental factors and the activity, abundance, fitness, and/or trait distribution for a species imply a causative role for those environmental factors but do not confirm it (Lande and Arnold 1983, Wiens 1989a,b). Experimentation is thus necessary to distinguish underlying causative forces which produce observed patterns from those noncausative factors which are simply correlated with patterns.

On the other hand, experiments can lead to results not suggested (or, at least, not widely believed) by known natural history. In this case, the role of observation is to confirm relationships revealed by experimentation not previously suspected. Previous natural history information may be available, but not just the right information. For example, de la Fuente and Marquis (forthcoming) have recently demonstrated experimentally that predatory ants reduce pathogen attack to leaves. This result suggests the need for further observation of the interaction between ants and fungal spores and/or between leaf damage by herbivores (which ants also reduce) and pathogen infection.

Finally, some experiments find their inspiration far afield, with little or no initial observation of the system. In these cases, postexperiment observation will serve to ground such experiments. Experimental (and observational) studies of the role of sexual selection in plants were inspired almost entirely by a wealth of theory, observation, and experimentation on this subject in the animal world (Willson 1979).

We thus caution that the interpretation of natural history observations can constrain unnecessarily the scope of experiments. For ecology to progress, accepted interpretation of natural history observations will need to be challenged continually. Often this will be accomplished most efficiently with an appropriate experiment.

In the field of plant–animal interactions, natural history has served as a source of inspiration for theory. Natural history also is used to judge the validity of experiments. It is impossible to evaluate the relevance of results from experimentation whose parameters are not grounded in basic natural history data. For example, part of the controversy concerning the impact of elk browsing on fitness of *Ipomopsis aggregata* (Paige and Whitham 1987, Bergelson and Crawley 1992, Paige 1994, Bergelson et al. 1996) centers, on one hand, on knowing the degree that elk browse the plants to the ground and, on the other, on whether artificial herbivory simulates the effect of such browsing. To quantify the impact of herbivory in natural systems, we need to simulate damage produced by the appropriate herbivores. It is as simple as that, but numerous published studies of experimental herbivory include manipulations that do not simulate or explicitly control for the kind of damage produced by herbivores. Pattern of removal among (e.g., Marquis 1992a) and within leaves (e.g., Coleman and Leonard 1995), timing (e.g., Maschinski and Whitman 1989), and method of removal (Baldwin 1990) are known to influence plant response. Notwithstanding, exploration beyond the observed range of natural herbivory (or other processes) can be extremely informative, if used properly. For example, it is likely that the investigator has not observed all possible levels and types of herbivory and that other levels and types occur at other times and other places. An experiment that goes beyond the presently observed levels would be useful in defining the boundaries of the response.

Despite the utility of the experimental approach, we described limitations inherent in each of our studies. By purposefully identifying limitations, investigators can avoid the danger of overgeneralizing the results of experiments. But overgeneralization is not unique to experimental studies. Indeed, we feel that much of the often shrill debate over the importance of competition in natural systems during the 1970s and 1980s arose from a perceived overgeneralization of correlational and observational field studies that sought confirmatory evidence for competition theory, rather than critical experimental tests of the theory and its assumptions. Careful consideration of the results of any scientific study is requisite for good science, regardless of technique or even field of inquiry.

With respect to plant–animal interactions, some limitations need to be addressed. For example, benefits and detrimental effects of interactions between plants and animals are more easily quantified for plants than for their animal counterparts. Plants can be easily marked, their aboveground growth measured, their fruits collected, and their seeds counted and weighed and tested for germinability. In some cases, it has even been possible to determine male fitness by examining the distribution of unique genotypes among the progeny of a plant population (though, more typically, only female fitness is quantified). Thus, for plants we have verified experimentally the influence of variable seed dispersal distance on the probability that a seedling will establish (e.g., Howe et al. 1985), the relative importance that different pollinator behaviors have for fruit set (Schemske and Horvitz 1984), and the relationship between amount of leaf area removed by herbivores and resulting plant seed production (Marquis 1992b).

In contrast, it is more difficult to experimentally manipulate animals to determine the role that plant resources have on population dynamics and fitness of animals. Rarely do we know how phenotypic variation in a plant population affects the resource base of the interacting animals and, in turn, serves as a basis for natural selection on those animals. For example, determining if a change in floral-tube length and an associated change in nectar availability have a selective effect on the visiting hummingbird depends on quantifying the relative importance of that particular plant species in the hummingbird's resource budget and ascertaining the availability of alternative resources that provide comparable nutrition. Both are obviously difficult tasks. Some systems are more amenable to investigate. For instance, Cushman and colleagues (Cushman and Beattie 1991, Cushman et al. 1994) have begun to quantify the importance of extrafloral plant nectar for mutualistic ant species with the goal to understand the reciprocal selective forces operating in ant–plant interactions. Likewise, herbivorous insect species that have specialized on a single host plant species provide a reasonable starting point for experimenting in the area of plant–herbivore interactions. For the most part, selective impacts of the plant on the herbivore remain unquantified (but see Berenbaum and Zangerl 1992). Until we begin to experiment with both plants and animals, we will not understand fully the factors that influence the evolution of plant–animal interactions.

Some problems require the invention of unique experimental techniques—for example, Thompson and Willson's (1978) invention of artificial fruit displays to examine the effect of disturbance on seed dispersal, Sork's (1984) use of metal-tagged acorns to investigate postdispersal seed fate, Schupp's (1988) innovation of seeds glued to fishing line to determine the fate of seeds secondarily dispersed by mammals, and Brown's (1988) development of giving up density of food in artificial foraging patches to investigate foraging, predation, and missed opportunity costs of foragers. However, invention is not essential to reveal new relationships. We feel that the experimental techniques we describe are revealing, but not because the techniques were not used previously. For example, Whelan and Willson's experiments using artificial displays adapted techniques (behavioral titrations) that were developed previously in enclosed arenas (Moermond and Denslow 1983, Levey et al. 1984). But by conducting the experiments in the wild, choices were made within the context that selection of fruit actually occurs. Similarly, numerous artificial defoliation experiments preceded that of Marquis (1984), particularly in agricultural systems. The insight gained from the experiment with *Piper* derived again from conducting the experiment in the wild, but also through careful natural history conducted prior to the experiment to establish the parameters of the experiment. Finally, the technique of excluding birds from trees via cages in our experiment on tri-trophic interactions had been used previously in a number of systems. The revelation of that experiment derived from conducting the experiment from a phytocentric, rather than a zoocentric, perspective. Our most important message is that experiments can provide insights that cannot be gleaned by other approaches. Much can be learned by knowing the range of experimental techniques that have already been used, as well as their inherent limitations, and by taking advantage of opportunities and ideas that call for new experimental techniques. Careful and creative statistical analysis can reveal significant results which are not predicted or expected, in addition to those predicted. Experiments have clearly contributed to the maturation of the field of plant–animal interactions, and we believe they will continue to do so.

ACKNOWLEDGMENTS   Research described was supported by a National Geographic grant, National Science Foundation grants BSR-86–00207 and DEB-81–10197, and a University of Missouri research grant to RJM, the Emily Rodgers Davis Endowment to the Morton Arboretum, and a Whitehall grant to M. Willson. We thank J. Bernardo, M. Bowles, C. Dunn, R. Flakne, C. Hochwender, J. Lill, R. Medina, K. Mothershead, J. Le Corff, G. Polis, K. Stowe, M. Willson, and E. Wold for their comments and cheerful discussion.

Literature Cited

Atlegrim, O. 1989. Exclusion of birds from bilberry stands: impact on insect larval density and damage to bilberry. Oecologia 79:136–139.

Baker, H. G. 1970. Evolution in the tropics. Biotropica 2:101–111.

Baldwin, I. T. 1990. Herbivory simulations in ecological research. Trends in Ecology and Evolution 5:91–93.

Belsky, A. J. 1986. Does herbivory benefit plants? A review of the evidence. American Naturalist 127:870–892.

Berenbaum, M. R., and A. R. Zangerl. 1992. Genetics of physiological and behavioral resistance to host furanocoumarins in the parsnip webworm. Evolution 46:1373–1384.

Bergelson, J., and M. J. Crawley. 1992. Herbivory and *Ipomopsis aggregata*: the disadvantages of being eaten. American Naturalist 139:870–882.

Bergelson, J., T. Juenger, and M. J. Crawley. 1996. Regrowth following herbivore attack in *Ipomopsis aggregata*: compensation but not overcompensation. American Naturalist 148:744–755.

Brown, J. S. 1988. Patch use as an indicator of habitat preference, predation risk, and competition. Behavioral Ecology and Sociobiology 22:37–48.

Coleman, J. S., and A. S. Leonard. 1995. Why it matters where on a leaf a folivore feeds. Oecologia 101:324–328.

Cushman, J. H., and A. J. Beattie. 1991. Mutualisms: assessing the benefits to hosts and visitors. Trends in Ecology and Evolution 6:193–195.

Cushman, J. H., V. K. Rashbrook, and A. J. Beattie. 1994. Assessing benefits to both participants in a lycaenid-ant association. Ecology 75:1031–1041.

Denslow, J. S. 1987. Fruit removal rates from aggregated and isolated bushes of the red elderberry, *Sambucus pubens*. Canadian Journal of Botany 65:1229–1235.

Dial, R., and J. Roughgarden. 1995. Experimental removal of insectivores from rain forest canopy: direct and indirect effects. Ecology 76:1821–1834.

Doutt, R. L. 1960. Natural enemies and insect speciation. Pan-Pacific Entomologist 36:1–14.

Ehrlich, P. R., and P. H. Raven. 1964. Butterflies and plants: a study in coevolution. Evolution 18:586–608.

Gilbert, L. E. 1980. Food web organization and the conservation of Neotropical diversity. Pages 11–33 in M. Soule and B. H. Wilcox (eds.), Conservation Biology. Sinauer, Sunderland, Massachusetts.

Gomez, J. M., and R. Zamora. 1994. Top-down effects in a tritrophic system: parasitoids enhance plant fitness. Ecology 75:1023–1030.

Hagan, J. M., III, and D. W. Johnston (eds.). 1992. Ecology and Conservation of Neotropical Migrant Landbirds. Smithsonian Institution Press, Washington, D.C.

Hairston, N. G., F. E. Smith, and L. B. Slobodkin. 1960. Community structure, population control, and competition. American Naturalist 94:421–425.

Herrera, C. M. 1982a. Seasonal variation in the quality of fruits and diffuse coevolution between plants and avian dispersers. Ecology 63:773–785.

———. 1982b. Some comments on Stiles' paper on temperate bird-disseminated fruits. American Naturalist 120:819–822.

Holmes, R. T., and J. C. Schultz. 1988. Food availability for forest birds: effects of prey distribution and abundance on bird foraging. Canadian Journal of Zoology 66:720–728.

Holmes, R. T., and F. W. Sturges. 1975. Avian community dynamics and energetics in a northern hardwoods ecosystem. Journal of Animal Ecology 44:175–200.

Holmes, R. T., J. C. Schultz, and P. Nothnagle. 1979. Bird predation on forest insects: an exclosure experiment. Science 206:462–463.

Howe, H. F. 1977. Bird activity and seed dispersal of a tropical wet forest tree. Ecology 58: 539–550.

Howe, H. F., E. W. Schupp, and L. C. Westley. 1985. Early consequences of seed dispersal for a Neotropical tree (*Virola surinamensis*). Ecology 66:781–791.

Jermy, T. 1984. Evolution of insect-host plant relationships. American Naturalist 124:609–630.

Johnson, R. A., M. F. Willson, J. N. Thompson, and R. I. Bertin. 1985. Nutritional values of wild fruits and consumption by migrant frugivorous birds. Ecology 66:819–827.

Krischik, V. A., and R. F. Denno. 1983. Individual, population, and geographic patterns in plant defense. Pages 463–512 in R. F. Denno and M. S. McClure (eds.), Variable Plants and Herbivores in Natural and Managed Systems. Academic Press, New York.

Kulman, H. 1971. Effects of insect defoliation on growth and mortality of trees. Annual Review of Entomology 16:289–324.

Lande, R., and S. J. Arnold. 1983. The measurement of selection on correlated characters. Evolution 37:1210–1226.

Levey, D. J., T. C. Moermond, and J. S. Denslow. 1984. Fruit choice in neotropical birds: the effect of distance between fruits on preference patterns. Ecology 65:844–850.

Levin, D. A. 1976a. Alkaloid-bearing plants: an ecogeographic perspective. American Naturalist 110:261–284.

———. 1976b. The chemical defenses of plants to pathogens and herbivores. Annual Review of Ecology and Systematics 7:121–159.

Lewis, A. 1984. Plant quality and grasshopper feeding: effects of sunflower condition on preference and performance in *Melanoplus differentialis*. Ecology 65:836–843.

MacArthur, R. 1969. Patterns of communities in the tropics. Biological Journal of the Linnean Society 1:19–30.

Marquis, R. J. 1984. Leaf herbivores decrease fitness of a tropical plant. Science 226:537–539.

———. 1987. Variation en la herbivoria foliar y su importancia selectiva en *Piper arieianum* (Piperaceae). Revista Biologia Tropical 35 (Suppl. 1): 133–149.

———. 1988. Phenological variation in the Neotropical understory shrub *Piper arieianum*: causes and consequences. Ecology 69:1552–1565.

———. 1992a. A bite is a bite is a bite? Constraints on response to folivory in *Piper arieianum* (Piperaceae). Ecology 73:143–152.

———. 1992b. Selective impact of herbivores. Pages 301–325 in R. S. Fritz and E. L. Simms (eds.), Plant Resistance to Herbivores and Pathogens: Ecology, Evolution, and Genetics. University of Chicago Press, Chicago, Illinois.

Marquis, R. J., and C. J. Whelan. 1994. Insectivorous birds increase growth of white oak through consumption of leaf-chewing insects. Ecology 75:2007–2014.

Maschinski, J., and T. G. Whitham. 1989. The continuum of plant responses to herbivory: the influence of plant association, nutrient availability and timing. American Naturalist 134:1–19.

McKey, D. 1975. The ecology of coevolved seed dispersal systems. Pages 159–191 in L. E. Gilbert and P. H. Raven (eds.), Coevolution of Animals and Plants. University of Texas Press, Austin.

McLaren, B. E., and R. O. Peterson. 1994. Wolves, moose, and tree rings on Isle Royale. Science 266:1555–1558.

Moermond, T. C., and J. S. Denslow. 1983. Fruit choice in Neotropical birds: effects of fruit type and accessibility on selectivity. Journal of Animal Ecology 52:407–420.

Nafus, D. 1991. Biological control of *Penicillaria jocosatrix* (Lepidoptera: Noctuidae) on mango on Guam with notes on the biology of its parasitoids. Environmental Entomology 20:1725–1731.

Owen, D. F., and R. G. Wiegert. 1981. Mutualism between grasses and grazers: an evolutionary hypothesis. Oikos 36:376–378.

Paige, K. 1994. Herbivory and *Ipomopsis aggregata*: differences in response, differences in experimental protocol: a reply to Bergelson and Crawley. American Naturalist 143: 739–749.

Paine, R. T. 1980. Food webs: linkage, interaction strength and community infrastructure. Journal of Animal Ecology 49:667–685.

Peck, K. M. 1989. Tree species preferences shown by foraging birds in forest plantations in northern England. Biological Conservation 48:41–57.

Power, M. E. 1990. Effects of fish in river food webs. Science 250:811–814.

Power, M. E., W. J. Matthews, and A. J. Stewart. 1985. Grazing minnows, piscivorous bass, and stream algae: dynamics of a strong interaction. Ecology 66:1448–1465.

Price, P. W. 1991. The plant vigor hypothesis and herbivore attack. Oikos 62:244–251.

Robinson, S. K., and R. T. Holmes. 1984. Effects of plant species and foliage structure on foraging behaviour of forest birds. Auk 101:672–684.

Root, R. B., and N. Cappuccino. 1992. Patterns in population change and the organization of the insect community associated with goldenrod. Ecological Monographs 62:393–420.

Schemske, D. W., and C. C. Horvitz. 1984. Variation among floral visitors in pollination ability: a precondition for mutualism specialization. Science 225:519–521.

Schupp, E. W. 1988. Factors affecting post-dispersal seed survival in a tropical forest. Oecologia 76:525–530.

Sinclair, A. R. E. 1979. Dynamics of the Serengeti ecosystem. Pages 1–30 in A. R. E. Sinclair and M. Norton-Griffiths (eds.), Serengeti, Dynamics of an Ecosystem. University of Chicago Press, Chicago, Illinois.

Snow, D. W. 1971. Evolutionary aspects of fruit-eating by birds. Ibis 113:194–202.

Snow, B. K., and D. W. Snow. 1988. Birds and Berries. Poyser, Carlton, England.

Sork, V. L. 1984. Examination of seed dispersal and survival in red oak, *Quercus rubra* (Fagaceae), using metal-tagged acorns. Ecology 65:1020–1022.

Spiller, D. A., and T. W. Schoener. 1990. A terrestrial field experiment showing the impact of eliminating top predators on foliage damage. Nature 347:469–472.

Stamp, N. E., and M. D. Bowers. 1996. Consequences for plantain chemistry and growth when herbivores are attacked by predators. Ecology 77:535–549.

Stiles, E. W. 1980. Patterns of fruit presentation and seed dispersal in bird-disseminated woody plants in the eastern deciduous forest. American Naturalist 116:670–688.

———. 1993. The influence of pulp lipids on fruit choice by birds. Pages 226–236 in T. H. Fleming and A. Estrada (eds.), Frugivory and Seed Dispersal: Ecological and Evolutionary Aspects. Kluwer Academic Publishers, Dordrecht.

Strong, D. R. 1992. Are trophic cascades all wet? Differentiation and donor-control in speciose ecosystems. Ecology 73:747–754.

Terborgh, J. 1989. Where Have All the Birds Gone? Princeton University Press, Princeton, New Jersey.

Thompson, J. N., and M. F. Willson. 1978. Disturbance and the dispersal of fleshy fruits. Science 200:1161–1163.

Wheelwright, N. T. 1983. Fruits and the ecology of resplendent quetzals. Auk 100:286–301.

Whelan, C. J., and M. F. Willson. 1994. Fruit choice in migrating North American birds: field and aviary experiments. Oikos 71:137–151.

Whitham, T. G., J. Maschinski, K. C. Larson, and K. N. Paige. 1991. Plant responses to herbivory: the continuum from negative to positive and underlying physiological mechanisms. Pages 227–256 in P. W. Price, T. M. Lewinsohn, G. W. Fernandes, and W. W. Benson (eds.), Plant-Animal Interactions: Evolutionary Ecology in Tropical and Temperate Regions. Wiley, New York.

Wiens, J. A. 1973. Pattern and process in grassland bird communities. Ecological Monographs 43:237–270.

———. 1989a. The Ecology of Bird Communities: Vol. 1. Foundations and Patterns. Cambridge University Press, Cambridge.

———. 1989b. The Ecology of Bird Communities: Vol. 2. Processes and Variations. Cambridge University Press, Cambridge.

Wiens, J. A., R. G. Cates, J. T. Rottenberry, N. Cobb, B. Van Horne, and R. A. Redak. 1991. Arthropod dynamics on sagebrush (*Artemisia tridentata*): effects of plant chemistry and avian predation. Ecological Monographs 61:299–321.

Willson, M. F. 1979. Sexual selection in plants. American Naturalist 113:777–790.

Willson, M. F., and J. N. Thompson. 1982. Phenology and ecology of color in bird-dispersed fruits, or why some fruits are red when they are ''green.'' Canadian Journal of Botany 60:701–713.

Willson, M. F., and C. J. Whelan. 1989. Ultraviolet reflectance of fruits of vertebrate-dispersed plants. Oikos 55:341–348.

———. 1993. Variation of dispersal phenology in a bird-dispersed shrub, *Cornus drummondii*. Ecology 63:151–172.

Willson, M. F., A. K. Irvine, and N. G. Walsh. 1989. Vertebrate dispersal syndromes in some Australian and New Zealand plant communities, with geographic comparisons. Biotropica 21:133–147.

22

Experimental Approaches to the Study of Evolution

JOSEPH TRAVIS & DAVID N. REZNICK

The study of evolution, a historical phenomenon, can be frustrating; historical subjects are seldom amenable to experimentation, and inferences about them depend on arguments of plausibility and likelihood. Further, evolutionary change is usually slow relative to the rate at which human patience is exhausted, so the phenomena under scrutiny take a long time to unfold before they become recognizable history. Nonetheless, the power of the experimental method can be used to study, in real time, some of what has unfolded and why. Our goal in this chapter is to focus on such experimental approaches; we argue that these methods, when combined with historical or comparative approaches, allow direct tests of certain kinds of hypotheses about evolution. Specifically, we examine how experimental approaches can be used to test hypotheses about the action of selection in natural populations. First, we distinguish a variety of comparative and historical methods, discuss how experimental approaches ought to dovetail with those methods, and illustrate our points with a specific example. We then place these experimental approaches in a larger context and categorize the variety of experimental approaches by their goals and methods. These experiments and methods are designed to measure the extent of genetic variation for the traits under scrutiny and to elucidate the action of natural selection. For clarity, we focus on the action of selection on morphological, physiological, or life history traits.

The Relationship of Comparative and Experimental Methods

Macro-and Microevolutionary Approaches to the Study of Evolution

One productive approach uses phylogenetic information to examine patterns of trait evolution and to construct hypotheses for those patterns. The issues examined include how often putatively adaptive changes have arisen (Dunham et al. 1988), whether particular trait values arise in association with particular environments (Block et al. 1993, McPeek 1995), whether evolution can produce reversals in complex life histories (Siddall et al. 1993), whether associations between traits in extant taxa reflect some form of correlated evolution (Sillen-Tullberg 1988, Donoghue 1989, Losos 1990, Wickman 1992), whether groups of interacting species have coevolved with each other (Mitter and Brooks 1983), and whether certain trait values have evolved more rapidly than neutral evolution would predict (McPeek 1995). The specific methods used to test these hypotheses depend on the nature of the traits under examination and the amount and quality of phylogenetic information available (Felsenstein 1985, Maddison 1990, Miles and Dunham 1993, Martins 1996).

Macroevolutionary studies make two contributions. First, they reveal what all evolutionary biologists wish to know—which historical phenomena have occurred. Second, they often narrow the range of viable hypotheses about the causes of those phenomena (e.g., Sillen-Tullberg 1988 and modifications by Maddison 1990). Macroevolutionary studies have two obvious limitations. First, in many cases they cannot distinguish between competing ecological hypotheses about the nature of selection on suites of traits and the constraints that have channeled the path of adaptive evolution. Second, in most cases they cannot evaluate how much genetic change underlies the amount of phenotypic change that has been measured.

These limitations can be overcome by microevolutionary studies, which we define as studies in conspecific populations. Indeed, a program that combines experimental studies at the microevolutionary scale with comparative approaches at that scale can address all of the macroevolutionary questions listed here. Thoughtfully executed microevolutionary studies thus offer a means of overcoming the limitations of macroevolutionary studies by addressing the same questions at a different, more immediate scale. Singer's study of the evolution of diet in *Euphydryas editha* (Singer 1994, Radtkey and Singer 1995) illustrates this thesis quite elegantly.

To be sure, some phenomena cannot be approached at the microevolutionary scale. Not all interesting patterns of trait variation show sufficient intraspecific variation, and many cases of intraspecific variation are not amenable to experimentation. Moreover, issues of genetics aside, the question asked at the microevolutionary scale is not necessarily the same question addressed at the macroevolutionary scale. In cases like industrial melanism in Lepidoptera, antibiotic resistance in bacteria, or morphological changes during the spread of introduced species, significant evolutionary change has been observed directly or can be inferred from historical records, and the microevolutionary questions of "how" and "why" that surround such change apply directly to the larger pattern. In most cases, however, the student of microevolution examines the end point of a recent history that has not been observed directly, such as persistent differentiation among conspecific populations. In these cases, the questions that can be asked are really variants of "what maintains the present pattern" and not "what caused the present pattern." This distinction is important; whether the answers to one question are the same as those to the other is itself an interesting issue, but one that is transcended by the value of the microevolutionary approach.

A Paradigm for Microevolutionary Studies

The virtues and limitations of microevolutionary studies have been recognized at least since Weldon (1893). The basic tenet of such studies is that evolution is a contemporary phenomenon that can be studied experimentally. We prefer the paradigm adopted in the 1930s by the ecological geneticists (Ford 1971). Ford's original proposal was to study evolution by natural selection through focusing on differences among phenotypes within a population and differences among populations within a species. The classic starting point required a discrete polymorphism such as shell banding in snails or wing color in moths. This choice made it possible to evaluate correlations between the frequency of a trait in a population and features of the environment that might offer clues about the functional significance of the trait. The resulting hypothesis is that differences in one or more environmental factors selected for the trait differences observed among populations. Given such clues, it is possible to evaluate the relative fitness of phenotypes under alternative conditions. With some methodological changes, the paradigm applies readily to continuously distributed trait variation (Reznick and Travis 1996). Kettlewell's (1955, 1973) study of industrial melanism is the classic application of this method, because it began with a historical trend and comparative pattern, then progressed to experiments that addressed hypotheses suggested by that pattern. It is this thoughtful combination of natural history, quantitative observation, and experiment that we endorse as a paradigm for investigating contemporary patterns.

The Comparative Method in Ecological Genetics: A Case Study

One of us (DR) applied this approach to natural populations of guppies (*Poecilia reticulata*) from the island of Trinidad. Previous investigators had found that these populations co-occur with different ensembles of predators that can be classified as either "high"- or "low"-predation communities (see Reznick and Endler 1982 for references). High-predation communities are generally found in the lower reaches of streams and are characterized by the presence of predators like the pike cichlid *Crenicichla alta* and fish in the characin family. These species are often excluded from upstream habitats by rapids or waterfalls, yielding communities that contain guppies plus fewer, smaller species of predators, such as the killifish *Rivulus hartii. Rivulus* is also less likely to prey on guppies. Such breaks in the distribution of guppy predators occur at different places in different streams, so there is considerable heterogeneity in the nature of the physical environment that is classified as high or low predation and considerable overlap among the environments of each type of community. These discontinuities in the distribution of predators also mean that the contrast between high and low predation is often present within a stream between adjacent or nearby guppy populations.

Before initiation of this work, three investigators (Gadgil and Bossert 1970, Law 1979, Michod 1979) considered the potential consequences of changes in age-specific mortality for the evolution of life history patterns. All three predicted that increased adult mortality rates would select for lower age at maturity and increased reproductive effort. Subsequent theory (e.g., Charlesworth 1994) often supported this prediction, but the fabric of life history theory has also become far more complex and now admits alternative predictions (e.g., those of Orzack and Tuljapurkar 1989). Nevertheless, existing theory provided a basis for making predictions about how guppy life histories would evolve in response to differences in predation.

We first made a comparative study to estimate the correlation between guppy life history phenotypes and predator ensembles in seven high-predation and five low-predation localities (Reznick and Endler 1982). Guppies were collected and preserved in the field. Variables such as size-specific fecundity, offspring size, and reproductive allotment (the percentage of total body weight that consists of developing embryos) were scored in the laboratory to characterize the life history. Reproductive allotment estimates the investment in each litter. When combined with laboratory estimates of the frequency of reproduction, it yields an index of reproductive effort (the comparative rate of investment of resources in reproduction). We also evaluated two variables that serve as indices of the age at maturity: average size of mature males and minimum size of reproducing females. Because males stop growing at maturity, the average size in a collection is a direct estimate of the average size at maturity. Because females continue to grow throughout their lives, minimum size is an index of the size at which they begin to reproduce. We assumed that growth rates were comparable at the different types of localities.

We found that guppies from high-predation localities have higher reproductive allotments (devote more resources to each brood) and also reproduce more frequently (Reznick and Endler 1982). The combination of more frequent reproduction with larger investments in each brood suggests that these fish had higher reproductive effort than guppies from low-predation localities. Male guppies from high-predation localities are also consistently smaller than their low-predation counterparts. Females from high-predation localities begin reproduction at a smaller size than their low-predation counterparts. These two observations suggest that guppies from high-predation localities reach maturity at an earlier age. Finally, guppies from high-predation localities produced more offspring per brood, and the size of individual offspring was smaller. These initial observations suggested that guppy life histories had evolved in response to differences in predation in accord with the predictions of life history theory.

We have since extended these observations to more localities with communities similar to those initially considered (Reznick 1989) and found that these patterns are robust. At the same time, we demonstrated that they are not influenced by year-to-year variation in the environment or by wet and dry seasons. We have also extended these comparisons to a new series of localities in Trinidad, Tobago, and Venezuela (Reznick et al. 1996b; Reznick, unpublished data). These communities are similar to the original series in Trinidad in having a contrast between high and low predation, but they contain a very different suite of predators. *Crenicichla* and all of the other predators associated with the south slope of the Northern Range of Trinidad are absent, replaced by species of gobies and mullets derived from a marine environment. Nevertheless, the life history patterns are the same as on the south slope and thereby strengthen the idea that it is mortality rates, rather than other variables, that matter.

Genetics

Genetics of Trait Variation

Almost every microevolutionary study involves investigation of how much of the phenotypic variation under scrutiny is genetically based. An ecologically based study of the agents of phenotypic selection need not include an estimate of genetic variation in the target traits to be valuable, but the implications of such studies for evolutionary issues must be considered cautiously (e.g., Núñez-Farfán and Dirzo 1994).

Unfortunately, genetic studies of trait variation in natural populations are more easily designed than executed. Several issues must be considered before a specific genetic study is designed, from how much genetic information is desired to whether other sources of variation can mimic genetic effects. We have discussed many of these issues elsewhere (Reznick and Travis 1996); here we focus on two issues that surround the use of experimental genetic studies.

In many cases, observations on natural populations have served as substitutes for experiments. For example, observations of mating and offspring production can yield pedigrees from which genetic parameters can be estimated (Smith and Zach 1979, van Noordwijk et al. 1980). These parameters will be accurate insofar as the resemblance among relatives is not caused by nongenetic effects or by genetic effects with a lag between generations. When such sources of resemblance among relatives are plausible, an experimental approach is necessary to eliminate those sources or estimate their effects. In natural populations of birds, nestling transplants and cross-fostering have been employed to estimate the effects of variation in parental care or nest environment on the resemblance among relatives (Smith and Dhondt 1980, Dhondt 1982). The problem is particularly vexing in plants; in many cases, elaborate experiments are necessary to estimate the impact of subtle sources of resemblance (Mazer and Gorchov 1996). Such approaches are often the only ones available because of the difficulties of mating and rearing many organisms.

An experimental approach to genetic inference usually involves controlled crosses among parents and rearing of offspring in an environment that allows full trait expression. Each component requires careful consideration in experimental design. If one is interested only in establishing that variation in a trait has some genetic basis, then simple mating designs and small numbers of family groups may suffice (Travis and Trexler 1984). If one is interested in estimating quantitative genetic parameters, then experimental crosses must capture the distribution of extant genetic variation through random sampling of a sufficient number of dams and sires and matings that are at random with respect to the trait values.

The choice of ontogenetic environment is critical (see Reznick and Travis 1996 for a fuller discussion). In the ideal world, the progeny of controlled crosses would be raised in the natural environments under scrutiny. The usual impediment to this ideal is that experimental progeny can die in the natural environment, and mortality rates may either be too high for an acceptable sample size or cause biases in parameter estimation if mortality is not random (Bennington and McGraw 1995). The usual solution is to raise progeny in more protected circumstances such as the laboratory, experimental garden, or greenhouse, where, unfortunately, a variety of factors may combine to alter the levels of phenotypic and genetic variance observed. For example, Service and Rose (1985) describe the effects of novel laboratory environments on trait expression in *Drosophila melanogaster*, and Travis et al. (1989) illustrate how the laboratory environment can induce rates of postmaturation growth in adult male sailfin mollies (*Poecilia latipinna*) that are never realized in nature.

Two lines of evidence can be marshaled to support the validity of a specific "protected rearing." First, if the distribution of phenotypic variation among experimental progeny is comparable to that seen in natural populations, untoward phenotypic expression, at least, is ruled out (e.g., Travis et al. 1987). No inference about untoward genetic variance is possible with this comparison.

Second, subsets of experimental progeny can be reared under protected and natural conditions so that the distributions of genetic and phenotypic variance can be compared where possible (see Groeters and Dingle 1996 for a fuller discussion). For example, Dudash (1990) raised plants from selfed matings, crossings to near neighbors, and crossings to far neighbors under greenhouse, garden, and natural conditions; she found that the mating-system effects on fitness were underestimated under the more protected conditions. Analogously, one of us (Travis 1983) raised individuals from the same seven full-sib families of tadpoles under two laboratory conditions and in experimental enclosures in the field; the ratio of among-family variance to within-family variance was one or two orders of magnitude higher in the field-raised animals, which suggests that protected conditions cause genetic variance to be underestimated in these animals (JT has repeated this experiment three times with comparable results). Whether protected rearings always produce conservative estimates of genetic variation is an unresolved empirical issue.

Genetic Basis of Interpopulation Variation: A Case Study with Guppies

We evaluated the genetic basis of interpopulation differences in guppy life histories in the second-generation laboratory-reared descendants from two high-predation and two low-predation localities (Reznick 1982b). We collected adult females from the wild, then isolated them in the laboratory; the females produced a series of broods that were maintained as separate pedigrees in subsequent experiments. After rearing the lab-born descendants through two generations in a common environment, we compared the life histories of individuals reared on strictly controlled levels of food. This protocol eliminates the possible influences of the environment on the life history phenotype. We assume that differences among population means that persist after two generations in a common environment have a genetic basis. We can thus quantify the extent to which genetic differences underlay the phenotypic differences we observed in natural populations; in the case of offspring size, we also used hybridization studies (Reznick 1981, 1982a). Guppies from high-predation sites matured at a lower age and had higher levels of reproductive effort (Reznick 1982b). They achieved higher reproductive efforts by initiating reproduction earlier, devoting more resources to each brood, and reproducing more frequently. As in the field, they also produced more offspring per brood, and each offspring was smaller than those produced by their counterparts from low-predation localities.

Because they have been extended to 10 new localities, including the fish assemblages found on the north slope of the Northern Range, Tobago, and parts of Venezuela (Reznick and Bryga 1996, unpublished), these genetic comparisons appear to be as widespread as implied by the comparative studies. It thus appears that the life history phenotypes, in this case, are a good index of genetic differences, in spite of the potential confounding effects of the environment. The laboratory studies also allow us to quantify more variables characterizing the life history, including direct estimates of age at maturity and reproductive effort. The disadvantage is that they are very labor-intensive; each study takes at least 9 months to complete, and they can only be applied to a subset of the localities covered by the initial comparative study. This trade-off between the extent of geographical study by the comparative method and the added detail but narrower scope of evaluating genetic differences is likely to apply to many systems. The combination of the two allows one to exploit the virtues of both.

Methods of Evaluating Natural Selection

Finding a genetic difference between populations that is correlated with some important feature of the environment still does not demonstrate a cause-and-effect relationship between the environmental effect and the trait. Establishing such a relationship requires critical evaluation of the presumed mechanism of natural selection. The method will depend entirely on the nature of the study system. In this section, we outline four classes of approaches to this question.

Mark-Recapture

In the case of guppies, the hypothesis derived from life history theory was based on the relative mortality rates of guppies in the two types of environments. Differences in the types of predators that cooccur with guppies presumably caused differences in age-specific mortality rates. The next step was to evaluate mortality rates directly by mark-recapture studies. Through successive recaptures or observations of marked individuals or cohorts, one can evaluate dependent variables associated with fitness. If such methods are extended for multiple generations, it is also possible to estimate lifetime reproductive success (Clutton-Brock 1988). Although mark-recapture studies are not experiments, we include them here for two reasons. First, they provide critical information on the performance of individuals. Second, as we will illustrate with guppies, when combined with experimental methods, mark-recapture studies can be an integral part of a convincing argument for adaptation.

When this method was applied to guppies, the goal was to evaluate the relative mortality rates of guppies from high-and low-predation localities. If differences in predation caused differences in guppy mortality rates, then recapture probabilities of guppies from high-predation localities should be lower. If these predators preyed selectively on adults, then the difference in mortality should be more dramatic for the adult age classes.

To address these hypotheses, we performed mark-recapture studies on populations in streams with discrete pool-riffle structure. These streams are divided into small pools with slow currents. Pools are bounded up- and downstream by steeper, narrower gradients called riffles. We found initially that guppies tend to congregate in pools and rarely cross riffles to neighboring pools. We also found that it was possible to collect every guppy in a pool very quickly. We were thus able to perform a series of mark-recapture assays in which we collected the entire population in a pool, marked individuals to indicate their initial size class, released them, then recollected them approximately 2 weeks later. When we recollected the fish, we also collected from pools up- and downstream from the release site to look for émigrés and re-collected each site on subsequent days to confirm that all marked individuals had been recaught. If the assumption is satisfied that virtually all marked fish that remain alive are recaught, then the probability of recapture is an estimate of survival. Indeed, recapture probabilities were lower at high-predation sites (Reznick et al. 1996a).

Two complications arise in interpreting such differential mortality rates as the sole direct mechanism that has selected for life history evolution in these guppies. First, although mortality rates are higher in high-predation localities, these higher rates apply equally to all size classes, rather than being concentrated on adults. Some treatments of life history theory predict that no differences in life histories will evolve under such circumstances (Law 1979, Michod 1979), whereas others predict the evolution of patterns similar to those we have observed (Kozlowski and Uchmanski 1987, Charlesworth 1994). Second, there is some evidence that these differences in predation have also caused differences in the pattern of habitat use, population density, and resource availability. When density dependence is present, the way the life history evolves will depend on how density effects are manifested in different age classes (Charlesworth 1994; see following discussion).

The more usual application of mark-recapture methods to natural populations does not involve the assumption that all individuals can be recollected (see LeBreton et al. 1990). Instead, all individuals are uniquely marked and populations are resampled repeatedly. The data consist of the pattern of recapture of individuals across sampling surveys. Maximum-likelihood techniques yield the most likely mortality pattern that is consistent with the pattern of recaptures over time.

Population Manipulations

Another method for evaluating mechanisms of natural selection involves short-term manipulations of natural populations to evaluate the response of phenotypes either to key features of the environment or to sets of experimental conditions. For example, Jones (1996) introduced experimentally produced plants with two flower-color morphs into three natural populations that exhibited consistent differences in the relative frequencies of each morph. The resultant fitness effects that she estimated were those of each color morph independent of background genotype; the introduction into three populations allowed a specific test of frequency-dependent fitness through differential fertility.

The classic reciprocal transplant experiment (e.g., Berven 1982) is one of many variations of this approach, and it can be quite powerful when integrated with comparative data, genetic analysis, and theory. Dudley (1996a,b) examined plant populations that differed in physiological traits apparently related to their soil-moisture regimes. Through comparative observations, reciprocal transplants, estimating genetic parameters, and measuring selection gradients (sensu Lande and Arnold 1983), she accumulated convincing evidence that the differentiation was adaptive. Moreover, she showed that the divergent trait values of the two populations she examined in detail could be combined with genetic parameters to yield estimated selection gradients that were strikingly similar to the gradients she measured directly in her field studies. This result strongly suggests that the forces maintaining the present differentiation among these populations are the ones that produced the differentiation in the first place.

Another important population manipulation is the experimental arrangement of genetic variation to examine how selection acts on breeding or mating systems. In the best of these studies, experimental populations are set in the natural environment and the investigators measure the fitness of individuals whose genotypes have been manipulated through controlled crosses to provide predefined levels of inbreeding and outbreeding (Waser and Price 1994) or sexual and asexual reproduction (Strauss and Karban 1994). These experiments can be viewed legitimately as population manipulations because in each one the investigator is merely rearranging extant genetic variation from a well-defined population. Antonovics and his colleagues have reported a remarkable series of field experiments of this type that manipulated the genotypes of individuals and their neighbors to test a variety of hypotheses about the adaptive significance of sexual reproduction (Schmitt and Antonovics 1986 and prior work cited therein).

Experiments in Artificial Environments

Experimental "mesocosms," or artificial environments, can be used in microevolutionary studies of selection. In this context, a mesocosm or artificial environment can range from a series of artificial streams to a greenhouse, and it is employed for three types of studies. First, an investigator uses a mesocosm to manipulate a specific environmental factor. For example, Norman et al. (1995) placed individual plants from inbred and outcrossed matings in artificial environments that differed in water availability to examine the effects of inbreeding on physiological traits and how those effects influenced fitness under alternative environmental conditions. Second, in some cases experimental manipulations of natural populations are not feasible and some type of artificial environment is the only reasonable recourse (e.g., Semlitsch and Wilbur 1989). Third, an investigator who combines a genetic manipulation with measurements of fitness may resort to a mesocosm as an alternative to a field study, where the risk of high mortality might compromise the desired genetic inference (e.g., Willis 1993).

Mesocosms of various types have been used in both short-term (one generation or less: Campbell et al. 1994) and long-term (multiple generations: Endler 1980, Spitze 1991, Scribner and Avise 1994) studies. Their value for short-term studies hinges on how short-term experiments are combined and integrated into a realistic empirical description of population processes, a problem that we discuss in more detail in the next section. In either short-term or long-term situations, the success of the mesocosm approach will depend upon whether the experimental design can re-create realistic distributions of the germane features of the natural environment. The extent to which realistic distributions can be maintained over time determines whether a mesocosm approach is useful for a long-term study and depends in turn on a variety of factors such as whether the mesocosms can support realistic densities; whether, for animal studies, they allow normal movement patterns and the development of a normal social system; and whether normal values for reproductive traits will be realized.

Enclosure Studies

Long a favorite method of ecologists, enclosure studies place organisms within enclosures, which themselves are placed in natural habitat. In some cases, the enclosures are designed to keep the study animals where they can be found; in others, they are designed to exclude a variety of unwanted visitors. In all cases, the enclosure is a fundamental, replicable experimental unit that allows the study of short-term processes in the natural environment. The value of these studies is that, like some mesocosm studies, they allow experimental study of selective processes that would not be possible in natural populations. The disadvantage of enclosure studies, beside the obvious one of their short-term nature, is that they illustrate only what *can* happen, not what does happen in natural populations; their interpretability depends heavily on their design and execution.

We illustrate the important considerations in enclosure studies by reviewing two different experimental studies. Travis et al. (1985a) manipulated tadpole density, familial origin, and predator composition in a large number of enclosures to answer two questions. First, could we verify prior results (Travis 1983) that more slowly growing tadpoles suffer higher cumulative risks of mortality from size-limited predators? Second, is the selective effect of combining different types of predators additive or synergistic?

To answer those questions we designed enclosures that were suitable for the species of tadpole we used, *Hyla gratiosa*. These animals migrate vertically in response to day and night, so the enclosures needed to hold sufficient water depth to allow vertical migration. Tadpoles also adjust their microhabitat use between open water and cover in response to predator presence, so each enclosure held vegetation, litter, and debris for cover. The two predators we employed, dragonfly naiads and larval salamanders, use different foraging strategies and use different microhabitats. Tadpoles could escape the salamanders, which forage on pond bottoms and in open water, by moving into cover provided by vegetation and debris, but this was the habitat in which the insects resided. Thus, when only one type of predator was present, the tadpoles had a spatial refuge from predation; when both predators were present, there was no easy escape. Therefore, if ever two predators exerted synergistic selective effects, these were the best candidates; if we could not find the phenomenon in this system, we were unlikely to see it in other combinations.

Two additional factors required careful consideration. We used densities of prey and predators that were at the lower end of the documented natural range (reviewed by Warner et al. 1991, Gascon and Travis 1992). This consideration is especially critical in examination of selective predation in an enclosure study; at high densities in a finite amount of space, predator–prey contact rates will be unnaturally high, and the overall effect of predation might be exaggerated. We also matched the size distribution of predators to the natural distribution as best we could, given practical limitation. Although dragonfly predation is only weakly related to dragonfly size (Travis et al. 1985b), salamander predation is strongly dependent on predator body size. Cumulative predation rates could be altered substantially by changes in the sizes of predators used in the enclosures.

The design considerations and certain features of the results that were divorced from the major goal of the experiment gave us confidence that we were studying a realistic process. The levels of tadpole mortality we documented in this study were below many estimates from natural populations and at the lower end of some other experimental estimates, so, if anything, we minimized the probability of overpredation. We found the same statistical effects of varying population density on tadpole life history traits that almost every other worker had found. The insect and salamander predators metamorphosed from the enclosures at normal sizes, suggesting they were not malnourished. The surviving tadpoles expressed a normal range of phenotypic variation in larval period and body size at metamorphosis; indeed, more slowly growing tadpoles suffered higher cumulative mortality risks, and the selective effect of the predators on growth rate was synergistic, even though overall mortality levels showed no such synergism.

In the second study, Trexler et al. (1992) examined mortality rates for adult and juvenile sailfin mollies (*Poecilia latipinna*) in the absence of predation. Our goals were to compare predation-free mortality rates of juveniles and adults in two locations, to determine whether mortality rates were size-specific, and to estimate annual variation in mortality rates. The results were used in conjunction with an enclosure-based study of wading-bird predation (Trexler et al. 1994) to examine the selective forces affecting body-size variation.

In the predator-free study, we wished to examine natural mortality patterns in two locations that represented end points of the types of habitats in which mollies are found in north Florida; an enclosure study was appropriate because fish were exposed to natural conditions in every way with the exception of predation risk. Mollies are herbivorous and graze on diatoms, bacteria, and algae in the sediment; the mesh enclosures settled well into the sediment and allowed the fish access to a normal distribution of food resources without introducing undesired effects of an altered physical environment. If predatory animals are the objects of study, special consideration must be given to the enclosure design; normal feeding can be hampered by mesh enclosures that exclude larger prey (Chambers 1984). We used small enclosures and began each experiment with four individuals per enclosure (five in one juvenile experiment); this initial density was well within the natural range, and lowered densities due to mortality did not affect life history parameters or vital rates (this was not designed as a study of density effects). The analysis of mortality rates in enclosure studies such as this one provokes several statistical considerations that we will not address here (see Trexler et al. 1992, 1994; Brodie and Janzen 1996).

Although short-term studies in either mesocosms or enclosures reveal what *can* happen in natural populations, their users usually wish to infer what *does* happen in nature. To do so, they must use some form of empirical modeling to extrapolate experimentally derived, phenotype-based vital rates into a fuller description of population processes (cf. Streifer 1974, Gurney and Nisbet 1985). A variety of approaches to this problem can be found in the literature (e.g., Ryan et al. 1992, Walters et al. 1992). For such modeling to have meaning, however, the experimental data must be derived from realistic re-creations of the distributions of germane variables so that the empirical models will be applicable to natural distributions of selective agents.

The requirement for realistic estimates of selection imposes a rigorous standard for experiments that often has subtle implications. For example, a study of selective predation might contrast mortality rates in the absence and presence of predation, but the impact of predators may be extremely high because of a high density of predators. If so, two sources of bias can be created. First, the mean effect of predators is overestimated. Second, the experiment also overestimates the *variance* in predator effect, because the variance in predator density is inflated by the single extreme difference between no predators and high predator density. This inflated variance, when used in empirical modeling, translates into a bias in evaluating the importance of selective predation for fitness differences. In general, such experimental studies of selection potentially yield biased estimates of the effects of selective agents, in this case predators. This potential source of bias has been discussed in ecological contexts by Gascon and Travis (1992) and Petraitis et al. (1996). The obvious protocol for avoiding these problems is to integrate carefully designed short-term experimental approaches with quantitative observation and comparative data from natural populations.

Selection Experiments

Types of Selection Experiments

Most readers are familiar with what might be termed *laboratory selection experiments* of two general types. In the first, the investigator uses artificial selection on a trait or traits to estimate quantitative genetic parameters, diagnose a suite of genetically covarying traits, or examine the reciprocity of correlated responses. In the second, the investigator manipulates populations and/or environmental factors to determine whether selection and response will occur over two (or more) generations as a function of experimental treatment. These include studies such as those on sexual selection among competing male genotypes (Partridge and Farquhar 1983) and the evolution of competitive ability (Joshi and Thompson 1995).

Here we highlight selection experiments on a larger scale, those performed in natural populations or mesocosms. Such experiments represent variations on two previous themes: population manipulations and long-term mesocosm studies. They can be categorized similarly to laboratory selection experiments into "artificial selection studies" and "manipulative studies of natural selective processes."

The work of Semlitsch and Wilbur (1989; see also Semlitsch et al. 1990) illustrates the first type. They exercised artificial selection on the tendency for salamander larvae to undergo paedomorphosis (gonadal maturation and reproduction), using stocks derived from two natural populations that differed in their observed rates of paedomorphosis and in the environmental factor postulated to be the selective agent for paedomorphosis (reliability of permanent water). Their experimental work was designed to investigate whether artificial selection would be differentially successful in the different stocks and allow some insight into the likely cause of the evolutionary divergence of the natural populations from which they were derived.

The second type of selection experiment, like its laboratory analogue, involves either manipulation of conditions or movement of novel populations into existing conditions at a new location. These experiments offer the opportunity to watch evolution occur in the field or mesocosm. The prevailing view of evolution as a process that requires long periods to unfold has probably inhibited us from considering this perspective in empirical studies. The truth is that we know very little about how strong natural selection can be and hence how quickly organisms can evolve. Endler's (1986) review of the subject suggests that natural selection can often be very intense. If so, then it might often be possible to study adaptation in nature from an experimental perspective.

Selection and Evolution in Introduced Guppy Populations

We (DR and colleagues) have exploited the discontinuities imposed by natural barriers on the distribution of guppies and their predators to evaluate the effects of predators more directly (Reznick and Bryga 1987, Reznick et al. 1990). In two experiments, we worked on streams with barrier waterfalls that stopped the upstream dispersal of all species of fish except *Rivulus*. Guppies were only found below the barrier waterfall in high-predation communities. We introduced these guppies above the barrier waterfall, thus moving them from a high-predation environment to a low-predation one. If predators are responsible for the evolution of the life-history differences between guppies from high- and low-predation populations, then the life histories of introduced guppies should evolve to match their new mortality patterns.

We evaluated the evolution of the life histories in the introduced populations by comparing the performance of second-generation laboratory-reared guppies from the introduction site and the population of guppies below the barrier waterfall (the control). We use the same protocol described previously to evaluate the life history "genotype."

We have found that the life histories of the introduced populations of guppies evolve in a fashion that is consistent with the life histories of guppies from low-predation environments. In one replicate of this experiment, executed on a tributary of the Aripo River, we evaluated the guppies 11 years (approximately 16 generations) after their introduction. In this case, male and female guppies from the introduction site matured at significantly higher ages than their counterparts from the high-predation control. The females also produced larger and fewer offspring in their first broods. In a second replicate, on a tributary to the El Cedro River, we have evaluated the guppies 4, 7, and 9 years after the introduction. After 4 years, the introduction-site males were older and larger at maturity than the controls, but there were no differences in any aspect of the female life history. After 7 and 9 years, females also showed delayed maturity, but there still were no changes in other variables, such as offspring size or fecundity.

These experiments have allowed us to follow the dynamics of the evolution of the life history. We have also used mark-recapture to demonstrate that the guppies from the introduction sites experience lower mortality rates than those from the high-predation controls. Evolution of these patterns in response to such a manipulation lends strength to our interpretation of the interpopulation differences in life histories as adaptations to predator-induced mortality.

These experiments support Endler's (1986) conclusions about the potential strength of natural selection. If we think of these experiments as episodes of directional selection, then it is possible to apply the same kinds of statistics to the rates and patterns of change that others have used to characterize either artificial selection or differential mortality and reproductive success observed within a generation (Haldane 1957, Lande and Arnold 1983). The results of these experiments, in combination with laboratory estimates of genetic variance-covariance matrices and field estimates of generation time, demonstrate that the rate of evolution attainable under natural selection is of the same order of magnitude as that obtained under artificial selection and five to seven orders of magnitude greater than what has been inferred from the fossil record (Reznick, Rodd, Shaw, and Shaw, unpublished data).

Phenotypic Engineering

The Argument for Altering Phenotypes

"Phenotypic engineering" is the experimental manipulation of genotypes or phenotypes to alter the distribution of phenotypic variances and covariances from their natural values. It is not a class of experiments separate from those described previously but a technique that can be used profitably within them. The appeal of phenotypic engineering is in its ability to overcome two limitations commonly encountered in studies of selection. First, normal levels of phenotypic variance may be too small to provide satisfactory power to detect certain types of selective effects, especially when selection is optimizing (Manly 1985, Travis 1989). Second, phenotypic variation in one subset of traits may be so confounded with variation in another that statistical power is insufficient to separate the causal effects of the subsets on fitness (Travis 1994).

Advocating phenotypic engineering might seem incompatible with our earlier emphasis on re-creating natural distributions of factors in experiments. A fair engineering of phenotypic variance, or a fair manipulation of a putative ecological factor, allows the nature of selection to be adduced throughout the range within which the trait or factor is varied. The creation of extreme values, or extreme treatments, is acceptable when it is done in conjunction with the use of more typical intermediate values; the goal is to capture the statistical power of enhanced variance without sacrificing the confidence of reliable interpolation between the extremes.

Phenotypic engineering is a novel term for a set of diverse experimental methods that have been in use for decades. Each method has its advantages and disadvantages; in general, the more advantageous methods carry the more potentially difficult consequences. For example, the simplest experimental method to increase the variance in the distribution of a trait that is hypothesized to be under selection is using large and small individuals in excess of their natural relative frequencies. This method's virtue, simplicity of execution, is tempered by its limited effectiveness (the availability of large and small individuals will set a practical upper limit for the phenotypic variance), its requirement for specialized statistical analysis (Trexler et al. 1992), and its inability to provide insights into how selection might act outside the natural range of phenotypic variation. In addition, such nonrandom choice of individuals cannot help but dissect the relative importance of correlated characters; indeed, it will be counterproductive because it will also maximize the covariance between the characters whose separate roles are in question. A variety of more complicated methods have been used to overcome these limitations; the choice among them will depend on whether their liabilities can be accommodated.

Genetic Molding of Phenotype Distributions

Genetic manipulation of phenotypic variance is a more complicated but highly informative method of phenotypic engineering. By using a variety of genetic crosses, an investigator can manipulate the mean and variance of a phenotype distribution independently. This method is most profitably used in studies of selection and putative local adaptation in two or more populations or locations.

Two examples illustrate the range of potential contributions that genetic manipulations can make. Schluter (1995) used F_1 hybrids between benthic and littoral forms of sticklebacks, along with parentals, to investigate the relationship between variation in morphology and variation in performance in each habitat. His results suggested that the differences between the forms were adaptive; moreover, he suggested that the hybrid's lack of intermediacy in performance, despite its intermediate values of morphological traits, indicated a natural reinforcing mechanism for reproductive isolation between the benthic and littoral forms.

Jordan (1991) created F_2 hybrids between two plant populations and transplanted hybrid and parental stock into each parental habitat. The expanded phenotypic variance characteristic of F_2 hybrids increased substantially the power to detect selection, but the use of F_2 hybrids offered Jordan another advantage; in the absence of tight linkage, F_2 hybrids will exhibit covariances among characters that are lower than those in either parental population. This reduced covariance, combined with the increased variance in every character, provided Jordan with higher statistical power to separate the causal effects of each character on fitness. In particular, the expanded variances and reduced covariances helped elevate the power of cross-product terms in regression analyses for detecting whether there was functional integration in the phenotypes favored at each location. The lack of statistical significance in these terms, despite the increased power, was a strong argument that the adaptive differentiation between populations that Jordan uncovered was an additive effect of divergent selection for a few morphological characters and did not involve any synergistic reinforcement across character values.

These examples illustrate the advantages of genetic manipulations in studies of selection, but this approach has two disadvantages. First, it is impractical for many organisms. Second, hybrid forms may exhibit fitness differences for reasons other than their particular phenotypic values for the traits under scrutiny, and those fitness effects may provoke a spurious interpretation of the experimental results. Of course, this liability is irrelevant if one is interested only in whether hybrids are intermediate with respect to fitness, regardless of cause, but if genetic manipulations are used for phenotypic engineering, then this issue must be addressed with either additional statistical analyses or additional studies of hybrid viability. For example, if hybrids suffered from other causes of reduced fitness, selection gradients within hybrid stocks, even if they are consistent with those in parental stocks, should account for less of the fitness variance than they do within parental stocks.

Direct Manipulation of the Phenotype

The most widespread method of phenotypic engineering is the direct manipulation of individuals to create altered trait values. In fact, three types of direct manipulation can be found in the literature, each with its own advantages and disadvantages.

The direct manipulation most often employed is mechanical or surgical manipulation of the phenotype. The advantage of this method is that it captures the full power of the experimental approach, particularly for distinguishing direct from indirect selection (Campbell et al. 1994). Trait values are altered so that a trait or set of traits is studied against a randomized background of all other potential effects. The disadvantages include the labor involved in developing and implementing the manipulation and the need to incorporate appropriate controls for the manipulation. Of course, some traits may not be amenable to manipulation.

Mechanical manipulation in the study of selection has a long history, beginning with Brower et al.'s (1964) manipulation of butterfly wing patterns and Semler's (1971) manipulation of male stickleback throat coloration. Semler (1971) pointedly discussed the importance of manipulating the throat coloration to separate the effect of the color from the effects of other behaviors or attributes, perhaps unnoticed by the investigator, with which it may naturally covary. Indeed, direct manipulation continues to play a major role in studies of sexual selection of both discrete (Wiernasz 1989, Scheffer et al. 1996) and continuous variation (Bischoff et al. 1985, Moller and de Lope 1994). In particular, mechanical manipulations have allowed investigators to examine how sexual selection would act on character states not normally seen in natural populations and thereby test broader hypotheses for their evolution (Andersson 1982, Basolo 1990). Direct manipulations of characters have been used less often to examine natural selection, despite the success of some exemplary studies (e.g., Sinervo 1990, Sinervo et al. 1992).

Another type of direct manipulation, which is enjoying increased popularity, is the use of endocrine treatments to alter an individual's hormonal state. Most of these studies have manipulated the testosterone level of males to alter their behavior patterns and examine the consequences for survivorship (Marler and Moore 1988, Dufty 1989) or reproduction (Hegner and Wingfield 1987). The virtues and liabilities of endocrine manipulations have been reviewed elegantly elsewhere (Ketterson and Nolan 1994). The advantages include expansion of phenotypic variances through a natural physiological process and allowing that expansion to be performed ''on demand,'' which permits investigators to test a multitude of ecological and evolutionary hypotheses. The disadvantages include the considerable difficulty of developing and implementing an acceptable delivery system for the hormone and, more important, the fact that many phenotypic traits will be simultaneously altered. The weight given to the latter issue will depend on the hypotheses under scrutiny.

Still another variety of direct manipulation is ''ecological manipulation.'' With this method an investigator exposes growing or developing individuals to one or more ecological factors, which they encounter normally in the natural habitat, and uses the effects of those factors to alter the levels of variance and covariance among traits. Obviously, the investigator employs extreme levels of such factors in order to expand phenotypic variances. For example, Anholt (1991) manipulated density and food level in a population of nymphal damselflies to expand dramatically the phenotypic variance in body size and morphology displayed by the emerging adults. McManus (1993) manipulated food level and photoperiod to expand the variance in lipid storage in sailfin mollies and reduce the covariance found in natural populations between body size and size-specific lipid concentrations.

The advantage of this method is that the manipulations act through mechanisms normally encountered in natural populations, so they merely exaggerate naturally induced variation. As always, the use of extreme levels of those factors is sensible provided that a reasonable number of intermediate levels are also employed. The most obvious disadvantages include the level of effort required to perform the manipulations and that this method works only for traits that exhibit substantial phenotypic plasticity. The more subtle potential disadvantage is that the manipulation may have other effects on the individuals besides simply altering the values of the characters under scrutiny. Anholt (1991) was able to discount this possibility by incorporating the specific manipulative history of each individual into elegant statistical tests with general linear models.

The most subtle potential disadvantage of this method is that the ecological manipulation may cause substantial mortality in addition to its effects on plastic characters. This effect can have two ramifications. First, the level of mortality must be anticipated so that the survivors represent a sufficiently large sample size for the subsequent study of phenotypic selection. Second, if mortality is not random, then the survivors may represent a group that has passed through a selective filter that the manipulation has made more stringent than normal. If the genes controlling the characters under scrutiny have pleiotropic effects that influence the probability of passing through that selective filter, then the evolutionary interpretation of the selective process studied afterward may be compromised. Whether such extensive pleiotropy exists across the life cycle is an open empirical question; the lessons from study of optimizing selection on bristle number in *Drosophila* (see Travis 1989) suggest that this ought to be a cause for concern in some organisms.

A potential problem for direct manipulations of any kind is that the manipulation may not do exactly what the investigator expects. For example, in some cases a manipulation may have an effect that is not anticipated by the investigator: Waldbauer and Sternburg (1975) suggested that Brower et al.'s (1964) manipulation of butterfly color patterns inadvertently created experimental animals that resembled palatable moth species, which compromised the original interpretation of the results. In other cases, the manipulation may not serve exactly the purpose desired by the investigator: Bernardo (1996) has suggested that some manipulations of egg size in animals also manipulated egg quality. In this case, the effect of the manipulation on fitness is not in dispute but rather the interpretation of that effect.

Conclusions

Our goal has been to review a spectrum of methods that are used to study the process of selection experimentally in natural populations. We have tried to present a convincing argument in support of the thesis that experimental methods, used in conjunction with comparative or observational ones, allow a direct study of the process of adaptation and permit the testing of hypotheses about its causes.

Two subtexts emerge. First, not every method is possible or perhaps even desirable in every investigation. Each method has its advantages and disadvantages, and the choice of method should be dictated by the nature of the question being asked and the pragmatic considerations involved.

Second, although these experimental methods allow adaptive evolution to be studied in real time, "real time" can entail at least half of any investigator's useful professional lifetime (cf. Cain and Provine 1992). There are two reasons why this is the case. The first, and less interesting, is practical: only so much work can be done in a day. The second, and more important, is that, even at the microevolutionary scale, evolution is a complicated phenomenon. The complex mechanics of almost any natural system will generate competing hypotheses that must be distinguished and tested. Moreover, almost all systems will show their students significant temporal variation in process. Sometimes this variation makes the "signal" harder to detect; sometimes it is, in fact, part of the "signal."

These subtexts argue for the importance of long-term studies and programs that support them. We hope to have argued successfully that the best long-term studies have a firm experimental orientation. Implicit in our thesis is that, if evolution can be studied through experiments, there is no reason to argue that other areas of environmental biology should not.

ACKNOWLEDGMENTS   Our work has been supported generously by the National Science Foundation. During preparation of this manuscript we were supported by grants DEB-94-19823 (DR) and DEB-92-20849 (JT). We thank J. Bernardo and W. J. Resetarits for their amazing patience.

Literature Cited

Andersson, M. 1982. Female choice selects for extreme tail lengths in widowbirds. Nature 299:818–820.

Anholt, B. R. 1991. Measuring selection on a population of damselflies with a manipulated phenotype. Evolution 45:1091–1106.

Basolo, A. L. 1990. Female preference predates the evolution of the sword in swordtail fish. Science 250:808–810.

Bennington, C. C., and J. B. McGraw. 1995. Phenotypic selection in an artificial population of *Impatiens pallida*: the importance of the invisible fraction. Evolution 49:317–324.

Bernardo, J. 1996. The particular maternal effect of propagule size, especially egg size: patterns, models, quality of evidence and interpretations. American Zoologist 36:216–236.

Berven, K. A. 1982. The genetic basis of altitudinal variation in the wood frog *Rana sylvatica*: I. An experimental analysis of life history traits. Evolution 36:962–983.

Bischoff, J. A., J. L. Gould, and D. I. Rubenstein. 1985. Tail size and female choice in the guppy (*Poecilia reticulata*). Behavioral Ecology and Sociobiology 17:253–255.

Block, B. A., J. R. Finnerty, A. F. R. Stewart, and J. Kidd. 1993. Evolution of endothermy in fish: mapping physiological traits on a molecular phylogeny. Science 260:210–214.

Brodie, E. D., III, and F. J. Janzen. 1996. On the assignment of fitness values in statistical analyses of selection. Evolution 50:437–442.

Brower, L. P., J. V. Z. Brower, F. G. Stiles, H. J. Croze, and A. S. Hower. 1964. Mimicry: differential advantage of color-patterns in the natural environment. Science 144:183–185.

Cain, A. J., and W. B. Provine. 1992. Genes and ecology in history. Pages 3-28 in R. J. Berry, T. J. Crawford, and G. M. Hewitt (eds.), Genes in Ecology. Blackwell, Oxford.

Campbell, D. R., N. M. Waser, and M. V. Price. 1994. Indirect selection of stigma position in *Ipomopsis aggregata* via a genetically correlated trait. Evolution 48:55–68.

Chambers, R. C. 1984. Competition and predation in temporary habitats. Ph.D. dissertation, Duke University, Durham, North Carolina.

Charlesworth, B. 1994. Evolution in Age-structured Populations. Cambridge University Press, Cambridge.

Clutton-Brock, T. (ed.). 1988. Reproductive Success. University of Chicago Press, Chicago, Illinois.

Dhondt, A. A. 1982. Heritability of blue tit tarsus length from normal and cross-fostered broods. Evolution 36:418–419.

Donoghue, M. J. 1989. Phylogenies and the analysis of evolutionary sequences, with examples from seed plants. Evolution 43:1137–1156.

Dudash, M. R. 1990. Relative fitness of selfed and outcrossed progeny in a self-compatible, protandrous species, *Sabatia angularis* L. (Gentianaceae): a comparison in three environments. Evolution 44:1129–1139.

Dudley, S. A. 1996a. Differing selection on plant physiological traits in response to environmental water availability: a test of adaptive hypotheses. Evolution 50:92–102.

———. 1996b. The response to differing selection on plant physiological traits: evidence for local adaptation. Evolution 50:103–110.

Dufty, A. M., Jr. 1989. Testosterone and survival: a cost of aggressiveness? Hormones and Behavior 23:185–193.

Dunham, A. E., D. B. Miles, and D. N. Reznick. 1988. Pages 441–522 in C. Gans and R. B. Huey (eds.), Biology of the Reptilia: Vol. 16. Ecology B, Defense and Life History. Liss, New York.

Endler, J. A. 1980. Natural selection on color patterns in *Poecilia reticulata*. Evolution 34:76–91.

———. 1986. Natural Selection in the Wild. Princeton University Press, Princeton, New Jersey.

Felsenstein, J. 1985. Phylogenies and the comparative method. American Naturalist 125:1–15.

Ford, E. B. 1971. Ecological Genetics, 3rd ed. Chapman and Hall, London.

Gadgil, M., and W. H. Bossert. 1970. Life historical consequences of natural selection. American Naturalist 104:1–24.

Gascon, C., and J. Travis. 1992. Does the spatial scale of experimentation matter: a test with tadpoles and dragonflies. Ecology 73:2237–2243.

Groeters, F. R., and H. Dingle. 1996. Heritability of wing length in nature for the milkweed bug, *Oncopeltus fasciatus*. Evolution 50:442–447.

Gurney, W. S. C., and R. M. Nisbet. 1985. Fluctuation periodicity, generation separation, and the expression of larval competition. Theoretical Population Biology 28:150–180.

Haldane, J. B. S. 1957. The cost of natural selection. Journal of Genetics 55:511–524.

Hegner, R. E., and J. C. Wingfield. 1987. Effects of experimental manipulation of testosterone levels on parental investment and breeding success in male house sparrows. Auk 104:470–480.

Jones, K. N. 1996. Fertility selection on a discrete floral polymorphism in *Clarkia* (Onagraceae). Evolution 50:71–79.

Jordan, N. 1991. Multivariate analysis of selection in experimental populations derived from hybridization of two ecotypes of the annual plant *Diodia teres* W. (Rubiaceae). Evolution 45:1760–1772.

Joshi, A., and J. N. Thompson. 1995. Alternative routes to the evolution of competitive ability in two competing species of *Drosophila*. Evolution 49:616–625.

Ketterson, E. D., and V. Nolan, Jr. 1994. Hormones and life histories: an integrative approach. Pages 327–353 in L. A. Real (eds.), Behavioral Mechanisms in Evolutionary Ecology. University of Chicago Press, Chicago, Illinois.

Kettlewell, H. B. D. 1955. Selection experiments on industrial melanism in the Lepidoptera. Heredity 10:287–301.

———. 1973. The Evolution of Melanism. Clarendon Press, Oxford.

Kozlowski, J., and J. Uchmanski. 1987. Optimal individual growth and reproduction in perennial species with indeterminate growth. Evolutionary Ecology 1:214–230.

Lande, R., and Arnold, S. J. 1983. The measurement of selection on correlated characters. Evolution 37:1210–1226.

Law, R. 1979. Optional life histories under age-specific predation. American Naturalist 114:399–417.

LeBreton, J.-D., G. Hemery, J. Clobert, and H. Coquillart. 1990. The estimation of age-specific breeding probabilities from recaptures or resightings in vertebrate populations: I. Transversal models. Biometrics 46:609–622.

Losos, J. B. 1990. The evolution of form and function: morphology and locomotor performance in West Indian *Anolis* lizards. Evolution 44:1189–1203.

Maddison, W. P. 1990. A method for testing the correlated evolution of two binary characters: are gains or losses concentrated on certain branches of a phylogenetic tree? Evolution 44:539–557.

Manly, B. F. J. 1985. The Statistics of Natural Selection on Animal Populations. Chapman and Hall, London.

Marler, C. A., and M. C. Moore. 1988. Evolutionary costs of aggression revealed by testosterone manipulations in free-living male lizards. Behavioral Ecology and Sociobiology 23:21–26.

Martins, E. P. 1996. Conducting phylogenetic comparative studies when the phylogeny is not known. Evolution 50:12–22.

Mazer, S. J., and D. L. Gorchov. 1996. Parental effects on progeny phenotype in plants: distinguishing genetic and environmental causes. Evolution 50:44–53.

McManus, M. G. 1993. Differential resource allocation in the sailfin molly, *Poecilia latipinna*: laboratory and field experiments. Ph.D. dissertation, Florida State University, Tallahassee.

McPeek, M. 1995. Testing hypotheses about evolutionary change on single branches of a phylogeny using evolutionary contrasts. American Naturalist 145:686–703.

Michod, R. E. 1979. Evolution of life histories in response to age-specific mortality factors. American Naturalist 113:531–550.

Miles, D. B., and A. E. Dunham. 1993. Historical perspectives in ecology and evolutionary biology: the use of phylogenetic comparative analyses. Annual Reviews of Ecology and Systematics 24:587–619.

Mitter, C., and D. R. Brooks. 1983. Phylogenetic aspects of coevolution. Pages 65–98 in D. J. Futuyma and M. Slatkin (eds.), Coevolution. Sinauer, Sunderland, Massachusetts.

Moller, A. P., and F. de Lope. 1994. Differential costs of a secondary sexual character: an experimental test of the handicap principle. Evolution 48:1676–1683.

Norman, J. K., A. K. Sakai, S. G. Weller, and T. E. Dawson. 1995. Inbreeding depression in morphological and physiological traits of *Schiedea lydgatei* (Caryophyllaceae) in two environments. Evolution 49:297–306.

Núñez-Farfán, J., and R. Dirzo. 1994. Evolutionary ecology of *Datura stramonium* L. in central Mexico: natural selection for resistance to herbivorous insects. Evolution 48: 423–436.

Orzack, S. H., and S. Tuljapurkar. 1989. Population dynamics in variable environments: the demography and evolution of iteroparity. American Naturalist 133:901–923.

Partridge, L., and M. Farquhar. 1983. Lifetime mating success of male fruitflies (*Drosophila melanogaster*) is related to their size. Animal Behaviour 31:871–877.

Petraitis, P. S., A. E. Dunham, and P. H. Niewiarowski. 1996. Inferring multiple causality: the limitations of path analysis. Functional Ecology 10:421–431.

Radtkey, R. R., and M. C. Singer. 1995. Repeated reversals of host-preference evolution in a specialist insect herbivore. Evolution 49:351–359.

Reznick, D. N. 1981. ''Grandfather effects'': the genetics of interpopulation differences in offspring size in the mosquito fish. Evolution 35:941–953.

———. 1982a. Genetic determination of offspring size in the guppy (*Poecilia reticulata*). American Naturalist 120:181–188.

———. 1982b. The impact of predation on life history evolution in Trinidadian guppies: genetic basis of observed life history patterns. Evolution 36:1236–1250.

———. 1989. Life-history evolution in guppies: 2. Repeatability of field observations and the effects of season on life histories. Evolution 43:1285–1297.

Reznick, D. N., and H. Bryga. 1987. Life history evolution in guppies (*Poecilia reticulata*): phenotypic and genetic changes in an introduction experiment. Evolution 41:1370–1385.

———. 1996. Life history evolution in guppies (*Poecilia reticulata*, Poeciliidae): V. Genetic basis of parallelism in life histories. American Naturalist 147:339–359.

Reznick, D. N., and J. A. Endler. 1982. The impact of predation on life history evolution in Trinidadian guppies (*Poecilia reticulata*). Evolution 36:160–177.

Reznick, D., and J. Travis. 1996. The empirical study of adaptation in natural populations. Pages 243–289 in M. R. Rose and G. Lauder (eds.), Adaptation. Academic Press, San Diego.

Reznick, D. N., H. Bryga, and J. A. Endler. 1990. Experimentally induced life-history evolution in a natural population. Nature 346:357–359.

Reznick, D. N., M. J. Butler IV, F. H. Rodd, and P. Ross, 1996a. Life-history evolution in guppies (*Poecilia reticulata*): VI. Differential mortality as a mechanism for natural selection. Evolution 50:1651–1660.

Reznick, D. N., F. H. Rodd, and M. Cardenas. 1996b. Life history evolution in guppies (*Poecilia reticulata*: Poeciliidae): IV. Parallelism in life history phenotypes. American Naturalist 147:319–338.

Ryan, M. J., C. M. Pease, and M. R. Morris. 1992. A genetic polymorphism in the swordtail *Xiphophorus nigrensis*: testing the prediction of equal fitnesses. American Naturalist 139:21–31.

Scheffer, S. J., G. W. Uetz, and G. E. Stratton. 1996. Sexual selection, male morphology, and the efficacy of courtship signalling in two wolf spiders (Araneae: Lycosidae). Behavioral Ecology and Sociobiology 38:17–23.

Schluter, D. 1995. Adaptive radiation in sticklebacks: trade-offs in feeding performance and growth. Ecology 76:82–90.

Schmitt, J., and J. Antonovics. 1986. Experimental studies of the evolutionary significance of sexual reproduction: IV. Effect of neighbor relatedness and aphid infestation on seedling performance. Evolution 40:830–836.

Scribner, K. T., and J. C. Avise. 1994. Population cage experiments with a vertebrate: the temporal demography and cytonuclear genetics of hybridization in *Gambusia* fishes. Evolution 48:155–171.

Semler, D. E. 1971. Some aspects of adaptation in a polymorphism for breeding colours in the threespine stickleback (*Gasterosteus aculeatus*). Journal of Zoology 165:291–302.

Semlitsch, R. D., and H. M. Wilbur. 1989. Artificial selection for paedomorphosis in the salamander *Ambystoma talpoideum*. Evolution 43:105–112.

Semlitsch, R. D., R. N. Harris, and H. M. Wilbur. 1990. Paedomorphosis in *Ambystoma talpoideum*: maintenance of population variation and alternative life-history pathways. Evolution 44:1604–1613.

Service, P. M., and M. R. Rose. 1985. Genetic covariation among life-history components: the effect of novel environments. Evolution 39:943–945.

Siddall, M. E., D. R. Brooks, and S. S. Desser. 1993. Phylogeny and the reversibility of parasitism. Evolution 47:308–313.

Sillen-Tullberg, B. 1988. Evolution of gregariousness in aposematic butterfly larvae: a phylogenetic analysis. Evolution 29:293–305.

Sinervo, B. 1990. The evolution of maternal investment in lizards: an experimental and comparative analysis of egg size and its effects on offspring performance. Evolution 44:279–294.

Sinervo, B., P. Doughty, R. B. Huey, and K. Zamudio. 1992. Allometric engineering: a causal analysis of natural selection on offspring size. Science 258:1927–1930.

Singer, M. C. 1994. Behavioral constraints on the evolutionary expansion of insect diet: a case history from checkerspot butterflies. Pages 279–296 in L. A. Real (ed.), Behavioral Mechanisms in Evolutionary Ecology. University of Chicago Press, Chicago, Illinois.

Smith, J. N. M., and A. A. Dhondt. 1980. Experimental confirmation of heritable morphological variation in a natural population of song sparrows. Evolution 34:1155–1158.

Smith, J. N. M., and R. Zach. 1979. Heritability of some morphological characters in the song sparrow. Evolution 33:460–467.

Spitze, K. 1991. *Chaoborus* predation and life-history evolution in *Daphnia pulex*: temporal pattern of population diversity, fitness, and mean life history. Evolution 45:82–92.

Strauss, S. Y., and R. Karban. 1994. The significance of outcrossing in an intimate plant-herbivore relationship: I. Does outcrossing provide an escape from herbivores adapted to the parent plant? Evolution 48:454–464.

Streifer, W. 1974. Realistic models in population ecology. Advances in Ecological Research 8:199–266.

Travis, J. 1983. Variation in growth and survival of *Hyla gratiosa* larvae in experimental enclosures. Copeia 1983:232–237.

———. 1989. The role of optimizing selection in natural populations. Annual Review of Ecology and Systematics 20:279–296.

———. 1994. Size-dependent behavioral variation and its genetic control within and among populations. Pages 165–187 in C. M. Boake (ed.), Quantitative Genetic Approaches to Animal Behavior. University of Chicago Press, Chicago, Illinois.

Travis, J., and J. C. Trexler. 1984. Investigations on the control of the color polymorphism in *Pseudacris ornata*. Herpetologica 40:252–257.

Travis, J., W. H. Keen, and J. Juilianna. 1985a. The effects of multiple factors on viability selection in *Hyla gratiosa* tadpoles. Evolution 39:1087–1099.

———. 1985b. The role of relative body size in a predator-prey relationship between dragonfly naiads and larval anurans. Oikos 45:59–65.

Travis, J., S. Emerson, and M. Blouin. 1987. A quantitative genetic analysis of larval life history traits in *Hyla crucifer*. Evolution 41:145–156.

Travis, J., J. A. Farr, M. McManus, and J. C. Trexler. 1989. Environmental effects on adult growth patterns in the male sailfin molly (*Poecilia latipinna*). Environmental Biology of Fishes 26:119–127.

Trexler, J. C., J. Travis, and M. McManus. 1992. Effects of habitat and body size on mortality rates of *Poecilia latipinna*. Ecology 73:2224–2236.

Trexler, J. C., R. C. Tempe, and J. Travis. 1994. Size-selective predation of sailfin mollies by two species of heron. Oikos 69:250–258.

van Noordwijk, A. J., J. J. van Balen, and W. Scharloo. 1980. Heritability of ecologically important traits in the great tit. Ardea 68:193–203.

Waldbauer, G. P., and Sternburg, J. G. 1975. Saturniid moths as mimics: an alternative interpretation of attempts to demonstrate mimetic advantage in nature. Evolution 29:650–658.

Walters, J. R., P. D. Doerr, and J. H. Carter III. 1992. Delayed dispersal and reproduction as a life-history tactic in cooperative breeders: fitness calculations from red-cockaded woodpeckers. American Naturalist 139:623–643.

Warner, S. C., W. A. Dunson, and J. Travis. 1991. Interaction of pH, density and priority effects on the survivorship and growth of two species of hylid tadpoles. Oecologia 88:331–339.

Waser, N. M., and M. V. Price. 1994. Crossing-distance effects in *Delphinium nelsonii*: outbreeding and inbreeding depression in progeny fitness. Evolution 48:842–852.

Weldon, W. F. R. 1893. On certain correlated variations in *Carcinus moenas*. Proceedings of the Royal Society of London B 5:318–329.

Wickman, P.-O. 1992. Sexual selection and butterfly design—a comparative study. Evolution 46:1525–1536.

Wiernasz, D. 1989. Female choice and sexual selection of male wing melanin pattern in *Pieris occidentalis* (Lepidoptera). Evolution 43:1672–1682.

Willis, J. H. 1993. Effects of different levels of inbreeding on fitness components in *Mimulus guttatus*. Evolution 47:864–876.

Index